Stem Cell Biology and Regenerative Medicine

Series Editor
Kursad Turksen, Ph.D.
kursadturksen@gmail.com

Our understanding of stem cells has been growing rapidly over the last decade. While the apparently tremendous therapeutic potential of stem cells has not yet been realized, their routine use in regeneration and restoration of tissue and organ function is greatly anticipated. To this end, many investigators continue to push the boundaries in areas such as the reprogramming, the stem cell niche, nanotechnology, biomimetics and 3D bioprinting, to name just a few. To capture and consolidate some of these developments in a timely way is the objective of the volumes in this series. We want each volume to be thought-provoking in identifying problems, offering solutions and providing ideas to excite further innovation in the stem cell and regenerative medicine fields.

More information about this series at http://www.springer.com/series/7896

Gerald Brandacher
Editor

The Science of Reconstructive Transplantation

Foreword by Dr. Thomas E. Starzl
Foreword by Dr. Raimund Margreiter

Humana Press

Editor
Gerald Brandacher
Department of Plastic and Reconstructive Surgery
The Johns Hopkins University School of Medicine
Baltimore, MD, USA

ISSN 2196-8985 ISSN 2196-8993 (electronic)
Stem Cell Biology and Regenerative Medicine
ISBN 978-1-4939-2070-9 ISBN 978-1-4939-2071-6 (eBook)
DOI 10.1007/978-1-4939-2071-6

Library of Congress Control Number: 2015930838

Springer New York Heidelberg Dordrecht London

Printed on acid-free paper

Springer is part of Springer Science+Business Media (www.springer.com)

For Barbara and Mo

Foreword: The Science of Reconstructive Transplantation

Reconstructive transplantation is an emerging area of transplant medicine that has become a viable option for patients with large and devastating tissue defects. Over the past decade, a rapidly growing number of face and upper extremity transplantations have been performed worldwide with highly encouraging outcomes. Advances in microsurgical techniques, transplant immunology, and immunosuppression have enabled such operations.

It has been a pleasure to see the life-changing impact of the hand, face, and other composite allografts. Recipients of these grafts represent a new generation of transplant recipient pioneers. The uniqueness of their grafts, which include donor bone marrow, could help further elucidate the mechanisms by which transplanted organs and tissues are accepted. In turn, novel strategies to facilitate these mechanisms may be developed.

The Science of Reconstructive Transplantation presents a comprehensive overview of the latest advances in basic and translational research in the field. Many of its leaders have contributed their expertise to the inspiring book. Important topics include reconstructive animal models, skin rejection, immune monitoring, stem cell-based immunomodulation strategies, costimulatory blockade, tolerance induction, chronic rejection, ischemia–reperfusion injury, nerve regeneration, and cerebrocortical reintegration.

The textbook should spark the interest of physicians, scientists, and surgeons, while serving as a valuable reference for students and scholars engaged in this novel and emerging area of transplantation.

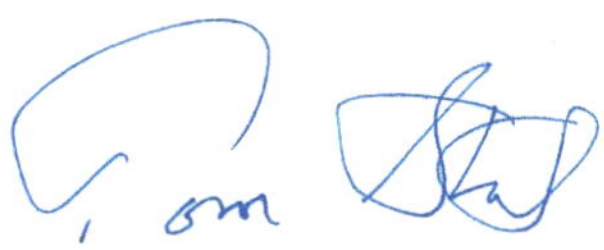

Thomas E. Starzl, M.D., Ph.D.
Professor of Surgery
Distinguished Service Professor
Professor Emeritus
Thomas E. Starzl Transplantation Institute
University of Pittsburgh

Foreword: The Science of Reconstructive Transplantation

Although the outcome of the initial series of hand and face transplantations has been better than any other field of transplantation, fewer than 100 upper extremity and only 29 face transplants have been performed since the first successful hand transplant in Lyon, France, in 1998. Wider application of reconstructive transplantation is hampered to a large extent by the lifelong need for immunosuppression. Minimization of immunosuppression or even induction of a specific immune tolerance has to be considered a prerequisite for further propagation of this most fascinating field. This book covers the various aspects of reconstructive transplantation with special emphasis on immunology. All chapters are written by top experts in the field.

The Science of Reconstructive Transplantation is a must for every clinician and scientist working in the field, but is also worthwhile for anyone with an interest in new developments in medical science.

Raimund Margreiter MD
Professor Emeritus
Department of Visceral, Thoracic and Transplant Surgery
University Hospital
Innsbruck—Austria

Preface

Reconstructive transplantation of vascularized composite allografts, such as hand, face, and abdominal wall transplants, has become a clinical reality and a viable treatment option for the patients suffering from complex tissue injuries or defects not amenable to conventional reconstruction. Despite the fact that early and intermediate outcomes are highly encouraging, skin rejection, the need for chronic immunosuppressive treatment and the relatively slow pace of nerve regeneration continue to hamper broader clinical application and further expansion of indications. Therefore, a thorough understanding of the unique biological features and basic mechanisms related to immunogenicity and neuroregeneration in this novel and emerging field of transplantation is key to establishing future treatment protocols that allow to favorably balance the risks and benefits for such *non-life-saving* but *life-changing* types of transplants. This was the impetus to embark on this project and create a book entirely devoted to research in reconstructive transplantation.

Written by renowned scientists and leaders and pioneers in the field, *The Science of Reconstructive Transplantation* thus presents a comprehensive overview of the latest advances in basic and translational research in the field of reconstructive transplantation with a particular emphasis on its potential therapeutic implications.

The book has been structured into two parts. Part I gives an overview of the history and development of reconstructive transplantation, discusses what can be learned from the experiences and successes over the past 60 years in solid organ transplantation, and provides insights from a recipient perspective describing the experience and daily life of the first US patient receiving a combined forearm and hand transplant. Part II concentrates on individual research areas, spanning topics such as the use of small and large animal models for reconstructive transplantation research, mechanism and diagnosis of skin rejection, immune monitoring concepts, cell-based immunomodulatory strategies, tolerance induction, chronic rejection, ischemia–reperfusion injury, models and tools to assess nerve regeneration, and cortical reintegration of vascularized composite allografts. While this volume is certainly not inclusive of all areas of transplantation research, it contains topics that are of major current interest and have significant potential for translation and future clinical applications.

Thus, the audience for this book includes biomedical researchers and basic scientists in the field of reconstructive transplantation, transplant immunology, and regenerative medicine, as well as clinicians, surgeons, and multidisciplinary specialists, interested in this novel and exciting field.

I am extremely grateful to Drs. Thomas E. Starzl and Raimund Margreiter for writing the forewords to this book. Dr. Starzl has taught me about the fundamental and natural laws of immunology. In addition, by sharing his personal reflections on one of the most exceptional transplant journeys, he taught me that through relentless pursuit of a goal and vision, eventually stunning successes can be achieved in a novel discipline. My surgical teacher and mentor, Dr. Margreiter, has involved me in reconstructive transplantation in the very early days of the field as a young resident and has supported my efforts throughout my entire professional career. He has been and remains to be the ultimate role model of a surgeon-scientist.

Finally, I would like to express my sincere appreciation to all the contributors for sharing their experience and knowledge in this book and for their excellent manuscripts. I would also like to thank Springer for supporting this exciting project and the opportunity to edit this volume.

Baltimore, MD, USA Gerald Brandacher, MD

Contents

Contributors

W. P. Andrew Lee Department of Plastic and Reconstructive Surgery, The Johns Hopkins University School of Medicine, Baltimore, MD, USA

Kodi Azari Department of Orthopedic Surgery, David Geffen School of Medicine at University of California-Los Angeles, Los Angeles, CA, USA

Lionel Badet Department of Transplantation, Edouard Herriot Hospital, Lyon, France

Rolf N. Barth Department of Surgery, University of Maryland School of Medicine, Baltimore, MD, USA

Stephen T. Bartlett Department of Surgery, University of Maryland School of Medicine, Baltimore, MD, USA

Brenda Blair Christine M. Kleinert Institute for Hand and Microsurgery, Kleinert Kutz Hand Care Center, Louisville, KY, USA

Pedro Bolado Division of Plastic and Reconstructive Surgery, La Paz University Hospital, Madrid, Spain

Gerald Brandacher Department of Plastic and Reconstructive Surgery, Vascularized Composite Allotransplantation (VCA) Laboratory, The Johns Hopkins University School of Medicine, Baltimore, MD, USA

Philip S. Brazio Department of Surgery, University of Maryland School of Medicine, Baltimore, MD, USA

Justin M. Broyles Department of Plastic and Reconstructive Surgery, The Johns Hopkins University School of Medicine, Baltimore, MD, USA

Cesar Casado-Sanchez Division of Plastic and Reconstructive Surgery, La Paz University Hospital, Madrid, Spain

Daniel J. Ceradini Institute of Reconstructive Plastic Surgery, New York University School of Medicine, New York, NY, USA

Curtis L. Cetrulo Vascularized Composite Allotransplantation Laboratory, Transplantation Biology Research Center & Division of Plastic & Reconstructive Surgery, Massachusetts General Hospital/Harvard Medical School, Boston, MA, USA

Jeff Chang Department of Surgery, Division of Plastic and Reconstructive Surgery, School of Medicine, University of Washington, Seattle, WA, USA

Damon S. Cooney Department of Plastic and Reconstructive Surgery, The Johns Hopkins University School of Medicine, Baltimore, MD, USA

Joanna Cwykiel Department of Orthopaedics, University of Illinois at Chicago, Chicago, IL, USA

Eric A. Elster Walter Reed National Military Medical Center, Bethesda, MD, USA

Uniformed Services University of Health Sciences, Bethesda, MD, USA

Naval Medical Research Center, Silver Spring, MD, USA

Jonathan A. Forsberg Walter Reed National Military Medical Center, Bethesda, MD, USA

Uniformed Services University of Health Sciences, Bethesda, MD, USA

Naval Medical Research Center, Silver Spring, MD, USA

Scott H. Frey Brain Imaging Center & Rehabilitation Neuroscience Laboratory, University of Missouri, Columbia, MO, USA

Vijay S. Gorantla Department of Plastic Surgery, University of Pittsburgh Medical Center, Pittsburgh, PA, USA

Johanna Grahammer Center of Operative Medicine, Department of Visceral, Transplant and Thoracic Surgery, Innsbruck Medical University, Innsbruck, Austria

Theresa Hautz Center of Operative Medicine, Department of Visceral, Transplant and Thoracic Surgery, Innsbruck Medical University, Innsbruck, Austria

Rahul M Jindal Walter Reed National Military Medical Center, Bethesda, MD, USA

Uniformed Services University of Health Sciences, Bethesda, MD, USA

George Washington University, Washington, DC, USA

Jean Kanitakis Department of Dermatology, Edouard Herriot Hospital, Hospices Civils de Lyon, Lyon, France

Christina L. Kaufman Christine M. Kleinert Institute for Hand and Microsurgery, Louisville, KY, USA

Barbara Kern Department of Visceral, Transplant, and Thoracic Surgery, Innsbruck Medical University, Innsbruck, Austria

Saami Khalifian Department of Plastic and Reconstructive Surgery, The Johns Hopkins University School of Medicine, Baltimore, MD, USA

Jerzy W. Kupiec-Weglinski The Dumont-UCLA Transplant Center, Division of Liver and Pancreas Transplantation, Department of Surgery, David Geffen School of Medicine at University of California-Los Angeles, Los Angeles, CA, USA

Josef M. Kurtz Transplantation Biology Research Center, Massachusetts General Hospital/Harvard Medical School and Emmanuel College, Boston, MA, USA

Joseph E. Kutz Christine M. Kleinert Institute for Hand and Microsurgery, Kleinert Kutz Hand Care Center, Louisville, KY, USA

Luis Landin Division of Plastic and Reconstructive Surgery, La Paz University Hospital, Madrid, Spain

David A. Leonard Transplantation Biology Research Center & Division of Plastic & Reconstructive Surgery, Massachusetts General Hospital/Harvard Medical School, Boston, MA, USA

Angelo A. Leto Barone Department of Plastic and Reconstructive Surgery, The Johns Hopkins University School of Medicine, Baltimore, MD, USA

Denver Lough Department of Plastic and Reconstructive Surgery, The Johns Hopkins University School of Medicine, Baltimore, MD, USA

Maria Madajka Department of Plastic Surgery, Cleveland Clinic, Cleveland, OH, USA

Yorell Manon-Matos Christine M. Kleinert Institute for Hand and Microsurgery, Kleinert Kutz Hand Care Center, Louisville, KY, USA

Kacey G. Marra Department of Plastic Surgery, University of Pittsburgh Medical Center, Pittsburgh, PA, USA

Michael R. Marvin Department of Medicine, University of Louisville, Louisville, KY, USA

Department of Surgery, Jewish Hospital Transplant Center, University of Louisville, Louisville, KY, USA

David W Mathes Department of Surgery, Division of Plastic and Reconstructive Surgery, University of Colorado Denver, Denver, CO, USA

Emmanuel Morelon Department of Transplantation, Nephrology and Immunology, Edouard Herriot Hospital, Lyon Civil Hospital, University of Lyon, Lyon, France

Rosemary Ouseph Department of Medicine, University of Louisville, Louisville, KY, USA

Palmina Petruzzo Department of Transplantation, Edouard Herriot Hospital, Lyon, France

Department of Surgery, University of Cagliari, Cagliari, Italy

Jan A. Plock Division of Plastic and Hand Surgery, University Hospital Zurich, Zurich, Switzerland

Christopher Pollock Double Hand Transplant Patient, Transplant Composite Tissue Allograph, Harrisburg, PA, USA

Johann Pratschke Center of Operative Medicine, Department of Visceral, Transplant and Thoracic Surgery, Innsbruck Medical University, Innsbruck, Austria

Department of General, Visceral and Transplant Surgery, Charité, Berlin, Germany

Kadiyala V. Ravindra Division of Transplantation, Duke University, School of Medicine, Durham, NC, USA

Sudheer K. Ravuri Vascularized Composite Allotransplantation (VCA) Laboratory, Department of Plastic Surgery, University of Pittsburgh, Pittsburgh, PA, USA

Eduardo D. Rodriguez Department of Plastic Surgery, New York University School of Medicine, New York, NY, USA

Stefan Schneeberger Center of Operative Medicine, Department of Visceral, Transplant and Thoracic Surgery, Innsbruck Medical University, Innsbruck, Austria

Department of Plastic and Reconstructive Surgery, The Johns Hopkins University School of Medicine, Baltimore, MD, USA

Riccardo Schweizer Adipose Stem Cell Center, Department of Plastic Surgery, University of Pittsburgh, Pittsburgh, PA, USA

Jaimie T. Shores The Johns Hopkins Outpatient Center, Baltimore, MD, USA

Maria Siemionow Department of Orthopaedics, University of Illinois at Chicago, Chicago, IL, USA

Marc A. Soares Institute of Reconstructive Plastic Surgery, New York University School of Medicine, New York, NY, USA

Kristin A Stevens Walter Reed National Military Medical Center, Bethesda, MD, USA

Robert Sucher Department of General, Visceral and Transplant Surgery, Charité, Berlin, Germany

Bruce Swearingen Department of Surgery, Division of Plastic and Reconstructive Surgery, School of Medicine, University of Washington, Seattle, WA, USA

Olivier Thaunat Department of Transplantation, Nephrology and Immunology, Edouard Herriot Hospital, Lyon Civil Hospital, University of Lyon, Lyon, France

Huey Y. Tien Christine M. Kleinert Institute for Hand and Microsurgery, Kleinert Kutz Hand Care Center, Louisville, KY, USA

Tsu-Min Tsai Christine M. Kleinert Institute for Hand and Microsurgery, Kleinert Kutz Hand Care Center, Louisville, KY, USA

Sami H. Tuffaha Department of Plastic and Reconstructive Surgery, The Johns Hopkins University School of Medicine, Baltimore, MD, USA

Victor W. Wong Department of Plastic and Reconstructive Surgery, The Johns Hopkins University School of Medicine, Baltimore, MD, USA

Part I
Introduction, History, and Clinical Impact

Chapter 1
Reconstructive Transplantation: From Scientific Dream to Clinical Reality

Gerald Brandacher, Saami Khalifian and W.P. Andrew Lee

Introduction

Transplantation of foreign or allogeneic tissues has captivated human imagination since antiquity, and what began as the medicine of mythology now spans a timeline of three millennia. Indeed, Greek mythology and religious texts are rich with examples of xeno- and allotransplantation. Icarus and his father, Daedalus, for example, attempted to fly across the sea from Crete to Greece with the help of bird wings attached to their arms—one of the earliest writings illustrating xenotransplantation [1]. Ancient folklore in many cultures describes the amalgamation of physical attributes from multiple species into one, such as the Chimera in Homer's *The Odyssey*, which is the fusion of a goat, lion, and a dragon, or the deity Zu in Babylonian myth, which is a lion-headed eagle with human arms. Perhaps the most famous of these chimeric heroes or gods is Ganesha, a Kumar child whom the Hindu god Shiva xenografted the head of an elephant, transforming him into the god of intellect and wisdom, the patron of art and science, and the remover of obstacles [2].

Interestingly, the concept of transplantation transcended temporospatial boundaries and has been reported in Egyptian, Chinese, Indian, and early Christian mythology as early as 2000 BC [3]. In ancient Chinese texts, the physician Pien Ch'iao exchanged the hearts of two warriors to cure the unbalanced equilibrium between the two men's energies [4]. In the Old Testament, the prophet Ezekiel also refers to cardiac transplantation, stating "a new heart also I will give you… I will take away the stony heart out of your flesh, and I will give you a heart of flesh," perhaps the

W. P. Andrew Lee (✉) · S. Khalifian
Department of Plastic and Reconstructive Surgery, The Johns Hopkins University School of Medicine, Johns Hopkins Outpatient Center 8152F, 601 North Caroline Street, 21287 Baltimore, MD, USA
e-mail: WPAL@jhmi.edu

G. Brandacher
Department of Plastic and Reconstructive Surgery, Vascularized Composite Allotransplantation (VCA) Laboratory, The Johns Hopkins University School of Medicine, Baltimore, MD, USA

G. Brandacher (ed.), *The Science of Reconstructive Transplantation*,
Stem Cell Biology and Regenerative Medicine, DOI 10.1007/978-1-4939-2071-6_1

earliest reference of a cardiac allograft. However, the most famous legend highlighting transplantation in antiquity is the "miracle of the black leg" in the third century AD by Saint Cosmas, a physician, and Saint Damian, a surgeon [5]. These twin brothers removed the malignant and gangrenous leg of an aged Roman deacon of the church, and successfully transplanted the leg of a recently deceased Ethiopian Moor to the patient while he slept. Upon waking, the Roman discovered he had a new, healthy leg, albeit a black one. Although it is unlikely that any of these legends are derived from historical fact, it underscores human fascination with the practice of transplantation. In particular, the legend of Cosmas and Damian describes for the first time a new concept: cadaveric transplantation—where the body of the dead can help the living.

Transplantation in Antiquity

Experimentation with various types of transplants continued in the following centuries. Although the earliest recorded human surgery only dates back to the Bronze Age, archaeological evidence suggests that Proto-Neolithic [6] and Neolithic [7] civilizations performed trephination, the removal of a circular disc of calvarium to relieve intracranial pressure, and this practice continued well into recorded history. Yet, the first reference to organ transplantation does not occur until AD 200, when the Chinese physician Hua Tuo is said to have replaced diseased organs with healthy ones [8]. Unfortunately, all of Tuo's medical texts were destroyed and he was executed, causing his surgical practices and expertise to fall into disuse.

Although it seems unlikely that proper surgical techniques were available to facilitate successful performance of solid organ transplantation (SOT) at that time, the procedure of skin grafting had already been performed for many centuries. In the second century BC, the Indian surgeon Sushruta pioneered skin grafting and rotational pedicle flaps for nasal reconstruction and described over a dozen ways to reconstruct ears and lips. His forward-thinking concepts were compiled in the *Sushruta Samhita,* which served as the primary treatise on plastic and reconstructive surgery for centuries [9]. The prognoses for skin grafts improved considerably through the fifteenth century AD as these surgical techniques continued to be optimized.

In the sixteenth century, Gaspare Tagliacozzi emerged as a popular figure in transplantation, and he, like Sushruta, was a specialist in rhinoplasty. Tagliacozzi successfully performed nasal reconstruction on a patient who had lost his nose using a flap from the patient's upper arm—one of the earliest records of a human autograft. Although the practice of donor consent did not exist, tissue transfer from slave to master had become common during this time, and as the Italian Poet Calenzio wrote, slaves would often "donate" their noses to their masters [10]. Possibly due to sectarian and secular objections, nasal reconstruction via allografting became the subject of satire, and was criticized by writers such as Voltaire. Similarly, in Samuel Butler's play, *Hudibras* [11], he highlights the injustice of such practices in a scene that ends with, "When the date of Nock was out, off dropt the sympathetic

Snout." Although allografts were commonly attempted during this era, they did not enjoy the same success as autografts, and invariably failed. Tagliacozzi and other surgeon-scientists of the era became acutely aware that the limitations and failure of allografting were likely due to "the force and power of individuality" [12]. This was one of the earliest recorded indications that individual differences precluded tissue transfer between genetically disparate individuals; however, very little was known about genetics at this time, and the pioneering work of Mendel would not be conducted for another 250 years.

A Renaissance in Transplantation: The Nineteenth Century

Reports of successful allografts began to circulate around the turn of the nineteenth century, when the Scottish surgeon, John Hunter, successfully transplanted the testes of a chicken into a hen "without altering the disposition of the hen" [13]. These experiments led Hunter to conclude that "transplantation is founded on a disposition in all living substances to unite when brought into contact with each other;" one of the earliest revelations regarding chimerism and allograft acceptance, which seems compatible with current philosophy on the matter. In 1804, the Milanese surgeon Baronia claimed to perform successful autogenous and xenogeneic skin transplants as well as free tendon allografts in sheep [.14]. However, Baronia's results were unable to be duplicated, and were subsequently disputed by Paul Bert in his 1863 thesis, *De la Greffe Animale* [15]. In his manuscript, Bert describes his own animal experiments with many kinds of allogeneic and xenogeneic skin transplants and notes that the results of Baronia seemed unlikely. Nevertheless, experimental transplants continued and by the end of the century, stable corneal transplants were performed in animals and humans, and skin, tendon, nerve, tooth, and cartilage-free grafts had all been reported.

The 1800s were truly a Renaissance for transplantation after the darkness of the Middle Ages. Indeed, the concept of cadaveric transplantation reemerged in popular culture during this time, which is highlighted by Mary Shelley's popular novel *Frankenstein* [16]. The novel describes a physically and morally superior creature constructed with organs and parts taken from a graveyard. Unfortunately, the creature turns to violence after his creator rejects him, thereby providing one of the earliest positive and negative depictions of transplantation in literature.

At this point, organ transplantation was still not technically feasible due to limitations in suturing techniques for vascular anastomosis. This obstacle was overcome in 1902 when Alexis Carrel introduced his vascular anastomosis technique [17], which unlocked a continuum of research in experimental organ transplantation, including an orthotopic canine head transplant. Within the same year, Austrian surgeon Emmerich Ullmann performed the first experimental kidney transplantation in animals in Vienna [18], which was followed a few years later by reports of unsuccessful attempts in humans by Mathieu Jaboulay [19]. In 1905, Carrel and Charles Guthrie performed the first cardiac transplant in animals [20]; however, the graft was rejected early, which was hypothesized to be due to malnutrition of the grafting tissue by Paul Ehrlich in the following year [21]. Renal transplantation re-emerged as the most promising model of clinical transplantation, and allogeneic canine transplants, xenogeneic transplants into humans, and cadaveric human kidney transplants (performed in 1933 by Ukrainian surgeon Voronoy) were attempted—all of which were invariably unsuccessful [22].

Transplantation Immunology in the Twentieth Century: The Key to Success

The mechanisms of rejection were nebulous at this time and presented a major hurdle to successful organ transplantation. In the early 1900s, major advances were made in the understanding of humoral immunity; however, very little was known about lymphocyte function and cell-mediated immunity. The discovery of ABO blood groups in 1901 by Austrian biologist and physician Karl Landsteiner was a major advance for transplantation and led to the introduction of clinical blood transfusion [23]. In 1912, Murphy and Rous described the predominance of lymphocytes in a tumor rejection model, but several decades would pass before the activation of T cells was understood to be the molecular basis for acute allograft rejection.

Unfortunately, the 1930s marked a period of decline for transplant immunology research, due to limited success in skin and organ transplants secondary to rejection. Then came World War II and the bombings of cities led to a significant increase in burn victims in need of skin allografts. At this time, long-term outcomes for skin allografts were still plagued by a high failure rate due to the rejection response. This led to the pioneering work of Sir Peter Medawar, who transplanted skin onto badly burned soldiers in London, although the procedure was only successful when performed between identical twins [21]. Medawar concluded that the rejection of human skin allografts is a result of actively acquired immune reactions—work that was later summarized by Billingham, Brent, and Medawar in their manuscript "Actively Acquired Tolerance," which became the preeminent treatise on engineering the immune system [24]. In this manuscript, they implicitly discuss the importance of chimerism for tolerance induction, noting the development of donor-specific tolerance by injection of donor cells into neonatal animals. Soon thereafter, this concept was explicitly demonstrated by Main et al., when it was shown that an immature or immunologically weakened (irradiated) organism was prone to tolerance through chimerism induction [25]. This knowledge of immunology facilitated one of the most remarkable advances in medicine in the twentieth century: the advent of successful SOT.

Prior to the 1950s, early outcomes after organ transplantation were poor; however, the incorporation of the aforementioned findings led Dr. Joseph Murray to perform the first successful living-related kidney transplantation between identical twins in 1954 [26]. Shortly thereafter, a rapid development and utilization of chemical immunosuppressive drugs took place. The use of agents such as azathioprine,

6-mercaptopurine, and steroids in the 1960s allowed for the success of cadaveric renal transplants in 1962, and ushered in a new era in SOT [27]. Thereafter, many organs were successfully transplanted for the first time. In 1963, Dr. James Hardy successfully transplanted the first lung at the University of Mississippi, and in the following year, he attempted to transplant a chimpanzee heart into a critically ill man; unfortunately, the heart only beat for 90 min. By 1966, the first successful pancreas transplant was performed at the University of Minnesota in a patient with uncontrolled diabetes and kidney failure, which resulted in stabilization of blood glucose levels. Dr. Thomas E. Starzl performed the first liver transplant in the following year on a 3-year-old child, and 20 years later, he performed the first multivisceral transplant [28]. The first successful heart transplant was carried out by Christian Barnard in 1967, followed by the first successful bone marrow transplant by E. Donnall Thomas in 1968, and first successful small bowel transplant in 1988 by Goulet et al. In 1972, Jean-François Borel discovered cyclosporine, which further improved survival outcomes after transplantation [29]. Since then, many potent and more selective immunosuppressive agents have been developed, which has enabled SOT to become the standard of care for patients with end-stage organ disease.

History and Current Status of Reconstructive Transplantation

Given the early success of SOT and the development of immunosuppressive drugs to combat rejection in the 1960s, human hand transplantation was first attempted in Ecuador in 1964 [30]. Although the procedure was technically successful from a surgical standpoint, the currently available immunosuppressive agents at the time (steroids and azathioprine) were insufficient to prevent rejection, and the limb was ultimately amputated 2 weeks after transplantation [31].

The failure of this first hand transplant underscored the previously described immunological challenges of skin transplants between genetically different individuals. As a result, it was felt that the transplantation of any skin-bearing allograft would be an insurmountable hurdle and a second attempt was not carried out for over 30 years [32]. During this long hiatus, accumulating evidence suggested that the skin of a vascularized composite allotransplant behaved differently than an isolated skin graft, and, therefore, would not be rejected as stringently [33]. Furthermore, animal models of vascularized composite allotransplantation (VCA) used to investigate newer generations of immunosuppressive agents such as calcineurin inhibitors (e.g., tacrolimus) and antiproliferative agents (e.g., mycophenolate mofetil), indicated that the loss of vascularized composite allografts could now be prevented [34].

These findings led to the organization of independent clinical hand transplantation teams in Lyon, France, under Jean-Michel Dubernard, and in Louisville, Kentucky, led by Warren Breidenbach. The group of Dubernard successfully performed the first unilateral hand transplantation in September 1998 [35], followed shortly thereafter by Breidenbach et al. in January 1999 [36], and many more around the world since then. In January 2000, Dubernard's group successfully performed the first bilateral hand transplantation, and within 3 years, upper extremity transplantation was carried out at the level of the forearm for the first time performed at the Innsbruck Medical University in Austria by Raimund Margreiter's team [37]. Due to concerns over the capacity for nerve regeneration to occur over long distances, upper arm transplantation did not occur in adults until 2008, when the first double arm transplant was performed in Munich, Germany. These early cases provided ample evidence that graft survival could be achieved, and underscored the importance of appropriate immunosuppression management, patient compliance, and close follow-up. In the past decade, over 100 hand/forearm/arm transplants have been performed with encouraging aesthetic, functional, and immunologic outcomes that have exceeded all initial expectations [38].

The success of clinical hand transplantation, however, also heralded another new era for reconstructive transplantation. Seven years after the first hand transplant, Dubernard and Devauchelle performed the first successful face transplant in Amiens, France, in November 2005 [39]. Since then 28 face transplants have been performed at multiple centers around the world [40]. In addition, many other types of vascularized composite allografts have successfully been transplanted with highly encouraging functional and immunological outcomes including larynx [41], trachea [42], vascularized knee [43], femur, abdominal wall [44], tongue [45], penis, and uterus [46]. Although the surgical techniques to perform these complex procedures have been optimized, the widespread application of this reconstructive modality is still limited by the risks of lifelong, high-dose, multidrug, systemic immunosuppression needed to prevent graft rejection. Thus, continued progress in the field of transplant immunology is critical to the continued success of reconstructive transplantation, as minimization or elimination of immunosuppressive agents is a key goal in bringing this life-changing procedure to routine clinical applicability.

Indeed, VCA centers and researchers around the world have successfully attempted to develop novel concepts of immunomodulation to prevent rejection after reconstructive transplantation or to induce donor antigen-specific tolerance. Advances in immunosuppressive drug development and cell-based therapies combined with the unique elements and biology of a vascularized composite allograft (e.g., the vascularized bone marrow component/niche), which may supply a continuous source of donor-derived stem cells, have shown the most favorable results in achieving this goal [47–50].

Outlook

Many breakthroughs in medicine, science, surgery, and drug development facilitated the development of VCA into a critically important reconstructive modality and clinical realty for patients not amenable to conventional treatment options. In 2013, the US Department of Health and Human Services ultimately acknowledged the importance of reconstructive transplantation as a vital treatment option and successfully established to classify vascularized composite allografts as organs, with their allocation, data collection, and reporting falling under the purview of the Organ Procurement and Transplantation Network (OPTN) and United Network for Organ Sharing (UNOS). Along those lines, implementation of guidelines and regulatory oversight by professional societies and governing agencies such as UNOS will be critically important to further streamline any issues related to organ donation, allocation, and data management unique to reconstructive transplantation in the future. Although more and more centers are embarking on reconstructive transplantation and implementing VCA programs, the number of procedures performed to date is still too small to conduct any randomized clinical trials. Therefore, establishing standardized outcome measures such as uniform criteria and assessment tools for motor and sensory recovery or improvement in quality of life that will allow the comparison of data across institutions and across protocols will be an important task in the years to come.

The future for reconstructive transplantation seems promising and bright, with rapid scientific developments continuously expanding indications, improving outcomes, and inching ever closer to minimization of immunosuppression and tolerance induction. Indeed, Dr. Cesar Milstein may have said it best during his acceptance of the Nobel Prize in 1984 for his discovery of the principles for the production of monoclonal antibodies:

> Although the way ahead [for immunology] is full of pitfalls and difficulties, this is indeed an exhilarating prospect. There is no danger of a shortage of forthcoming excitement in the subject. Yet, as always, the highlights of tomorrow are the unpredictabilities of today.

References

1. Reemtsma K. Xenotransplantation—a brief history of clinical experiences: 1900–1965. In: Cooper DKC, Kemp E, Reemtsma K, White DJG, editors. Xenotransplantation. Berlin: Springer; 1991. pp. 9–22.
2. Brown RL. Ganesh: studies of an Asian god. Albany: State University of New York. SUNY; 1991.
3. Conolly BW, Benanzio M. Cosmas and damian revisited. Hand Transplantation. Milan Berlin Heidelberg New York. Springer; 2007. pp. 3–10.
4. Kahan BD. Pien Ch'iao, the legendary exchange of hearts, traditional Chinese medicine, and the modern era of cyclosporine. Paper presented at: Transplantation proceedings 1988.
5. da Varagine J. Leggenda aurea, trad. it. di C. *Lisi, Firenze*. 1952.
6. Solecki RS, Solecki RL, Agelarakis AP. The proto-neolithic cemetery in shanidar cave. Vol. 7. Texas: Texas A & M University Press; 2004.
7. Skvarilova B, Povýsil C, Horak M, Kyzlinkova K. Trephinations—old surgical intervention. Acta chirurgiae plasticae. 2007;49(4):103.
8. Tubbs RS, Riech S, Verma K, Chern J, Mortazavi M, Cohen-Gadol AA. China's first surgeon: Hua Tuo (c. 108–208 AD). Child's Ner Syst. 2011;27(9):1357–60.
9. Bhishagratna KL. The Sushruta Samhita. Calcutta, Vol. 1. 1907.
10. Hauben DJ, Baruchin A, Mahler D. On the history of the free skin graft. Ann Plast Surg. 1982;9(3):242–6.
11. Butler S. Hudibras. Frederick London: Warne; 1887.

12. Tagliacozzi G. De Curtorum Chirurgia per insitionem. London: Bindonus; 1596.
13. Martin C. John Hunter and tissue transplantation. Surg Gynecol Obstet. 1970;131(2):306.
14. Baronio G. Degli innesti animali. Stamperia e Fonderia del Genio, Milan; 1804.
15. Bert P, Martinet IdÉ. De la greffe animale. Paris: J.-B. Baillière et Fils; 1863.
16. Shelley M. Frankenstein. Broadview Press, Peterborough, ON Broadview; 1999.
17. Carrel A. La technique opératoire des anastomoses vasculaires et de la transplantation des viscères. Lyon: Association typographique; 1902.
18. Schultheiss D, Bloom DA, Wefer J, Jonas U. Tissue engineering from Adam to the zygote: historical reflections. World J Urol. 2000;18(1):84–90.
19. Jaboulay M. Greffe de reins au pli du coude par soudures arterielles et veineuses. Lyon Med. 1906;107(39):575–83.
20. Carrel A, Guthrie CC. Anastomosis of blood vessels by the patching method and transplantation of the kidney. JAMA 1906;47(20):1648–51.
21. Baar H. From Ehrlich-Pirquet to Medawar and Burnet; a revolution in immunology. J Maine Med Assoc. 1963;54:209.
22. Hamilton D. Kidney transplantation: a history. In: Morris PI, editor. Kidney transplantation: principles and practice. 4th ed. London: Saunders; 1994. S. 1.
23. Landsteiner K, Miller CP Jr. Serological studies on the blood of the primates: II. The blood groups in anthropoid apes. J Exp Med. 1925;42(6):853.
24. Billingham RE, Brent L, Medawar PB. Quantitative studies on tissue transplantation immunity. II. The origin, strength and duration of actively and adoptively acquired immunity. Proc R Soc Lond B. 1954;143(910):58–80.
25. Main JM, Prehn RT. Successful skin homografts after the administration of high dosage X radiation and homologous bone marrow. J Natl Cancer Inst. 1955;15(4):1023–9.
26. Guild WR, Harrison JH, Merrill JP, Murray J. Successful homotransplantation of the kidney in an identical twin. Trans Am Clin Climatol Assoc. 1955;67:167–73.
27. Merrill JP, Murray JE, Takacs FJ, Hager EB, Wilson RE, Dammin GJ. Successful transplantation of kidney from a human cadaver. JAMA. 1963;185:347–53.
28. Starzl TE. History of clinical transplantation. World J Surg. 2000;24(7):759–82.
29. Borel J. Comparative study of in vitro and in vivo drug effects on cell-mediated cytotoxicity. Immunology. 1976;31(4):631.
30. Gilbert R. Transplant is successful with a cadaver forearm. Med Trib Med News. 1964;5:20.
31. Gilbert R. Hand transplanted from cadaver is reamputated. Med Trib Med News. 1964;5:23.
32. Schneeberger S, Khalifian S, Brandacher G. Immunosuppression and monitoring of rejection in hand transplantation. Tech H and Up Extrem Surg. 2013;17(4):208–14.
33. Lee WP, Yaremchuk MJ, Pan YC, Randolph MA, Tan CM, Weiland AJ. Relative antigenicity of components of a vascularized limb allograft. Plast Reconstr Surg. 1991;87(3):401–11.
34. Jones JW Jr, Ustuner ET, Zdichavsky M, et al. Long-term survival of an extremity composite tissue allograft with FK506-mycophenolate mofetil therapy. Surgery. 1999;126(2):384–8.
35. Dubernard JM, Owen E, Herzberg G, et al. Human hand allograft: report on first 6 months. Lancet. 1999;353:1315–20.
36. Jones JW, Gruber SA, Barker JH, Breidenbach WC. Successful hand transplantation—one-year follow-up. New Engl J Med. 2000;343(7):468–73.
37. Schneeberger S, Ninkovic M, Gabl M, et al. First forearm transplantation: outcome at 3 years. Am J Transplant. 2007;7(7):1753–62.
38. Petruzzo P, Lanzetta M, Dubernard JM, et al. The international registry on hand and composite tissue transplantation. Transplantation. 2010;90(12):1590–4.
39. Devauchelle B, Badet L, Lengele B, et al. First human face allograft: early report. Lancet. 2006;368(9531):203–9.
40. Khalifian S, Brazio PS, Mohan R, et al. Facial transplantation: the first 9 years. Lancet. 2014 Dec 13;384(9960):2153–63.
41. Strome M, Stein J, Esclamado R, et al. Laryngeal transplantation and 40-month follow-up. N Engl J Med. 2001;344:1676–9.

42. Delaere P, Vranckx J, Verleden G, De Leyn P, Van Raemdonck D. Tracheal allotransplantation after withdrawal of immunosuppressive therapy. N Engl J Med. 2010;362(2):138–45.
43. Hofmann GO, Kirschner MH. Clinical experience in allogeneic vascularized bone and joint allografting. Microsurgery. 2000;20(8):375–83.
44. Levi DM, Tzakis AG, Kato T, et al. Transplantation of the abdominal wall. Lancet. 2003;361(9376):2173–6.
45. Birchall M. Tongue transplantation. Lancet. 2004;363(9422):1663.
46. Diaz-Siso JR, Bueno EM, Sisk GC, Marty FM, Pomahac B, Tullius SG. Vascularized composite tissue allotransplantation—state of the art. Clinical transplantation. 2013;27(3):330–7.
47. Leonard DA, Kurtz JM, Cetrulo CL Jr. Achieving immune tolerance in hand and face transplantation: a realistic prospect? Immunotherapy. 2014;6(5):499–502.
48. Leonard DA, Kurtz JM, Mallard C, et al. Vascularized composite allograft tolerance across MHC barriers in a large animal model. Am J Transplant. 2014;14(2):343–55.
49. Brandacher G, Lee WP, Schneeberger S. Minimizing immunosuppression in hand transplantation. Expert Rev Clin Immunol. 2012;8(7):673–83; quiz 684.
50. Schneeberger S, Gorantla VS, Brandacher G, et al. Upper-extremity transplantation using a cell-based protocol to minimize immunosuppression. Ann Surg. 2013;257(2):345–51.

Chapter 2
Clinical Pearls and Pitfalls in Reconstructive Transplantation

Huey Y. Tien, Yorell Manon-Matos, Tsu-Min Tsai, Christina L. Kaufman and Joseph E. Kutz

Historical Perspective

The field of limb replantation and subsequently hand transplantation is founded on the work done by Jacobson and Suarez [31], Kleinert et al. [21], Buncke and Schultz [9], Tamai et al. [25], Chen and Bao [10], who developed techniques for the anastomosis of blood vessels as small as 1 mm in diameter. Based on decades of experience at our center in replantation of digits and the hand and forearm [11, 20, 22–24] as well as at that of our colleagues [33], an approach to hand transplantation was developed that has been modified over the past 15 years, based on our outcomes and the types of patients encountered. As presented by others, a transplant is much like a replant, but there are distinct differences [14]. The field of reconstructive transplantation encompasses more than hand and arm transplantation, and can be applied to the restoration of other composite tissues such as the face [30] and the larynx [6]. This technique has application anywhere the surgeon needs to replace "like with like," and conventional reconstructive surgery fails to restore function and cosmesis. Potential applications include vascularized joint transplantation such as the wrist and the knee. Trials have been initiated in vascularized knee transplants, which have demonstrated some of the challenges that come with transplantation of less well-vascularized tissue [12, 17].

This chapter discusses the technique of hand transplantation and highlights some of the challenges with respect to the harvest of the donor organ, the preparation of the recipient, with a focus on the sequence of surgical procedures of transplantation of the donor hand to its new recipient.

J. E. Kutz (✉) · H. Y. Tien · Y. Manon-Matos · T.-M. Tsai · C. L. Kaufman
Christine M. Kleinert Institute for Hand and Microsurgery, Kleinert Kutz Hand Care Center, 225 Abraham Flexner Way, Louisville, KY 40202, USA
e-mail: jkutz@cmki.org

G. Brandacher (ed.), *The Science of Reconstructive Transplantation*,
Stem Cell Biology and Regenerative Medicine, DOI 10.1007/978-1-4939-2071-6_2

Indications/Contraindications

The process of considering a patient for hand transplantation is an extensive one. There are multiple and complex medical, social, and financial considerations. Over the past 15 years, inquiries from more than 600 potential hand transplant candidates have been screened at this center. To date, eight of these patients have received a transplant. Current requirements of chronic systemic immunosuppression to maintain the graft require the patient be in good health. Rigorous compliance with respect to medication and rehabilitation therapy demands persons of strong character and work ethic, with plenty of social support. Finally, it is imperative to ensure that the patient will have the insurance coverage or means to pay for immunosuppressive medication and the extensive clinical care required to monitor and treat side effects of the immunosuppression. Hand transplantation should only be considered for patients who have failed conventional reconstructive surgery and prosthetic therapy, and meet all of the previous criteria.

An additional contraindication that must be considered for a hand transplant is the presence of preformed donor-specific antibodies between the donor and recipient. In two of our patients, preformed antibodies against specific human leukocyte antigen (HLA) types were present, and as such, one donor was turned down for the first recipient and ten donors were incompatible for the second candidate, who has a high level of panel-reactive antibody (PRA). For patients with a PRA of more than 0% and anti-HLA antibodies of known specificity, this center has an exclusion criterion that the recipient must have a negative crossmatch against donor cells before the operation can proceed. In one case, an inadequate donor sample resulted in a technically difficult crossmatch assay. In turn, this delayed the start of the operation by an additional 2 h. Situations like this must be taken into consideration for coordinating the start of the team for the preparation of the recipient, and the back table preparation of the donor hand. Care must be taken that neither team starts too early. Of note, the structures in the donor arm will be of known length. Conversely, in the recipient, the soft tissue structures are almost always proximal to the level of bone amputation, and may be encased in scar. As such, the recipient dissection is usually a much longer procedure than the 1–2 h required for dissection of the donor extremity.

Other issues that may delay the start of the procedure are issues with the donor arteries, especially as the arteries in the arm may have been used for the placement of intravenous or intra-arterial lines in the donor (Fig. 2.1). Once a patient is determined to be a potential donor for hand transplantation, intravenous access lines in the upper extremities should be removed and relocated to another site such as the groin. No further needle sticks in the donor graft should be allowed. In order to avoid warm ischemia time, the donor graft should be kept at 4 °C for as long as possible during the dissection process, and while waiting for the completion of the recipient dissection.

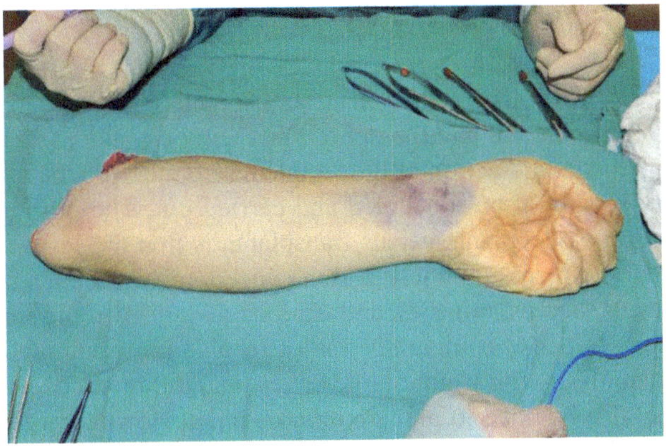

Fig. 2.1 Donor arm on the back table prior to dissection. Note the bruising after the removal of an intravenous line used in the donor prior to donation. The use of lines in the donor arm may compromise the vessels

Technique

Some surgeons have referred to hand transplantation as nothing more than a replant under ideal conditions. This might be true if you compare the allogeneic reconstruction of a guillotine amputation with ample pristine tissue from the donor to the replantation of a crushed and avulsed extremity with significant contamination and warm ischemia time. This highlights the major differences between the two techniques. In transplantation, there is plenty of donor tissue, and the surgeon is under less pressure to make a decision whether the damaged recipient tissue should be conserved. Another primary difference is that in transplants bone and tendon length must be carefully adjusted to match the recipient for a proper biomechanical function of the graft. Often in replantation, the bones and tendons are already at or near the appropriate length.

The specifics of the techniques of hand transplantation revolve around five different areas: (1) preoperative planning, (2) harvest and preparation of the donor graft, (3) preparation of the recipient and transplant of the donor graft, (4) immunosuppression and management of graft rejection, and finally (5) rehabilitation and therapy of the allograft to ensure the best possible function. The last two areas are beyond the scope of this chapter, and are mentioned only briefly. This chapter concentrates on the preoperative planning, harvest, and preparation of the donor, as well as the preparation of the recipient and attachment of the donor graft for transplantation of the upper extremity.

Preoperative Planning for the Operation

Once a patient has been accepted for hand transplantation, preoperative planning with respect to the approach and the donor graft will be required. If the transplant will require an unusual dissection of the recipient or donor, practice sessions in the cadaver laboratory should be scheduled and carried out. The primary differences between the surgery of replants versus transplants is that the level of amputation is different in the recipient and donor and the uncertainty of the length and quality of recipient vessels, nerves, muscles, and tendons remain until the surgery actually starts. Some centers perform imaging studies to obtain a roadmap of the recipient vascular anatomy by ultrasound [14]. Other imaging methods such as MRI or angiograms can also be used to prepare this roadmap. However, regardless of the imaging studies of the recipient prior to transplant, our experience has been that extent and quality of the patient's vasculature and other soft tissues may look different from expected upon dissection. The surgeon will not know for certain what there is to work with until the recipient has been explored intraoperatively.

As part of the planning process, a transplant algorithm should be developed. This algorithm is used by the nurse coordinator and the lead surgeon to identify who should be called and when they should be called when a donor becomes available. At minimum, there should be two teams of surgeons, with at least two surgeons in each team who are experienced in performing replants. The first team's primary responsibility is to go on the procurement run and harvest the donor graft and perform the dissection on the back table in the recipient's operating room (OR). The second team's primary responsibility is to perform the recipient dissection in preparation for the donor graft. If a bilateral transplant is being performed, four teams are needed. Hand transplantation is a long procedure and while a large team can reduce fatigue, care must be taken to avoid confusion and miscommunication about what has been completed when multiple surgeons are entering and leaving the OR. In our experience, teams of two surgeons rotated with each step, i.e., two for the dissection of the recipient, two for the osteosynthesis, two for the artery repair, etc. Each step takes approximately 2 h to complete (the osteosynthesis less time, and tendon and vein repair more time). Therefore, while in most cases multiple teams of two surgeons rotated throughout the transplant, once the donor dissection is completed, two teams could alternate between the steps of the procedure. This would give each team ample rest breaks and reduce surgeon fatigue. With this setup, a team of four surgeons would be sufficient to perform a hand transplant.

In addition, skilled anesthesiologists, nurses, and support staff are critical. Hypovolemia, metabolic acidosis, and reperfusion syndrome are risks of replantation [5, 7, 13, 32] and transplant patients must be carefully monitored by both surgical and anesthesiology teams. Recently, Caterson et al. have hypothesized that these ischemia reperfusion injury events may be a primary force causing graft injury and initiating rejection of the graft [8]. Preoperative planning, coordination, and communication with the surgical teams are of paramount importance to reduce ischemia time and operative complications.

Preparation of the Recipient

The recipient is prepped for surgery, has repeat laboratories, and has blood drawn for the donor crossmatch. The recipient is dosed with Benadryl and Solumedrol prior to infusion of thymoglobulin or alemtuzumab. Infusion of these induction agents may continue perioperatively. It is the experience at our center that the surgery is unaffected by these medications. Surgery is performed under general anesthesia and axillary block. Good collaboration with the anesthesiologist is necessary to maintain adequate circulatory status throughout the case. Quite often the patient comes to the OR relatively hypovolemic. The adjunctive use of alpha-mimetic agents for circulatory support should be avoided.

Harvest and Preparation of the Donor Graft

Acceptance of the donor occurs in stages; first, the parameters are communicated to the organ procurement organization (OPO) with respect to age, sex, skin tone, and general size. This information is used by the OPO coordinators to determine which donor families should be approached about hand donation. Additionally, the OPO reviews the donor with respect to infectious disease markers and serologic compatibility. Our program matches for blood type and the recipient must not have preformed antibodies against the donor. Both the donor and the recipient are HLA typed to determine Digital Signature Algorithm (DSA), but HLA type is not currently a consideration in accepting a donor. Finally, a radiograph of the entire donor upper extremity is sent to the hand surgery team for evaluation of whether the bone size is a good match with respect to length of the donor and the recipient. The hand surgeons also rule out any congenital or acquired bone or joint problems in the donor before accepting the graft.

The details of the recovery of the donor hand have been previously reported [2]. In most cases, recovery of the donor hand was performed prior to the recovery of the solid organs. A sterile tourniquet was placed on the donor upper extremity above the elbow, and a fishmouth incision is made at or slightly proximal to the elbow, and the elbow was disarticulated.

The order of dissection and tagging is as follows:

- Cephalic and basilic veins (obtaining sufficient extra length for vein grafts, vessels should be procured as proximally as possible, the level of the tourniquet being the limiting factor).
- Medial antebrachial cutaneous nerve (MACN) and the lateral anebrachial cutaneous (LACN) are dissected and tagged.
- The brachial artery and veins are dissected and tagged.
- Ulnar, median, and radial nerves are dissected and tagged.
- The medial and lateral epicondyle muscles are elevated in a subperiosteal plane. The biceps, brachialis, and triceps tendons are transected.

- An elbow capuslotomy is performed and the joint disarticulated. In the case of an above-the-elbow transplant, the arm would be disarticulated at the shoulder.

The donor graft is then prepared for transport to the recipient. If the donor and the recipient are in the same hospital, the donor graft would not be harvested until immediately before the recipient is taken to the OR to reduce ischemia time as much as possible. More frequently, the graft has to be prepared for transport at a separate hospital, which is often located in a different state. Currently, the graft is perfused with cold (4 °C) preservative solution. Our center prefers University of Wisconsin (UW) solution [3], but histidine–tryptophan–ketoglutarate (HTK) solution (Custodiol) was used in one case. The preservation solution serves a dual purpose of inducing hypothermia, which slows metabolism of the cells in the graft, and maintaining the intracellular electrolyte balance, thus extending the time before irreversible cell death occurs. The preservative is a proprietary mix that comes in bags and is connected to the graft via a cannulated artery (brachial artery in the cases harvested to date at our center) and then cold solution is flushed through the graft until the back flow is clear and the blood has been flushed from the graft. A drip of the UW solution is then attached and allowed to back drip through the veins. The graft is then dressed in gauze moistened with sterile saline and placed in a sterile plastic bag. The entire graft is placed in a cooler with ice and transported to the recipient site.

Our protocol allows for up to 12 h of cold ischemic time, starting from disarticulation of the extremity if our team is the first to procure, or from cross-clamp of the donor if the hand recovery team is following one of the solid organ recovery teams. Our group has harvested extremity grafts from a hospital a few blocks from the recipient OR, to as far away as Texas (about 950 miles from recipient hospital). The longest cold ischemic time from disarticulation to preparation of the donor hand in the OR has been 11 h. Of note, experimental evidence suggests a safe cold ischemia time might be much longer than 12 h. In a model of canine forelimb replantation, it was demonstrated that grafts could be held for up to 3 days and remain viable after replantation [4]. In this study, one forelimb of eight puppies and seven dogs was amputated, perfused with iced Collins solution, maintained at 4 °C for 72 h (78.5 h total anoxia), and replanted. Five animals were followed for 1 year to assess bone growth. Additional animals underwent bone labeling on days 1, 8, and 15, and were sacrificed at 22 days to assess osteocyte survival. Osteocytes survived replantation in all dogs and one puppy; most osteocytes died in two puppies. In five long-term puppies, central epiphyseal growth was disturbed, but the peripheral portions maintained nearly normal growth, with almost normal bone length being achieved at 1 year [4]. This data with carefully preserved limb replants with long cold ischemia time is in contrast to warm ischemia times which are extremely detrimental and contribute to ischemia reperfusion injury (IRI) and metabolic acidosis, especially when large muscles are reperfused [33]. Sabapathy et al. suggest that mid-forearm to wrist level replantation not be performed at all if warm ischemia exceeds 8 h, and that replantation is contraindicated with more than 6 h of warm ischemia for more proximal level replantation [28]. While keeping the allograft for 3 days before transplantation is not advocated, it is likely that cold ischemia time of longer than 12 h is possible in clinical VCA procedures. Reperfusion injury is a risk of limb transplantation. Proximal limb amputations with larger muscle mass are at higher risk of reperfusion syndrome, which can include multiorgan system failure and death [5, 7, 13, 32]. These risks are increased by more warm ischemic time. In a study of 14 patients who underwent replantation of an upper or lower limb, with 6 h or more of warm ischemia time, multiple reperfusion syndrome events were noted [33]. Four of the 14 patients suffered intraoperative hypotension with or without acidosis, and one child suffered acute bronchospasm following revascularization of the limb. One female patient died as a result of the hypotension and acute metabolic acidosis [33]. These risks will be mitigated in hand transplantation as there is some control of the donor warm ischemia time. It is the warm ischemia time that is most damaging to the neuronal and muscle tissues of the graft [5]. In the case of bilateral transplantation, the potential large amount of tissue and muscle mass may increase this risk. This is one rationale for in support of unilateral transplantation in clinical trials.

Surgical Dissection of the Recipient

Once the graft has been provisionally accepted by the transplant team, the recipient is notified, made nil per os (NPO), and arrangements are made to get the recipient to the hospital. At the time of graft procurement, the recovery team needs to ensure that lymph nodes are obtained from the donor so that a donor crossmatch can be performed. This test takes about 4 h to perform once the tissue typing laboratory has both the donor tissue and a sample of the recipient's serum. If the clinical protocol lists the presence of donor-specific antibodies as an exclusion criterion, the surgery cannot proceed. With new procedures that allow the tissue typing laboratory to perform "virtual crossmatches," some centers do not require a donor crossmatch to be performed prior to starting the dissection of the recipient. Our center has adopted the policy of requiring a prospective donor crossmatch if the recipient has known preformed specificities and a panel-reactive antibody (PRA). If the recipient is at low risk, i.e., has a PRA of 0 % and no known specificities identified on the virtual crossmatch, a retrospective donor crossmatch is requested to be performed on the next business day, and the recipient dissection can begin as soon as the donor graft is at the recipient hospital.

Upon arrival to the hospital, the recipient is rescreened for infectious disease markers, re-consented by the lead hand surgeon, and prepared for surgery. When it is clear that the graft will be acceptable for transplantation, the patient is given the required peri-transplant medications and immunosuppression, placed supine on the operating table and anesthetized.

Preparation of the recipient limb or limbs should occur at the same time as preparation of the donor graft on the back table. In cases of crush or burn injury, there may be significant scarring. Some centers use ultrasound imaging studies to map the veins prior to transplant [14], but this has not been found to be necessary in our

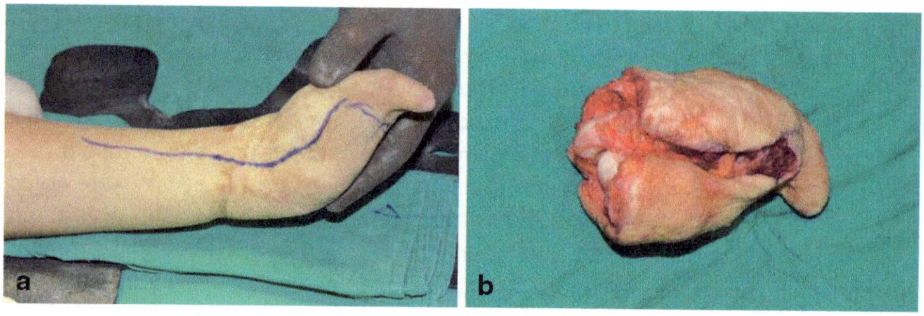

Fig. 2.2 Preparation of the dissection of the recipient, taking care to preserve as much of the recipient tissue as possible. Note in Fig. 2.2b that the digital nerve has been dissected as far as possible

experience. A tourniquet is placed on the recipient's extremity that will receive the transplant. The incision used has been the mosaic or zigzag incision, to allow for tissue swelling, or in case that more skin is needed to close at the end of the case.

Careful dissection and identification of structures are paramount, and it is at this point the real status of recipient tissue is identified. In several of our cases, once the recipient tissues were fully exposed, it was found that the recipient arteries were smaller and shorter than expected; the tendons had retracted significantly, and were of insufficient length for straightforward reconstruction. Many times, fewer tendons were available than expected. In one case, despite a mid-forearm amputation, limited extensor tendons were available and the patient required a weave to give tenodesis. The surgeon should be aware that anatomic landmarks may be distorted due to the injury to the recipient, and the exact anatomy may be difficult to predict. Significant variation in anatomy and quality of the tissue should be expected. Versatility is required of the surgical team during this portion of the operation. As with the donor dissection, it is critical to label the identified structures. Our center keeps sterilized waterproof labels in the OR specifically for this purpose. Once blood flow is restored, the surgical field changes dramatically and blood and edema can make identification of small structures very difficult. Other centers use indelible ink on Esmarch bandages sewn to the structure with 2-0 silk [14].

Depending on the level of the amputation, the recipient's extremity is dissected in such a manner that will preserve as much native tissue as possible for use in the transplant. This includes all vessels and nerves to the largest extent possible (Fig. 2.2). In general, the nerves are dissected and labeled first, then tendons, and then arteries and veins that will be used. The stump is then turned over and any available tissue such as the sensory branch of the ulnar and radial nerves is dissected and labeled (Fig. 2.3). Once all of the structures have been dissected and labeled, the radius and ulna in the case of a mid-forearm amputation are prepared for fixation with the donor graft. Communication between the teams is paramount, especially as length, and even availability of structures within the recipient may not be known until the actual surgery. Care must be taken that nerves, tendons, and vessels are of adequate length in both directions prior to bone fixation of the recipient and donor.

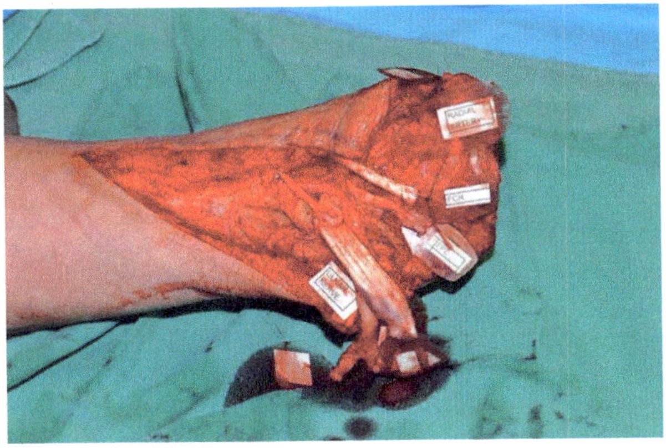

Fig. 2.3 Labeling of the recipient structures

Preparation of the Donor Graft

The donor graft is brought to the recipient's OR and is dissected and prepared for transplantation on a sterile back table. In the example of a mid-forearm transplant, radial and ulnar skin flaps are created taking care to preserve the subcutaneous veins. Dissection is routinely started from the volar aspect. Care is taken to identify and label the relevant structures such as the cephalic veins, flexor tendons, median and ulnar nerves, radial and ulnar arteries, etc. Preserving more length than is needed is essential in order to compensate for a lack of recipient tissue. Attention is then turned to the dorsal aspect in a similar fashion. The extensor tendons are then isolated and labeled, as well as any available blood vessels. At the mid-forearm level, only the superficial branch of the radial nerve needs to be identified in the donor. The extensor tendons are isolated and labeled as well as any blood vessels (Fig. 2.4). Following identification of these structures, the donor muscle bellies are detached and discarded. The osteotomy is usually performed at the middle third of the forearm, in communication with the recipient team. The entire dissection is carried out using cool conditions and irrigation fluids.

In several of our transplants, a long donor brachial artery was used. The brachial artery was dissected, taking care to not skeletonize and minimally disturb the peri-adventitial tissues. In the case of mid-forearm transplantation, this technique reduces ischemia time, allowing the team to restore arterial blood flow with a single anastomosis, and reduces the risk of complications related to the anastomosis. However, there is debate as to whether the dissection of significant length of the artery from surrounding tissue may increase the risk of intimal hyperplasia, as has been observed in some vascular autografts [27]. More recent studies suggest that skeletonization of thoracic artery grafts may produce better results than pedicle grafts [1]. While still a matter of discussion at our center, in the most recent cases, the approach in mid-forearm procedures has been to repair and connect the donor and recipient at the level of the radial and ulnar arteries, rather than use a long brachial

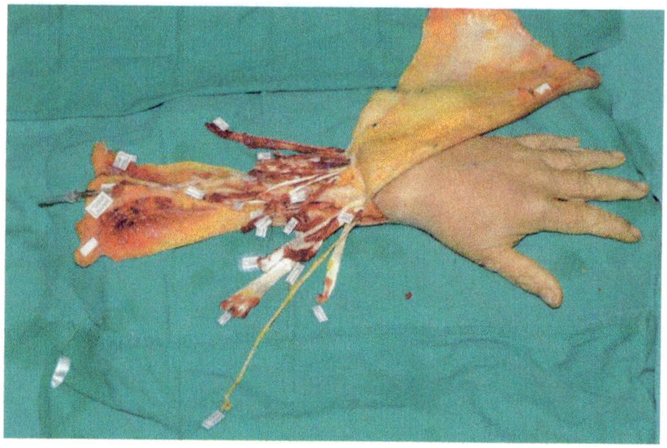

Fig. 2.4 Labeling the donor structures

artery. This was done for reasons of technical ease and to prevent additional dissection into the recipient extremity.

As with replants, the order of repair at any center is at the discretion of the attending surgeons and is planned with the best possible outcome as the goal. In general, our center repairs bone, arteries, veins, flexor tendons, nerves, extensor tendons, and then skin. The sequential repair of the flexor tendons before the extensor tendons allows for better adjustment of proper tendon tension. Some centers have chosen to repair nerves before the arteries, as the bloodless field reduces technical difficulty [14, 18], but this is the choice of the surgeon.

Bone Fixation

Osteosynthesis is the first step of transplantation. Achieving skeletal appropriate length and stability will set the foundation for the subsequent steps of the surgery (Fig. 2.5). Our goal is to achieve a limb length that is identical to native, but shortening may be accepted in order to allow ease of the arterial and tendon reconstruction. Even minimal extra length may mandate the use of vein grafts to perform arterial and venous repairs. Depending on the level of amputation, the radius and the ulna of the donor are measured to ensure matching recipient length of the contralateral arm, and positioned to allow proper alignment and the best chance for bony union. The osteotomy is created perpendicular to the longitudinal axis of the radius and ulna and in most cases a low-profile 3.5-mm locking compression plate was used for fixation.

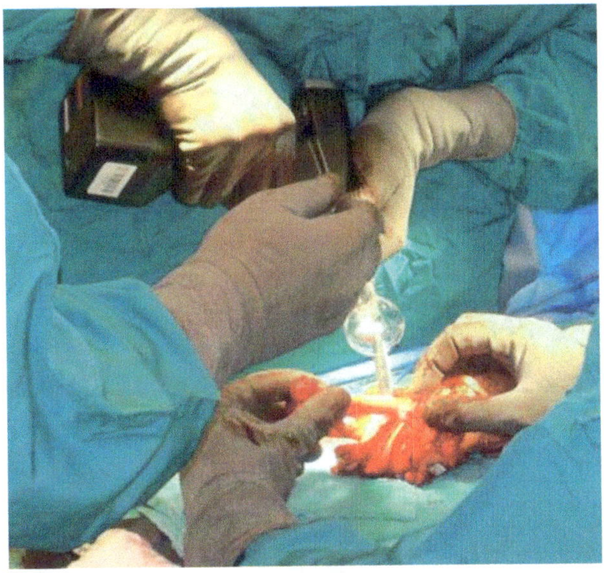

Fig. 2.5 Osteotomy prior to osteosynthesis

Restoration of Blood flow

Following osteosynthesis the arterial anastomosis are created. Because of the high concentration of potassium in UW solution, care has been taken to flush the donor graft with physiologic solution such as Ringer's saline or plasma prior to the reperfusion. This is in order to avoid cardiovascular complications, such as hyperkalemic cardiac arrest or bradyarrhythmia. The radial and ulnar arteries are most often the first to be repaired, followed by the venae comitantes veins with simple end-to-end anastomosis using the microscope. As the recipient may have significant scarring, in most cases simple end-to-end repairs are performed. When the vessels have retracted proximally into the recipient arm the use of end-to-side anastomosis is preferred. The end-to-side anastomosis may provide a stronger repair and better preserve blood flow to the graft (Fig. 2.6). Following arterial repair, at least some of the venae comitantes are also repaired in the same fashion. The skin is then provisionally repaired to protect the underlying anastomoses before proceeding with the tendon repairs. These repairs are followed by the flexor tendons, the nerves, and then the extensor tendons.

The use of vein grafts should be avoided in favor of primary anastomosis, as vein grafts add significantly to operative time and increase the risk of anastomotic complications. Also, it is very important to limit the amount of soft tissue retraction during the anastomosis to avoid kinking when the surrounding soft tissue and skin flaps are released. Finally, in order to flush the system of toxic products developed during the warm and cold ischemia times, once arterial blood flow is established (Fig. 2.7),

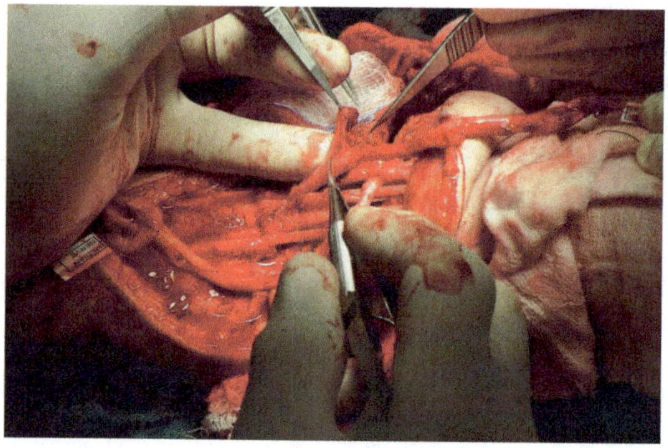

Fig. 2.6 End-to-side arterial repair

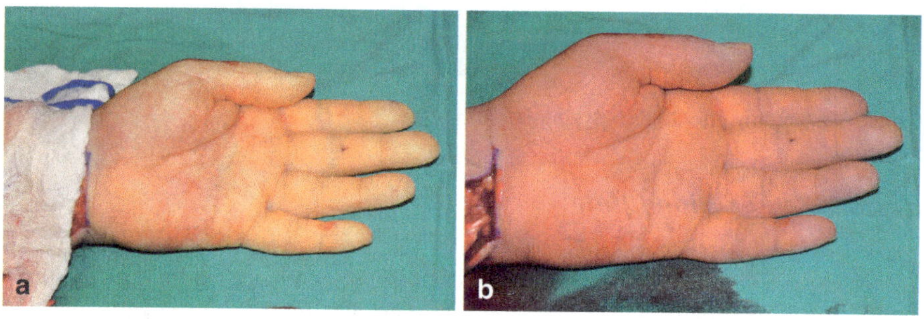

Fig. 2.7 Establishment of blood flow to the hand (2.7a and 2.7b)

the surgeon should clamp the large veins but allow blood to flow to flush out the toxic products of warm and cold ischemia from the donor graft. This flushing is allowed to continue for approximately 5 min before continuing with the procedure.

At this point, the surgical team and anesthesiologist need to collaborate in monitoring the patient's circulatory system. It is very easy to underestimate the extent of hypovolemia due to hemorrhage from the venous system. The use of alpha-mimetic agents to increase blood pressure can increase the technical difficulty of vessel repairs and decrease perfusion of the allograft. It is critical to have packed red blood cells and other blood products available.

Nerve Repair

Following restoration of blood flow, the nerves are repaired next, more as matter of technical ease. This is at the discretion of the surgical team. In all cases, the anastomosis is done as distal to the graft as possible to reduce the length that axons must

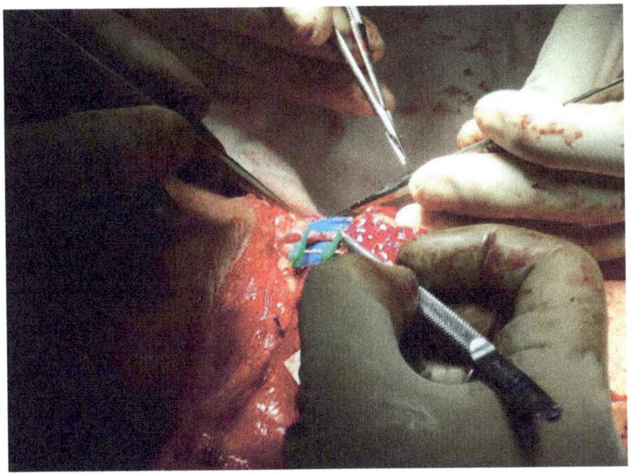

Fig. 2.8 End-to-end epineurial repair of nerves

regenerate and speed up the nerve recovery time post transplant. The anastomosis is done as a simple end-to-end epineurial repair (Fig. 2.8). Typically, the anastomosis is at the level of the wrist. For the median nerve, this is proximal to the motor branch while for the ulnar nerve it may require a separate anastomosis of the dorsal sensory branch. If desired, a collagen-based nerve wrap can be used to protect the repair, but this is at the discretion of the surgeon.

In preparation of the recipient, great care is taken to preserve as much of the recipient nerve as possible, to reduce the length of nerve regeneration as much as possible. In the case of more distal transplants, a more detailed connection of sensory and motor nerves can be performed. The branches of the sensory and motor nerves of the ulnar and medial nerves can be mapped and/or stained to achieve the best possible functional outcome. The goal is to make the nerve repair as close to the target tissues (muscle, skin) as much as possible.

Tendon Repair

Repair of the tendons can be challenging. This stage of the procedure will require versatility and patience on the part of the surgeon because of unexpected distortions in the anatomy and landmarks within the injured recipient extremity. In general, flexor tendons are repaired first. This will make it easier to achieve proper tension with the extensor tendons. When all tendons are present, a primary repair is performed end to end with a core and epitendinous suture. In all of the cases transplanted here, a strong repair of tendons was performed so that active movement can be started as soon as possible after transplant. In general, the flexor tendons are repaired first, with either a Pulvertaft weave, or a Tsai six-strand suture. The Pulvertaft weave is preferred if tendon length allows. The tension of repair is adjusted to

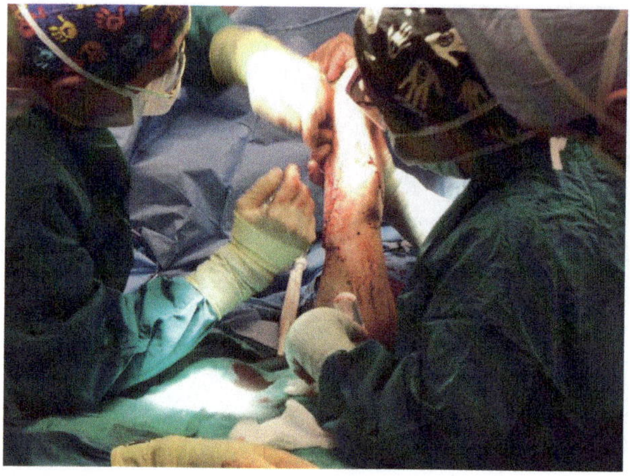

Fig. 2.9 Assessment of the tension and function of the tendon repairs

the natural flexion arcade of the fingers. In some cases where insufficient recipient tendons are present, a tendon graft is added. For example, one recipient required an extension to be able to fasten the extensor to the thumb, which was through the brachial radialis to the thumb extensor. A key element of the hand transplant is to ensure proper tension between the extensor and flexor tendons so that good function is achieved. Once the repair of the flexor tendons is complete, the volar skin is provisionally closed. However, before attaching the skin to the dorsal structure, a vein is repaired quickly to facilitate venous drainage of the allograft. It must be emphasized that versatility is required of the surgical team during the repair of the extensor tendons. Anticipate that all extensors or sufficient muscle will not be present. For good function, at minimum, the following should be attached: one wrist flexor tendon; one flexor digitorum profundus (FDP) tendon to the middle, ring, and small finger; one FDP to the index finger; and one flexor pollicis longus (FPL) to the thumb. With respect to the extensor tendons, a minimum of one tendon to the extensor carpi radialis brevis (EcRB) muscle; one *extensor* digitorum communis (EPC) to the middle, ring, and small finger; and one EPC to the index finger should be attached. A minimum of one tendon to the abductor pollicis longus (APL) muscle and the extensor pollicis longus muscle/tendon should be repaired. Of course, the ideal situation is a one-to-one recipient to donor tendon repair, but if this is not possible, good function can be obtained with these repairs at minimum. All repairs are tensioned to the normal tenodesis of the hand (Fig. 2.9).

Venous Repair and Outflow

The majority of the venous outflow occurs through the deep veins, most of which have been repaired at this point. However, the superficial veins, such as the cephalic vein, provide significant contribution to outflow. Once the arteries, nerves,

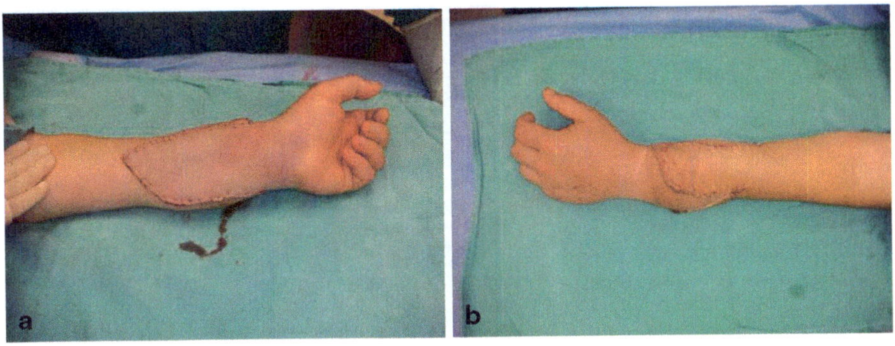

Fig. 2.10 a volar side of hand transplant, b dorsal side of hand transplant

and tendons have been repaired, significant time and energy are now devoted to the venous repair. The challenge is that there may be significant fatigue on the part of the surgeon. Rotating teams can be very helpful at this point. While the veins to be anastomosed are several millimeters in diameter, surrounding soft tissue edema and bleeding can make the repair difficult. Some centers advocate a prophylactic carpal tunnel release as they have dealt with significant post-transplant edema [14]. Our center has not chosen this approach.

Skin Closure

As the hand transplant procedure is lengthy, significant swelling of the graft can occur and there is often not enough skin to loosely close. Another consideration is that the orientation of venous repairs within the skin flap may require larger flaps that expected. If very long skin flaps (highly encouraged) from the donor were obtained, revision of the flaps may be required prior to closing. In the process of closing, more dorsal veins can be anastomosed to further improve drainage of the graft. (As many additional veins as possible should be repaired.) A zigzag or mosaic four-flap interposing incision results in a closure that reduces scar contracture (Fig. 2.10). The skin is closed over an implantable thermocouple, with a light pressure dressing. The hand is stabilized and protected in an above-elbow splint with the elbow at 90°, with the hand in a functional position. The inner phalangeal joints are placed in extension, the wrist is in slight extension (10°) and the metacarpophalangeal joints are in slight flexion.

Early Postoperative Care

Postoperatively, the hand is kept just above heart level. After the first few days, the plaster splint is removed and the recipient is fitted with a crane outrigger brace [16, 29] (Fig. 2.11). The patient is encouraged to get out of bed as soon as possible and

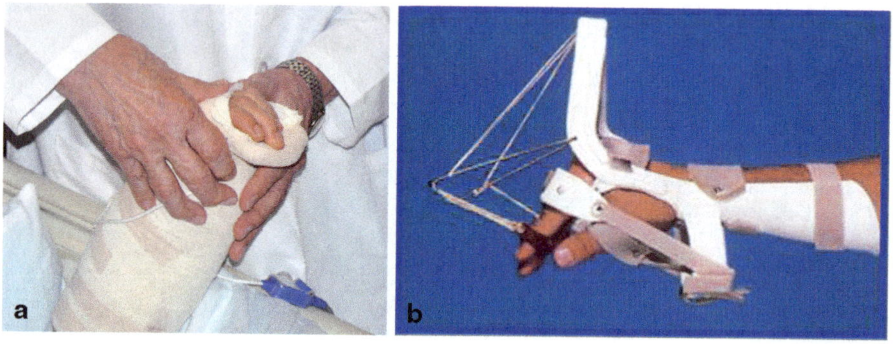

Fig. 2.11 **a** Plaster cast after skin closure and **b** Crane outrigger brace placed at day 3–5 post transplant

start active motion of the graft. Physical therapy sessions are started by day 3–4 post transplant. The crane outrigger brace allows early therapy to begin focusing on range of motion (ROM) and edema management.

Discussion

In the opinion of the authors, the important considerations for optimal outcome of hand transplantation surgery are:

- Patient selection, with an emphasis on work ethic, realistic expectations, and family support.
- There should be sufficient recipient tissue to power the donor graft. In repair of old amputations, the soft tissue is often not equal to the recipient bone in length.
- Appropriate bone fixation, which may require shortening to facilitate soft tissue reconstruction.
- Use a microscope for anastomosis of arteries, veins, and nerves.
- Create nerve anastomosis as distally as possible.
- After flexor tendon repair, provisionally close volar skin and anastomose a vein to facilitate drainage.
- Venous repair will take the most time and takes place at the end of the operation when surgeon fatigue is high. Plan accordingly.
- Minimize soft tissue retraction during vessel repairs to avoid kinking after skin closure
- Versatility in the hand surgery team is required to handle these unexpected situations. Rotating teams of hand surgeons to minimize fatigue is highly encouraged.
- Support of transplant, anesthesia, critical care specialties is paramount—preventing under resuscitation, compensating for underestimation of blood loss
- Physical therapy with certified hand therapists familiar with replants, and early movement is critical to good functional outcomes

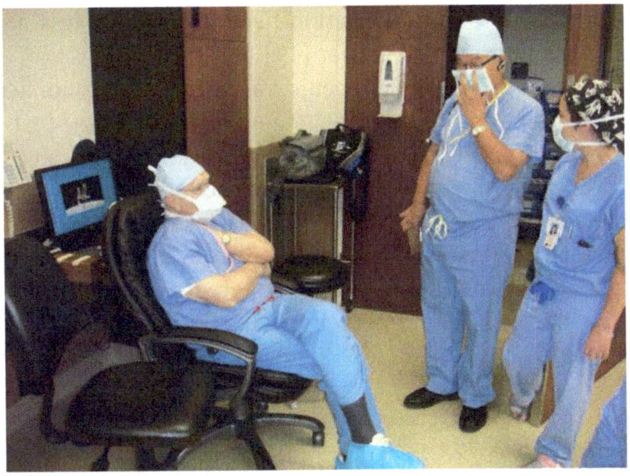

Fig. 2.12 Dr. Tsai and Dr. Kutz discussing the case

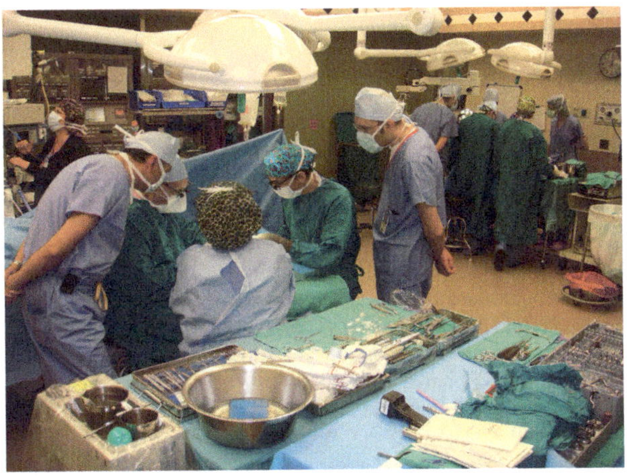

Fig. 2.13 Concurrent surgical teams working on recipient and donor dissection

The field of VCA is still young, with less than 150 transplants of hands and faces worldwide. Nonetheless, outcomes to date have exceeded early expectations [15], and the number of centers who are starting hand transplant programs is expanding. About 25% of the cases performed to date are more than 5 years post transplant (www.handregistry.com (Oct 2013), [26]). Clinical follow-up ranges from a few months to nearly 15 years. With the successes, there have been both graft losses and patient mortality. These losses have occurred primarily in the first year after the transplant. Graft losses that occur after the first year have been related primarily to patient compliance. As such, patient selection is extremely important. Programs are strongly encouraged to have potential candidates speak to someone who has had a hand or face transplant. Chronic rejection-like sequelae are less frequent than in solid organ transplantation, but do appear [19].

In successful cases, VCA recipients enjoy a quality of life not achievable with conventional reconstruction or currently available prosthetics. Most graft loss has occurred relatively early, suggesting that efforts to improve should focus on this time period. More follow-up is needed to determine the rates and targets of chronic rejection, and the characteristics of VCA unique to face versus hand transplantation. Ultimately, tolerance induction will be needed to allow the widespread application of this treatment alternative. However, adherence to the considerations listed above should give the hand transplant recipient the best outcomes currently possible (Figs. 2.12 and 2.13).

Acknowledgments This work has been supported by grants from the Department of Defense, Office of Army Research and Office of Navy Research, as well as institutional support from Jewish Hospital, part of Kentucky One Health, the Jewish Hospital Foundation, the Christine M. Kleinert Institute and the Kleinert Kutz Hand Care Center.

Disclosure: The authors do not have any conflicts to disclose regarding this manuscript.

References

1. Ali E, Saso S, Ashrafian H, Athanasiou T. Does a skeletonized or pedicled left internal thoracic artery give the best graft patency? Interact Cardiovasc Thorac Surg. 2010;10(1):97–104.
2. Banegas RN, Moreno R, Duggal A, Breidenbach WC, III. Surgical aspects of donor hand recovery for transplantation. J Reconstr Microsurg. 2012;28(1):21–6.
3. Belzer FO. Clinical organ preservation with UW solution. Transplantation. 1989;47(6):1097–8.
4. Beppu M, Takahashi F, Tsai TM, Ogden L, Sharp JB Jr. Experimental replantation of canine forelimbs after 78.5 hours of anoxia. Plast Reconstr Surg. 1989;84(4):642–8.
5. Beyersdorf F, Schlensak C. Controlled reperfusion after acute and persistent limb ischemia. Semin Vasc Surg. 2009;22(1):52–7.
6. Birchall MA, Lorenz RR, Berke GS, Genden EM, Haughey BH, Siemionow M, et al. Laryngeal transplantation in 2005: a review. Am J Transplant. 2006;6(1):20–6.
7. Blaisdell FW. The pathophysiology of skeletal muscle ischemia and the reperfusion syndrome: a review. Cardiovasc.Surg. 2002;10(6):620–30.
8. Caterson EJ, Lopez J, Medina M, Pomahac B, Tullius SG. Ischemia-reperfusion injury in vascularized composite allotransplantation. J Craniofac Surg. 2013;24(1):51–6.
9. Buncke H, Schulz W. Total ear reimplantation in the rabbit utilising microminiature vascular anastomoses. Br J Plast Surg. 1966;19(1):15–22. PMID 5909469.
10. Chen ZW, Bao YS. Microsurgery in China. Clin Plast Surg. 1980;7(4):437–74.
11. Chew WY, Tsai TM. Major upper limb replantation. Hand Clin. 2001;17(3):395–410, viii.
12. Diefenbeck M, Nerlich A, Schneeberger S, Wagner F, Hofmann GO. Allograft vasculopathy after allogeneic vascularized knee transplantation. Transpl Int. 2011;24(1):e1–5.
13. Eltzschig HK, Collard CD. Vascular ischaemia and reperfusion injury. Br Med Bull. 2004;70:71–86.
14. Hartzell TL, Benhaim P, Imbriglia JE, Shores JT, Goitz RJ, Balk M, et al. Surgical and technical aspects of hand transplantation: is it just another replant? Hand Clin. 2011;27(4):521–30, x.

15. Hatrick NC, Tonkin MA. H and transplantation: a current perspective. ANZ J Surg. 2001;71(4):245–51.
16. Hodges A, Chesher S, Feranda S. Hand transplantation: rehabilitation: case report. Microsurgery. 2000;20(8):389–92.
17. Hofmann GO, Kirschner MH. Clinical experience in allogeneic vascularized bone and joint allografting. Microsurgery. 2000;20(8):375–83.
18. Jablecki J, Kaczmarzyk L, Patrzalek D, Domanasiewicz A, Boratynska Z. First Polish forearm transplantation: report after 17 months. Transplant Proc. 2009;41(2):549–53.
19. Kaufman CL, Ouseph R, Blair BW, Kutz JE, Tsai TM, Scheker LR, et al. Graft vasculopathy in clinical hand transplantation. Am J Transplant. 2012;12(4):1004–16.
20. Kleinert HE, Tsai TM. Microvascular repair in replantation. Clin Orthop Relat Res. 1978;(133):205–11. http://journals.lww.com/corr/pages/issuelist.aspx?year=1978.
21. Kleinert HE, Kasdan M, Romero JL. Small blood vessel anatomosisi for salvage of severely injured upper extremity. J Bone Joint Surg Am. 1963;45:788.
22. Kleinert HE, Serafin D, Kutz JE, Atasoy E. Reimplantation of amputated digits and hands. Orthop Clin North Am. 1973;4(4):957–67.
23. Kleinert HE, Jablon M, Tsai TM. An overview of replantation and results of 347 replants in 245 patients. J Trauma. 1980a;20(5):390–8.
24. Kleinert HE, Jablon M, Tsai TM. An overview of replantation and results of 347 replants in 245 patients. J Trauma. 1980b;20(5):390–8.
25. Komatsu S, Tamai S. Successful replantation of a completely cut-off thumb. Plast Reconstr Surg. 1968;42:374–7.
26. Murphy BD, Zuker RM, Borschel GH. Vascularized composite allotransplantation: An update on medical and surgical progress and remaining challenges. J Plast Reconstr Aesthet Surg. 2013a;66(11):10. p. 1449–55.
27. Purcell C, Tennant M, McGeachie J. Neo-intimal hyperplasia in vascular grafts and its implications for autologous arterial grafting. Ann R Coll Surg Engl. 1997;79(3):164–8.
28. Sabapathy SR, Venkatramani H, Bharathi RR, Bhardwaj P. Replantation surgery. J Hand Surg Am. 2011;36(6):1104–10.
29. Scheker LR, Hodges A. Brace and rehabilitation after replantation and revascularization. Hand Clin. 2001;17(3):473–80.
30. Siemionow M, Ozturk C. Face transplantation: outcomes, concerns, controversies, and future directions. J Craniofac Surg. 2012;23(1):254–9.
31. Suarez EL, Jacobson JH. Results of small artery endarterectomy-microsurgical technique. Surg Forum. 1961;12:256–7.
32. Tantry TP, Kadam D, Shenoy SP, Bhandary S, Adappa KK. Perioperative evaluation and outcomes of major limb replantations with ischemia periods of more than 6 hours. J Reconstr Microsurg. 2013;29(3):165–72.
33. Wood MB. Finger and hand replantation. Surgical technique. Hand Clin. 1992;8(3):397–408.

Chapter 3
Reconstructive Transplantation: What Can We Learn from Solid Organ Transplantation?

Philip S. Brazio, Eduardo D. Rodriguez, Stephen T. Bartlett and Rolf N. Barth

Reconstructive transplantation is a relatively new field with as-yet uncharted boundaries, while solid organ transplantation (SOT) has been developing for over half a century. Although many lessons from SOT may be applied to vascularized composite allotransplantation (VCA), novel challenges also arise. Until a larger body of evidence can be built to guide reconstructive transplantation, it will be crucial to distinguish which lessons translate effectively, and which problems demand unique solutions (Table 3.1).

Patient and Center Selection

Patient selection is the starting point for transplantation. Patients who may benefit from transplantation may be still deemed inappropriate by patient selection committees because of inability to take care of an allograft subsequent to confounding medical and social factors. Patient selection committees are composed of surgeons, physicians experienced in immunosuppression (IS), transplant coordinators, social workers, psychiatrist, nutritionist, financial counselors, and others each with a respective voice in selecting individuals to be acceptable for transplantation and placed on a transplant waiting list.

The selection of medically appropriate candidates has obvious similarities between organ and VCA. Patients must not have major cardiovascular, cerebrovascular, infectious, or malignant contraindications to both major surgery and lifelong IS. The technical considerations of VCA include vascular, neurologic, skeletal, and soft

R. N. Barth (✉) · P. S. Brazio · S. T. Bartlett
Department of Surgery, University of Maryland School of Medicine, 29 S. Greene Street, Suite 200, Baltimore, MD 21201, USA
e-mail: rbarth@smail.umaryland.edu

E. D. Rodriguez
Department of Plastic Surgery, New York University School of Medicine, New York, NY, USA

G. Brandacher (ed.), *The Science of Reconstructive Transplantation*,
Stem Cell Biology and Regenerative Medicine, DOI 10.1007/978-1-4939-2071-6_3

Table 3.1 Summary of similarities and differences between solid organ transplantation and vascularized composite allotransplantation

Area	Common characteristics	Considerations unique to VCA
Patient selection	Selection against contra-indications to surgery or IS Minimize risk of noncompliance	Awareness of PTSD Availability of long-term psychiatry/counseling Expectations management Higher standards for informed consent given non-lifesaving
Center selection	Centers with multidisciplinary expertise	More restricted volume and number of centers
Donor selection	Immunological matching Donor health/age	Increased recipient sensitization Additional anatomic/cosmetic requirements Current need for personal examination of donor
Allograft procurement, preservation	Goal of decreasing ischemic time, especially warm ischemia	Variable graft anatomy, margins Possibly increased complexity, time of in situ dissection Unknown acceptable cold ischemia time
Organ allocation, sharing	Regulation by OPTN/UNOS under HRSA, DHHS Possibility for shipping under cold ischemia	Increased stringency of procurement possibly causing geographical limitations Lower demand
Immunosuppression	Standard 3-drug regimen as baseline for most protocols Active progress in IS minimization protocols employing cellular therapies	Heightened graft immunogenicity Non-lifesaving graft decreases acceptability of IS complications
Rejection	Treatable with steroid boluses, increased IS	Availability of topical therapies Unknown role of chronic Ab-mediated rejection Unknown long-term graft survival
Complications	High risk of local, systemic infection	Technical complications, including cosmetic issues, less robust blood supply to graft Goals of sensory/motor functional restoration

VCA vascularized composite allotransplantation, *PTSD* posttraumatic stress disorder, *IS* immunosuppression, *HRSA* Health Resources and Service Administration; *DHHS* Department of Health and Human Services, *OPTN* Organ Procurement and Transplantation Network *UNOS* United Network for Organ Sharing, *Ab* antibody

tissue matching that has substantial variation between every patient. The technical approach for each patient is highly unique and planned to an extent not necessary for most solid organ transplant recipients.

Psychosocial criteria center on recipients who must be carefully chosen to minimize the risk of noncompliance causing rejection. Liver transplantation for alcoholic hepatitis in particular adds the risk of graft failure from recidivism. In VCA, changes in self-perception and extensive rehabilitation demands should increase the stringency of the selection process.

The SOT literature has provided extensive analysis of risk factors for medication nonadherence. Histories of substance abuse or pretransplant nonadherence, poor understanding of the transplant and IS, and low perceived self-efficacy [1–3] should raise concern for nonadherence. Beyond medication compliance, VCA (and especially extremity) recipients must participate actively in potentially grueling physical therapy programs. Because of this, VCA recipient selection committees should further evaluate recipient motivation level for rehabilitation.

While all SOT recipients do not generally require psychiatric assessment, VCA recipients require both psychiatric assessment and possible plans for long-term management. VCA is unique in that the mechanism of disease is usually traumatic; significant psychiatric trauma may also manifest in the form of posttraumatic stress disorder. This may require specific attention that will be new to transplant professionals. Furthermore, the principles of SOT do not adequately address the subjective satisfaction of VCA recipients. Function of a solid organ is essentially binary: the graft functions or it does not. "Perfect" graft function in a VCA (total restoration of native form and function) may never be feasible, and grafts fall along a continuum of partial restoration. Because of these difficulties, reconstructive surgeons commonly stress management of patient expectations.

Appreciation of a defect by living for a period of time with deformity can increase satisfaction with reconstruction [4, 5], and is considered by some essential to providing informed consent for VCA. A parallel with regard to timing may be seen in fulminant hepatic failure. Patients may undergo the entire selection and transplantation process while intubated and sedated, unexpectedly emerging with a lifesaving allograft that demands lifetime medication compliance. While these recipients largely have good outcomes even in cases of fulminant alcoholic hepatitis [6], VCA are unlikely to be lifesaving with a few possible exceptions such as chest or abdominal wall. Because of this, an acute stage transplant will likely require higher justification. One face transplant has been described in which multiple failed salvage attempts and a large exposed defect and threatened a patient's life, leading to urgent transplantation [7]. The outcomes from this case are likely to be scrutinized to determine whether the parallels of emergent liver transplantation truly apply.

The selection of transplant center for VCA will have unique challenges in comparison to SOT. Currently, transplant centers are supervised by specific government regulations and required for public disclosure of patient and graft outcomes on a biannual basis. VCA has not achieved a place as a standardized therapy, and while the government has recently issued guidelines for the oversight of VCA in parallel to SOT [8], most VCA have been performed as parts of institutional review board (IRB)-approved clinical trials. Outcomes from individual centers are not accessible for either patients or referring physicians to make decisions. Just as in SOT, the surgeon must be part of a multidisciplinary team, including transplant physicians, psychiatrists, social workers, immunologists, pathologists, and nurse coordinators. The recipient must be able to count on lifelong follow-up for IS and its potential complications. Finally, the size and technical expertise of the surgical team itself cannot be discounted. Unlike SOT, reconstructive transplants may violate conventional anatomic boundaries and require a team of multiple surgeons working in tandem for over 24 h. The expertise of one or two surgeons will not be sufficient in these cases. It is likely that reconstructive transplantation, even more so than SOT, will remain the province of only select few academic centers with the necessary infrastructure and multidisciplinary resources.

Donor Selection

VCA requires an equivalent level of immunological stringency as SOT, with additional constraints imposed by extensive recipient sensitization as well as needs for anatomical compatibility not encountered in SOT.

As with SOT, current immunosuppressive regimens largely allow transplantation to occur regardless of human leukocyte antigen (HLA) type. Although a VCA itself does not demand more stringent crossmatching, certain characteristics of VCA recipients may impose severe constraints on donor selection: VCA recipients are likely to have undergone multiple reconstructive procedures, possibly with allogeneic biological materials and multiple blood product transfusions over time. The virtual crossmatch, as in kidney transplantation, is useful for narrowing the donor pool appropriately in these patients.

The assessment of tissue quality for potential VCA donors includes donor characteristics such as age, and comorbidities, including vascular disease that may shorten the life span of the graft. As opposed to recipient tobacco use, donor tobacco use is not a significant donor factor in SOT and should not be expected to affect VCA outcomes.

In addition to screening techniques translated from SOT, physical parameters must also be accounted for in VCA. Skin color, gender, and basic morphometric parameters such as intercanthal distance, mandibular width, and upper anterior facial height can be easily checked against the potential recipient. Unlike SOT, however, many of these parameters are malleable according to surgeon and recipient preference. Physical selection stringency can be adjusted without affecting immunological outcomes, for example, by considering opposite-gender face donors for a highly sensitized recipient.

Unlike SOT, in which remotely transmitted data can be used to judge organ quality and suitability for a given recipient, to date no VCA has been reported in which the allograft was accepted by the transplant team without the opportunity for physical examination. While purely objective criteria may one day be developed using color matching, cephalometry, and computed tomography, these data currently serve only as adjuncts to the personal examination of a prospective donor.

Allograft Procurement and Preservation

Surgical teams typically recover solid organs in a tightly choreographed multiteam operation that begins with thoracoabdominal incision and ends 2–3 h later with removal of all visceral organs in immediate sequence [9, 10]. In contrast, the variability of VCA between organ type and recipient defect extent prevents prescription of a standard protocol.

If a donor is a candidate for VCA but not solid organ donation, the sequence is simple. For a multiorgan donor, however, some level of coordination is necessary. Beginning VCA procurement first may threaten solid organ integrity if donor perfusion is impaired by prolonged anesthesia or blood loss. Because they are potentially lifesaving, solid organs must ultimately be given priority. Organs such as extremities, abdominal walls, or smaller face allografts, which can be procured quickly and are unlikely to impact donor physiology, can continue to be procured before the start of thoracoabdominal dissection. VCA such as trachea and uterus, which lie deeper and can be quickly isolated after cardiac death, are more suited to procurement following cold cardioplegia infusion and cardiac explantation, in sequence with abdominal organ procurement. Abdominal vessel cannulation for rapid cooling of VCA is unlikely to interfere significantly with solid organ procurement.

Complex face allografts represent a unique challenge in procurement. Isolation of multiple vessels, nerves, muscles, and osseous structures in situ [11] may require well over 12 h. The ideal approach would begin with face procurement and coordinate teams' actions so that each procurement could conclude immediately following systemic heparinization of the donor: a "face-first, concurrent completion" approach. It would also incorporate mechanisms for proceeding quickly to procurement of solid organs if their integrity is threatened [12].

Clinical and basic research has provided solid organ transplant surgeons guidelines regarding the amount of time each type of organ can withstand cold ischemia. Extrapolating from free tissue transfer and hand transplantation, ischemia times under 4 h should be well tolerated by VCA [11, 13, 14]. The skeletal muscle component is likely to be less tolerant of cold ischemia than solid organs [15]. Prolonged subcritical ischemia time may furthermore increase risk of rejection [16].

Unlike solid organs, however, VCA do not have an established acceptable cold ischemia time. The greatest difference compared to SOT is an unprecedented level of variability in the graft itself, which may never allow "hard" temporal limits on VCA cold ischemia. Tolerable ischemia time is heavily influenced by type and proportion of tissues. SOT typically are constant in size and tissue content, encapsulated within a fascial layer and attached only by their vascular pedicles and ductal structures. The margins of a VCA, on the other hand, vary following the recipient defect not only in breadth but also in depth and in types of tissue.

Despite this variability, surgeons performing VCA would benefit significantly from guidelines establishing maximum cold ischemia limits for defined categories of tissue types, for example, soft tissue only, musculocutaneous, osteocutaneous, and osteomyocutaneous. Further research may additionally help to identify whether size of the allograft influences speed of initial cooling and therefore tolerance to further ischemia.

Substantial research has been conducted in both solid organs and VCA regarding the optimum means of organ preservation. While various strategies, including modification of preservation solutions and pulsatile perfusion, have been attempted to decrease cold ischemic damage, the standard approach remains simple cold storage in a fluid such as University of Wisconsin solution.

Organ Sharing and Allocation

Standardized techniques for organ procurement allow the sharing of organs on regional and national scales. If VCA preservation strategies are optimized, it is conceivable that logistics could allow sharing between different centers. Allograft type and complexity are likely to represent the largest limit to organ sharing. Even an extremity allograft could potentially be procured with generous margins, shipped, and customized to the recipient defect on the back table.

Organ sharing and allocation is a perpetual topic of discussion in SOT, while, VCA may never require this type of scheme. Such strategies in SOT are motivated by the chronic shortage of donors for a large population of patients in need of transplantation. This is illustrated by 2012 UNet data from the US Organ Procurement and Transplantation Network. The median adult waiting time for in the past decade was 4–5 years for a kidney and 4–14 months for a liver. These wait times reflect high transplant volume, and higher demand for organs: In 2012, 34,000 kidney and 11,000 liver recipients were added to waiting lists, compared to 14,000 donors in the same year. In contrast, the yearly volume of potential VCA recipients is unlikely to exceed 0.1–1 % of these figures in the near future.

To this date, the longest reported waiting period for a VCA has been 8 months, not due to competition for the same graft but rather by factors, including recipient sensitization, strict anatomical criteria, limited geographical search area, and low public awareness leading to reticence to approaching every potential donor family about VCA donation. Allocation of VCA in the future will likely be limited by these factors as well as small overall pools of both donors and recipients. Initial work will need to focus on increasing the donor pool by boosting public awareness of VCA donation and transitioning to a standard-of-care approach. Subsequent efforts should attempt to establish a system for defining objective VCA allograft requirements for a given recipient, maximizing the donor pool by expanding geographic range. Allocation schemes that determine priority based on severity of disease (such as Model for End-Stage Liver Disease (MELD) for liver) or time on waiting list (as in kidney transplantation) are less likely to be relevant or necessary.

The 2013 decision to regulate VCA as organs under the auspices of the Health Resources and Service Administration, Department of Health and Human Services [8], will have significant impact in allocation decisions. Although specific schemes and time lines have not yet been established, the Organ Procurement and Transplantation Network/United Network for Organ Sharing will have the power to regulate VCA. The experience and organization added by this oversight may accelerate and expand the possibilities for organ sharing.

Immunosuppression

VCA adopts IS management regimens directly from well-established organ transplant protocols. Induction has performed with T cell depletion consisting of either rabbit anti-thymocyte globulin (Thymoglobulin) or alemtuzumab (Campath). T cell depletion is utilized for the majority of SOT, but rarely required for liver transplantation [17, 18]. While almost all VCA have been performed as clinical research, most maintenance IS medications are the same triple therapy as in SOT: a calcineurin inhibitor (tacrolimus), antimetabolites (mycophenolate mofetil), and steroids. VCA still have not achieved widespread success with steroid elimination protocols. Experimental adjuncts, including infused donor bone marrow (BM), have been added in some protocols as an attempt to minimize rejection and reduce immunosuppressive requirements [19–21]. The addition of infused BM has still resulted in multiple rejection episodes and some graft losses [22]. The use of BM outside of tolerance protocols utilizing nonmyeloablative conditioning in SOT has likewise not achieved reliable success as an immunomodulatory approach.

VCA research has been motivated with similar goals to SOT to achieve immunologic tolerance [23]. The rationale for tolerance in VCA has been based on the near universal rejection episodes and concern for a high requirement of lifelong IS in a younger patient population. The belief that tolerance would decrease the risks of VCA to a more acceptable level justified numerous mixed chimerism approaches that have been utilized in renal transplantation [24]. While studies in rodent models have achieved immunologic tolerance to VCA [25], no reproducible large-animal model has demonstrated tolerance induction to disparate tissue elements, including skin and mucosal tissues (which appear to be among the most immunogenic). The additional challenge for VCA compared to successful tolerance approaches is that all VCA will come from deceased donors. SOT tolerance protocols have utilized living donors with the ability to plan for a transplant date and perform necessary preconditioning prior to surgery. While some studies have attempted to develop protocols for the induction of tolerance to deceased donors that would be accomplished at the time of transplant and afterwards, these have not been successfully applied in clinical trials.

One of the central strategies to establish tolerance in experimental and clinical SOT has been the establishment of recipient chimerism—defined as either transient or durable detection of donor lymphohematopoietic cells in the peripheral blood of recipients. This requires infusion of donor BM in SOT. VCA is unique in that many VCA grafts contain vascularized BM that may function as a self-contained and self-renewing reservoir of BM cells [26]. This has demonstrated immunomodulatory properties in nonhuman primate studies that have associated the presence of VBM with significant immunomodulatory capacity to prevent graft rejection and prolong graft survival compared to grafts without vascularized BM, when combined with immunosuppressive medications [27]. Additionally, the skin may function as a source of dendritic cells that may facilitate chimerism and tolerance [28]. These unique elements of VCA, as compared to SOT, may contribute to the better than expected half-life of VCA, which remains undefined but does not appear any shorter than for SOT despite early and near universal rejection episodes.

Rejection

Vascularized composite allografts appear to be at higher risk of rejection than solid organs, with first-year rejection of any grade approaching 100 %. The first-year rejection rate for renal allografts treated with lymphocyte depletion and weaned off steroids as part of IS minimization can be under 10 % [29]. Two factors likely contribute to this contrast: higher immunogenicity of tissues, especially skin, and higher detection of even subtle rejection by directly visualizing the graft. This is a very unique aspect of VCA compared to SOT—that daily monitoring for rejection is possible. SOT depends on significant clinical dysfunction for the development of clinical signs, and subsequent laboratory and histologic analysis of biopsy specimens.

The mainstays of treatment for acute cellular rejection have translated successfully from SOT to VCA. Most rejection is treated with intravenous boluses of corticosteroids, followed by tapers of varying length. Initial rejection episodes when moderate to severe have been treated with additional T cell depletion provided by thymoglobulin. Unique to VCA is the ability to locally treat acute rejection with topical formulations of corticosteroids and tacrolimus. These have been reported to be successful for Banff grade I rejection, but more commonly are added to systemic agents. Studies in small-animal models suggest efficacy may be comparable to systemic tacrolimus, but this has not been verified clinically [30]. Antibody-mediated rejection has not yet appeared with the same frequency as SOT, but one reported case necessitated treatments similar to SOT, including eculizumab and bortezomib (both without success), followed by plasmapheresis, thymoglobulin, and alemtuzumab [31]. Chronic rejection has become increasingly recognized in both research and clinical models of VCA and similar to the vascular manifestations in every transplanted solid organ. Chronic allograft vasculopathy, as described in several clinical extremity transplants and a nonhuman primate model, involves severe intimal hyperplasia and vasculopathy without donor-specific antibodies or C4d deposition [32]. Unlike SOT, vasculopathy in VCA is focused on vessels that may not be accessible by routine biopsy. Thus, alternate modalities, including high-resolution ultrasound, may allow for detection and quantification of vasculopathy [32].

The long-term survival of VCA is the clinical question looming largest at this time. VCA losses to date have been due to acute rejection, infection and overwhelming sepsis, medication noncompliance, and malignancy. This is similar to SOT graft losses. Chronic rejection results in 18 % of all renal allograft losses at 5 years [33] and medication noncompliance is an additional major etiology. SOT experience has defined each allograft with a half-life: 3.6, 5.2, 8.5, 8.8, 11, and 16.7 years for bowel, lung, liver, kidney, heart, and pancreas [34, 35] in order of increasing longevity. This has not been defined for VCA. It is conceivable that the ability to immediately detect and treat even subtle rejection, leading to shorter and less severe rejection episodes, may decrease cumulative injury to the graft resulting in longer half-lives.

Complications of Immunosuppression

Direct effects of IS include renal failure and diabetes. Renal failure is a well-known and major complication, with rates between 7 and 21 % for heart, lung, liver, and intestinal transplant recipients on conventional calcineurin inhibitor-based immunosuppressive regimens [36]. Acute renal insufficiency has been described in multiple VCA patients, including one who subsequently required hemodialysis. Posttransplant diabetes mellitus can affect all immunosuppressed patients. Follow-up performed on renal transplant patients at of 8.3 years demonstrated that 20 % had developed diabetes as a result of IS [37]. There is no reason to expect any difference in rates or implications of these complications for SOT versus VCA.

Malignancy is another well-described and potentially lethal consequence of IS. SOT transplant recipients have increased rates of all malignancies, including a cumulative rate of skin malignancies of 7.4 % at 3 years and a rate of 7.5 % for nonskin malignancies [38] Posttransplant lymphoproliferative disorder (PTLD) is one of the most lethal malignancies associated with transplantation, with reported rates of between 1 % for renal allografts and much higher rates of approximately 10% and 20 % for lung and intestine transplant recipients, respectively. VCA also has reported cases of PTLD that have been severe and life threatening [19]. Nonhuman primate research suggests that PTLD in VCA may more closely resemble BM transplantation than SOT in that the malignant B cell transformation occurs in donor BM cells [39].

Bacterial, viral, and fungal infections increase in prevalence with IS and mandate specific prophylaxis in the first months post transplant. Nonetheless, rates of infection in the first year after renal transplantation are between 8.9 % for fungal infection and 37.5 % for bacterial infections [40]. General infectious prophylaxis for SOT includes antifungal treatments often with fluconazole for 30 days, antiviral therapy with valganciclovir or acyclovir for 90 days, and antibacterial therapy with trimethoprim/sulfamethoxazole for 6 months. VCA have resulted in serious and even fatal infectious complications, and may have even greater infectious risks given the colonization of mucosal surfaces and constant exposure to the external environment [41]. Nevertheless, pharmacologic prophylactic strategies for VCA have been similar to SOT [42].

Technical complications vary significantly between different types of SOT, and are more closely related to complications of nontransplant procedures in their respective anatomical domain than they are to each other. Thus, it is difficult to derive specific lessons regarding technical complications. Nevertheless, it is worth noting that IS increases the probability of wound-related complications, including dehiscence and wound infection [40, 43], and usual techniques for approximating tissue and maintaining wound sterility may not meet with the degree of success normally expected by reconstructive surgeons.

Unlike SOT, one of the essential functions of many VCA is motor and sensory restoration. The closest parallel may be seen in heart transplantation, which demands only intrinsic pacemaking—and may be easily supplanted by a pacemaker

[44, 45]. Instead, VCA surgeons must be guided by experience in digit and limb replantation and rehabilitation, as well as the emerging body of evidence in VCA itself.

Conclusions

VCA, like SOT, is a complex medical science in evolution. Multidisciplinary teams worldwide are called upon to tackle a myriad of technical, immunological, psychosocial, logistical, and legal issues. Certain challenges are unique to VCA, including heightened immunogenicity, psychological identity changes, and the non-lifesaving nature of the graft. Nevertheless, this science did not arise from a vacuum. Decades of refinement in SOT have been the indispensable foundation for the emergence of VCA. Large gaps still exist in the availability and consistency of both types of transplantation. The holy grail of tolerance remains a distant target. Moving forward, the cross-application of insights from each field will accelerate the pace of advancement in both.

References

1. Kung M, Koschwanez HE, Painter L, Honeyman V, Broadbent E. Immunosuppressant nonadherence in heart, liver, and lung transplant patients: associations with medication beliefs and illness perceptions. Transplantation. 2012;93(9):958–63. Epub 2012/02/22.
2. De Geest S, Borgermans L, Gemoets H, Abraham I, Vlaminck H, Evers G, et al. Incidence, determinants, and consequences of subclinical noncompliance with immunosuppressive therapy in renal transplant recipients. Transplantation. 1995;59(3):340–7. Epub 1995/02/15.
3. Lieber SR, Volk ML. Non-adherence and graft failure in adult liver transplant recipients. Dig Dis Sci. 2013;58(3):824–34. Epub 2012/10/12.
4. Guyomard V, Leinster S, Wilkinson M, Servant JM, Pereira J. A Franco-British patients' and partners' satisfaction audit of breast reconstruction. JPRAS. 2009;62(6):782–9. Epub 2008/04/09.
5. Haekens CM, Enajat M, Keymeulen K, Van der Hulst RR. Self-esteem and patients' satisfaction after deep inferior epigastric perforator flap breast reconstruction. Plast Surg Nurs. 2011;31(4):160–6. Epub 2011/12/14.
6. Mathurin P, Moreno C, Samuel D, Dumortier J, Salleron J, Durand F, et al. Early liver transplantation for severe alcoholic hepatitis. New Engl J Med. 2011;365(19):1790–800. Epub 2011/11/11.
7. Press TA. Emergency face transplant successfully performed in Poland. New York: CBS News; 2013. http://www.cbsnews.com/8301-204_162-57585664/emergency-face-transplant-successfully-performed-in-poland/. cited 2013 June 1
8. Health Resources and Services Administration DoHaHS. Organ procurement and transplantation network, final rule. Fed Regist 2013;78(128):40033–42.
9. Starzl TE, Miller C, Broznick B, Makowka L. An improved technique for multiple organ harvesting. Surg Gynecol Obstet. 1987;165(4):343–8. Epub 1987/10/01.

10. Starzl TE, Hakala TR, Shaw BW Jr., Hardesty RL, Rosenthal TJ, Griffith BP, et al. A flexible procedure for multiple cadaveric organ procurement. Surg Gynecol Obstet. 1984;158(3):223–30. Epub 1984/03/01.
11. Pomahac B, Papay F, Bueno EM, Bernard S, Diaz-Siso JR, Siemionow M. Donor facial composite allograft recovery operation: Cleveland and Boston experiences. Plast Reconstr Surg. 2012;129(3):461e–7e. Epub 2012/03/01.
12. Brazio PS, Barth RN, Bojovic B, Dorafshar AH, Garcia JP, Brown EN, et al. Algorithm for total face and multiorgan procurement from a brain-dead donor. Am J Transpl. 2013. Epub 2013/08/07.
13. Lanzetta M, Petruzzo P, Dubernard JM, Margreiter R, Schuind F, Breidenbach W, et al. Second report (1998–2006) of the International Registry of Hand and Composite Tissue Transplantation. Transpl Immunol. 2007;18(1):1–6. Epub 2007/06/23.
14. Villamaria CY, Rasmussen TE, Spencer JR, Patel S, Davis MR. Microvascular porcine model for the optimization of vascularized composite tissue transplantation. J Surg Res. 2012;178(1):452–9. Epub 2012/06/02.
15. Blaisdell FW. The pathophysiology of skeletal muscle ischemia and the reperfusion syndrome: a review. Cardiovasc Surg (London, England). 2002;10(6):620–30. Epub 2002/11/28.
16. Pradka SP, Ong YS, Zhang Y, Davis SJ, Baccarani A, Messmer C, et al. Increased signs of acute rejection with ischemic time in a rat musculocutaneous allotransplant model. Transpl Proc. 2009;41(2):531–6. Epub 2009/03/31.
17. Lang KS, Georgiev P, Recher M, Navarini AA, Bergthaler A, Heikenwalder M, et al. Immunoprivileged status of the liver is controlled by Toll-like receptor 3 signaling. J Clin Invest. 2006;116(9):2456–63. Epub 2006/09/07.
18. Knechtle SJ, Kwun J. Unique aspects of rejection and tolerance in liver transplantation. Semin Liver Dis. 2009;29(1):91–101. Epub 2009/02/25.
19. Morelon E, Testelin S, Petruzzo P, Badet L, Michallet M, Dubois V, et al. Face transplantation with combined hematopoietic stem cell infusion and vascularized bone marrow: first year of follow-up. Transpl Intl. 2011;24:75.
20. Schneeberger S, Gorantla VS, Brandacher G, Zeevi A, Demetris AJ, Lunz JG, et al. Upper-extremity transplantation using a cell-based protocol to minimize immunosuppression. Ann Surg. 2013;257(2):345–51. Epub 2012/09/25.
21. Brandacher G, Lee WP, Schneeberger S. Minimizing immunosuppression in hand transplantation. Expert Rev Clin Immunol. 2012;8(7):673–83; quiz 84. Epub 2012/10/20.
22. Morelon E, Dubois V, Testelin S, Petruzzo P, Badet L, Michallet M, et al. Face transplantation with combined hematopoietic stem cell infusion and vascularized bone marrow transplantation is not associated with mixed chimerism in humans. Am J Transpl. 2013;13:202–3.
23. Armed Forces Institute of Regenerative Medicine Annual Report 2009. Army Medical Research and Materiel Command, Armed Forces Institute of Regenerative Medicine, 2009.
24. Kawai T, Cosimi AB, Spitzer TR, Tolkoff-Rubin N, Suthanthiran M, Saidman SL, Shaffer J, Preffer FI, Ding R, Sharma V, Fishman JA, Dey B, Ko DS, Hertl M, Goes NB, Wong W, Williams WW, Jr., Colvin RB, Sykes M, Sachs DH. HLA-mismatched renal transplantation without maintenance immunosuppression. N Engl J Med. 2008;358(4):353–61.
25. Siemionow M, Klimczak A. Chimerism-based experimental models for tolerance induction in vascularized composite allografts: Cleveland clinic research experience. Clin Dev Immunol. 2013;2013:831410. Epub 2013/04/11.
26. Brazio PS, Bojovic B, Ha JS, Brown EN, Panda A, Munivenkatappa R, et al. Infused bone marrow fails to prolong vascularized composite allograft survival in nonhuman primates: 2572. Transplantation. 2012;94:573.
27. Barth RN, Rodriguez ED, Mundinger GS, Nam AJ, Ha JS, Hui-Chou H, et al. Vascularized bone marrow-based immunosuppression inhibits rejection of vascularized composite allografts in nonhuman primates. Am J Transpl. 2011;11(7):1407–16. Epub 2011/06/15.
28. Siemionow M, Madajka M, Cwykiel J. Application of cell-based therapies in facial transplantation. Ann Plast Surg. 2012;69(5):575–9. Epub 2012/10/10.

29. LaMattina J, Goldschen L, Munivenkatappa R, Gurk-Turner C, Powell J, Klassen D, et al. Decreased rejection and equivalent survival with selective alemtuzumab induction for renal transplantation. Am J Transpl. 2013;13:84.
30. Gharb BB, Rampazzo A, Altuntas SH, Madajka M, Cwykiel J, Stratton J, et al. Effectiveness of topical immunosuppressants in prevention and treatment of rejection in face allotransplantation. Transplantation. 2013;95(10):1197–203. Epub 2013/03/28.
31. Pomahac B. Facial Transplantation—The Boston, Massachusetts Clinical Experience. Presented at State of the Art: Facial Reconstruction & Transplantation. 2013.
32. Kaufman CL, Ouseph R, Blair B, Kutz JE, Tsai TM, Scheker LR, et al. Graft vasculopathy in clinical hand transplantation. Am J Transpl. 2012;12(4):1004–16. Epub 2012/02/14.
33. Matas AJ, Humar A, Gillingham KJ, Payne WD, Gruessner RW, Kandaswamy R, et al. Five preventable causes of kidney graft loss in the 1990s: a single-center analysis. Kidney Intl. 2002;62(2):704–14. Epub 2002/07/12.
34. Lodhi SA, Lamb KE, Meier-Kriesche HU. Solid organ allograft survival improvement in the United States: the long-term does not mirror the dramatic short-term success. Am J Transpl. 2011;11(6):1226–35. Epub 2011/05/14.
35. Lamb KE, Lodhi S, Meier-Kriesche HU. Long-term renal allograft survival in the United States: a critical reappraisal. Am J Transpl. 2011;11(3):450–62. Epub 2010/10/27.
36. Ojo AO, Held PJ, Port FK, Wolfe RA, Leichtman AB, Young EW, et al. Chronic renal failure after transplantation of a nonrenal organ. New Engl J Med. 2003;349(10):931–40. Epub 2003/09/05.
37. Cosio FG, Pesavento TE, Kim S, Osei K, Henry M, Ferguson RM. Patient survival after renal transplantation: IV. Impact of post-transplant diabetes. Kidney Intl. 2002;62(4):1440–6. Epub 2002/09/18.
38. Kasiske BL, Snyder JJ, Gilbertson DT, Wang C. Cancer after kidney transplantation in the United States. Am J Transpl. 2004;4(6):905–13. Epub 2004/05/19.
39. Barth RN, Nam AJ, Stanwix MG, Kukuruga D, Drachenberg CI, Bluebond-Langner R, et al. Prolonged survival of composite facial allografts in non-human primates associated with posttransplant lymphoproliferative disorder. Transplantation. 2009;88(11):1242–50. Epub 2009/12/10.
40. Snyder JJ, Israni AK, Peng Y, Zhang L, Simon TA, Kasiske BL. Rates of first infection following kidney transplant in the United States. Kidney Intl. 2009;75(3):317–26. Epub 2008/11/21.
41. Gordon CR, Avery RK, Abouhassan W, Siemionow M. Cytomegalovirus and other infectious issues related to face transplantation: specific considerations, lessons learned, and future recommendations. Plast Reconstr Surg. 2011;127(4):1515–23. Epub 2011/04/05.
42. Hammond SP. Infections in composite tissue allograft recipients. Infect Dis Clin North Am. 2013;27(2):379–93. Epub 2013/05/30.
43. Roine E, Bjork IT, Oyen O. Targeting risk factors for impaired wound healing and wound complications after kidney transplantation. Transpl Proc. 2010;42(7):2542–6. Epub 2010/09/14.
44. Herre JM, Barnhart GR, Llano A. Cardiac pacemakers in the transplanted heart: short term with the biatrial anastomosis and unnecessary with the bicaval anastomosis. Curr Opin Cardiol. 2000;15(2):115–20. Epub 2000/08/30.
45. Melton IC, Gilligan DM, Wood MA, Ellenbogen KA. Optimal cardiac pacing after heart transplantation. Pacing Clin Electrophysiol. 1999;22(10):1510–27. Epub 1999/12/10.

Chapter 4
The Daily Life of a Hand Transplant Recipient

Christopher Pollock

November 28, 2008, is a day I will never forget. On that day, over 5 years ago, I was in a traumatic farm accident that changed my life forever. It was the day after Thanksgiving, and I was helping my friend on his farm by harvesting field corn. Around 5:15 pm, while using the one row corn picker, I stopped the tractor because I saw that the wagon was overflowing and corn was falling to the ground. I wanted to move the corn further back in the wagon. I kept the machine running at normal operating speed as I continued around to the front of the tractor. As I came to the snout, or front of the corn picker, I noticed that there was an ear of corn on it. I tried to knock the ear of corn into the running machine, but it did not move. So, on the second attempt, I tapped on the ear with my left hand. I was wearing my army jacket which had a loose cuff. Right then, the chain snagged my loose cuff pulling my hand into the snapping rollers. I screamed as I tried to pull my left hand out with my right hand. That was a tragic decision, because my right hand also got pulled into the machine. For 30 min, I was trapped in the loudly running machine yelling for help while continuing to try and dislodge my hands. Fighting to get loose, I asked the LORD to let me die. After the third time, I had a warm sensation that overcame me. I knew that I was fighting for my life, but I could feel God sustaining me through all of this.

Around 5:45 pm, help finally arrived. I was surprisingly calm as I explained how to shut the machine off. When the EMTs arrived on site, they quickly concluded that the Lifeline helicopter needed to be called. It took approximately a half hour to dislodge me from the machine. Through all of this, I never lost consciousness. My wife eventually showed up at the scene and reassured me that everything would be fine, and that I would be safe. For the 30 min I was trapped, my life had changed forever. I no longer had hands. It seemed that this accident might even bring my family back together since my wife and I had been separated at the time.

First US patient receiving a combined forearm and hand transplantation

---

C. Pollock (✉)
Double Hand Transplant Patient, Transplant Composite Tissue Allograph, 3225 Meadow lane, Harrisburg, PA 17109, USA
e-mail: cpollock68@gmail.com

G. Brandacher (ed.), *The Science of Reconstructive Transplantation*, Stem Cell Biology and Regenerative Medicine, DOI 10.1007/978-1-4939-2071-6_4

Once I was dislodged from the machine, the medics put me into the helicopter and attended to my needs during the 8-min flight to the Hershey medical facility. Finally, I was on the stretcher going to the trauma center when a nurse said, "Mr. Pollock, you are at the trauma unit and you will be treated." That is when I knew it was okay to let go, and at 6:30 pm, I closed my eyes and finally passed out. I found out later that while I was unconscious, I continued to answer questions for about 20 min prior to going into surgery.

On Sunday, I awoke like I had been resurrected to a new life. I had almost died. After multiple surgeries, I had become a bilateral upper extremity amputee, however, I had a whole new outlook on my life. I feel like the experience made a chemical and spiritual change for me, and to this day, I have a much more positive outlook on life. My relationships are better, especially with my twin children. Though my wife and I eventually divorced, she was extremely helpful through the months following the accident. I now live my life one day at a time, and I am grateful for each day of life that I have been given. I can neither change the past, nor the future, so I try to get the most out of each day.

As time passed, I experienced no phantom pains, only sensations. I am grateful for how well my recovery progressed. I did not have a problem dealing with what could be viewed as a major loss. I thank God for each day, and for the peace I felt after my bilateral amputations. Eventually, I was fitted with a prosthesis. I adapted to using the prosthetics as well as my new life. I was learning to do things such as driving and normal daily tasks. I was thankful and did not view my life as a hardship but as a challenge that I would overcome!

While going through occupational therapy, the possibility of hand transplantation was introduced to me, but I, for some reason, did not give it a lot of thought. More time passed and the idea was brought up again, but I still did not see it as an option for me. Then, about 6–7 months later, I walked into my parent's dining room and saw a *People* magazine with a story about the first bilateral transplant patient in the USA. As I read the article, "New Hands for Jeff," I was inspired to look into the procedure. One quote really struck home for me, "Just recently, Valarie (Jeff's wife) intertwined her fingers with his. 'That's the first time I've held your hand in 10 years,' she told her husband. 'Just wait,' Jeff replied, 'until I can squeeze yours back [1].'" I guess the timing was just right, because the very next day, I called the team at the University of Pittsburgh Medical Center (UPMC) that has performed Jeff's transplant and inquired about the program. After speaking with someone about the program, I chose to get a screening for the surgery. My thought was, "What do I have to lose?"

The UPMC website along with the article in *People* magazine was also my first introduction to the "Pittsburgh Protocol" for immunomodulation after the hand transplant. The protocol works to help reduce the number of medications to prevent rejection episodes. The protocol as compared to current protocols used worldwide was designed in the hopes to reduce the number of side effects and occurrences of complications like drug toxicity.

The screening involved several days of testing and interviews with psychologists and social workers as well as the surgeons and doctors heading the program. Once the screening was complete, the surgery, treatment protocol, and recovery were discussed in detail. The most important point to me was that I only needed to be on one type of antirejection medicine. I asked many questions. I knew it was going to be a long road; however, I believed the end result would be a better quality of life for me as well as much more independence.

Before I was even considered a viable candidate for this surgery and research protocol, I had to realize that it would take time, hard work, and the right frame of mind, in other words a positive outlook and attitude. I also knew that I had to be regimented and methodical with my therapy and my medications as well as continually communicating with my doctors.

It was actually a fairly easy decision for me to go through with the transplants. I had to wait for a bit to be put on the transplant wait list since it had been less than a year since my accident, but once on the list, I only had to wait a couple of months until a matching donor was identified. On February 3, 2010, at 11:20 pm, I received the phone call I had been waiting for—there was a donor and I needed to get to Pittsburgh as soon as possible. On February 5, 2010, I was no longer a double upper extremity amputee. I am now a bilateral hand transplant recipient. My hand was replaced on my left side, and I was the first in the USA to have my entire forearm replaced above the elbow on the right. It is a beautiful gift and I thank God as well as the donor and his family for this miracle.

When I awoke from the surgery, I was on a ventilator and numb. I was quickly removed from the ventilator. After 14 days, I received the bone marrow of the donor (a crucial step in the Pittsburgh Protocol). I spent the next few weeks recovering until I could start moving my hands myself. Until then, occupational therapists came in daily to work with my hands to keep them pliable and to prevent scarring with the tendons and muscles bundling together.

In the months that followed, I had therapy Monday through Friday for 6 h; this quickly added up to over 2300 h of hand therapy. Additionally, I was monitored closely by the transplant team. I was finally able to return home to Harrisburg in August 2010. I continued my weekly therapy regimen, and began to work toward a life of normalcy with my hands becoming more functional each week. In January of 2011, I decided to enroll in college, so I enrolled at Harrisburg Area Community College (HACC). During my second semester, I was assigned a research paper. I had to pose a question and research to find the answer or to support for my thesis. Before and after my transplants and during my medical follow-ups at UPMC, I found that I had so many questions related to hand transplantation such as wondering about what type of recovery I was expected to make, about rejection, and whether it could be physically harmful to me or even fatal. So, I decided to make these the focus of my paper. The research that I did helped to reduce my anxiety and to learn even more about my outlook for the future of my hands and my overall health.

I found out that hand transplantation is one example of vascularized composite allotransplantation (VCA). VCA, or reconstructive transplantation, is a broader term for transplantations that involve tissues such as skin, muscle, tendon, nerve, blood vessels, lymph nodes, cartilage, bone, and bone marrow. VCA appears to be much more intensive and complex than solid organ transplants [2, 3].

I learned that the International Registry on Hand and Composite Tissue Transplantation (IRHCTT) publishes data annually regarding "functional recovery, patient and graft survival, adverse events, and complications" [2, 3]. This was a great resource and answered some of my questions about what to expect from my recovery and my transplants in the years to come. Here is an excerpt from what I read.

> In follow-up of at least 1 year, 100% of recipients developed protective sensibility [the ability to feel hot and cold], 90% developed tactile sensibility [the ability to feel touch, pressure, edges, etc…], and 84% developed discriminative sensibility [the ability to differentiate the feeling of different objects and the details of the object]. Seventy-five percent of recipients reported improved quality of life, and many have returned to employable status. [3, 4]

Rejection was another area that concerned me. My body needed time to heal and adjust to the new tissue, and to recover from the trauma the body goes through from such extensive surgery. Rejection is when the body's immune system attacks the transplant as if it is a "foreign" body. With allotransplantation of extremities, rejection is easily noticed and caught in the early stages. I would notice a rash, swelling, and redness of the skin. To control this, I would have my immunosuppression medicine dose increased, blood work drawn, and usually I would have to apply medicine (a greasy, staining ointment) externally to the transplanted tissue.

During the first year of post-transplants, I recall having several rejection episodes. During the first few months post transplant, there were some major episodes that had the medical team readmitting me to the hospital. Due to my compliance and diligence, these episodes were caught early and with the help of modern medicine and the medical team, they could be quickly reversed and I recovered completely. However, since May 2011, I had not had an episode of rejection.

I think the following excerpt from the book, *Plastic and Reconstructive Surgery,* best sums it up:

> That the ultimate goal of HT (Hand Transplantation) is to attain functional motor recovery of the transplanted hand superior to myoelectrical prosthesis and to achieve sensory function for discrimination and tactile sensation, while at the same time, maintaining minimal adverse effects secondary to unavoidable immunosuppression. [5]

I realized several points through my research and from my personal experiences with hand transplantation. The most significant point that I have come to appreciate is that some questions I have could not be answered, as we are still learning as the field of VCA continues to develop. I now see that the doctors are most likely looking at my outcome in order to better predict the outcome of future upper extremity transplant patients. Thus far, my hand transplants are progressing successfully and continually. I am very patient, organized, motivated, and determined. I follow the protocol to the letter, and, I believe, this is why I am doing so well. Secondly, there are still a lot of unknowns in the science of reconstructive transplantation. I am a patient they consider to be successful because my nerves continue to respond showing regeneration and my hands continue to become more useful and responsive. As far as how well will my hands work, considering I am the first bilateral hand transplant with the right extremity above the elbow joint, I have to realize that many people are looking to me for that answer.

Fig. 4.1 Chris Pollock performing garden work in his backyard 4 years after his combined hand/forearm transplant

Bottom line: would I do it again? No hesitation for me—YES! The transplants have opened up my world. I am now attending college and pursuing a degree. I can see progress in my abilities monthly. I can eat with chopsticks, do home repairs and garden work (Fig. 4.1), feel someone's handshake and touch my kids again. I am able to manage and function far better than I ever could with my prostheses and, as far as normal, I just take it one-day-at-a-time. Being positive keeps me motivated to continue doing well, and I am excited about the fact that I am writing maybe a small piece of history with my double hand transplant experience.

Acknowledgments I would like to thank Alysoun Goebel for proofreading and assistance with this manuscript.

References

1. Egan NW. New hands for Jeff. People Mag. 2009 Jul 27;72(4):70.
2. Shores JT, Brandacher G, Schneeberger S, Gorantla VS, Lee WP. Composite tissue allotransplantation: hand transplantation and beyond. J Am Acad Orthop Surg. 2010 Mar;18(3):127–31.
3. Lovasik D, Foust DE, Losee JE, Lee WP, Brandacher G, Gorantla VS. Helping hands: caring for the upper extremity transplant patient. Crit Care Nurs Clin North Am. 2011 Sep;23(3):505–17.
4. Ninkovic M, Weissenbacher A, Gabl M, Pierer G, Pratschke J, Margreiter R, et al. Functional outcome after hand and forearm transplantation: what can be achieved? Hand Clin. 2011 Nov;27(4):455–65.
5. Gordon CR, Siemionow MZ. Clinical experience with hand transplantation. In: Siemionow MZ, Eisenmann-Klein M, editors. Plastic and reconstructive surgery. New York: Springer; 2010. pp. 729–49.

Part II
Specific Areas of Research

Chapter 5
Small Animal Models for Reconstructive Transplantation

Barbara Kern and Robert Sucher

Introduction

Significant advances in the field of reconstructive transplantation over the past decade have propelled the evolution of hand and face transplantation from an experimental surgical procedure to an effective treatment option for those suffering from disfiguring tissue defects and limb amputations [1]. Several factors have contributed to the clinical success of reconstructive transplantation, including the development of immunomodulatory therapies, the perfection of microsurgical techniques, and a profound understanding of the immunological mechanisms involved in chronic tissue rejection and vascularized composite allograft tolerance [2]. A large number of these medical advances have been achieved through the repeated utilization of experimental animal models [3]. In addition, and reproducible data generated from these unremitting *in vivo* experiments have been key to the development of reliable immunomodulatory protocols that could not have been solely discovered by *in vitro* experiments alone. Conversely, a plethora of additional transplant related *in vitro* tests form today's backbone of a modern transplant research unit that is devoted to make the next giant leap in immunology.

The selection of an animal model for a specific transplant procedure usually depends on (1) the nature of the experimental study, (2) the type of tissue component to be evaluated, and (3) the financial resources available.

Although short-term functional outcomes after reconstructive transplantation are promising, rejection and conventional high-dose immunosuppressive treatment continue to limit broader clinical application of this innovative surgical procedures

B. Kern (✉)
Department of Visceral, Transplant, and Thoracic Surgery, Innsbruck Medical University, Anichstrasse 35, 6020 Innsbruck, Austria
e-mail: barbara.kern@i-med.ac.at

R. Sucher
Department of General, Visceral and Transplant Surgery, Charité - Universitätsmedizin Berlin, Berlin, Germany

G. Brandacher (ed.), *The Science of Reconstructive Transplantation*,
Stem Cell Biology and Regenerative Medicine, DOI 10.1007/978-1-4939-2071-6_5

[4]. Therefore, animal model-based research in this field mainly focuses on the investigation of the basic mechanisms of the alloimmune response during rejection [5]; novel strategies and protocols for immunomodulation and tolerance induction after reconstructive transplantation are tested extensively in animals before they can be applied to humans [6]. A second significant research interest focuses on the improvement of nerve regeneration and functional outcome after reconstructive transplantation [7]. It is therefore important to distinguish between animal models designed for either basic immunologic or specific functional studies.

Vascularized composite allografts are composed of a wide variety of different tissue types, including bone and bone marrow, cartilage, muscle, nerve, blood vessels, lymphatic tissue, and skin; each of these tissues carry distinct immunological and functional properties. In this context, it has been shown that the alloimmune response and antigenicity vary substantially depending on the individual tissue component. It is interesting and important to note, however, that the entire limb allograft consisting of multiple tissues generates a much weaker immune response than allografts of its individual components [8]. The development of various animal surgical models containing different combinations of tissues is therefore crucial for the investigation of the immunological processes at hand, as no one model can adequately dissect every individual mechanism from the role it plays in concert.

Funding uncertainty is an inescapable aspect of modern research; economic recession and governmental budgetary restraints have resulted in a worldwide pattern of constrained funding [9]. This competitive funding environment demands that every laboratory works as efficiently as possible to maximize data output while keeping expenses low. Small animal models, therefore, are optimal for *in vivo* experiments as they strike the best balance between translatability and cost-effectiveness.

It is also important to keep in mind that animal experimentation by its very nature takes a considerable toll on animal life [10], and scientists have an ethical obligation to minimize the pain and distress of experimental animals.

Animal Models for Transplantation Research

The complexity of acute/chronic allograft rejection and functional recovery as well as the urgent need for improved patient outcomes provide the impetus to employ small animal models for reconstructive transplantation. Primarily, small animals provide comparative living systems that allow researchers to investigate both the orchestrated course of the immune response after allotransplantation and how various kinds of preventive interventions and treatment protocols can be used to modulate the recipient's immune system and halt allograft rejection.

The utility of the animal model varies by species, available reagents and tools to support the study, and level of homogeneity with the human body. The most commonly used animal models for transplant research are rats and mice. Other less frequently employed transplant models include larger animals such as hamsters, guinea pigs, rabbits, dogs, pigs, and nonhuman primates. Although large animal experiments have been and will always be the best way to study safety and efficacy of novel therapeutic regiments that are qualified for human phase 1 studies [6], small animal research has and will always remain the cornerstone of *in vivo* basic science. In rodents, techniques such as inbreeding, the expression of foreign genes, and the alteration of germ lines have advanced to make remarkable precision around the genomic locations that can be modified for investigation. In other species, similar tools still have yet to be incorporated into the researcher's toolbox.

Studies with the goal to investigate questions regarding alloimmune response, ischemia-reperfusion injury (IRI), as well as nerve regeneration can be carried out in small animal models, which might improve the knowledge about transplantation in the clinical setting.

Rat Hind Limb Transplant Models

The rat is a major animal model for the study of human health and disease. The large number of inbred strains and vast amount of physiological, behavioral, biochemical, cellular, pharmacological, toxicological, and immunological data gathered over the past 40 years make the rat a superb model for research efforts in the field of reconstructive transplantation.

The first vascularized composite allotransplantation (VCA) model describing orthotopic hind limb transplantation in rats was introduced by Shapiro and Cerra [11]. The surgical procedure involved time-consuming reattachments of bones, muscles, nerves, and blood vessels that demanded delicate microsurgical skills. Since its publication in 1978, the model has been successfully used in a multitude of functional and immunological studies and must therefore be considered the current gold standard animal model in reconstructive transplantation research. The orthotopic hind limb transplant model in the rat can be used for functional, immunological, and ischemia-reperfusion-related studies.

Walking track analysis that analyzes the dynamic foot placement of the animal was the first reliable technique that truly allowed noninvasive assessment of motor function and nerve regeneration in the rat hind limb [12]. Further quantification of neuroregeneration based on measurements from walking tracks can be performed by the sciatic function index, which allows a functional assessment of the sciatic nerve [13]. A second quantitative evaluation of hind limb motor function in rats with selective sciatic, tibial, and peroneal nerve injury can be performed using the Bain–Mackinnon–Hunter index. The rather simple method of quantifying the size and distribution of the animal's footprints on paper, however, has shown that animals with selective nerve injuries have walking tracks that are consistent, predictable, and based on known neuromuscular deficits [14]. More recent, highly sensitive tools for the detailed assessment of gait changes in rodents rely on computerized models. The CatWalk XT™, for example, delivers objective quantitative measurements of footfalls and gait in rodents and has successfully been used in various rat models of osteoarthritis [15], spinal cord injury [16], peripheral nerve regeneration [17], and hind limb transplantation [18]. Due to the partly striking inconsistencies

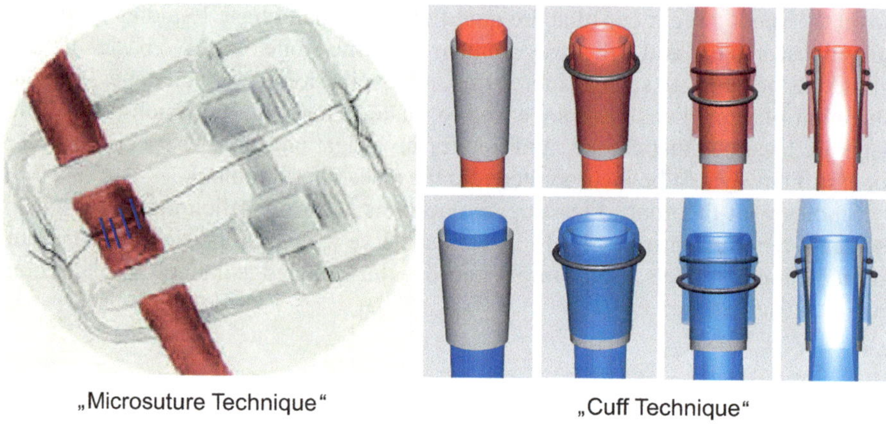

Fig 5.1 Conventional suture technique for microvascular anastomosis (*left*) and non-suture cuff technique (*right*)

between the results of walking track and electrophysiologic examinations of regenerating nerves, however, it is important to be careful with the interpretation of its recordings [19].

Although the orthotopic rat hind limb model has been performed with great success over the past decades, several attempts have been made to make it less time consuming and technically demanding. Since the suture connection of blood vessels seemed to be the surgically most demanding step in the whole procedure, more feasible anastomosis techniques have been applied. One such technique uses a polyimide or polyethylene cylinder, slightly larger in diameter than the blood vessel itself, which is pulled over the donor vessel so that the vessel wall can finally be everted over the cuff. For anastomosis, the recipient's blood vessel is simply pulled over the donor vessel and cuff, which finally results in an end-to-end anastomosis of both vessels (Fig. 5.1). This delicate technique, which establishes endothelial contact between both donor and recipient vessel, was first applied by Alexis Carrel, a French surgeon and biologist who was awarded the Nobel Prize in Physiology or Medicine in 1912 for his pioneering vascular suturing techniques [20]. For investigation of ischemia-reperfusion-related injuries, where prolonged cold storage leads to extremely fragile vessel walls and renders the conventional suture anastomosis nearly impossible, the cuff technique was able to simplify and shorten the whole surgical procedure and thus provide a rapid and reliable surgical approach to hind limb transplantation in rats [21].

Since orthotopic hind limb transplantation is a traumatic surgical procedure that includes the resection of the recipient limb to make room for an allograft, there are multiple concerns, including intra- and postoperative bleeding, insufficient fixation and stabilization of the graft bone to the recipient bone, and loss of normal gait; these concerns have fueled the development of less traumatic and less time-consuming VCA models. The most frequently used alternatives to the orthotopic hind limb model are heterotopic transplants of myocutaneous or osteomyocutaneous flaps to the recipient's groin [22]. One of the unique biologic features of a VCA is that its composition as a whole determines immunological outcome [8]. Therefore, heterotopic transplant models bear the advantage that relative sizes of individual graft tissue components can be easily manipulated in order to examine their impact on overall immunogenicity. One example of a heterotopic transplant model using a hind limb that contains the entire femur bone was described by Ulusal et al. [23]. In this case, an epigastric skin flap was used to cover the surgical defect at the recipient site. The authors also reported a decreased risk for postoperative bleeding and embolism since the integrity of the graft's femur bone could be preserved when compared to the "original" orthotopic hind limb model where osteotomy at the mid-femur level has to be performed using an intramedullary splint to connect both the donor with the recipient femur bone. A similar model containing a smaller allograft which only consisted of the lower leg was developed by Nazzal et al. [24]. A third modified model was described by Tobin et al. [25] in which the osteomyocutaneous graft consisted of the distal femur and proximal tibia and used the femoral artery and vein as the vascular pedicle. Since the recipient's femoral vessels were used for graft reperfusion, the animal's foot was solely perfused by its deep femoral vessels. Using this approach, the authors reported a significant reduction in operation time and recipient mortality. These heterotopic allografts are extremely well suited for immunological and ischemia-reperfusion-related studies; however, their functional assessment after transplantation is severely limited because they are not connected to afferent nor efferent nerves.

When it comes to the study of immunologic outcomes after VCA besides skin and muscle, it is crucial to include other tissue components like vascularized bone marrow that may have a substantial impact on the posttransplant immunological behavior of the graft. Zamifrescu et al. provided evidence that the bone component of a VCA, which comprises a permanent source of vascularized bone marrow, is capable of inducing cellular microchimerism at postoperative days 30 and 60 in a heterotopic rat hind limb model. In contrast, animals which only received intravenous suspensions of foreign bone marrow cells were unable to generate a stable donor cell population together with its own [26]. A similar study by Kubitskiy et al. confirmed improved survival rates of hind limbs transplanted along with vascularized bone marrow; however, their experimental setting was unable to generate stable chimerism in the recipient [27].

Another cellular-based approach to impact on the recipient's immune system is achieved by a repeated intravenous administration of donor adipose-derived stem cells. In combination with transient conventional immunosuppression, these immunomodulatory cells are capable of suppressing alloreactive T cells while increasing the CD4/CD25/Foxp3 T regulatory cell population *in vitro* and *in vivo*, resulting in a significant prolongation of rat hind limb allograft survival [28]. In addition, the authors report significantly elevated levels of donor cell chimerism and upregulation of transforming growth factor-β and interleukin-10 levels, all accounting for the beneficial outcome.

Transplantation of foreign tissues is inevitably paired to a transient stop of blood and nutrient supply to the grafted tissue. Paradoxically, the restoration of circulation results in a profound inflammatory response is commonly referred to as IRI. Another interesting scientific emphasis is therefore centered around the investigation of IRI-related injuries since it might critically influence the outcome of graft and patient survival after reconstructive transplantation [29]. Using the cuff technique for vascular anastomosis, Sucher et al. have established a reliable rat hind limb transplant model to study different aspects of IRI [21]. However, more detailed information about how IRI might contribute to the immunological and functional outcome of a VCA are still under intensive investigation [see also Chapter 22].

Mouse Models

Although mice are only one-tenth the size of rats, they provide several advantages when it comes to basic immunologic *in vivo* experiments. The widespread use of genetically defined inbred and knockout strains, as well as animals that have either had their DNA sequences modified to resemble those found in human beings, or that have human cells incorporated into them, can improve the congruence between the animal model and the human counterpart [30, 31]. This modern technology, which is predominantly available in mice, is widely used when computer models, cell cultures, or other animal models do not provide sufficient accuracy.

Murine nonvascularized skin grafts have been used for decades in organ transplant research; however, due to remarkable differences of the immune response of vascularized versus nonvascularized grafts, Jiang et al. established an ear transplant model which could be considered as the first murine VCA model [32]. In this case, the donor operation consisted of harvesting the ear with intact arterial and venous pedicles, which was subsequently transplanted orthotopically to the recipient. The applicability of this model for reconstructive transplantation is limited, however, since it predominantly consists of skin, cartilage, and blood vessels and lacks tissue components such as muscle, nerve, bone, and bone marrow. Subsequently, Tung et al. developed both heterotopic [33, 34] and orthotopic [35] hind limb transplant models which paved the way for pioneering studies of acute and chronic rejection, as well as tolerance induction in murine composite allografts. However, rate-limiting factors for the widespread distribution of this model were again challenging supermicrosurgical anastomoses of blood vessels. Eventually, studies by Foster and Liu [36] applying a nonsuture cuff technique for the femoral vein anastomosis and by Sucher and Lin et al. [18] using again the cuff technique for both arterial and venous femoral vessels have resulted in further optimized surgical outcomes.

Similar to the rat model, heterotopic murine VCA models have been developed using the groin or the cervical region as graft recipient sites, all bearing the advantages of lower intra- and postoperative mortality rates but facing disadvantages of impaired functional graft assessment [18].

Face Transplant Models

Apart from the classical hind limb transplant model in rodents, several groups have focused on other types of VCA such as face, larynx, and vascularized knee joint transplants. The first experimental model for full face and scalp transplantation was described by Ulusal et al. in 2003 [37]. In this case, the upper face and scalp of the donor was transplanted orthotopically using both common carotid arteries and jugular veins for revascularization. The major problem this model is confronted with, however, is the missing trigeminal nerve anastomosis, making it invaluable for functional and sensational studies on nerve regeneration. One important aspect of functional recovery after reconstructive transplantation is cortical reintegration of the graft, which occurs due to the newly regained sensory input from the periphery posttransplant. Since each individual whisker of the rat can be correlated to a certain anatomical area in the cortex, Washington et al. developed a rat hemiface transplant model in which nerve conduction studies of the reanastomosed facial nerve could be performed with great accuracy [38]. The "mystical flap pad model" described by Landin and Cavadas again used the rats' vibrissal system to study cortical reafferentiation through simple stimulation of the whiskers [39]. When model systems are designed to assess the immunologic outcome after face transplantation, vascularized bone marrow components such as the mandible are critical elements of the graft. Siemionow et al. were the first to study the influence of vascularized bone marrow on the immunologic outcome after face transplantation in rats [40].

Discussion

With new developments in genetics, drug discovery, stem cell research, and bioengineering, small animal models in reconstructive transplantation are employed to deliver new groundbreaking insights into the immunological and functional behavior of vascularized composite tissue allografts. Nevertheless, a combination of both *in vivo* and *in vitro* tests is necessary for profound research in this novel emerging surgical field. The advancement of biotechnology may eliminate the further use of animal research in the long term; however, in the short term, it is more likely to present us with new troubling questions.

References

1. Dubernard JM. Hand and and face allografts: myth, dream, and reality. Proc Am Philos Soc. 2011;155(1):13–22.
2. Brandacher G. Composite tissue transplantation. Methods Mol Biol. 2013;1034:103–15.

3. Brandacher G, Grahammer J, Sucher R, Lee WP. Animal models for basic and translational research in reconstructive transplantation. Birth Defects Res C Embryo Today. 2012;96(1):39–50.
4. Sucher R, Hautz T, Brandacher G, Lee WP, Margreiter R, Schneeberger S. [Immunosuppression in hand transplantation: state of the art and future perspectives]. Handchir Mikrochir Plast Chir. 2009;41(4):217–23.
5. Unadkat JV, Schneeberger S, Goldbach C, Solari MG, Washington KM, Afrooz PN, et al. Investigation of antibody-mediated rejection in composite tissue allotransplantation in a rat limb transplant model. Transpl Proc. 2009;41(2):542–5.
6. Kirk AD. Crossing the bridge: large animal models in translational transplantation research. Immunol Rev. 2003;196:176–96.
7. Yan Y, MacEwan MR, Hunter DA, Farber S, Newton P, Tung TH, et al. Nerve regeneration in rat limb allografts: evaluation of acute rejection rescue. Plast Reconstr Surg. 2013;131(4):499e–511e.
8. Lee WP, Yaremchuk MJ, Pan YC, Randolph MA, Tan CM, Weiland AJ. Relative antigenicity of components of a vascularized limb allograft. Plast Reconstr Surg. 1991;87(3):401–11.
9. Nurse P. US biomedical research under siege. Cell. 2006;124(1):9–12.
10. Animals in research: "Can they suffer?". Lancet. 2011;378(9788):289.
11. Shapiro RI, Cerra FB. A model for reimplantation and transplantation of a complex organ: the rat hind limb. J Surg Res. 1978;24(6):501–6.
12. Hruska RE, Kennedy S, Silbergeld EK. Quantitative aspects of normal locomotion in rats. Life Sci. 1979;25(2):171–9.
13. de Medinaceli L, Freed WJ, Wyatt RJ. An index of the functional condition of rat sciatic nerve based on measurements made from walking tracks. Exp Neurol. 1982;77(3):634–43.
14. Bain JR, Mackinnon SE, Hunter DA. Functional evaluation of complete sciatic, peroneal, and posterior tibial nerve lesions in the rat. Plast Reconstr Surg. 1989;83(1):129–38.
15. Ferreira-Gomes J, Adães S, Castro-Lopes JM. Assessment of movement-evoked pain in osteoarthritis by the knee-bend and CatWalk tests: a clinically relevant study. J Pain. 2008;9(10):945–54.
16. Hendriks WT, Eggers R, Ruitenberg MJ, Blits B, Hamers FP, Verhaagen J, et al. Profound differences in spontaneous long-term functional recovery after defined spinal tract lesions in the rat. J Neurotrauma. 2006;23(1):18–35.
17. Vogelaar CF, Vrinten DH, Hoekman MF, Brakkee JH, Burbach JP, Hamers FP. Sciatic nerve regeneration in mice and rats: recovery of sensory innervation is followed by a slowly retreating neuropathic pain-like syndrome. Brain Res. 2004;1027(1–2):67–72.
18. Sucher R, Lin CH, Zanoun R, Atsina KK, Weinstock M, Pulikkotil B, et al. Mouse hind limb transplantation: a new composite tissue allotransplantation model using nonsuture supermicrosurgery. Transplantation. 2010;90(12):1374–80.
19. Rupp A, Dornseifer U, Fischer A, Schmahl W, Rodenacker K, Jütting U, et al. Electrophysiologic assessment of sciatic nerve regeneration in the rat: surrounding limb muscles feature strongly in recordings from the gastrocnemius muscle. J Neurosci Methods. 2007;166(2):266–77.
20. Dutkowski P, de Rougemont O, Clavien PA. Alexis Carrel: genius, innovator and ideologist. Am J Transpl. 2008;8(10):1998–2003.
21. Sucher R, Oberhuber R, Rumberg G, Hautz T, Zelger B, Glodny B, et al. A rapid vascular anastomosis technique for hind-limb transplantation in rats. Plast Reconstr Surg. 2010;126(3):869–74.
22. Chang KP, Huang SH, Lin CL, Chang LL, Lin SD, Lai CS. An alternative model of composite tissue allotransplantation: groin-thigh flap. Transpl Int. 2008;21(6):564–71.
23. Ulusal AE, Ulusal BG, Hung LM, Wei FC. Heterotopic hindlimb allotransplantation in rats: an alternative model for immunological research in composite-tissue allotransplantation. Microsurgery. 2005;25(5):410–4.
24. Nazzal JA, Johnson TS, Gordon CR, Randolph MA, Lee WP. Heterotopic limb allotransplantation model to study skin rejection in the rat. Microsurgery. 2004;24(6):448–53.

25. Adamson LA, Huang WC, Breidenbach WC, Rahhal D, Xu H, Huang Y, et al. A modified model of hindlimb osteomyocutaneous flap for the study of tolerance to composite tissue allografts. Microsurgery. 2007;27(7):630–6.
26. Zamfirescu D, Popovicu C, Stefanescu A, Bularda A, Popescu M, Zegrea I, et al. Vascularized bone marrow transplantation model in rats as an alternative to conventional cellular bone marrow transplantation: preliminary results. Transpl Proc. 2011;43(9):3549–51.
27. Kubitskiy A, Li J, Lanzetta M, Owen E, McCaughan GW, Bishop GA. Simultaneous vascularized bone marrow transplantation to promote acceptance of hind limb allografts and its effects on central and peripheral chimerism. Transplantation. 2007;83(9):1273–6.
28. Kuo YR, Chen CC, Goto S, Lee IT, Huang CW, Tsai CC, et al. Modulation of immune response and T-cell regulation by donor adipose-derived stem cells in a rodent hind-limb allotransplant model. Plast Reconstr Surg. 2011;128(6):661e–72e.
29. Kern B SR. Ischemia and reperfusion injury in reconstructive transplantation. Anaplastology. 2013;2:108. doi:10.4172/2161-1173.1000108.
30. Klysik J. Mice and humans: chromosome engineering and its application to functional genomics. Acta Biochim Pol. 2002;49(3):553–69.
31. van der Weyden L, Bradley A. Mouse chromosome engineering for modeling human disease. Annu Rev Genomics Hum Genet. 2006;7:247–76.
32. Jiang J, Humar A, Gracia B, Zhong R. Surgical technique for vascularized ear transplantation in mice. Microsurgery. 1998;18(1):42–6.
33. Tung TH, Mohanakumar T, Mackinnon SE. Development of a mouse model for heterotopic limb and composite-tissue transplantation. J Reconstr Microsurg. 2001;17(4):267–73.
34. Tung TH, Mohanakumar T, Mackinnon SE. Effect of age on thrombosis and morbidity in the mouse-limb transplantation model. J Reconstr Microsurg. 2002;18(1):47–52.
35. Tung TH, Mackinnon SE, Mohanakumar T. Long-term limb allograft survival using anti-CD40 L antibody in a murine model. Transplantation. 2003;75(5):644–50.
36. Foster RD, Liu T. Orthotopic hindlimb transplantation in the mouse. J Reconstr Microsurg. 2003;19(1):49–52; discussion 3–4.
37. Ulusal BG, Ulusal AE, Ozmen S, Zins JE, Siemionow MZ. A new composite facial and scalp transplantation model in rats. Plast Reconstr Surg. 2003;112(5):1302–11.
38. Washington KM, Solari MG, Sacks JM, Horibe EK, Unadkat JV, Carvell GE, et al. A model for functional recovery and cortical reintegration after hemifacial composite tissue allotransplantation. Plast Reconstr Surg. 2009;123(2 Suppl):26S–33S.
39. Landin L, Cavadas PC. The mystacial pad flap: a functional facial flap in rats. Ann Plast Surg. 2006;56(1):107–8.
40. Siemionow M, Nasir S. Impact of donor bone marrow on survival of composite tissue allografts. Ann Plast Surg. 2008;60(4):455–62.

Chapter 6
Use of Large-Animal and Nonhuman Primate Models for Reconstructive Transplantation

Bruce Swearingen, Jeff Chang and David W. Mathes

Introduction

The field of reconstructive transplantation is now a clinical reality with multiple reports documenting the successful transplantation of vascularized composite allografts (VCA) to restore lost hands and damaged faces. Since the first hand transplant performed in 1998 and the first face transplant performed in 2005, there are now more than 59 hand transplants and more than 24 face transplants performed worldwide [1–5]. While the benefits of these transplants are clear, the application of this technique is currently limited to experimental protocols due largely to the need for chronic immunosuppression to maintain these nonlife-saving transplants.

The administration of immunosuppression leads to long-term survival of these transplants but is also accompanied with unwanted effects such as hypertension, diabetes, Cushing's disease, nephrotoxicity, and infections. Other complications that have been reported include avascular necrosis of the hip and malignancy [1, 3, 6]. Despite the use of modern immunosuppressive regimes, patients still experience episodes of acute rejection and more recently the emergence of signs of chronic rejection [7, 8]. The use of lifelong immunosuppression can be seen in itself as a chronic disease characterized by its own set of risks. The morbidity of these drugs affects the quality of life, alters the risk profile, and can jeopardize the benefits gained from a successful transplant. Immunologic tolerance, however, would allow for the long-term survival of these organs without the need for chronic immunosuppression. This would significantly impact the risk–benefit ratio and allow for the more widespread use of VCA to reconstruct lost limbs and facial deformities.

D. W. Mathes (✉)
Department of Surgery, Division of Plastic and Reconstructive Surgery,
University of Colorado Denver, Denver, CO, USA
e-mail: david.mathes@ucdenver.edu

B. Swearingen · J. Chang
Department of Surgery, Division of Plastic and Reconstructive Surgery,
School of Medicine, University of Washington, Seattle, WA, USA

G. Brandacher (ed.), *The Science of Reconstructive Transplantation,*
Stem Cell Biology and Regenerative Medicine, DOI 10.1007/978-1-4939-2071-6_6

The development of protocols to induce tolerance to VCA is critical to the future of the field of reconstructive transplantation. While there have been reports of inducing tolerance to kidney organ allografts in select patients, these have been limited to living kidney donors [9–11]. The VCA can only be transplanted from a brain-dead donor and is likely more immunogenic than the kidney given that it has a cutaneous component. Thus, progress in the development of a clinically relevant protocol is dependent on animal models that can truly mimic the human condition.

The development of successful protocols to induce tolerance to VCA has historically been centered on rodent models. Gibson and Medawar performed the original studies in neonatal mice that first suggested that tolerance to skin could be achieved after the injection and persistence of donor cells [12].

This observation was confirmed in nature by Owen, who noted that when dizygotic Freemartin cattle twins shared a placenta, they demonstrated persistence of donor cells, and would tolerate skin grafts derived from the other twin [13].This tolerance to the skin grafts was not as robust as has been observed in the mouse model and some of these cattle went on to reject their donor skin. This observation highlights the importance of using large-animal models to develop tolerance protocols. Small-animal models are critical to explore multiple variations in the development of a tolerance protocol. However, results that have been demonstrated in these small-animal models have not translated directly to the clinic. In fact, these protocols often fail even when applied to large-animal models. In this chapter, we highlight both the critical results and key limitations of small-animal models and the importance of the use of large-animal models to develop clinically relevant protocols for VCA transplantation.

Small-Animal Models and Mixed Chimerism

The vast majority of experimental research in the field of VCA transplantation has been conducted in small-animal models. The advantages of using small animals such as rodents are based on the ease of handling, reliability of the model, lower cost, ability to use multiple different groups in each experiment, rapid return of results, and the availability of reagents. These experiments have provided the foundation for much of the more recent work in large-animal models. However, the success of a regimen in a rodent model has not directly translated into a large-animal model [14].

The majority of rodent models have sought to establish a state of mixed chimerism, where both the donor and the recipient immune systems are present. The VCA consists of a hind-limb transplant that included vascularized bone, muscle, fat, and skin. Initial studies utilized depletion of host T cells, total body irradiation (TBI), and bone marrow transplantation (BMT) to create stable chimeras. Foster et al. demonstrated that establishment of stable mixed chimerism in rats leads to tolerance to the hind-limb allografts [15]. The conditioning protocol in this experiment consisted of antilymphocyte serum (ALS), 500–700 centigray (cGy) TBI, and the use of T-cell-depleted donor bone marrow. The authors noted that the level of donor chimerism (> 19 %) appeared to determine whether or not the animal was tolerant to the donor hind limb. Unfortunately, those rats with high levels of donor cell chimerism went on to develop graft-versus-host disease (GVHD). In a subsequent study, they irradiated the hind limb before transplant and prevented the development of GVHD, while maintaining tolerance to the allograft.

Tolerance of a donor hind limb via mixed chimerism was confirmed in several other studies using marrow transplantation and TBI. Prabhune et al. employed a protocol that used high dose of TBI (950 cGy) combined with the simultaneous transplantation of T-cell-irradiated marrow and a donor hind limb [16]. This protocol led to high level of donor cell chimerism (average 85 %), but the majority of animals did not survive long term due to the toxicity of the radiation. In a similar study, Esumi et al. was able to establish a high level of donor cell chimerism (>98 %) and tolerance to the donor hind limb without evidence of GVHD using a combination of TBI (900 cGy) and fludarabine [17].

As the use of high-dose TBI would not be clinically relevant for clinical transplantation of VCA, Huang et al. attempted to determine the optimal dose of TBI delivered in combination with ALS and a donor hind limb transplant [18]. They found that at a dose 600 cGY TBI, the rats where chimeric and tolerant, but again the majority succumbed to GVHD. When the dose was decreased to 400 cGy, the level of donor cell chimerism decreased (mean of 15 %) with only one animal being tolerant. Finally, the use of 200 cGy TBI led to a very low level of donor cell chimerism (mean of 10 %) and no tolerant animals. In the rodent model, the use of lower doses of TBI did not lead to reliable tolerance.

In more recent work, Pan et al. added the administration of mesenchymal stem cells (MSC) and the use of longer duration of post-grafting immunosuppression to their tolerance-induction protocol [19]. Their protocol only required the use of 300 cGy TBI and all rats were noted to be chimeric with no evidence of GVHD. They did note that those rats with at least 30 % donor cell chimerism accepted their donor limb transplants while those with lower levels went on to reject their transplant. Again, it appears that there is a careful balance that must be established between donor and host cells.

Despite evidence from the above experiments that long-term engraftment of the donor cells is required for tolerance, it remains unclear if a certain level of donor cell chimerism correlates with a state of tolerance or if the presence of these donor cells is merely a surrogate for tolerance. Kuo et al. demonstrated that after infusing recipient dendritic cells exposed to donor antigens along with ALS, they could create tolerant animals with low donor cell chimerism (14 %) [20]. In a study by Siemionow et al. using a T cell depletion protocol via a T cell receptor antibody and a short course of post-graft immunosuppression, hind-limb tolerance was achieved with donor cell chimerism less than 10 % [20]. Using the same T cell receptor antibody and deoxyspergualin (agent that inhibits T cell maturation into Th1 pathway), Quatra et al. was able to induce tolerance in rats with no long-term donor cell chimerism [21]. Similar results were seen by Adamson et al. when they generated tolerant rats using 600-cGy TBI and ALS. Although tolerant to the VCA, these animals did not exhibit high levels of donor cell chimerism (less than 1 % 3 months postoperatively) [22].

Co-Stimulatory Blockade in Rodents

Early studies from Tung et al. attempting to induce tolerance in a murine model by using a CD40 antibody to block T cell co-stimulation resulted in mice exhibiting split tolerance [23–25]. These animals are tolerant of all components of a hind-limb allograft except for the skin. Tolerance towards the muscle component does not appear to require long-term chimerism, as these mice show no evidence of donor cell chimerism. In a later study by the same group, the authors added CTLA-4 Ig to block CD28(28). Again these mice show no long-term chimerism; they reject the skin component but remain tolerant towards the muscle component of the VCA. These studies suggest that complete tolerance towards composite tissues needs further blockade than just the co-stimulatory pathway.

Other groups have had further success with obtaining full tolerance towards the VCA. Zhong et al. were able to create tolerant mice using a combination of a CD45 antibody and an analogue of deoxyspergualin [26]. They demonstrate that these animals are tolerant to the hind limb and tolerance is donor specific as evident by acceptance of donor skin grafts. However, these mice show low levels of donor cell chimerism (1–2 %), again supporting the hypothesis that a high level of chimerism is not necessary to maintain tolerance towards VCA. Likewise, Li et al. blocked the CD40 pathway and after a short course of rapamycin, were able to induce tolerance in their mice with low levels of chimerism (less than 2 %) [27].

Large-Animal Models

The published studies using small-animal models provided us with important data in regards to many immunologic aspects of VCA transplantation. However, the translation of the data derived from these small-animal experiments to the clinic has been very limited. Tolerance-induction protocols that are successful in rodents frequently fail when applied in large-animal models such as swine, canine, and non-human primates. More often than not, attempts to induce tolerance in large animals require additional agents and even with these more complicated regimes, the result is only sporadic cases of tolerance. The skin component in large animals appears to be more difficult to induce tolerance to when compared to small-animal models [28]. While no animal model is a perfect human surrogate and each has its own idiosyncrasies and limitations, it does provide the best way to test possible tolerance protocols. It is clear that experiments in large animals remain the best way to gain sufficient experience to initiate ethically designed human trials. All new therapies in transplantation need to be studied for safety and efficacy in at least one large-animal model prior to phase 1 study in humans.

Common Large-Animal Models

Early experimental work in transplantation traditionally used the dog model to test the surgical techniques and the application of new immunosuppressive drugs. Many of the technical issues related to early solid organ transplantation were refined in the canine model and the first successful use of an immunosuppressive agent, proof of the efficacy of 6-MP and azathioprine, was demonstrated in a renal canine model [29]. Over time, however, the use of dogs in surgical research has declined, as a result of the social pressures. However, their role in bone marrow [30], pancreatic islet [31], and intestinal transplantation [32] and now VCA [33] has remained prominent. The canine major histocompatibility complex (MHC) is well characterized, and dog leukocyte antigen (DLA)-typed animals are available for studies requiring defined MHC haplotypes [34].

As the number of canine transplant models has declined, the swine has emerged as a good replacement for the large-animal model. The pig has two major advantages over dogs and primates. First, they are generally accepted for human consumption, and the public harbors relatively little resentment for porcine research. Second, pigs are easily bred and, thus, can be genetically manipulated. Although outbred pigs were used in many of the early experiments investigating cyclosporine (CSP) [35], this model has been used with far greater validity following the development of the MHC-defined mini-pigs [36–38]. Sachs has led the development of well-characterized inbred pigs bred to near-homozygosity at the swine leukocyte antigen (SLA) locus specifically for use in transplant experiments. It has been one of the more commonly employed large-animal models for VCA experimentation.

The nonhuman primates are the animal models most often used when evaluating biologics such as monoclonal antibodies and other agents with a high degree of human specificity [39]. Their relevance to human transplantation is well established [39]. Several species are used, including baboons, macaques (cynomologus, rhesus, and pigtail), and, rarely, the chimpanzees. Macaques are the most commonly used primate in transplantation as they are small in size and have sufficient homology to exhibit cross-reactivity with most immune molecules of the humans. The one notable exception is that antibodies to the macaque CD3 receptor do not cross-react with human CD3 and vice versa. Despite this finding, human-specific polyclonal agents are frequently used in macaques with a little regard for this difference [40]. Baboons have become a commonly used species in xenotransplantation, where their larger size is an advantage with porcine organs. Chimpanzees have the most homology with humans and have in fact served as donors for humans on historical occasions [41]. However, the evolutionary stature of chimps and their endangered status have made their use impractical, if not generally unethical.

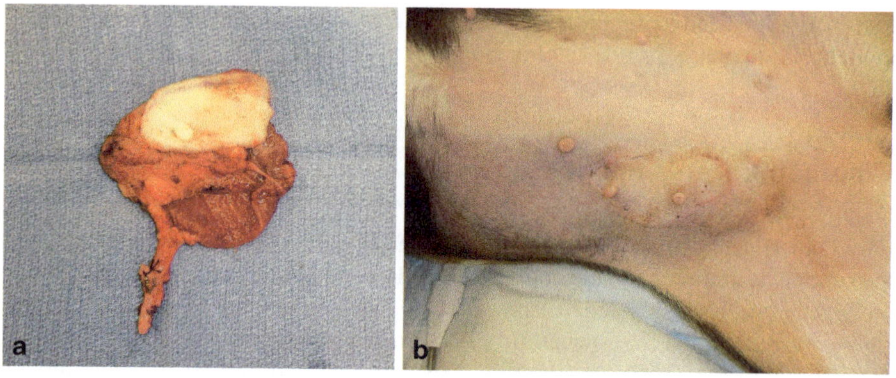

Fig. 6.1 Canine myocutaneous VCA model. **a** Myocutaneous flap harvested. **b** Flap inset and healed

Monkeys are not inbred, but the animals available for transplant research cannot be considered truly outbred either. Close attention to typing is mandatory if experiments are to be interpreted, particularly with the small numbers required in the modern research environment [39]. Common typing methods that must be employed to demonstrate genetic differences include mixed lymphocyte culture, serologic delineation of the class II loci (particularly the rhesus macaque DR region), one-dimensional isoelectric focusing (a biochemical characterization technique for both class I and class II), and polymerase chain reaction-based techniques for the highly polymorphic macaque exon 2 of MHC-DRB [39].

Canine Models in VCA

In our laboratory, we developed a preclinical canine model for VCA with a myocutaneous rectus abdominis allograft (Fig. 6.1) [42]. We then applied a clinically significant non-myeloablative hematopoietic stem cell transplant (HCT) regimen to induce tolerance to VCA allografts. This regimen was initially performed in DLA-identical littermate recipients and consisted of 200 cGy TBI before and mycophenolate mofetil (MMF)/ CSP given after HCT to control both GVHD and host-versus-graft (HVG) reactions. The regimen has been successfully translated into the clinic to treat human patients with both malignant and nonmalignant diseases by grafts from human leukocyte antigen (HLA)-matched related and unrelated donors without the need for testing in any additional animal models [43–45].

Using this regimen, we first demonstrated that donor-specific tolerance can be induced to a VCA after the establishment of mixed chimerism across a minor genetic barrier. All five animals transplanted with a vascularized myocutaneous tissue allograft demonstrated long-term acceptance of their transplant for greater than 1 year [33]. The allografts appeared normal with excellent hair growth with no evidence of rejection in the skin or muscle. In contrast, four control animals that were transplanted across the same barrier without immunosuppression rejected their allografts (15–30 days). The tolerant animals also demonstrated stable engraftment of their bone marrow transplants as demonstrated by the presence of donor granulocytes and lymphocytes.

We have also demonstrated that tolerance to the VCA is not dependent on the previous establishment of donor cell chimerism. Our clinically relevant model entailed the simultaneous marrow and VCA transplants using the same non-myeloablative HCT protocol. In this study, we observed 100 % acceptance of all components of the VCA, specifically skin and muscle, for all four dogs that underwent simultaneous transplantation of VCA and hematopoietic stem cells (HSC) for periods greater than 1 year. In addition, all of the tolerant dogs went on to accept a second non-vascularized skin graft while they rejected a third-party skin graft. These results confirmed that tolerance is not dependent on the previous establishment of stable mixed chimerism. In fact, tolerance may not be dependent on long-term engraftment of the donor bone marrow as evidenced by one of the four dogs that demonstrated initial engraftment of the HCT but had no detectable donor cells in the blood or bone marrow after week 12. Despite the loss of the HSC allograft, the VCA remained without any evidence of rejection. To control for the influence of the conditioning regimen, we also performed four transplants with the same non-myeloablative regimen but without any HSC infusion. All animals demonstrated acute rejection after the completion of the post-grafting immunosuppression. Currently, our focus has been modifying this regimen to use across greater genetic disparities and examining the need for long-term engraftment of donor cells for the maintenance of tolerance.

Swine Models in VCA

The swine model has been frequently used for the exploration and development of clinical protocols for the transplantation of VCA (Fig. 6.2). Ustuner et al. reported transplantation of a radial forelimb osteomyocutaneous flap between size-matched outbred swine with the use of a daily CSP, MMF, and prednisone oral regimen

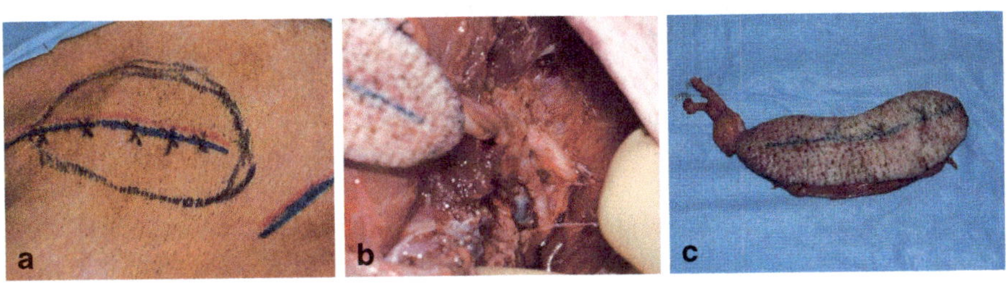

Fig. 6.2 Fasciocutaneous VCA swine model. a Flap design; outline, femoral and medial saphenous vessels marked. b Dissection of medial saphenous vessels to junction with femoral artery and vein. c Isolated fasciocutaneous VCA ready for transplantation. (Photographs courtesy of Dr. Curt Cetrulo)

[46]. Of the eight swine, two sustained severe rejection, three demonstrated mild-to-moderate rejection, and three were free of rejection at the termination of the experiment at 90 days. No drug toxicity was evident in serum hematologic and chemical parameters of immunosuppressed animals. The Louisville group also examined the use of tacrolimus, MMF, and prednisone in the same swine model. Five of nine animals that survived to the study end at 90 days were noted to be free of rejection [47]. This work served as a basis for their human hand transplants performed under the same chronic immunosuppression regimen. However, this work did not attempt to induce tolerance to these allografts.

The swine model has also served to explore techniques to induce tolerance to the VCA allograft. Initial work by Lee et al. sought to achieve host tolerance to musculoskeletal allografts through matching of the MHC antigens between donor and host swine with only a 12-day course of cyclosporine [48]. Allografts from MHC-mismatched donors treated with cyclosporine and allografts from MHC-matched (minor antigen mismatched) donors not treated with cyclosporine were rejected. However, allografts from MHC-matched donors treated with 12 days of cyclosporine showed no evidence of rejection until sacrifice up to 47 weeks after transplantation. This protocol did not induce tolerance to skin and when a cutaneous portion was added to the transplant, only one out of the six animals transplanted maintains tolerance to the entire allograft [28]. The majority demonstrated "split tolerance" where the skin was rejected but the muscle and bone survived long term. Attempts to extend this work utilizing high-dose tacrolimus rather than cyclosporine failed to induce tolerance to myocutaneous grafts across greater genetic disparities.

Hettiaratchy et al. sought to apply a mixed chimerism protocol to the same VCA model across greater genetic barriers (haploidentical and fully mismatched MHC) [49]. In these experiments, he used a nonmyeolablative protocol that combined anti-CD3 antibody, 150 cGY thymic radiation, and the infusion of donor hematopoietic cells with the VCA transplant [49]. These animals received either bone marrow or cytokine-mobilized peripheral blood mononuclear cells with 30 days of cyclosporine for post-grafting immunosuppression. As was noted in the previous experiments, split tolerance was again observed with rejection of the skin noted by day 60. In addition, all chimeric animals developed cutaneous GVHD approximately 70 days post transplant. While these cases did respond to treatment with immunosuppression, this complication limits clinical applicability.

HSC engraftment and stable mixed chimerism can be achieved in Massachusetts General Hospital (MGH) miniature swine conditioned with 100 cGY TBI and CD3-immunotoxin before transplant with 15×109 conditioned media-peripheral blood mononuclear cell (CM-PBMC) per kilogram and 45 days treatment with CSP A [42]. Two animals on this protocol received primarily vascularized skin flaps transplanted from the original donor on a sapheno-femoral vascular pedicle. Unfortunately, one animal died 46 days post transplant from unrelated complications, but another accepted this skin transplant indefinitely with follow-up of more than 1 year and no gross or histologic evidence of rejection at any time. This animal maintained stable, multilineage mixed chimerism, detectable in peripheral blood, thymus, and bone marrow and did not develop GVHD [43]. This was the first reported induction of skin tolerance in a large-animal model across an MHC barrier and provided proof-of-principle for the induction of tolerance of skin-bearing VCA using the mixed chimerism approach.

Recent work in the swine model from Kuo et al. demonstrated acceptance of a VCA using a combination of low-dose TBI (150 cGy), intrathymic irradiation, bone marrow infusion, and MSC transplant, followed by a short duration of post-grafting immunosuppression [50]. This protocol appeared to lead to prolonged acceptance of the allograft. The authors then modified the protocol to exclude administration of donor bone marrow, and again noted prolonged survival of the allograft [51]. This work suggested that the use of MSC alone may enhance survival of the transplant. However, neither donor chimerism nor donor-specific tolerance was addressed in their studies.

Nonhuman Primates

Trials to induce tolerance to VCA in nonhuman primates have been ineffective as well. There is no published report of a successful induction of tolerance to a VCA in the literature. In fact, until the advent of improved immunosuppression, even attempts to maintain the allograft with chronic immunosuppression failed. Daniel et al. and Stark et al. each published their experience with hand transplantation in a baboon model (Fig. 6.3) [52, 53]. They found thateven with the administration of high-dose CSP and steroids failed to prevent rejection. However, in a more recent publication, Gold et al. reported the ability to maintain a transplant for up to 65 days in select animals with CSP alone [54].

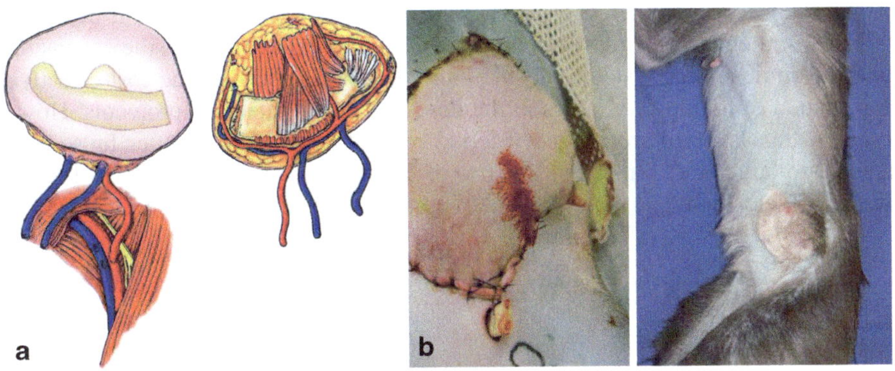

Fig. 6.3 Primate osteomyocutaneous VCA model. **a** Osteomyocutaneous flap design. **b** Flap inset and healed. (Photographs courtesy of Dr. Rolf Barth)

Barth et al. employed a similar mandibular allograft primate model and demonstrated long-term survival (177 days) with high-dose tacrolimus [55]. Unfortunately, in all of these monkeys, the VCA was inevitably rejected and the majority of the monkeys developed post-transplant proliferative disorder (PTLD). The addition of MMF to the same protocol allowed for the long-term survival of the allografts [56]. This study also evaluated the role of the inclusion of a vascularized bone marrow segment. Facial segments containing bone enjoyed prolonged survival until the immunosuppression was withdrawn, afterwards all allograft went on to reject. In contrast, transplants devoid of the mandibular component were all lost to acute rejection by day 15 [57]. The histologic evaluation of the five long-term surviving animals that later underwent immunosuppression withdrawal all demonstrated features consistent with chronic rejection, including neointimal proliferation and transplant vasculopathy [56]. The group also examined these animals for the presence of T-regulatory cells in the biopsy specimens and in the peripheral blood and found no correlation between presence of T-regulatory cells and the presence or the absence of rejection [58].

There have been limited studies using co-stimulatory blockade in primate tolerance protocols. Barth et al. found that the addition of anti-CD28 to their protocol led to prolongation of the allograft but not tolerance [59]. In a recent unpublished experiment by Cendales et al. attempting to block both T cell co-stimulation with CTLA 4-Ig and antigen-presenting cells via LFA 3-Ig, it was unable to prevent rejection of a transplanted myocutaneous forearm flap.

Conclusions

Vascularized composite allotransplantation represents a paradigm shift in the reconstruction of complex facial defects and extremity loss. However, the application of VCA is currently limited by the need for chronic immunosuppression. The development for strategies to decrease or eliminate this requirement is key to the expansion of this technique. Initial studies in small-animal models allow for the cost-effective exploration of multiple variables in the development of tolerance protocols. However, the translation of protocols developed in small-animal models often fails when applied to large-animal models. The use of large-animal models is critical for the further development and advancement of clinical VCA transplantation.

References

1. Breidenbach WC 3rd. Update on hand transplantation: Louisville. In: American Society of Reconstructive Transplantation: 2010. Chicago; 2010.
2. Lantieri L. Face transplant: a paradigm change in facial reconstruction. J Craniofac Surg. 2012;23(1):250–53.

3. Lanzetta M, Petruzzo P, Dubernard JM, Margreiter R, Schuind F, Breidenbach W, Nolli R, Schneeberger S, van Holder C, Gorantla VS et al. Second report (1998–2006) of the International Registry of Hand and Composite Tissue Transplantation. Transpl Immunol. 2007;18(1):1–6.
4. Gordon CR, Siemionow M, Papay F, Pryor L, Gatherwright J, Kodish E, Paradis C, Coffman K, Mathes D, Schneeberger S, et al. The world's experience with facial transplantation: what have we learned thus far? Ann Plast Surg. 2009;63(5):572–78.
5. Pomahac B, Bueno EM, Sisk GC, Pribaz JJ. Current principles of facial allotransplantation: the Brigham and Women's Hospital experience. Plast Reconstr Surg. 2013;131(5):1069–76.
6. Landin L, Cavadas PC, Ibanez J, Roger I. Malignant skin tumor in a composite tissue (bilateral hand) allograft recipient. Plast Reconstr Surg. 2010;125(1):20e–21e.
7. Diefenbeck M, Nerlich A, Schneeberger S, Wagner F, Hofmann GO. Allograft vasculopathy after allogeneic vascularized knee transplantation. Transpl Int. 2011;24(1):e1–5.
8. Kaufman CL, Ouseph R, Blair B, Kutz JE, Tsai TM, Scheker LR, Tien HY, Moreno R, Ozyurekoglu T, Banegas R, et al. Graft vasculopathy in clinical hand transplantation. Am J Transplant. 2012;12(4):1004–16.
9. Kawai T, Cosimi AB, Spitzer TR, Tolkoff-Rubin N, Suthanthiran M, Saidman SL, Shaffer J, Preffer FI, Ding R, Sharma V, et al. HLA-mismatched renal transplantation without maintenance immunosuppression. N Engl J Med. 2008;358(4):353–61.
10. Millan MT, Shizuru JA, Hoffmann P, Dejbakhsh-Jones S, Scandling JD, Grumet FC, Tan JC, Salvatierra O, Hoppe RT, Strober S. Mixed chimerism and immunosuppressive drug withdrawal after HLA-mismatched kidney and hematopoietic progenitor transplantation. Transplantation. 2002;73(9):1386–91.
11. Leventhal J, Huang Y, Xu H, Goode I, Ildstad ST. Novel regulatory therapies for prevention of Graft-versus-host disease. BMC Med. 2012;10:48.
12. Billingham R, Brent L, Medawar P. Actively acquired tolerance of foreign cells. Nature. 1953;172:603.
13. Owen RD. Immunogenetic Consequences of Vascular Anastomoses between Bovine Twins. Science. 1945;102(2651):400–1.
14. Leonard DA, Cetrulo CL Jr, McGrouther DA, Sachs DH. Induction of tolerance of vascularized composite allografts. Transplantation. 2013;95(3):403–9.
15. Foster RD, Fan L, Neipp M, Kaufman C, McCalmont T, Ascher N, Ildstad S, Anthony JP, Niepp M. Donor-specific tolerance induction in composite tissue allografts. Am J Surg.1998;176(5):418–21.
16. Prabhune KA, Gorantla VS, Perez-Abadia G, Francois CG, Vossen M, Laurentin-Perez LA, Breidenbach WC, Wang GG, Anderson GL, Pidwell DJ, et al. Composite tissue allotransplantation in chimeric hosts part II. A clinically relevant protocol to induce tolerance in a rat model. Transplantation. 2003;76(11):1548–55.
17. Esumi T, Inaba M, Ichioka N, Kushida T, Iida H, Ikehara S. Successful allogeneic leg transplantation in rats in conjunction with intra-bone marrow injection of donor bone marrow cells. Transplantation. 2003;76(11):1543–48.
18. Huang WC, Lin JY, Wallace CG, Chuang WY, Wei FC, Liao SK. Vascularized bone grafts within composite tissue allotransplants can autocreate tolerance through mixed chimerism with partial myeloablative conditioning: an experimental study in rats. Plast Reconstr Surg. 2010;125(4):1095–103.
19. Pan H, Zhao K, Wang L, Zheng Y, Zhang G, Mai H, Han Y, Yang L, Guo S. Mesenchymal stem cells enhance the induction of mixed chimerism and tolerance to rat hind-limb allografts after bone marrow transplantation. J Surg Res. 2010;160(2):315–24.
20. Kuo YR, Huang CW, Goto S, Wang CT, Hsu LW, Lin YC, Yang KD, Chen CL, Lee WP. Alloantigen-pulsed host dendritic cells induce T-cell regulation and prolong allograft survival in a rat model of hindlimb allotransplantation. J Surg Res. 2009;153(2):317–25.

21. Quatra F, Lowenberg DW, Buncke HJ, Romeo OM, Brooks D, Buntic RF, Baxter-Lowe LA. Induction of tolerance to composite tissue allograft in a rat model. Microsurgery. 2006;26(8):573–78.
22. Adamson LA, Huang WC, Breidenbach WC, Rahhal D, Xu H, Huang Y, Pidwell DJ, Wei FC, Tobin G, Ildstad ST. A modified model of hindlimb osteomyocutaneous flap for the study of tolerance to composite tissue allografts. Microsurgery. 2007;27(7):630–36.
23. Tung TH, Mackinnon SE, Mohanakumar T. Long-term limb allograft survival using anti-CD40 L antibody in a murine model. Transplantation. 2003;75(5):644–50.
24. Tung TH, Mackinnon SE, Mohanakumar T. Prolonged limb allograft survival with CD 40 costimulation blockade, T-cell depletion, and megadose donor bone-marrow transfusion. Microsurgery. 2005;25(8):624–31.
25. Tung TH, Mackinnon SE, Mohanakumar T. Combined treatment with CD40 costimulation blockade, T-cell depletion, low-dose irradiation, and donor bone marrow transfusion in limb allograft survival. Ann Plast Surg. 2005;55(5):512–18.
26. Zhong T, Liu Y, Jiang J, Wang H, Temple CL, Sun H, Garcia B, Zhong R, Ross DC. Long-term limb allograft survival using a short course of anti-CD45RB monoclonal antibody, LF 15–0195, and rapamycin in a mouse model. Transplantation. 2007;84(12):1636–43.
27. Li Z, Benghiat FS, Charbonnier LM, Kubjak C, Rivas MN, Cobbold SP, Waldmann H, De Wilde V, Petein M, Schuind F, et al. CD8 + T-Cell depletion and rapamycin synergize with combined coreceptor/stimulation blockade to induce robust limb allograft tolerance in mice. Am J Transplant. 2008;8(12):2527–36.
28. Mathes DW, Randolph MA, Solari MG, Nazzal JA, Nielsen GP, Arn JS, Sachs DH, Lee WP. Split tolerance to a composite tissue allograft in a swine model. Transplantation. 2003;75(1):25–31.
29. Calne RY. The initial study of the immunosuppressive effects of 6-mercaptopurine and azathioprine in organ transplantation and a few words on cyclosporin A. World J Surg. 1982;6(5):637–40.
30. Ladiges WC, Storb R, Thomas ED. Canine models of bone marrow transplantation. Lab Anim Sci. 1990;40(1):11–15.
31. Shapiro AM, Geng Hao E, Lakey JR, Finegood DT, Rajotte RV, Kneteman NM. Defining optimal immunosuppression for islet transplantation based on reduced diabetogenicity in canine islet autografts. Transplantation. 2002;74(11):1522–28.
32. Pirenne J. Contribution of large animal models to the development of clinical intestinal transplantation. Acta Gastroenterol Belg. 1999;62(2):221–5.
33. Mathes DW, Hwang B, Graves SS, Edwards J, Chang J, Storer BE, Butts-Miwongtum T, Sale GE, Nash RA, Storb R. Tolerance to vascularized composite allografts in canine mixed hematopoietic chimeras. Transplantation. 2011;92(12):1301–8.
34. Wagner JL, Burnett RC, Storb R. Organization of the canine major histocompatibility complex: current perspectives. J Hered. 1999;90(1):35–8.
35. Calne RY, White DJ, Rolles K, Smith DP, Herbertson BM. Prolonged survival of pig orthotopic heart grafts treated with cyclosporin A. Lancet. 1978;1(8075):1183–85.
36. Singer DS, Parent LJ, Ehrlich R. Identification and DNA sequence of an interspersed repetitive DNA element in the genome of the miniature swine. Nucleic Acids Res. 1987;15(6):2780.
37. Gustafsson K, LeGuern C, Hirsch F, Germana S, Pratt K, Sachs DH. Class II genes of miniature swine. IV. Characterization and expression of two allelic class II DQB cDNA clones. J Immunol. 1990;145(6):1946–51.
38. Hirsch F, Sachs DH, Gustafsson K, Pratt K, Germana S, LeGuern C. Class II genes of miniature swine. III. Characterization of an expressed pig class II gene homologous to HLA-DQA. Immunogenetics. 1990;31(1):52–6.
39. Kirk AD. Transplantation tolerance: a look at the nonhuman primate literature in the light of modern tolerance theories. Crit Rev Immunol. 1999;19(5–6):349–388.

40. Preville X, Flacher M, LeMauff B, Beauchard S, Davelu P, Tiollier J, Revillard JP. Mechanisms involved in antithymocyte globulin immunosuppressive activity in a nonhuman primate model. Transplantation. 2001;71(3):460–68.
41. Reemtsma K. Renal heterotransplantation from nonhuman primates to man. Ann N Y Acad Sci. 1969;162(1):412–18.
42. Mathes DW, Noland M, Graves S, Schlenker R, Miwongtum T, Storb R. A preclinical canine model for composite tissue transplantation. J Reconstr Microsurg. 2010;26(3):201–7.
43. Maris M, Woolfrey A, McSweeney PA, Sandmaier BM, Nash RA, Georges G, Maloney DG, Molina A, Chauncey T, Yu C, et al. Nonmyeloablative hematopoietic stem cell transplantation: transplantation for the 21st century. Front Biosci. 2001;6:G13–16.
44. Maris MB, Niederwieser D, Sandmaier BM, Storer B, Stuart M, Maloney D, Petersdorf E, McSweeney P, Pulsipher M, Woolfrey A, et al. HLA-matched unrelated donor hematopoietic cell transplantation after nonmyeloablative conditioning for patients with hematologic malignancies. Blood 2003;102(6):2021–30.
45. Sandmaier BM, McSweeney P, Yu C, Storb R. Nonmyeloablative transplants: preclinical and clinical results. Semin Oncol. 2000;27(2 Suppl 5):78–81.
46. Ustuner ET, Majzoub RK, Ren X, Edelstein J, Maldonado C, Perez-Abadia G, Breidenbach WC, Barker JH. Swine composite tissue allotransplant model for preclinical hand transplant studies. Microsurgery. 2000;20(8):400–6.
47. Ustuner ET, Zdichavsky M, Ren X, Edelstein J, Maldonado C, Ray M, Jevans AW, Breidenbach WC, Gruber SA, Barker JH, et al. Long-term composite tissue allograft survival in a porcine model with cyclosporine/mycophenolate mofetil therapy. Transplantation. 1998;66(12):1581–87.
48. Lee WP, Rubin JP, Bourget JL, Cober SR, Randolph MA, Nielsen GP, Ierino FL, Sachs DH. Tolerance to limb tissue allografts between swine matched for major histocompatibility complex antigens. Plast Reconstr Surg. 2001;107(6):1482–90; discussion 1491–82.
49. Hettiaratchy S, Melendy E, Randolph MA, Coburn RC, Neville DM, Jr., Sachs DH, Huang CA, Lee WP. Tolerance to composite tissue allografts across a major histocompatibility barrier in miniature swine. Transplantation. 2004;77(4):514–21.
50. Kuo YR, Goto S, Shih HS, Wang FS, Lin CC, Wang CT, Huang EY, Chen CL, Wei FC, Zheng XX, et al. Mesenchymal stem cells prolong composite tissue allotransplant survival in a swine model. Transplantation. 2009;87(12):1769–77.
51. Kuo YR, Chen CC, Shih HS, Goto S, Huang CW, Wang CT, Chen CL, Wei FC. Prolongation of composite tissue allotransplant survival by treatment with bone marrow mesenchymal stem cells is correlated with T-cell regulation in a swine hind-limb model. Plast Reconstr Surg. 2011;127(2):569–79.
52. Daniel RK, Egerszegi EP, Samulack DD, Skanes SE, Dykes RW, Rennie WR. Tissue transplants in primates for upper extremity reconstruction: a preliminary report. J Hand Surg Am. 1986;11(1):1–8.
53. Stark GB, Swartz WM, Narayanan K, Moller AR. Hand transplantation in baboons. Transplant Proc. 1987;19(5):3968–71.
54. Gold ME, Randzio J, Kniha H, Kim BS, Park HH, Stein JP, Booth K, Gruber HE, Furnas DW. Transplantation of vascularized composite mandibular allografts in young cynomolgus monkeys. Ann Plast Surg. 1991;26(2):125–32.
55. Barth RN, Nam AJ, Stanwix MG, Kukuruga D, Drachenberg CI, Bluebond-Langner R, Hui-Chou H, Shipley ST, Bartlett ST, Rodriguez ED. Prolonged survival of composite facial allografts in non-human primates associated with posttransplant lymphoproliferative disorder. Transplantation. 2009;88(11):1242–50.
56. Mundinger GS, Munivenkatappa R, Drachenberg CB, Ha JS, Vaca EE, Shipley ST, Papadimitriou JC, Bartlett ST, Rodriguez ED, Barth RN. Histopathology of chronic rejection in a nonhuman primate model of vascularized composite allotransplantation. Transplantation. 2013;95(10):1204–10.

57. Barth RN, Rodriguez ED, Mundinger GS, Nam AJ, Ha JS, Hui-Chou H, Jones LS, Panda A, Shipley ST, Drachenberg CB, et al. Vascularized bone marrow-based immunosuppression inhibits rejection of vascularized composite allografts in nonhuman primates. Am J Transplant. 2011;11(7):1407–16.
58. Brazio PS, Munivenkatappa RB, Bojovic B, Ha JS, Brown EN, Hess AS, Bartlett ST, Rodriguez ED, Barth RN. Regulatory T cells are not predictive of outcomes in a nonhuman primate model of vascularized composite allotransplantation. Transplantation. 2013;96(3):267–73.
59. Mundinger GS, Nam AJ, Hui-Chou HG, Stanwix MG, Jones LS, Drachenberg CB, Kukuruga D, Shipley ST, Dorafshar AH, Panda A, et al. Nonhuman primate model of fibula vascularized composite tissue allotransplantation demonstrates donor-recipient bony union. Plast Reconstr Surg. 2011;128(6):1193–204.

Chapter 7
Unique Immunological Features of Vascularized Composite Allografts

Kadiyala V. Ravindra

Introduction

Skin is the largest organ in the body and immunologically the most intriguing. The early history of transplantation medicine is dominated by pursuit of skin transplantation. This was particularly crucial during the World War II—war victims sustained extensive burns that were often fatal. Pioneering researchers such as Sir Peter Medawar did extensive work on skin allotransplantation. Sir Peter Medawar was commissioned by the British Medical Council to explore skin transplantation. He performed some of the early experiments in this field. Skin allografting led to rejection in 7–10 days initially followed by a more rapid loss when a second graft from the same donor was attempted. He surmised that the allograft led to development of some humoral factors that then led to the accelerated loss subsequently.

The goal of skin allografting remained unfulfilled until 15 years ago. The first successful hand transplantation [1] in 1998 opened a new frontier in transplantation medicine that is slowly expanding. The success that the field of vascularized composite allotransplantation (VCA) currently enjoys did not come accidentally. It has been the result of painstaking research by many and the courage of the patients to undertake experimental procedures.

History

Skin was the tissue that was classically used in most early studies of transplantation. Ironically, it proved to be the most difficult tissue to transplant. Plastic surgeons have been at the forefront of transplantation from the beginning. World War II resulted in many patients with extensive burns succumbing due to lack of sufficient

K. V. Ravindra (✉)
Division of Transplantation, Duke University, Durham NC 27710, USA
e-mail: KV.Ravindra@duke.edu

G. Brandacher (ed.), *The Science of Reconstructive Transplantation*,
Stem Cell Biology and Regenerative Medicine, DOI 10.1007/978-1-4939-2071-6_7

unburned skin available for grafting. The obvious solution seemed to be use of allogeneic skin grafting.

In the 1930s, Padgett [2] reported the use of skin allografts from family and unrelated donors to cover severely burned patients. The duration of allograft survival was unpredictable but in some cases, the graft lasted long enough to enable control of infection and fluid loss. Gibson and Medawar [3] demonstrated the "second set" phenomenon where the second allograft from the same donor was rejected more rapidly than the first. This established that the "rejection process" was not immutable and implied an allergic or immunologic process that might be susceptible to manipulation.

The high degree of immunogenicity of skin has been known for many decades. Joseph Murray, who is credited with the first successful kidney transplant, studied the antigenicity of different tissues and organs in the 1960s. Experimenting with dogs, he demonstrated that with the same immunosuppression, renal grafts survived 175 days as compared to 19 days for skin grafts [4]. In an important paper on plastic and reconstructive surgery, Murray presented a relative scale of antigenicity of tissues and organs: skin and lung being the most antigenic followed by the liver, heart, kidney, and pancreas [5]. Murray's hypothesis was that skin being the biological barrier against the environment, it has evolved into the strongest barrier against foreign antigens [2]. Although the relative scale of antigenicity has been revised through subsequent progress in solid organ transplantation, successful transplantation of skin-bearing structures occurred nearly four and a half decades after the first renal transplant in 1954 [6]. This inability to transplant skin was a cause of immense frustration for the early transplant pioneers. Medawar [7] articulated this disappointment when he wrote: "The success of organ (kidney) transplantation has overthrown the doctrinal tyranny of skin grafts."

Transplantation of hand and facial structures has been long sought after. An attempt at hand transplant in 1964 in Ecuador ended into a failure. This early experience reflected experimental work at the time and was largely due to lack of potent immunosuppression. Steroids and Azathioprine were clearly inadequate to handle the immune challenge that a skin-bearing structure posed. Confidence in feasibility of skin-bearing transplants had to wait until the calcineurin inhibitors became the drivers of success in solid organ transplantation. Experimental data followed that clearly demonstrated that these agents were potent enough to quell the immunologic challenge of skin.

Despite the availability of laboratory success, many ethical and logistic challenges had to be overcome to translate the knowledge into the clinical realm. Dubernard led the way by performing the first successful hand transplant and many other centers have followed. The success in upper limb transplantation opened the door to an entirely new field now called "vascularized composite allotransplantation (VCA)." The original term for the field was composite tissue allotransplantation but was revised by the working group of the American Society of Transplant Surgeons. Innovative plastic surgeons have embraced this new technique and have achieved unparalleled success in facial reconstruction, transforming not only their

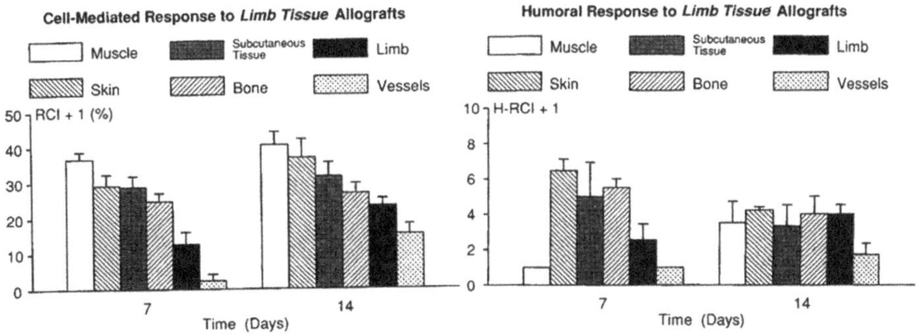

Fig. 7.1 Immune response to limb tissue allografts [5]

patients but also the entire field of reconstructive surgery. We are probably only at the start of a major shift in this subspecialty.

Components of Vascularized Composite Allografts

Solid organs have uniform histology through the organ while vascularized composite transplants are not composed of one tissue. They consist of an array of tissues including skin, subcutaneous tissue, muscles, nerves, tendons, bones and cartilage, vessels, etc. Each of these structures has a different immunogenic potential.

As mentioned earlier, Murray [5] found that the skin possessed the highest degree of antigenicity (Fig. 7.1). However, VCA is more complex. In pursuit of reducing the antigenicity of limb allografts by removing or suppressing the more antigenic portions, Lee[8] studied the relative antigenicity of the various limb tissues. Across a strong histocompatibility barrier in rats, he studied the humoral and cell-mediated immune response to transplanting vascularized skin flap, groin flap without skin, gastrocnemius muscle, knee joint, segment of femoral artery and vein, and the entire hind limb. The results were more complex than had been propounded by Murray (Figs. 7.1 and 7.2). The various limb tissues reacted with the host immune system in a predictable but differing timing and intensity. Vascularized muscle elicited the greatest cell-mediated response while skin, subcutaneous tissue, and bone produced a higher humoral response. Another interesting finding in this study was that when the entire hind limb was transplanted, the rejection response was slower and the rejection occurred more slowly. This finding corroborated an earlier report by Black [9] and different mechanisms were thought to be at play. These included "a consumption phenomenon," antigen competition, induction of enhancing antibodies, and an activation of suppressor T cells. These experiments also established that vascularized skin allografts elicited immune responses sooner: 1–2 weeks as against over 2 weeks in non-vascularized skin grafts.

Fig. 7.2 Relative antigenicity of tissues [5]

TABLE II

Relative Scale of Antigenicity of Tissues and Organs

Most antigenic:	Skin
	Lung
Less antigenic:	Liver
	Heart
Least antigenic:	Kidney
	Pancreas

Source: From Murray, J. E. Organ transplantation (skin, kidney, heart) and the plastic surgeon. *Plast. Reconstr. Surg.* 47: 425, 1971.

TABLE III

Relative Scale of Antigenicity of Limb Tissues

	Cellular Response	Humoral Response
Most antigenic:	Muscle	[Skin
		Subcutaneous tissue
	[Skin	Bone]
	Subcutaneous tissue	
	Bone]	Limb
	Limb	Muscle
Least antigenic:	Vessel	Vessel

Note: Based on immune responses measured for each limb tissue at 1 week after transplant.

The differential immune response generated by different organs/tissues from the same donor results in acceptance of one and rejection of another. This phenomenon of "split tolerance" has been extensively studied in the laboratory.

Reasons for High Immunogenicity of Skin

Skin is the largest organ of the body and is the first point of contact with the external environment. This has resulted in a specialized and potent immunological apparatus in the skin and surrounding tissues. Structurally, it consists of the epidermis and dermis. The epidermis is chiefly composed of ectodermal-derived keratinocytes and a smaller proportion of Langerhans cells, Merkle cells, and melanocytes. The dermis supports the epidermis and is composed of fibrous connective tissue (collagen and elastin) in a matrix of ground substance. Based on the difference in connective tissue density and arrangement, the dermis is divided into two layers: the superficial papillary dermis and the deeper reticular dermis which overlies the subcutaneous fat. Other important components of skin include the hair follicles, sebaceous, ecrine, and apocrine glands. The cutaneous vasculature is composed of two plexuses: the superficial and a deep plexus of arterioles and venules. Paralleling the blood supply of the skin is a lymphatic system that serves to allow Langerhans' cells to travel the regional lymph nodes.

The mechanisms of cutaneous defense consist of both innate and adaptive immune responses. The innate immunity consists of a physical barrier of stratum corneum and its various antimicrobial products including secreted lipids. In addition, antibacterial antibodies (immunoglobulin (Ig) A) are secreted by the sweat and sebaceous glands.

The adaptive cutaneous immunity is orchestrated through the "skin associated lymphoid tissue" (SALT) analogous to the gut-associated lymphoid tissue (GALT). This includes the Langerhans' cells, resident and migratory lymphocytes, keratinocytes, etc. The skin is rich in dendritic cells (DCs) which are the predominant antigen-presenting cells in the body. These Langerhans cells are abundant in the epidermal and dermal tissues and play a key role in the initiation of immune responses in the skin. They constitutively express major histocompatibility complex (MHC) class I antigens. Upon stimulation, the DCs and skin keratinocytes present class II antigens, intercellular adhesion molecule 1 (ICAM-1) and pro-inflammatory cytokines. The chief function of these cells is to process and present antigens encountered in the epidermis to naive T cells and thus initiate the adaptive immune response. Circulating skin-homing lymphocytes are a part of the cutaneous immune response and normal human skin demonstrates extravascular T lymphocytes—both CD4 and CD8. This is relevant to interpretation of skin biopsies in the transplant situation.

In the context of VCA transplantation, the cutaneous Langerhans' cells, which constitute 2–4 % of epidermal cells, are thought to play a crucial role in the priming of naive host T cells against the donor antigens. This results in the rejection response. Keratinocytes, which compose 90 % of epidermal cells, support the cell-mediated immune response. The DCs present in the dermis (dermal DCs) are able to migrate to the secondary lymphoid tissues. This property enables them to carry skin antigens to the host's lymphoid structures leading to initiation of the rejection phenomenon.

In addition, skin possesses tissue specific soluble antigens that are thought to play an important role in its antigenicity.

Immune Activation

Following transplantation, skin DCs migrate from the graft via the lymphatic vessels to the recipient's lymph nodes. They may present donor antigens to the recipient in two ways: the direct path where the host T cells recognize the donor MHC molecules present on the donor DCs and the indirect path where the donor peptides are bound to recipient DCs and then present them to the recipient T cells. Allo-recognition by either mechanism can trigger the rejection of the allogeneic skin graft. In addition, natural killer (NK) cells may be activated by the absence of self MHC class I molecules on allogeneic cells—this leads to direct killing of donor cells and the production of pro-inflammatory cytokines such as interferon (IFN)-γ and tumor necrosis factor (TNF)-α [10].

Role of Adhesion Molecules

The infiltration of alloantigen specific T cells into the skin of VCA grafts has been considered the key feature of acute rejection in composite tissue transplants. Adhesion molecules are vital for the function of immune cells. These molecules help activate leukocytes and convert them from an inactive, non-sticky state to an adhesive state [11]. This permits the activated cells to adhere to the vascular endothelium and thus migrate to inflamed tissues. The expression patterns of adhesion molecules are specific for each population of cells [12].

Various adhesion molecule types have been identified and broadly belong to two groups: selectins and integrins. It is currently thought that selectins (E, P, and L type) are important in leukocyte infiltration of skin in inflammation. Leukocytes also express integrins (lymphocyte function-associated antigen (LFA)-1 and macrophage-1 antigen (Mac-1)) which bind to ICAMs (ICAM-1 and 2) expressed on vascular endothelium.

Following transplantation, there is a unique injury related to restoration of blood flow following a period of ischemia: *ischemia–reperfusion injury*. This results in upregulation of adhesion molecules and generation of damage-associated molecular patterns (DAMPs). This leads to recruitment of lymphocytes [13] with potential adverse effect on graft—both acute and long-term injury [14]. Blocking adhesion molecules has been studied in experimental models to blunt the effect of ischemia–reperfusion injury. P-selectin blockade has been shown to reduce ischemia–reperfusion injury following liver transplantation in mouse models [15] and in a phase II clinical study [16]. In the study by Busuttil [16], the selectin antagonist known as recombinant P-selectin glycoprotein ligand IgG (rPSGL-Ig) was shown to significantly reduce the incidence of poor early graft function in liver transplant recipients, including those with a high donor risk index.

Adhesion Molecules in VCA

Study of skin biopsies from hand transplant recipients has demonstrated a strong correlation of LFA-1, ICAM-1, and E-selectin with the severity of acute rejection [17]. This finding was followed up with an elegant series of experimental studies by the Innsbruck group [17]. In a rat hind limb VCA model, Effomycine M was used subcutaneously to inhibit E and P-selectins. Long-term graft survival was demonstrated in five of six animals. Use of other agents to block ICAM-1 and LFA-1 was also reported to prolong graft survival significantly [11]. The ability to use these agents *locally* in the case of VCA transplants provides opportunities that are not feasible in solid organ transplantation. If this path can be further explored, it has the potential to target the early phase of immune activation with minimal systemic effects and thus facilitate lower systemic immunosuppression.

Rejection in VCA

By 2006, three different classification systems to grade acute rejection in VCA transplants had been published [18–20]. As the clinical volumes began to grow, there was a need to develop consensus on the issue. This was addressed at the Ninth Banff Conference on Allograft Pathology in 2007. The consensus group on composite tissue allotransplantation published its recommendation in 2008 [21]. Guidelines included the following: (a) specimen size—one "4-mm" punch biopsy of skin taken from the most red and indurated area of involved skin, (b) sample must have epidermis and adnexa, dermis, subcutaneous tissue, and vessels, and (c) slide preparation with hematoxylin–eosin (H&E) and periodic acid–Schiff (PAS) stains.

Clinical findings of rejection include mild pink discoloration, gradual erythema, macules progressing to red infiltrated lichenoid papules with or without limb edema, and onychomadesis in advance rejection [21]. Skin lesions can be either scattered over the allograft or present in a confluent pattern. In the Banff 2007 working classification, the area of graft involvement was graded as follows: <10 %, 10–50 %, and >50 % of the graft. However, this description has not been widely used in subsequent literature as this is not vital to grading the rejection event. The microscopic appearance is the basis for grading and is shown in Table 7.1.

Grade 1 rejection (mild) includes mainly lymphocytic perivascular aggregates in the dermis without epidermal involvement. With progression, the cellular infiltrate spreads into the epidermis. In severe cases, there is dense involvement of epidermis with apoptosis, dyskeratosis, and keratinolysis which could end up in frank necrosis of skin (Table 7.1 [21]).

While acute rejection in hand transplantation typically presents with maculopapular erythematous rash that is diffuse or patchy/focal over the forearms and dorsum of hands, a variant form has been described to occur rarely [22]. This atypical form involves the palmar skin and nails and has been attributed to repetitive mechanical stress of the palm. The features include red papules and lichenification of the palmar skin and dystrophy of the nail. The response to steroids was poor and resolution required the use of lymphocyte depleting agents Thymoglobulin and alemtuzumab.

Table 7.1 The Banff 2007 working classification of skin-containing composite tissue allograft pathology [21]

Grade 0	No or rare inflammatory infiltrates
Grade I	Mild. Mild perivascular infiltration. No involvement of the overlying epidermis
Grade II	Moderate. Moderate-to-severe perivascular inflammation with or without mild epidermal and/or adnexal involvement (limited to spongiosis and exocytosis). No epidermal dyskeratosis or apoptosis
Grade III	Severe. Dense inflammation and epidermal involvement with epithelial apoptosis, dyskeratosis, and/or keratinolysis
Grade IV	Necrotizing acute rejection. Frank necrosis of epidermis or other skin structures

Immunohistochemical Studies of Rejection in VCA

Cendales [23] studied the cellular infiltrate seen during acute rejection using immunohistochemical staining. The study of 29 specimens from both hand and abdominal wall transplants showed that the cells were predominantly CD4+ in milder cases and CD8+ in advanced cases. Hautz [17], in a more recent study, described a different pattern of infiltrate. In a study of 174 skin biopsy specimens collected over a 9-year time scale from five hand transplant recipients, the author showed that the perivascular infiltrate was predominantly CD3+T lymphocytes—a tendency for a predominance of CD8 positive lymphocytes in milder cases and CD4 positive cells in advanced cases was noted. During rejection, 10–50% of cells were identified to be CD68+ histiocytes/ macrophages. The numbers were increased during higher grades of acute rejection. CD 20+B cells were rarely detected (0–5%) in the skin of hand transplant recipients. In addition, Fox p3 and indoleamine 2, 3-dioxygenase expression correlated with the severity of rejection—suggesting a tendency toward self-limitation of the alloimmune response during the rejection process in VCA [24].

Humoral Immunity in VCA

The role of HLA antibodies in solid organ transplantation is well established. Since the landmark studies of Patel and Terasaki [25, 26], pre-transplant identification of donor-directed human leukocyte antigen (HLA) antibodies (donor-directed HLA-specific alloantibodies (DSAs)) has been a critical prelude to renal allotransplantation [27]. The presence of DSAs is largely a contraindication in renal transplantation. Recent innovations such as the use of desensitization techniques [28] have enabled successful transplantation in highly sensitized individuals. The development of DSAs following transplantation has been associated with episodes of acute rejection, chronic rejection, and graft loss in renal transplantation [29, 30].

The ability to detect DSAs has improved significantly in the past few years with the development of solid-phase detection assays [27]. This enhanced sensitivity and specificity in DSA detection has led to many unanswered questions regarding their relevance [31].

The role of preformed HLA antibodies in VCA transplantation has not been studied in depth. Although most VCA transplant centers use pre-transplant crossmatch testing [32, 33], the precise role DSAs play in hand or face transplantation is yet to be determined. A recent study [34] in Wistar Furth (WF) rats demonstrated that VCA grafts are rejected in an accelerated but not hyperacute fashion in the presence of allosensitization and preformed DSA. Additionally, this rejection was mainly cell-mediated and differed mechanistically from renal transplantation.

DSAs have developed during the follow-up of hand transplant recipients (de novo DSA) but have not been clearly shown to have adverse effects [33, 35]. C4d is a by-product of complement activation and the presence of staining for C4d on histopathology is considered a hallmark of acute antibody-associated injury of the allograft [36]. C4d deposition has been investigated in skin biopsies from VCA recipients with mixed results. Kanitakis [37] reported absence of C4d deposition in a study of 60 biopsy specimens obtained from four VCA recipients (three hand and one face). However, Landin [38] reported the occurrence of C4d deposits in the capillaries of skin biopsy specimens from two hand transplant recipients—both during and in the absence of clinical rejection episodes. Thus, further studies are needed with long-term follow-up of VCA recipients to further assess the role of preexisting and de novo DSAs in the field of VCA.

Regulatory T Cells

Regulatory T cells are thought to counter rejection and promote tolerance in the setting of transplantation. Many types of regulatory T cells have been identified in the recent past. These include CD8+T cells, CD4-CD8- double negative T cells, CD8+CD28-, natural killer (NK) T cells, and γδ T cells [39–42]. But the best studied are the CD4+regulatory T cells (T_{regs}). These have been characterized by high and stable expression of surface interleukin (IL)-2 receptor α chain (IL-2Rα, $CD25^{hi}$) and the transcription factor, fork-head box protein 3 (FoxP3) [43]. These cells are derived from thymus and are CD4+CD25+FoxP3+and are referred to as natural T_{regs} (nT_{regs}), in contrast to the induced T_{regs} (iT_{regs}) which are generated in the periphery and whose activation requires T cell receptor engagement and cytokines [44]. T_{regs} have been shown to prevent rejection of allogenic skin grafts in T cell deficient nude mice given $CD25^{-}$ T cells [43]. In a murine skin transplant model following thymectomy and partial T cell depletion, *in vitro* expanded T_{regs} have been shown to induce donor-specific transplantation tolerance [45]. Trials are currently in the pipeline to use adoptive T_{reg} cellular therapy in inducing transplantation tolerance [44].

Studies in human hand transplantation have demonstrated the presence of T_{regs} in transplanted skin. Intracellular staining of skin biopsy with highly specific monoclonal antibodies (mAbs) and measuring the FoxP3 messenger RNA (mRNA) expression has demonstrated the presence of FoxP3 positive cells in the grafted hand. In addition, these cells showed immunosuppressive properties when isolated in culture. These cells were found to be present as far out as 6 years posttransplantation [46]. It has been suggested that the presence of these cells could play a role in the long-term survival of VCA grafts [46].

Vascularized Bone Marrow Transplant

In the experimental setting, VCA grafts (face and limb) are thought to function as a vascularized bone marrow transplant (VBMT). Hewitt [47] established macrochimerism with vascularized limb allografts from Lewis X Brown-Norway F1 to Lewis rats and reported long-term survival in eight recipients treated with cyclosporine.

When immunosuppression was discontinued, two of eight animals did not show histologic evidence of acute rejection.

Based on laboratory data, it was widely anticipated that clinical hand transplantation would induce chimerism due to the viable bone marrow component of the graft. However, this expectation has not been fulfilled. Peripheral blood kinetic studies on two hand transplant recipients failed to demonstrate either donor macrochimerism or donor-specific hyporesponsiveness in mixed lymphocyte reaction [48]. Only a transient low level of peripheral microchimerism (1:75,000 cells) was detected. Another clinical study failed to demonstrate recipient-derived antigen-presenting cells (Langerhans' cells) in the epidermis of the graft beyond 77 days [49]. Upper limb allografts may not be a significant source of donor hematopoietic cells and the little that is engrafted will likely not survive without some conditioning.

Even when additional stem cell product is used along with VCA graft, there has been lack of success in engraftment of the stem cells. In the Pittsburgh study [33], unmodified donor stem cell transplant was added to hand transplantation as a way of minimizing immunosuppression. In the absence of additional conditioning other than alemtuzumab induction, the marrow graft failed to take and chimerism was not seen. Despite this, the five patients reported could be maintained on tacrolimus monotherapy with infrequent rejection episodes.

Clinical studies in renal transplantation have successfully utilized different conditioning regimens to facilitate chimerism following combined kidney and stem cell transplantation [50–52]. It appears that stem cell transplant survival will require more intense conditioning than is feasible with standard immunosuppression of VCA. This might explain why the VBMT that is thought to be part of a VCA graft fails to engraft. Thus, it seems unlikely that human hand allografts can be considered to a VBMT with the current immunosuppression protocols.

Conundrum of Chronic Rejection in VCA

Chronic rejection is a term that has been used to describe a slow decline of graft function over long-term follow-up. The mechanisms are poorly understood and manifest differently with each organ. The vasculature seems to be the main target—chronic obliteration and scarring of vessels in the organ. In the heart, this is seen as an "accelerated graft atherosclerosis" while in the kidney there is damage to the microscopic vasculature along with interstitial fibrosis and tubular atrophy. In the lung, this is manifest as "bronchiolitis obliterans" and in the liver with paucity of bile duct referred to as the "vanishing bile duct syndrome."

In renal transplantation, chronic renal allograft injury is multifactorial [53]. The etiology is both immunological (sensitization, HLA disparity, previous acute rejection episodes) and non-immunological (donor age, delayed graft function, calcineurin inhibitor toxicity, arterial hypertension, infections). The occurrence of acute rejection and the development of DSAs have been linked to chronic rejection in solid organ transplantation [29].

VCAs are subject to a higher prevalence of acute rejection episodes as compared to solid organ transplants [54]. Yet, these episodes do not seem to lead to chronic changes [55]. Development of DSA has been documented in hand transplant recipients [33, 35] but there is little correlation with immunological damage to the graft [38]. Whether the ability to detect acute rejection at an earlier point in the immunological cascade and thus treat it or the lack of adequate long-term follow-up are the reasons for this absence of clear cut long-term immunological damage to VCA grafts remains to be investigated.

There is evidence that vasculopathy does occur in VCA. An upper extremity graft was lost due to severe vasculopathy at 9 months from unclear reasons. Sophisticated techniques such as the ultrasound biomicroscopy are being investigated as a tool to detect changes in graft vessels as a means of detecting chronic damage to the graft in VCA [35]. The incidence of such injury and its etiology remain undefined at this time.

However, there is experimental data that supports chronic damage to VCA grafts. In a rat hind limb allotransplantation model, multiple episodes of acute rejection ultimately led to vasculopathy, skin and muscle atrophy, sclerotic bone, and an up-regulation of profibrotic gene expression resulting in fibrosis [56]. Although the study protocol may not truly reflect the clinical situation, there is need for further exploration.

Conclusions

VCA is immunologically unique due to the multiple tissues involved and the high immunogenicity of the skin component. It provides unique opportunities to study transplant immunology due to its visibility and ease of biopsy. The following areas provide opportunities for research in future:

1. The role of ischemia-reperfusion injury in VCA transplantation
2. The role of preformed and de novo DSAs
3. The significance of HLA disparity
4. Development of specific biochemical markers of immunological injury

References

1. Dubernard JM, Owen E, Herzberg G, et al. Human hand allograft: report on first 6 months. Lancet. 1999;353:1315–20.
2. Murray JE. The first successful organ transplants in man. J Am Coll Surg. 2005;200:5–9.
3. Gibson T, Medawar PB. The fate of skin homografts in man. J Anat. 1943;77:299–310 4.
4. Moseley RV, Sheil AG, Mitchell RM, Murray JE. Immunologic relationships between skin and kidney homografts in dogs on immunosuppressive therapy. Transplantation. 1966;4:678–87.
5. Murray JE. Organ transplantation (skin, kidney, heart) and the plastic surgeon. Plast Reconstr Surg. 1971;47:425–31.

6. Guild WR, Harrison JH, Merrill JP, Murray J. Successful homotransplantation of the kidney in an identical twin. Trans Am Clin Climatol Assoc. 1955;67:167–73.
7. Brent L, Medawar PB. Tissue transplantation: a new approach to the "typing" problem. Br Med J.1963;2:269–72.
8. Lee WP, Yaremchuk MJ, Pan YC, Randolph MA, Tan CM, Weiland AJ. Relative antigenicity of components of a vascularized limb allograft. Plast Reconstr Surg. 1991;87:401–11.
9. Black KS, Hewitt CW, Fraser LA, et al. Composite tissue (limb) allografts in rats. II. Indefinite survival using low-dose cyclosporine. Transplantation. 1985;39:365–8.
10. Benichou G, Yamada Y, Yun SH, Lin C, Fray M, Tocco G. Immune recognition and rejection of allogeneic skin grafts. Immunotherapy. 2011;3:757–70.
11. Hautz T, Wolfram D, Grahammer J, et al. Mechanisms and mediators of inflammation: potential models for skin rejection and targeted therapy in vascularized composite allotransplantation. Clin Dev Immunol. 2012;2012:757310.
12. von Andrian UH, Mackay CR. T-cell function and migration. Two sides of the same coin. New Eng J Med. 2000;343:1020–34.
13. Jaeschke H. Molecular mechanisms of hepatic ischemia-reperfusion injury and preconditioning. Am J Physiol Gastrointest Liver Physiol. 2003;284:G15–26.
14. Fondevila C, Busuttil RW, Kupiec-Weglinski JW. Hepatic ischemia/reperfusion injury–a fresh look. Exp Mol Pathol. 2003;74:86–93.
15. Dulkanchainun TS, Goss JA, Imagawa DK, et al. Reduction of hepatic ischemia/reperfusion injury by a soluble P-selectin glycoprotein ligand-1. Ann Surg. 1998;227:832–40.
16. Busuttil RW, Lipshutz GS, Kupiec-Weglinski JW, et al. rPSGL-Ig for improvement of early liver allograft function: a double-blind, placebo-controlled, single-center phase II study. Am J Transplant. 2011;11:786–97.
17. Hautz T, Zelger B, Grahammer J, et al. Molecular markers and targeted therapy of skin rejection in composite tissue allotransplantation. Am J Transplant. 2010;10:1200–9.
18. Bejarano PA, Levi D, Nassiri M, et al. The Pathology of full-thickness cadaver skin transplant for large abdominal defects: a proposed grading system for skin allograft acute rejection. Am J Surg Pathol. 2004;28:670–5.
19. Kanitakis J, Petruzzo P, Jullien D, et al. Pathological score for the evaluation of allograft rejection in human hand (composite tissue) allotransplantation. Eur J Dermatol. 2005;15:235–8.
20. Cendales LC, Kirk AD, Moresi JM, Ruiz P, Kleiner DE. Composite tissue allotransplantation: classification of clinical acute skin rejection. Transplantation. 2006;81:418–22.
21. Cendales LC, Kanitakis J, Schneeberger S, et al. The Banff 2007 working classification of skin-containing composite tissue allograft pathology. Am J Transplant. 2008;8:1396–400.
22. Schneeberger S, Gorantla VS, van Riet RP, et al. Atypical acute rejection after hand transplantation. Am J Transplant. 2008;8:688–96.
23. Cendales LC, Kirk AD, Moresi JM, Ruiz P, Kleiner DE. Composite tissue allotransplantation: classification of clinical acute skin rejection. Transplantation. 2005;80:1676–80.
24. Hautz T, Brandacher G, Zelger B, et al. Indoleamine 2,3-dioxygenase and foxp3 expression in skin rejection of human hand allografts. Transplant Proc. 2009;41:509–12.
25. Patel R, Terasaki PI. Significance of the positive crossmatch test in kidney transplantation. N Engl J Med.1969;280:735–9.
26. Patel R, Glassock R, Terasaki PI. Serotyping for homotransplantation. XIX. Experience with an interhospital scheme of cadaver-kidney sharing and tissue typing. JAMA. 1969;207:1319–24.
27. Gebel HM, Liwski RS, Bray RA. Technical aspects of HLA antibody testing. Curr Opin Organ Transplant. 2013;18:455–62.
28. Gloor J, Stegall MD. Sensitized renal transplant recipients: current protocols and future directions. Nat Rev Nephrol. 2010;6:297–306.
29. Lefaucheur C, Suberbielle-Boissel C, Hill GS, et al. Clinical relevance of preformed HLA donor-specific antibodies in kidney transplantation. Contrib Nephrol 2009;162:1–12.

30. Lefaucheur C, Loupy A, Vernerey D, et al. Antibody-mediated vascular rejection of kidney allografts: a population-based study. Lancet. 2013;381:313–9.
31. Taner T, Gandhi MJ, Sanderson SO, et al. Prevalence, course and impact of HLA donor-specific antibodies in liver transplantation in the first year. Am J Transplant. 2012;12:1504–10.
32. Ravindra KV, Buell JF, Kaufman CL, et al. Hand transplantation in the United States: experience with 3 patients. Surgery. 2008;144:638–43; discussion 43–4.
33. Schneeberger S, Gorantla VS, Brandacher G, et al. Upper-extremity transplantation using a cell-based protocol to minimize immunosuppression. Ann Surg. 2013;257:345–51.
34. Wu S, Xu H, Chen B, et al. Sensitized recipients exhibit accelerated but not hyperacute rejection of vascularized composite tissue allografts. Transplantation. 2011;92:627–33.
35. Kaufman CL, Ouseph R, Blair B, et al. Graft vasculopathy in clinical hand transplantation. Am J Transplant. 2012;12:1004–16.
36. Solez K, Colvin RB, Racusen LC, et al. Banff '05 Meeting Report: differential diagnosis of chronic allograft injury and elimination of chronic allograft nephropathy ('CAN'). Am J Transplant. 2007;7:518–26.
37. Kanitakis J, McGregor B, Badet L, et al. Absence of c4d deposition in human composite tissue (hands and face) allograft biopsies: an immunoperoxidase study. Transplantation. 2007;84:265–7.
38. Landin L, Cavadas PC, Ibanez J, Roger I, Vera-Sempere F. CD3 + -mediated rejection and C4d deposition in two composite tissue (bilateral hand) allograft recipients after induction with alemtuzumab. Transplantation. 2009;87:776–81.
39. Reibke R, Garbi N, Ganss R, Hammerling GJ, Arnold B, Oelert T. CD8 + regulatory T cells generated by neonatal recognition of peripheral self-antigen. Proc Natl Acad Sci U S A. 2006;103:15142–7.
40. Zhang ZX, Yang L, Young KJ, DuTemple B, Zhang L. Identification of a previously unknown antigen-specific regulatory T cell and its mechanism of suppression. Nat Med. 2000;6:782–9.
41. Haribhai D, Lin W, Relland LM, Truong N, Williams CB, Chatila TA. Regulatory T cells dynamically control the primary immune response to foreign antigen. J Immunol. 2007;178:2961–72.
42. Monteiro M, Almeida CF, Caridade M, et al. Identification of regulatory Foxp3 + invariant NKT cells induced by TGF-beta. J Immunol. 2010;185:2157–63.
43. Sakaguchi S, Sakaguchi N, Asano M, Itoh M, Toda M. Immunologic self-tolerance maintained by activated T cells expressing IL-2 receptor alpha-chains (CD25). Breakdown of a single mechanism of self-tolerance causes various autoimmune diseases. J Immunol. 1995;155:1151–64.
44. Safinia N, Leech J, Hernandez-Fuentes M, Lechler R, Lombardi G. Promoting transplantation tolerance; adoptive regulatory T cell therapy. Clin Exp Immunol. 2013;172:158–68.
45. Golshayan D, Jiang S, Tsang J, Garin MI, Mottet C, Lechler RI. *In vitro*-expanded donor alloantigen-specific CD4 + CD25 + regulatory T cells promote experimental transplantation tolerance. Blood. 2007;109:827–35.
46. Eljaafari A, Badet L, Kanitakis J, et al. Isolation of regulatory T cells in the skin of a human hand-allograft, up to six years posttransplantation. Transplantation. 2006;82:1764–8.
47. Hewitt CW, Black KS, Dowdy SF, et al. Composite tissue (limb) allografts in rats. III. Development of donor-host lymphoid chimeras in long-term survivors. Transplantation. 1986;41:39–43.
48. Granger DK, Briedenbach WC, Pidwell DJ, Jones JW, Baxter-Lowe LA, Kaufman CL. Lack of donor hyporesponsiveness and donor chimerism after clinical transplantation of the hand. Transplantation. 2002;74:1624–30.
49. Thaunat O, Badet L, El-Jaafari A, Kanitakis J, Dubernard JM, Morelon E. Composite tissue allograft extends a helping hand to transplant immunologists. Am J Transplant. 2006;6:2238–42.
50. Leventhal J, Abecassis M, Miller J, et al. Chimerism and tolerance without GVHD or engraftment syndrome in HLA-mismatched combined kidney and hematopoietic stem cell transplantation. Sci Transl Med. 2012;4:124ra28.

51. Kawai T, Cosimi AB, Spitzer TR, et al. HLA-mismatched renal transplantation without maintenance immunosuppression. New Eng J Med. 2008;358:353–61.
52. Scandling JD, Busque S, Dejbakhsh-Jones S, et al. Tolerance and withdrawal of immunosuppressive drugs in patients given kidney and hematopoietic cell transplants. Am J Transplant. 2012;12:1133–45.
53. Pascual J, Perez-Saez MJ, Mir M, Crespo M. Chronic renal allograft injury: early detection, accurate diagnosis and management. Transplant Rev. 2012;26:280–90.
54. Petruzzo P, Lanzetta M, Dubernard JM, et al. The International Registry on Hand and Composite Tissue Transplantation. Transplantation. 2010;90:1590–4.
55. Morelon E, Kanitakis J, Petruzzo P. Immunological issues in clinical composite tissue allotransplantation: where do we stand today? Transplantation. 2012;93:855–9.
56. Unadkat JV, Schneeberger S, Horibe EH, et al. Composite tissue vasculopathy and degeneration following multiple episodes of acute rejection in reconstructive transplantation. Am J Transplant. 2010;10:251–61.

Chapter 8
Immunological Similarities and Differences Between Extremity and Face Transplants

Palmina Petruzzo and Lionel Badet

Introduction

Since the first successful hand allotransplantation in 1998 and the first partial face allotransplantation in 2005, 61 patients have received a single or bilateral upper extremity transplantation, one patient a bilateral lower extremity transplantation, and 20 patients a partial or total face transplantation. Clinical reconstructive transplantation programs have been developed in several countries as a method of choice in plastic and reconstructive surgery for those patients suffering from complex injuries or malformations not amenable to conventional reconstruction. However, the principal factor limiting the widespread clinical application of reconstructive transplantation is the requirement for lifelong immunosuppression that carries many possible complications and side effects, especially in the setting of nonlife-saving transplants.

Improved understanding of the mechanisms of acute and chronic rejection in hand and face allotranplantation will allow for improved transplant management, as well as the future development of immunosuppression minimizing and tolerance-inducing protocols. The similarities and differences between the collective experiences of face and extremity transplantation are analyzed in this chapter.

Acute Rejection

Hand and face transplantations are called "composite tissue allotransplantations" (CTA) or "vascularized composite allotransplantations" (VCA) as they consist of histogenetically different tissues including skin, connective tissue, muscle, bone,

P. Petruzzo (✉) · L. Badet
Department of Transplantation, Edouard Herriot Hospital,
5, Place d'Arsonval, 69437 Lyon, France
e-mail: petruzzo@medicina.unica.it

P. Petruzzo
Department of Surgery, University of Cagliari, Cagliari, Italy

G. Brandacher (ed.), *The Science of Reconstructive Transplantation,*
Stem Cell Biology and Regenerative Medicine, DOI 10.1007/978-1-4939-2071-6_8

bone marrow, nerves, and blood vessels, that comprise a single functional unit transferred from a deceased donor to a recipient, much like a solid organ [1]. In other words, a VCA is a single vascularized unit comprised of different tissues with different antigenic loads.

Skin was historically considered the most antigenic tissue as proposed by Murray in his relative scale of the antigenicity of tissues and organs [2]. However, on the basis of their experimental studies, some authors concluded that no single tissue is dominant in primarily vascularized limb allografts. Moreover, they demonstrated that a whole limb allograft elicits a less intense immune response than each individual component of the composite tissue allografts [3].

Clinical experience, however, seems to confirm the contention that skin is the main target of acute rejection (AR). The first clinical signs of AR manifest on the skin: Suspicion is based on visual inspection, and then confirmed by histological examination. Whether the dominant immune response is really directed towards the skin is uncertain; however, since information on the involvement of the other components of composite tissue allografts is difficult to obtain. Indeed, much fewer data are available on the pathologic findings of deeper tissues during AR.

Experimental studies in a rat hind limb transplant model have shown an important involvement of muscles during acute rejection episodes that lead to fibrosis and impairment of muscular function and strength [4].

However, clinical experience shows that even during severe rejection, the changes found in underlying tissues (muscles, nerves bones, and tendons) are less severe than those present in the skin [5, 6].

Although the immunosuppressive drugs currently used in solid organ transplantation ensure VCA viability, the majority of extremity and face recipients experienced at least one episode of AR in the first year after transplantation (Figs. 8.1 and 8.2).

In extremities, as well as in face transplantations, AR reactions manifest clinically as erythematous macules, diffuse redness, or asymptomatic papules over the allografted skin [7]. Microscopically, they show characteristic, although nonspecific, changes involving mainly the dermis and the epidermis that may even extend

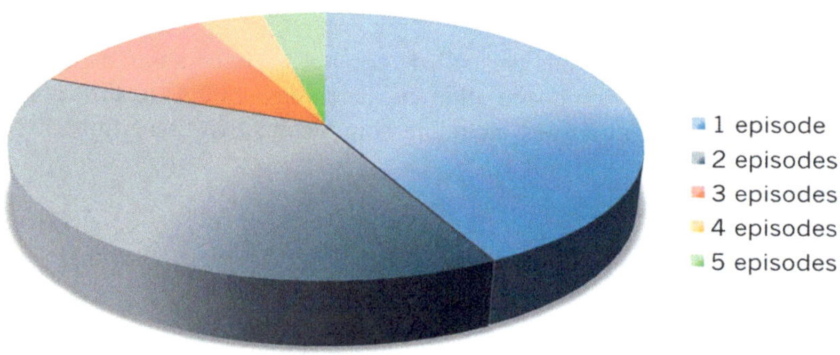

Fig. 8.1 Episodes of acute rejection within the first year in upper extremity transplantation

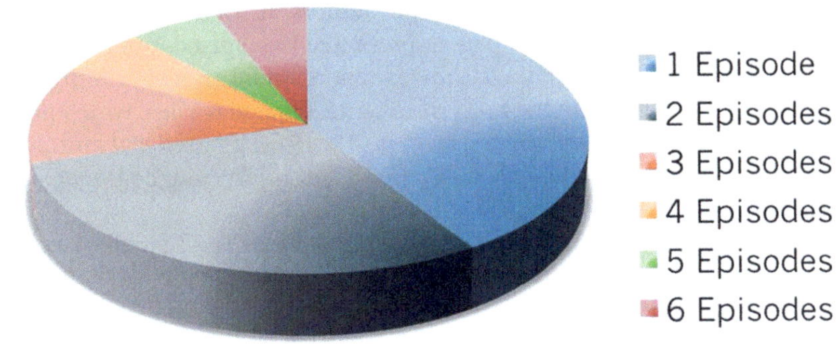

Fig. 8.2 Episodes of acute rejection within the first year in face transplantation

to the hypodermis in cases of severe rejection. These changes consist of a perivascular lymphocytic infiltrate in the superficial and mid-dermis predominantly made of CD3$^+$/CD4$^+$ T cells, with smaller proportions of CD8$^+$ and TIA-1$^+$ cytotoxic T cells, FoxP3$^+$ T regulatory cells, and occasional CD68$^+$ histiomonocytic cells of recipient's origin. The epidermis can show exocytosis and keratinocyte necrosis or apoptosis associated with basal keratinocyte vacuolization. More rarely, the epidermis shows spongiosis (intercellular edema) or lichenoid changes (orthokeratotic hyperkeratosis, hypergranulosis, acanthosis, band-like subepidermal infiltrate) similar to those observed in (lichenoid) graft-versus-host disease (GvHD). In the case of very severe rejection, the epidermis (and its appendages, hair follicles, and sweat glands) may show extensive necrosis. In those severe cases, the infiltrate may extend to the hypodermis and also contain eosinophils [5]. A specific score (Banff score 2007) consisting of five severity grades has been established in order to assess AR in hand and face transplantations [8, 9].

In the case of facial allotransplantation, biopsies obtained from the allografted oral mucosa show qualitatively similar changes, which are as a rule more pronounced than those found concomitantly on the skin [10–12]. The explanations for this discrepancy are unclear; however, one possible explanation could be the existence of a higher density of vessels and antigen-presenting cells (dendritic and endothelial cells) in mucosa versus skin.

The presence of C4d deposits in the skin and their significance is somewhat controversial. In our experience, such deposits are rarely (if ever) detected in the allografted hand and face skin and mucosae [13]. In some studies, such deposits have been reported in the skin of hand transplants with and without signs of AR, albeit in the absence of concomitant donor-specific antibodies [14, 15]. The role of humoral rejection in VCA is not clear. It has yet to be demonstrated [16] in experimental limb transplantation, although in clinical hand and face transplantations alloantibody production against donor human leukocyte antigen (HLA) has been reported. In the unique case of lower extremity transplantation [17], the patient presented two AR episodes that were graded on the basis of Banff score; C4d deposits were detected without donor-specific HLA antibodies, but they were considered to be "artifacts" by the authors. It is interesting to note that in the first posttransplant year the patient presented two episodes of AR, grade II and III, respectively. These findings showed that the number and severity of AR episodes are not correlated to the mass of the transplanted tissues, which primarily consists of muscle. Indeed, several studies [18, 19] suggested that muscles are the most sensible tissue to the ischemia/reperfusion injury and, consequently, to rejection.

The immune response in hand and face transplantations is essentially T-cell mediated, and the cytotoxic activity of these T cells is donor specific, as shown in the first face transplantation recipient [20]. It has also been shown that the rejection of a VCA in sensitized recipients is mainly cell mediated and differs mechanistically from that of solid organ transplantation [21]. The predominance of the T-cell response during VCA rejection probably explains the efficacy of the traditional triple therapy approach used in the majority of VCAs (steroid, tacrolimus, mycophenolate mofetil (MMF), which primarily targets the T-cell response.

Chronic Rejection

In solid organ transplantation, high rates of acute rejection are often associated with high incidence of chronic rejection and low organ survival rates. However, despite the high incidence of AR episodes to date (76.7 % of the hand recipients and 54.5 % of the face recipients presented at least one episode of AR in the first posttransplant year), no clear evidence of chronic rejection has been found in compliant patients on long-term follow-up [7] and the graft survival rate was 90.5 and 95 % at 1 year for upper extremity and face transplants, respectively.

Insufficient data are available to define specific changes of chronic rejection in VCA. The Banff 2007 classification, while useful in assessing acute rejection, has not yet included features of chronic rejection [8]. Clinicopathologic features suggestive of chronic rejection could include myointimal proliferation of arterioles, loss of adnexa, nail changes, skin and muscular atrophy, and fibrosis of deep tissues.

The small number of extremity and face allotransplantations performed to date, the relatively short follow-up, and the limited data on deep tissue biopsies make it difficult to evaluate the real incidence of chronic dysfunction of VCA.

The hallmark feature of chronic rejection is graft vasculopathy, which initiates with injury of vessels' endothelium, proliferation and migration of smooth muscle cells, deposited matrix protein, and finally concludes with arterial luminal narrowing.

An experimental murine study employing a rat hind limb transplantation model showed that graft vasculopathy was determined by multiple AR episodes and that it was the last lesion to occur after skin and muscular atrophy [22].

Another experimental model of face transplantation in nonhuman primates [23] showed graft vasculopathy involving both the external carotids as well as smaller vessels when immunosuppression was withdrawn. All grafts developed arteritis, intimal hyperplasia progressing to vessel occlusion, and tertiary lymphoid follicles.

The first clinical case of graft vasculopathy [24] occurred 275 days after transplantation and was characterized by acute ischemia of a grafted hand. On the basis of the pathological findings, which showed intimal hyperplasia and vessel occlusion similar to that reported in heart transplantation, the Louisville team introduced a sophisticated ultrasound technique for monitoring transplant vasculopathy. Although this technology remains experimental and not yet ready for clinical application, varying degrees of neointimal hyperplasia in other hand-grafted recipients were reported without correlation with the number of AR episodes, follow-up time, and the presence of acute skin rejection.

Additionally, one case of vasculopathy in a vascularized knee allograft has been reported [25]. In this case, necrosis of the skin sentinel flap occurred 36 months after transplantation, and the knee was removed 50 months posttransplantation after showing intimal hyperplasia and occlusion of the small vessels.

Although currently there are no reports of face allograft vasculopathy, careful evaluation of skin lesions and the treatment of all AR episodes are imperative for proper graft management.

A face loss was reported due to patient noncompliance with his immunosuppressive regimen. The patient died afterwards, but the circumstances concerning his death are not clear.

It is of paramount importance, particularly in facial transplant, to detect the early signs of graft vasculopathy due to the tremendous psychological impact of chronic rejection and graft loss on the recipients.

Graft-Versus-Host Disease, Chimerism, and Tolerance

Although bone marrow was present in the transplanted upper extremity bones and a substantial amount of active hematopoietic cells was in the bone marrow of the bilateral femoral transplantation, GvHD did not occur in any of these cases. Similarly, GvHD also failed to develop in partial and total face transplantations including the mandible, a bone that contains a rich marrow compartment.

It is even more remarkable that no GvHD occurred in the recipients of face or upper extremity transplantations who also received bone marrow infusions.

GvHD has never been observed in VCA to date; however, this possibility has to be considered prior to expanding immunosuppressive protocols including bone marrow infusion [26].

Peripheral blood microchimerism was detected for a very short period in two hand recipients in Louisville [27]. Although the long bones transferred as part of arm transplantation contained a substantial amount of active bone marrow, it was neither sufficient to induce chimerism nor allow for a reduction in immunosuppressive treatment.

Chimerism was not assessed in the case of bilateral femoral transplantation.

The team from Pittsburgh University and Johns Hopkins University [28] used a new protocol of "immunomodulation" or "minimizing immunosuppression" in upper extremity transplantation based on a regimen of donor bone marrow infusion 10–14 days posttransplantation and low-dose tacrolimus monotherapy after an induction with Campath-1H. Although they have obtained interesting and highly encouraging results, neither chimerism nor tolerance has been detected in these patients to date.

Similarly, face transplant recipients, including those receiving bone marrow infusions, did not show chimerism or tolerance. The only exception to this observation is first face transplant patient, who showed a transient microchimerism in a $CD34^{+}$-enriched cell population [20]. Despite this anomaly, there was no evidence for reduction in immunosuppressive requirements. In this case, the absence of durable donor chimerism and tolerance may be explained by an insufficient hematopoietic engraftment due to poor hematopoietic stem cell quality or an insufficient conditioning regimen.

At present, no VCA recipient has proved to be spontaneously tolerant. Indeed, it was noted in the first hand allotransplantation and in all recipients who discontinued immunosuppressive therapy that consequent rejection of the graft inevitably occurs [5, 7].

It is evident that the results obtained in rodent vascularized bone marrow transplant could not be transferred to the clinical practice in extremity as well as in face transplantation.

Immunosuppression and Complications

The large majority of upper extremity transplants [7] have been maintained on immunosuppression therapy similar to that used in solid organ transplantation, consisting of tacrolimus, steroids, and MMF. The induction therapy included antithymocyte globulin, basiliximab, and more recently, Campath-1H. All recipients received tacrolimus in the early postoperative period because of the stimulatory effect of this drug on the synthesis of axotomy-induced growth-associated protein (GAP-43) that seems to promote nerve regeneration [29]. Over the years, various modifications have been made to the initial maintenance treatment in order to decrease the risk for opportunistic infections, metabolic disorders, and malignancies. Such modifications include steroid-sparing maintenance, MMF-free treatment, and replacing tacrolimus with sirolimus.

The use of conventional immunosuppression in VCA is associated with the same complications commonly reported in solid organ transplantation [7], and these side effects have led to the creation of a novel protocol to minimize maintenance immunosuppression in upper extremity transplantation. This protocol is based on Campath-1H and prednisolone for induction, followed by tacrolimus monotherapy and infusion of donor bone marrow isolated by nine vertebral bodies on day 14 [28].

Five patients received this treatment with encouraging results. Although all of them presented with at least one episode of AR and developed donor-specific anti-HLA antibodies, only a few metabolic complications occurred and there were no reported infections.

The immunosuppressive therapy used in the case of lower extremity transplantation was similar to that used in upper extremity transplantation. It was based on Campath-1H as induction, and corticosteroid, tacrolimus, and MMF as maintenance treatment [17].

At present, it is difficult to demonstrate the superiority of one immunosuppressive regimen over another due to the lack of prospective randomized studies and the limited number of grafted patients.

The large majority of face transplant patients received antithymocyte globulins as immunosuppressive therapy, and all of them received corticosteroids, tacrolimus, and MMF in the immediate postoperative period. In addition, three patients also received topical immunosuppressants [7]. The maintenance therapy of multiple face transplant patients was modified during follow-up: Three patients were switched from tacrolimus to sirolimus, four patients were changed to steroid-free treatment, and another four cases were maintained on MMF-free treatment. In three cases, bone marrow cells were infused after transplantation in addition to induction therapy (antithymocyte globulins) and the conventional triple-drug maintenance therapy.

Immunosuppressive therapy in face transplantation is usually more aggressive than in extremity transplantation as graft loss is considered a more catastrophic event in the former group.

Graft loss was prevented in extremity as well as in face recipients compliant with the abovementioned protocols, but episodes of AR occurred in the majority of cases. Thus far, all episodes of skin rejection were reversible when treated promptly and effectively.

As reported by the International Registry on Hand and Composite Tissue Allotransplantation, the treatment of AR episodes in upper extremity transplantations was based on increase in oral dose or intravenous (IV) administration of corticosteroids at the first episode. Subsequent episodes were treated with IV corticosteroids or other drugs such as antithymocyte globulins or Campath-1H. The topical application of immunosuppressive drugs (steroid and tacrolimus ointments) was used for the first time in transplantation, and 95 % of patients received it as AR treatment with or without systemic therapy.

In face transplantation, the treatment of AR mainly involved IV corticosteroids and increases in the doses of oral immunosuppressants; other drugs such as antithymocyte globulins or Campath-1H were administered when the episodes were steroid resistant. Extracorporeal photochemotherapy was also used in several cases for the first time in VCA [30]. Although the mechanism of extracorporeal photochemotherapy has not been fully elucidated and at present is difficult to propose it as antirejection therapy for VCA, it has been successfully used after face transplantation to reverse AR episodes.

Extremity transplantations

Opportunistic infections
- CMV reactivation: 11
- Herpes virus: 3
- Herpes zoster: 1
- EBV infection: 1
- Clostridium difficilis infection: 2
- Cutaneous mycosis: 5
- Bacterial infection: 14 (1 osteitis and 3 infections of graft connective tissues)

Metabolic complications:
- Hyperglycemia: 19
- Increased creatinine values: 9
- End-stage renal disease (haemodialysis): 1
- Cushing Syndrome: 1
- Arterial hypertension: 5
- Avascular necrosis of the hip: 2

Malignancies:
- Basal cell carcinoma of nose: 1
- Post-transplant lymphoproliferative disease: 1

Face Transplantation

Opportunistic infections
- CMV reactivation: 3
- Herpes virus: 5
- EBV infection: 2
- Cutaneous mycosis: 2
- Bacterial infection: 12
  - 2 facial cellulitis
  - 1 pneumopathy
  - 1 pneumonia with sepsis
  - 1 sepsis

Metabolic complications:
- Hyperglycemia:3 (including 2 PTDM)
- Increased creatinine values: 4
- Arterial hypertension: 2
- Increase in γ-GT values: 1

Malignancies:
- Post-transplant lymphoproliferative disease: 1
- Basal cell carcinoma of recipient face: 1
- Uterus carcinoma: 1

Fig. 8.3 Complications

Local immunosuppressants have been used in both face and hand transplantation, yet their efficacy remains unproven. These local agents do not seem to be sufficient to reverse episodes of severe AR without additional systemic immunosuppressive treatment. A recent experimental model [31] showed the efficacy of clobetasol and tacrolimus ointment in the treatment of AR episodes as an adjunct agent to systemic therapy.

The use of conventional immunosuppression in both extremity and face transplants is associated with the complications usually reported in solid organ transplantation.

The main complications reported in the International Registry of Hand and Composite Tissue Transplantation [7] are metabolic ones, infections, and malignancies (Fig. 8.3).

Although the small number of patients, great number of variables, and short follow-up periods limit the statistical significance between the two groups, there seems to be a slight higher incidence of complications in face transplantation compared to extremities.

The higher incidence of bacterial, fungal, and viral infections could also be correlated to the presence of oral mucosa and/or paranasal sinuses in the grafts, as well as the higher incidence of burns among face recipients [32]. These patients are more exposed to bacterial infections as systemic immunosuppression may allow indolent, resistant bacteria to reemerge in a clinically significant manner with possible overwhelming sepsis after transplantation. This phenomenon occurred after a simultaneous face and bilateral hand transplantation and was followed by the recipient's consequent death [33].

Pertinent fungal infections may include candidal stomatitis, which can be difficult to distinguish from an episode of mucosal rejection in face transplantation.

Although no serious fungal infections have been reported thus far, the potential that fungal spores might colonize the graft (nasal and oral mucosa, paranasal sinuses) exists.

The most common viral infection or reactivation in face and upper and lower extremity transplantations was *Cytomegalovirus* (CMV). CMV infection or reactivation may not only condition the posttransplant period but also can increase the risk of AR.

Herpes viruses seem to be more common in face than in hand transplantation; the infection can be a reactivation of a latent infection or an infection through donor-derived transmission.

Face and extremity transplant patients are also at risk for Epstein–Barr virus (EBV) infection, which can lead to posttransplant lymphoproliferative disease (PTLD). Two such cases have been reported: one in a hand and one in a face transplant recipient [7].

These infectious risks can be reduced with a strategy based on antibiotic, fungal, and viral prophylaxis and avoiding the match EBV donor positive/EBV recipient negative. It is more complicated to use the same strategy for CMV as many regions have a considerable CMV-positive population, and in this case, prophylaxis and careful monitoring (CMV PCR) with antiviral therapy upon positive test results may be the best solution.

The primary metabolic complications associated with chronic immunosuppression are hyperglycemia, decrease in renal function, and arterial hypertension. Cancer of the skin, PTLD, and one case of uterus carcinoma were the specific malignancies reported in extremity and face transplant recipients.

The risks of the immunosuppression should not be underestimated; however, many of them are dose dependent and can be avoided with careful patient monitoring.

Conclusions

Extremity and face transplants are both composed of heterogeneous tissues and seem to show the same immunological features. The presence of mucosa in face transplantation could increase the susceptibility to infections, and in some cases may induce more severe AR lesions.

The immunosuppressive regimens used to maintain these two types of transplants are similar, although the more aggressive treatment used to prevent rejection in face transplantation also carries the risk of over-immunosuppression. This difference in the approach to immunosuppressive treatment is due to the fact that the loss of a facial allograft would likely have a more significant psychological impact on recipients than the loss of an extremity VCA.

A delicate balance between over- and under-immunosuppression is required to avoid infections, malignancies, and complications correlated to the toxicity of the drugs with graft vasculopathy, chronic rejection, and graft loss.

References

1. Pomahac B, Becker YT, Cendales L, et al. Vascularized composite allotransplantation research: the emerging field. Am J Transpl. 2012;12:1062.
2. Murray JE. Organ transplantation (skin, kidney, heart) and plastic surgeon. Plast Reconstr Surg. 1971;47:425.
3. Lee WP, Yaremchuk MJ, Pan YC, et al. Relative antigenicity of components of vascularized limb allograft. Plast Reconstr Surg. 1991;87:401.
4. Unadkat JV, Bourbaus D, Afrooz PN, et al. Functional outcomes following multiple acute rejections in experimental vascularised composite allotransplantation. Plast Reconstr Surg. 2013;131(5):720e.
5. Kanitakis J, Jullien D, Petruzzo P, et al. Clinicopathologic features of graft rejection of the first human hand allograft. Transplantation. 2003;76:688.
6. Landin L, Cavadas P, Nthumba P, Ibanez J, Vera-Sempere F. Preliminary results of bilateral arm transplantation. Transplantation. 2009;88:749.
7. Petruzzo P, Dubernard JM. The international registry on hand and composite tissue allotransplantation. Clin Transpl. 2011;27(4):411.
8. Cendales L, Kanitakis J, Schneeberger S, et al. The Banff 2007 working classification of skin-containing composite tissue allograft pathology. Am J Transpl. 2008;8:1396.
9. Kanitakis J, Cendales L. Classification of face rejection: Banff classification for CTA. In: Siemionow Maria, editor. The know-how of face transplantation. Springer Verlag London; 2011. pp. 199–205.
10. Kanitakis J, Badet L, Petruzzo P, et al. Clinicopathological monitoring of the skin and oral mucosa of the first human face allograft. Report on the first eight months. Transplantation. 2006;82:1610.
11. Lantieri L, Meningaud JP, Grimbert P, et al. Repair of the lower and middle parts of the face by composite tissue allotransplantation in a patient with massive plexiform neurofibroma: a 1-year follow-up study. Lancet. 2008;372:639.
12. Siemionow M, Papay F, Alam D, et al. First U.S. near-total human face transplantation-a paradigm shift for massive facial injuries. Lancet. 2009;374:203.
13. Kanitakis J, McGregor B, Badet L, et al. Absence of C4d deposition in human composite tissue (hands and face) allograft biopsies: an immunoperoxidase study. Transplantation. 2007;84:65.
14. Landin L, Cavadas PC, Ibañez J, Roger I, Vera-Sempere F. CD3+-mediated rejection and C4d deposition in two composite tissue (bilateral hand) allograft recipients after induction with alemtuzumab. Transplantation. 2009;87(5):776.
15. Hatz T, Zelger B, Brandacher G, et al. Histopathologic characterization of mild rejection (grade I) in skin biopsies of human hand allografts. Transpl Int. 2012;25(1):56.
16. Unadkat JV, Schneeberger S, Goldbach C, et al. Investigation of antibody-mediated rejection in composite tissue allotransplantation in rats. Transpl Proc. 2009;41(2):542.
17. Cavadas PC, Thione A, Carballeira A, Blanes M. Bilateral transfemoral lower extremity transplantation: result at 1 year. Am J Transpl. 2013 May;13(5):1343.
18. Khalil AA, Aziz FA, Hall JC. Reperfusion injury. Plast Reconstr Surg. 2006;117(3):1024.
19. Cendales LC, Kirk AD, Moresi JM, Ruiz P, Kleiner DE. Composite tissue allotransplantation: classification of clinical acute skin rejection. Transplantation. 2006;81(3):418.
20. Dubernard JM, Lengelé B, Morelon E, et al. Outcomes 18 months after the first human partial face transplantation. N Engl J Med. 2007;357(24):2451.
21. Wu S, Xu H, Chen B, et al. Sensitized recipients exhibit accelerated but not hyperacute rejection of vascularized composite tissue allografts. Transplantation. 2011;92(6):627.
22. Unadkat JV, Schneeberger S, Horibe EH, et al. Composite tissue vasculopathy and degeneration following multiple episodes of acute rejection in reconstructive transplantation. Am J Transpl. 2010;10:251.

23. Mundinger GS, Munivenkatappa R, Drachenberg CB, et al. Histopathology of chronic rejection in a nonhuman primate model of vascularized composite allotransplantation. Transplantation. 2013;295(10):1204.
24. Kaufman CL, Ouseph R, Blair B, et al. Graft vasculopathy in clinical hand transplantation. Am J Transpl. 2012;12(4):1004–16.
25. Diefenbeck M, Nerlich A, Schneeberger S, Wagner F, Hofmann GO. Allograft vasculopathy after allogeneic vascularized knee transplantation. Transpl Int. 2011;24(1):e1.
26. Granger DK, Briedenbach WC, Pidwell DJ, et al. Lack of donor hyporesponsiveness and donor chimerism after clinical transplantation of the hand. Transplantation. 2002;74:1624.
27. Ramsamooj R, Llull R, Black KS, Hewitt CW. Composite tissue allografts in rats: IV. Graft-versus-host disease in recipients of vascularized bone marrow transplants. Plast Reconstr Surg. 1999;104:1365.
28. Schneeberger S, Gorantla VS, Brandacher G, et al. Upper-extremity transplantation using a cell-based protocol to minimize immunosuppression. Ann Surg. 2013;257(2):345.
29. Gold BG, Yew JY, Zeleny-Pooley M. The immunosuppressant FK506 increases GAP-43 mRNA levels in axotomized sensory neurons. Neurosci Lett. 1998;241:25.
30. Hivelin M, Siemionow M, Grimbert P, et al. Extracorporeal photopheresis: from solid organ to face transplantation. Transpl Immunol. 2009;21:117.
31. Gharb BB, Rampazzo A, Altuntas SH, et al. Effectiveness of topical immunosuppressants in prevention and treatment of rejection in face allotransplantation. Transplantation. 2013 May 27;95(10):1197–203
32. Gordon CR, Avery RK, Abouhassan W, Siemionow M. Cytomegalovirus and other infectious issues related to face transplantation: specific considerations, lessons learned, and future recommendations. Plast Reconstr Surg. 2011;127(4):1515.
33. Carty MJ, Hivelin M, Dumontier C, et al. Lessons learned from simultaneous face and bilateral hand allotransplantation. Plast Reconstr Surg. 2013;132(2):423.

Chapter 9
Advances in Diagnosing Skin Rejection and Immune Monitoring

Emmanuel Morelon, Olivier Thaunat and Jean Kanitakis

Skin Rejection

Introduction

The fear that the skin would be ineluctably rejected after vascularized composite allotransplantation (VCA) was considered a major obstacle for hand transplantation prior to the first unilateral hand transplantation successfully performed in Lyon in 1998 [1]. The first human hand transplantation performed in 1964 by Gilbert et al. in Ecuador was indeed lost to rejection after only 2 weeks due to the lack of effective immunosuppressive therapy at that time [2]. As early as the late 1960s, experimental data suggested that the level of immunogenicity varied from one tissue to another. The skin was considered the tissue that carries the highest immunogenic potential because it was possible to induce tolerance to most allografted organs except the skin. This contention was further developed a few years later by Murray et al. [3], who proposed a scale of relative immunogenicity of tissues and organs, ranking the skin first, far above all the other tissues tested. The validity of this concept in the human setting was suggested by the histological examination of the first human hand allograft, removed during month 29 posttransplantation for uncontrolled rejection due to voluntary discontinuation of immunosuppression, showing that the most severe pathological changes were found in the skin, whereas

E. Morelon (✉) · O. Thaunat
Department of Transplantation, Nephrology and Immunology, Edouard Herriot Hospital, Lyon Civil Hospital, University of Lyon, Inserm U 1111, 69003 Lyon, France
e-mail: emmanuel.morelon@chu-lyon.fr

O. Thaunat
e-mail: olivier.thaunat@chu-lyon.fr

J. Kanitakis
Department of Dermatology, Edouard Herriot Hospital, Hospices Civils de Lyon, Lyon, France
e-mail: Jean.kanitakis@univ-lyon1.fr

G. Brandacher (ed.), *The Science of Reconstructive Transplantation*,
Stem Cell Biology and Regenerative Medicine, DOI 10.1007/978-1-4939-2071-6_9

only mild inflammation was found in muscles and tendons, and bone and joints were spared [4].

However, on the basis of experimental studies, Lee et al. concluded that no single tissue is dominant in primarily vascularized limb allografts; they further demonstrated that a whole limb allograft elicits a less intense immune response than each individual component of the composite tissue allograft [5]. Finally, one patient from the Louisville group who developed severe intimal hyperplasia and vasculopathy early after hand transplantation had normal skin at the time of graft loss, suggesting that vessels and skin are differentially targeted by the immune system and that skin can be spared despite its high immunogenicity while vessels are rejected [6]. Altogether, now more than 15 years of clinical experience under modern immunosuppressive therapy have confirmed that the skin is the main target of acute rejection (AR) in VCA.

Immunology of Skin Rejection

Cell Mediators of Skin Rejection

As for most cutaneous inflammatory reactions, the immune response against the donor skin is primarily T cell mediated. The skin contains at least two cell types that can recruit and activate T cells in dermal capillaries: (i) Langerhans cells (belonging to the dendritic cell family) and dermal dendritic cells and (ii) keratinocytes. Langerhans and dermal dendritic cells are professional antigen-presenting cells and act as a critical stimulus for sensitization of recipient's alloreactive T lymphocytes during the afferent phase of rejection, and keratinocytes can express major histocompatibility complex (MHC) class II molecules on their cell surface under inflammatory conditions [7] and may thereby contribute to the stimulation of $CD4^+$ T cells. The role of T cells in acute skin rejection is suggested by the composition of the cell infiltrate found in the skin during allograft rejection, which is made mainly by $CD3^+$/$CD4^+$ (and to a lesser degree $CD8^+$) T cells (Fig. 9.2, [8]) and by the donor-specific cytotoxic activity of these T cells, as shown in the first face transplant recipient [9]. In a large series of biopsies that has characterized cell infiltration in hand transplant patients over time, the majority of infiltrating cells were $CD3^+$ T lymphocytes [10]. The CD4/CD8 ratio seems to vary over time and with the severity of rejection [10]. Infiltration with $CD68^+$ macrophages observed in the dermis during skin rejection could also contribute to T cell activation [8, 10]. Besides $CD4^+$and $CD8^+$ and TiA-1+cytotoxic T cells, $CD4^+$/$CD25^+$/$FoxP3^+$T regulatory cells were also found in donor skin at different time points of follow-up in several patients [10–12]. T regulatory cells in the recipient lymphoid tissue may protect the allograft from an initial attack, while these cells, when present in the allograft, may help downregulate the effector cells that have infiltrated it. Similar to renal transplantation [13], the role of $CD4^+$/$CD25^+$/$FoxP3^+$T cells in the skin of composite tissue allografts remains un-

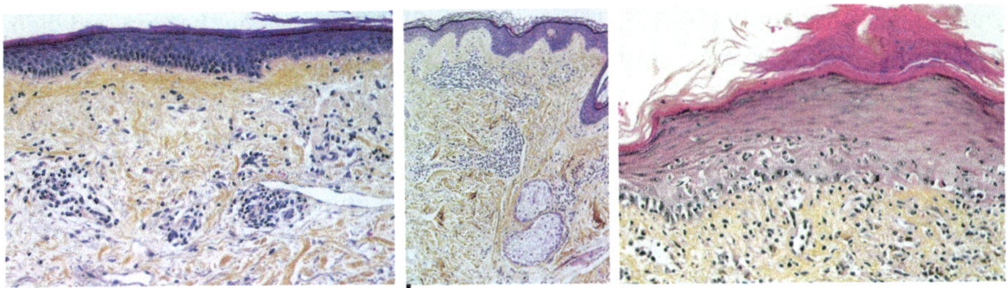

Fig. 9.1 Pathology of skin acute rejection. Illustration of the VCA Banff 2007 classification. **a** Grade 1, mild rejection. Pathological appearance of the sentinel skin graft of the first face transplant patient 6 years after transplantation (patient 7) shows mild perivascular infiltration without involvement of the overlying epidermis. **b** Grade 2, moderate rejection. Skin biopsy of the forearm in a bilateral hand transplant patient 2 years after transplantation (patient 5). Pathological appearance of the skin shows moderate-to-severe perivascular inflammation without epidermal involvement. **c** Grade 3, severe rejection. Skin biopsy on facial allograft 2.5 years after transplantation (patient 8). Pathological appearance of the skin shows lichenoid epidermal hyperplasia and lymphocyte infiltration in the upper dermis. The epidermis contains lymphocytic exocytosis. VCA vascularized composite allotransplantation

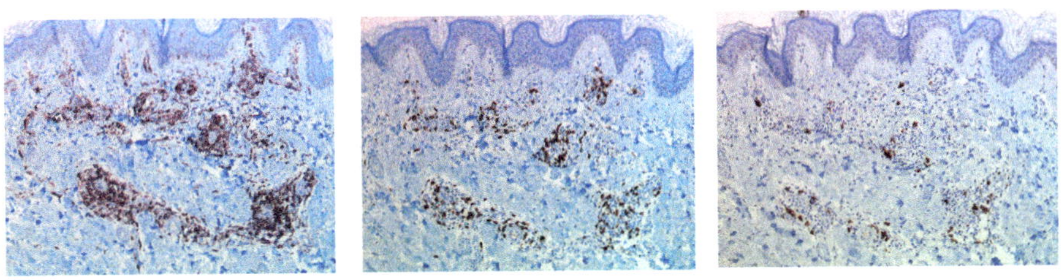

Fig. 9.2 Immunohistochemical analysis of skin rejection in hand transplantation. Immunohistochemical studies showed that dermal-infiltrating lymphocytes expressed predominantly a $CD3^+$/$CD4^+$ phenotype, with fewer cells expressing the $CD3^+$/$CD8^+$ phenotype. $CD20^+$ B cells are scarcely represented. (Hand transplant biopsy at 2 years posttransplantation, patient 5)

clear, including their clinical relevance in terms of immunoregulation and tolerance induction. Indeed, $CD4^+$/$CD25^+$/$FoxP3^+$T were not detected in skin biopsies in the absence of rejection, and FoxP3 expression was increased in rejection episodes at later time points after transplantation [10].

The role of B cells in acute skin rejection is uncertain. B cells are poorly represented within skin infiltrates (Fig. 9.2, [8, 10]). Circulating donor-specific anti-human leukocyte antigen (HLA) antibodies (DSA) have not been durably detected in hand transplant recipients, despite a high incidence of AR episodes [10, 14, 15]. The patient from the Louisville group who lost his grafted hand from vascular rejection had no DSA at the time of graft loss. Interestingly, DSA appeared 2 days after amputation [6], suggesting that anti-HLA antibodies could have been sequestered in the graft, as observed in kidney transplant recipients who developed DSA after transplant nephrectomy [16, 17]. The presence of C4d deposits in the skin and their significance as a marker of antibody-mediated rejection (AMR) remains doubtful. In kidney transplantation, deposition of the C4d complement degradation product in the capillary lumen is a relevant marker of AMR [18]. In our experience of hand and face transplantation, such deposits are exceptionally, if ever, detected in the allografted skin and mucosa [19]. Additionally, intraluminal capillary C4d deposits have been reported by other groups in the skin component both with and without concomitant signs of AR, DSA, and B cell infiltration [20–22]; this questions the usefulness of C4d detection in diagnosing VCA rejection. Besides producing antibodies, B cells might be involved in skin rejection through their ability to secrete cytokines and to act as antigen-presenting cells. B cells are indeed key components to generate memory $CD4^+$ T cells [23].

Altogether, these data suggest that the humoral arm of immune response does not play a significant role in skin AR in VCA.

Lymphocyte Migration to the Skin

Recruitment of immune cells to the skin graft involving chemokines and adhesion molecules is a key event in the process of AR as in other inflammatory cutaneous reactions [24, 25]. Expression of markers involved in cell trafficking has been extensively analyzed in protocol and for-cause hand transplant skin biopsies [10]. None of the markers investigated was found constitutively upregulated in the skin in the absence of histological signs of rejection. In contrast, the expression levels of lymphocyte function-associated antigen 1 (LFA-1), intercellular adhesion molecule 1 (ICAM-1), E-selectin, and P-selectin were upregulated upon rejection and correlated with the severity of rejection. This study also suggested that the superficial layers of the skin might be the primary sites of lymphocyte infiltration during the rejection process before their migration into the epidermis [10].

What Triggers Acute Skin Rejection in VCA?

Cellular rejection occurs as a result of an imbalance between the immunologic processes that maintain graft tolerance and those that promote graft rejection. One potential mechanism that could trigger such processes is activation of the innate immune response by ischemia/reperfusion injury (IRI) that subsequently initiates the adaptive alloimmune responses in the recipient. Emerging evidence has shown that innate immune activation as a consequence of IRI may occur in VCA [26, 27]. In addition, nonspecific stimuli such as mechanical trauma to the skin may also activate innate immune reactions [28]. Also viral infections, particularly cytomegalovirus (CMV) and herpesviruses, may trigger rejection episodes [9, 29].

Diagnosis of Skin Rejection

Frequency of Skin Rejection in the Clinical Setting

Although the immunosuppressive drugs currently used in solid organ transplantation usually ensure VCA viability, the majority of patients experience at least one episode of skin AR in the first posttransplant year. According to the International Hand and Composite Tissue Transplantation Registry [30], 85 % of hand-grafted patients and 54.5 % of face-grafted patients presented at least one episode of AR in the first posttransplant year, while multiple rejections developed in 56 % of them. Furthermore, steroid-resistant rejections are frequently observed, and require treatment with anti-thymocyte globulin (Thymoglobulin, Genzyme), basiliximab, or alemtuzumab (Campath-1H) [30]. Repeated episodes of skin rejection were observed in some patients beyond the first year after transplantation [30]. Table 9.1 shows the incidence of AR episodes in face and hand transplant patients in Lyon. All patients had at least one rejection episode in the first year, yet few patients experienced late AR episodes that required T-cell-depleting treatment by Thymoglobulin or Campath-1H.

Table 9.1 Incidence and severity of skin AR episodes in face and hand transplant patients. Lyon and Amiens experience

Patients	Sex	VCA	TR date (d/m/y)	No. of episodes	POD	Banff score	Initial IS treatment
Patient 1	M	Bilateral hands	13.01.2000	2	53, 72	2, 2	ATG, Tac, St, MMF
Patient 2	M	Bilateral hands	30.04.2003	3	57, 86, 2759	2, 2, 2	ATG, Tac, St, MMF
Patient 3	F	Bilateral hands	19.02.2007	7	16, 271, 635, 951, 1365, 1855, 2250	2, 2, 3, 2, 3, 3, 3	ATG, Tac, St, MMF
Patient 4	M	Bilateral hands	4.07.2008	1	65	2	ATG, Tac, St, MMF
Patient 5	M	Bilateral hands	11.07.2009	3	10, 350, 560	2, 2, 2	ATG, Tac, St, MMF
Patient 6	M	Bilateral hands	05.11.2012	3	20, 88, 154	2, 3, 3	ATG, Tac, St, MMF
Patient 7	F	Face	27.11.2005	2	23, 214	2, 3	ATG, Tac, St, MMF
Patient 8	M	Face	27.11.2009	8	41, 112, 186, 239, 474, 527, 540, 931	3, 3, 2, 2, 3, 3, 3, 3	ATG, Tac, St, MMF
Patient 9	F	Face	13.06.2012	1	12	3	ATG, Tac, St, MMF

TR transplantation, *POD* postoperative day, *IS* immunosuppressive, *ATG* Thymoglobulin, *Tac* tacrolimus, *St* steroids, *MMF* mycophenolate mofetil, *VCA* vascularized composite allotransplantation

The high rate of AR episodes reported in this field of transplantation might be due to the prompt diagnosis of AR, as the corresponding lesions are easily seen, and to the high immunogenicity of the skin, as discussed above. Despite the high incidence of AR episodes, the graft survival rate was 96 % at 1 year, and no convincing evidence of chronic (skin) rejection was found in compliant recipients on long-term follow-up [15].

Positive Diagnosis of Skin Rejection

Acute skin rejection is diagnosed by visual inspection as it manifests clinically with erythematous macules, diffuse skin redness (Figs. 9.3 and 9.4), or asymptomatic papules over the allografted skin with or without burning pain. Lesions are usually distributed on the dorsum of hands and forearms and may be bilateral when numerous [8, 31–33]. All changes can be associated with limb edema. Atypical rejection that affects the skin of the palm and nail beds has also been reported. Nail lesions include leukonychia and dystrophy, sometimes resulting in nail loss [28].

Interestingly, skin rejection in face transplantation may have a different clinical presentation form than that of hand transplantation. In our experience, face rejection usually appears as diffuse redness combined with diffuse edema (Fig. 9.3 and [9]). The different aspect might be related to venous and lymphatic vascularization that differs between the hands and face, as during the same episode of rejection, the patients may present macular lesion of AR on the sentinel skin flap and diffuse redness and edema on the facial graft skin (Fig. 9.3).

Given the importance of visual inspection for the diagnosis of skin rejection, patients need to be educated for routine daily inspection of the graft, at least during the first year posttransplant. Since the clinical appearance is not specific, diagnosis of skin rejection has to be substantiated by histological examination, even though the pathological findings alone are not totally specific.

Pathology of Skin Rejection

Skin Biopsy

In contrast to renal and liver transplantation, where rejection can be suspected by serological biomarkers of organ dysfunction, histological examination of skin biopsies remains the only established technique for assessment of skin rejection in VCA. Skin biopsy is a nonrisky procedure and easy to perform using a 4-mm punch scalpel. An adequate sample should contain the epidermis, dermis, and some quantity of subcutaneous tissue (hypodermis) [34]. The main drawback of skin biopsy is the resulting scar that can affect the aesthetic outcome in particular in face transplantation. Biopsy of the cheek oral mucosa or the sentinel skin flap of donor origin has been used as an alternative in order to limit damage to the grafted face by repeated

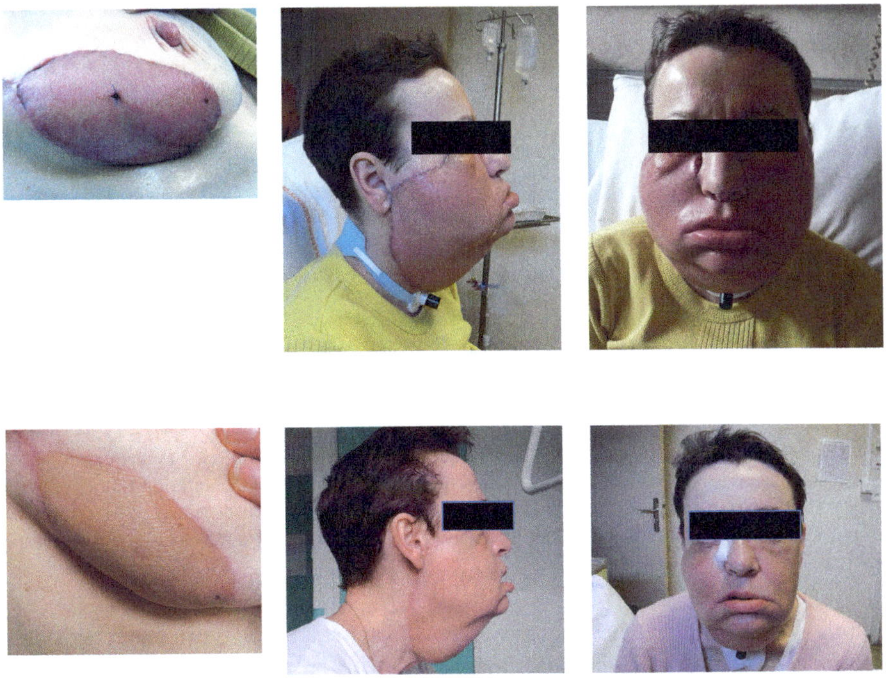

Fig. 9.3 Clinical aspects of skin rejection in face transplantation. Acute rejection grade 3, 12 days after face transplantation (patient 9): diffuse erythema on the sentinel skin graft (*upper left panel*); diffuse erythema and diffuse edema on the facial graft (*upper right panel*). The acute rejection episode was treated successfully with three boluses of intravenous methylprednisolone. The allograft skin of the sentinel skin graft (*lower left panel*) and the facial graft (*lower right panel*) shows normal appearance at 3 months posttransplantation

skin biopsies [35, 36]. The macroscopic features of the sentinel flap correspond well to those of the facial graft, and the pathological patterns of rejection are similar when skin biopsy specimens from both sites are compared [37]. Interestingly, the pathological patterns of rejection may appear earlier on the transplanted mucosa and also be more pronounced in the oral mucosa as compared with both the facial and the sentinel flap skin [37].

Pathology of Acute Skin Rejection

Microscopically, skin AR shows characteristic, although nonspecific, changes involving mainly the dermis and the epidermis that may also extend to the hypodermis in the case of severe rejection (Fig. 9.1). The earliest changes consist of a perivascular lymphocytic infiltrate in the superficial and mid dermis, predominantly made of $CD3^+$ $CD4^+$ and TIA-1^+ cytotoxic $CD8^+$ T cells, $FoxP3^+$ T regulatory cells, and $CD68^+$ histiomonocytic cells [8, 10]. In more severe rejection grades, this infiltrate may fill the dermis and invade the epidermis (exocytosis). The epidermis

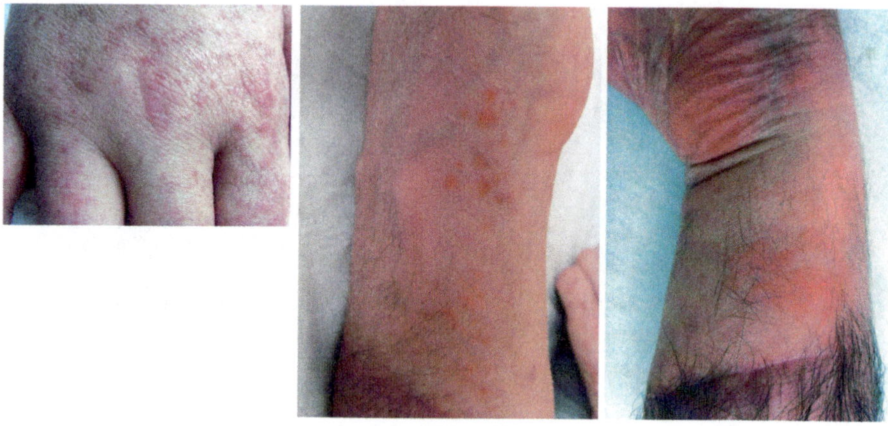

Fig. 9.4 Various clinical aspects of skin rejection in hand transplantation. Erythematous scaly papular lesions over the hand allograft, 5 years after bilateral hand transplantation (patient 3, *left panel*). Erythematous macula on the hand allograft during month 2 post-graft (patient 6, *middle panel*). Diffuse erythematous macula on the hand allograft during month 5 post-graft (patient 6, *right panel*)

is initially spared in the early phase of AR. At later stages, it shows exocytosis and keratinocyte necrosis or apoptosis associated with basal keratinocyte vacuolization. More rarely the epidermis shows spongiosis (intercellular edema) or lichenoid changes (orthokeratotic hyperkeratosis, hypergranulosis, acanthosis, band-like subepidermal infiltrate), similar to those observed in (lichenoid) graft-versus-host disease (GVHD). In the case of very severe rejection, the epidermis (and its appendages, hair follicles, and sweat glands) may show extensive necrosis. In those severe cases, the infiltrate may extend to the hypodermis and contain also eosinophils. On the basis of these changes, a specific score (Banff score 2007) has been established in order to assess the severity of AR (Table 9.2 and [33]).

Table 9.2 The BANFF 2007 working classification of skin-containing composite tissue allograft pathology. This system comprises the following five severity grades in order to assess the severity of acute rejection [33]

Grade 0 (no rejection)	No or rare inflammatory infiltrates
Grade 1 (mild rejection)	Mild perivascular infiltration. No involvement of the overlying epidermis
Grade 2 (moderate rejection)	Moderate-to-severe perivascular inflammation with or without mild epidermal and/or adnexal involvement (limited to spongiosis and exocytosis). No epidermal dyskeratosis or apoptosis
Grade 3 (severe rejection)	Dense dermal inflammation associated with epidermal involvement (basal keratinocyte vacuolization, keratinocyte apoptosis, and/or necrosis)
Grade 4 (necrotizing acute rejection)	Frank necrosis of epidermis or other skin structures

It should be noted here that the above pathological changes are not specific for AR as they can be found in a number of inflammatory, infectious, or proliferative dermatoses [38]. Ancillary techniques have been applied in an attempt to increase the specificity of AR diagnosis, such as immunophenotyping of infiltrating cells or detection of C4d in the allografted skin. The composition of the cell infiltrate is not very discriminative, since it is similar to that found in most inflammatory dermatoses. As discussed above, the presence of $FoxP3^+$ T regulatory cells, detectable in the allograft up to several years post-graft [11, 12] is interesting, although its prognostic significance remains so far unclear.

Chronic Skin Rejection

Surprisingly, despite a high incidence of skin AR, the occurrence of skin chronic rejection (CR) in VCA is rare and has not been reported so far in patients that were compliant with immunosuppressive therapy on long-term follow-up.

Experimental studies have shown that it is possible to induce CR in VCA. Data from a rat hind limb allograft model showed changes consistent with CR after 11 ± 3 episodes of AR. The skin was the main target of the immune response with progressive dermal atrophy and sclerosis and apoptosis of epithelial cells in the hair follicles with permanent hair loss. There are not yet sufficient data available in the clinical setting to define specific changes of CR in VCA. Indeed, the Banff 2007 classification did not include features of CR [33]. Clinicopathological features suggestive of skin CR could include loss of adnexa nail changes, skin atrophy, and fibrosis of deep tissues [33].

We recently investigated all allograft structures by histology, magnetic resonance imaging, ultrasonography, and high-resolution peripheral quantitative computed tomography scan in four bilateral hand transplant patients and one facial allotransplant recipient who all complied with the immunosuppressive treatment. We found no lesions that could suggest CR, such as dermal fibrosis and vascular stenosis [15].

Skin CR has been reported in the first hand transplant recipient after amputation of the graft following discontinuation of his immunosuppressive treatment. The rejected skin allograft showed a histological aspect that resembled chronic lichenoid GVHD [4]. Interestingly, the first patient who lost his hand allograft reportedly because of vascular rejection had no evidence of skin CR at the time of graft loss [6].

Finally, one face transplant patient from our group developed an Epstein–Barr virus (EBV)-positive B cell lymphoma and then a hepatic leiomyosarcoma that necessitated a drastic reduction of the immunosuppressive treatment resulting in multiple AR episodes. The face allograft, as well as the sentinel skin flap, showed some clinical and histological features consistent with CR beginning in the second year after transplantation. This patient showed exactly the predicted lesions of VCA CR, such as a sclerotic aspect of the graft, presence of dense dermal collagen fibers with hyalinosis, and atrophy of the adnexa. The large vessels remained unaffected while the dermal capillaries showed thickened walls and narrowed lumina (unpublished data, manuscript in preparation).

In summary, the experience obtained from 15 years of VCA has shown that chronic skin rejection is a rare event in patients who remain on immunosuppressive treatment. However, it may develop in noncompliant patients or in those where life-threatening side effects necessitate reduction of immunosuppression. One plausible explanation for the discrepancy between the high incidence of acute skin rejection and the absence of chronic skin rejection in compliant patients might be the exceptional healing potential of skin tissue as compared with other transplanted organs (such as the kidney).

Immune Monitoring

Overview on Immune Monitoring in Organ Transplantation

In the past 20 years, major progress has been made in prolonging graft and patient survival after organ transplantation as a result of the development of more efficient immunosuppressive drugs that have dramatically reduced the incidence of acute cellular rejection. However, the occurrence of AMR and subclinical rejection episodes, which markedly influence the long-term graft survival of transplanted organs, cannot be completely prevented with the current standard-of-care immunosuppressive protocols. Furthermore, long-term allograft survival requires lifelong immunosuppression that precipitates renal toxicity, opportunistic infections, diabetes, hypertension, and tumor formation. To avoid the toxic side effects caused by permanent immunosuppressive treatment, research in transplantation has focused on new treatment strategies, such as inducing tolerance or minimizing immunosuppression, in immunologically low-risk patients. The current challenge in transplantation is therefore to develop personalized treatment based on biomarkers at different biological levels that reflect the individual's immune reactivity and enable transplant clinicians to identify patients at risk for allograft rejection or, conversely, patients in whom immunosuppression could be safely minimized (review in [39]). Until now, monitoring of the immunosuppressive therapy received by solid organ transplant recipients has relied on the measurement of drug blood levels. Although this strategy has shown some efficacy in preventing drug toxicity, it is clearly inaccurate for the evaluation of the level of each patient's response against his graft, which remains a mandatory step in proposing a tailored immunosuppressive regimen.

Progress in transplant immunology has shown that the recipient's adaptive immune response relies on two effector arms that reject an allogenic transplant. The cellular cytotoxic response involves graft infiltration by activated T lymphocytes, which induces apoptosis of allogenic cells. The second arm is the humoral response, which relies on the generation of DSA by plasma cells in secondary lymphoid organs. Circulating DSA in turn bind to allogenic targets expressed by graft endothelial cells, which triggers the activation of the classical complement pathway and the recruitment of innate immune effectors responsible for antibody-dependent cell cytotoxicity (ADCC) and microvascular inflammation. Although these two processes usually function in collaboration, some rejection episodes can be exclusively cellular or humorally mediated. Accurate monitoring of allogenic immune responses therefore requires the combination of different techniques.

The question of the monitoring timing is equally important. When performed before transplantation, the tests allow practitioners to grade the a priori "immunological risk" and to adjust the immunosuppression accordingly. Immune monitoring can also be performed as part of the follow-up process posttransplantation, either in regularly predefined intervals or as indicated by graft dysfunction. The former option offers the theoretical possibility to identify recipients at risk for rejection before this becomes clinically apparent. While this strategy can improve long-term outcome, it also requires increased tests (and therefore costs) and defining the time points when the monitoring should be performed.

Another important aspect is the source of the samples available for the monitoring. There is an ongoing debate as to whether the peripheral blood accurately reflects the events occurring within the graft. Although the analysis of a transplant biopsy will undoubtedly give more direct information, sequential analysis, a prerequisite for accurate diagnosis, is unlikely to be possible. For kidney transplant recipients, urine is an easily accessible source of material that has proved to be informative [40]; however, urine biomarkers will not be relevant for patients receiving other types of transplants, particularly vascularized composite tissue allografts.

Finally, it should be noted that immune monitoring in transplantation is an emerging field. Most studies published so far are from single centers and the strength of their conclusions are limited by the inclusion of a small number of patients. Furthermore, most of them focused on renal transplant recipients; therefore, their conclusions may not be directly transposable to the field of VCA. For instance, accumulating evidence points toward a major role for DSA in the immune-mediated failure of kidney, lung, and heart transplants [41]; in contrast, evidence that DSA are capable of driving VCA rejection is still lacking (see above).

Monitoring of the Humoral Response

Circulating DSA

The humoral alloimmune response is monitored by quantification of serum DSA levels. In the past, antibody detection was performed exclusively with lymphocyte targets in complement-dependent cytotoxicity assays. These assays had a low sensitivity (they only detect high titers of complement-fixing antibodies, resulting in false-negative results) and were time consuming. Recent developments of solid-phase antibody detection assays and flow cytometric technology have revolutionized our ability to detect HLA antibodies with regard to both sensitivity and specificity [42].

False-negative reactions still occur, however, because of antibody adsorption by the graft [43] (which can sometimes leave no detectable anti-HLA antibody in the circulation) or because the humoral response against the graft is directed toward polymorphic non-HLA alloantigens and/or nonpolymorphic autoantigens. Such antibodies, that can be involved in graft rejection [23], are not detected by the assays described above. Most often, however, clinicians using these tests are facing the opposite problem of "excessive sensitivity," with anti-HLA antibodies being detected in asymptomatic patients. Efforts are currently being made to standardize laboratory procedures so that cutoff levels of antibody titers can be defined. Preliminary data also suggest that it is possible to refine the technology to improve the predictive value. An assay for complement-fixing C1q on HLA single-antigen beads has been developed and preliminary data suggest that it correlates better with rejection and failure of heart and kidney grafts [44].

Allospecific B Cells

Immunological dogma states that the production of a specific antibody should be preceded by the increase of B cell clones of the same specificity. Monitoring the frequency of HLA-specific B cells longitudinally may therefore offer the possibility to detect humoral alloimmune responses in their early phase, at the time of their development, when they can still be controlled using B-cell-depleting agents (i.e., anti-CD20 mAb, which are ineffective on plasma cells). Using recombinant monomeric HLA molecules as target molecules for the HLA antibodies to bind, a novel Enzyme-Linked Immuno Spot (ELISPOT) assay has been recently described which allows for quantification of HLA-specific B cells in sensitized individuals [45]. Whether the increase of peripheral HLA-specific B cells actually precedes the formation of antibodies is currently under investigation. As outlined above, it should be remembered that B cells are more than mere plasma cell precursors. They are endowed with several important antibody-independent functions, including the ability to present antigens to T cells and to orchestrate the immune response through the provision of a wide range of cytokines [46]. Monitoring B cell response might therefore provide interesting clues regarding the cellular response.

Monitoring of Cellular Response

Soluble CD30

In activated T cells, the membrane-bound CD30 molecule, a member of the tumor necrosis factor receptor superfamily, is proteolytically cleaved. This generates a soluble form (sCD30), which can be measured in the serum. It has been suggested that pre- or posttransplant levels of sCD30 represent a biomarker for graft rejection associated with an impaired outcome for transplanted patients [47].

Polyclonal Lymphocyte Stimulation Assays

The overall function of lymphocytes can be determined by their intracellular adenosine triphosphate (iATP) content after phytohemagglutinin (PHA) stimulation, the latter reflecting the "energy" levels available within the cell. A Food and Drug Administration–approved iATP assay is available commercially, the ImmuKnow assay (Cylex, Columbia, MD). High $CD4^+$ T cell iATP levels have been shown (albeit inconsistently) to be associated with a higher risk for AR in kidney, liver, heart, and small bowel transplantation, whereas low levels were a risk factor for infections [48]. A major drawback of this technique is that it determines a patient's general immune status rather than the T cell reactivity specifically directed toward the allograft.

Quantifying Donor-Specific Memory T Cell Reactivity

Memory T cells respond more rapidly to stimulation by alloantigens and are more difficult to suppress than their naive counterparts. Memory T cells generated in response to environmental stimuli (e.g., previous transplant, blood transfusion, viral infections) become more prevalent with age and can cross-react with alloantigens from the donor graft, despite no previous exposure to tissue from that donor. Donor-specific memory T cell reactivity can be quantified using interferon gamma (IFN-γ) ELISPOT [49].

Increased levels of donor-reactive IFN-γ-producing T cells pre-kidney and post-kidney transplantation were shown to be a risk factor for AR and a predictor of graft function, independent of the humoral response [50]. Based on these promising results, the European Reprogramming the Immune System for Establishment of Tolerance (RISET) consortium (www.risetfp6.org) has recently implemented a rigorous approach to optimize, standardize, and validate IFN-γ ELISPOT, enabling it to be used by multiple laboratories [51].

Flow Cytometry

Multiparameter flow cytometry is a flexible tool allowing simultaneous quantification and phenotyping of many immune cell subsets. Regulatory T cells (Treg) have been shown to prevent allograft rejection in many animal models of transplantation. Therefore, monitoring of Treg in the peripheral blood may detect unresponsiveness to the graft. Renal graft recipients with lower percentages of Treg in the peripheral blood are indeed at increased risk of chronic graft dysfunction [52].

Memory T cells can be identified thanks to their surface expression of CD45RA, CCR7, and CD62 L which differs from that of naive T cells. The increase in a circulating memory T cell subset has been associated with a higher risk for AR in kidney and heart transplant recipients [50]. Other studies have reached similar conclusions with different subsets, such as activated $CD8^+CD69^+$ T cells during both acute renal and cardiac rejection [53]. The discriminatory potential of assays relying on general activation markers is likely to be poor, however, as the same activation markers will increase on any T cell when responding to an antigen. In line with this observation, CMV infection in renal transplant patients coincided with upregulation of $CD69^+$ on CD8 T cells in the absence of acute renal allograft rejection [50]. To circumvent this limitation, new procedures are under development that will hopefully provide data on antigen specificity of T and B cells, signaling pathway activity, transcription factor expression, and cytokine production.

Overall, flow cytometry is a useful monitoring technique that provides a wealth of data. However, because flow cytometer configuration, setup, and data acquisition and analysis are expectedly different among centers, standardization to enable multiparameter phenotyping to be used in biomarker studies is crucial [54].

Gene Expression Profiling

Perforin, granzyme B, and FasL are molecules involved in cytotoxicity whose level of gene expression can be quantified by polymerase chain reaction (PCR). Upregulation of two or more of these molecules in the peripheral blood correlates with AR in kidney transplant recipients [55], a finding confirmed and extended in subsequent studies [50]. However, as already discussed with other biomarkers of immune activation, this tool lacks specificity since it does not make a distinction between rejection and viral infection.

Advances in molecular biology have led to the development of microarrays, which allow for the simultaneous quantification of the expression of a vast numbers of genes. Using this tool, it has been possible to identify a signature specific for operational tolerance to kidney allografts [56–58]. Current investigations will determine if this monitoring can be used to select recipients that can be weaned from immunosuppression.

Immune Monitoring in VCA

VCA differs from other types of organ transplantation on the following issues that impact patient care: (i) acute skin rejection does not induce graft dysfunction and has no biological signature, (ii) skin rejection is easy to detect by visual examination, (iii) skin biopsy is a nonrisky procedure that can be repeated as often as required, (iv) the limited number of patients limits validation of biomarkers by appropriate clinical trials, and, finally, (v) upper limb and face allografts contain bone marrow cells that could be involved in graft acceptance as described in rodent models of vascularized bone marrow transplantation, whereas lower hind limb transplantation is associated with tolerogenic posttransplant chimerism [59].

Pre-transplant Assessment of Sensitized Patients

Sensitization to nonself HLA prior to transplantation can occur through three main routes: blood transfusion, pregnancy, and previous transplantation. Patients who are candidates for VCA may have been exposed to allogenic HLA antigens by transfusions performed during surgery for hand amputation or face disfiguration. In addition, burned patients can be sensitized through allogenic skin grafted to cover the burned body areas. Therefore, even if the role of AMR has not been clearly defined in VCA, identification of anti-HLA sensitized patients must be systematically performed before transplantation by detecting anti-HLA class I and anti-HLA class II antibodies with highly sensitive techniques such as enzyme-linked immunosorbent assay (ELISA) or Luminex assay [42]. Development of anti-HLA antibodies might be a sign of thymo-dependent B cell activation through allogenic-activated $CD4^+$ T cells involved in skin rejection. In the near future, direct measurement of T cell sensitization with new cellular techniques, such as IFN-ELISPOT assay [49], should be of help in detecting alloantigen-specific memory T cells in patients awaiting a VCA and in adapting the induction and maintenance immunosuppressive treatment accordingly in high-risk recipients. As outlined above, kidney transplant recipients with high frequencies of donor-reactive memory T cells detected pre-transplantation by IFN-ELISPOT are at risk of severe AR episodes during the early posttransplant period [49, 60].

Posttransplant Assessment of Alloreactivity and Immunological Status

For-Cause and Protocol Biopsies

The ultimate goal of the posttransplant phase is to avoid both a state of over-immunosuppression and of under-immunosuppression. Because the skin is the main target for AR, the most reliable method to monitor alloimmune activation against the VCA graft after transplantation is the use of for-cause and protocol skin biopsies that can more specifically reveal the local ongoing anti-donor immune response.

As outlined above, for-cause biopsies are skin biopsies performed whenever new skin lesions appear on the graft. Patients have to be educated to examine their graft to detected early cutaneous changes that reveal clinical acute skin rejection. The alloimmune response against a VCA allograft is most stringent during the early post-operative period. Occurrence of one or two AR episodes in the first posttransplant year is common and should not prompt revision of the maintenance immunosuppressive treatment. In contrast, multiple episodes of AR later after transplantation could reveal either a noncompliant patient to immunosuppressive treatment, or a maintenance immunosuppressive treatment inadequate in controlling the alloimmune response. Occurrence of acute clinical skin rejection during the weaning process should also prompt (at least temporarily) discontinuation of the reduction of immunosuppression. Conversely, patients who experience less than two episodes of acute skin rejection during the first posttransplant year can be considered as candidates for weaning protocols.

Protocol biopsies, performed on well-functioning grafts at regular time points, are the gold standard for detecting subclinical intragraft processes in organ transplantation. Protocol biopsies can also be informative in revealing new mechanisms and cellular players of graft rejection. However, there is currently no clear consensus regarding the legitimacy of protocol biopsies in organ transplantation. Protocol biopsies could allow for detection of subclinical histological changes that may impact the management of the patient at early posttransplant stages, thus delaying graft loss [61]. However, opponents of protocol biopsies claim that analysis of the clinical benefit of protocol biopsies has not been formally assessed and that these benefits may not outweigh the risks associated with this invasive procedure [62]. These limitations of protocol biopsies in organ transplantation should not be considered in VCA because the risk of skin biopsies is negligible. In human hand and face transplantation, it is recommended that protocol biopsies be obtained weekly until the end of the first month. Then, skin specimens should be taken at 2, 3, 6, 9, 12, 18, 24, 30, and 36 months. After year 3, skin biopsies can be collected biannually [34]. In hand transplantation, protocol biopsies are usually performed on the volar surface of the forearm. In face transplantation, protocol biopsies should preferably be obtained from the sentinel skin graft or alternatively from the oral mucosa so as to avoid damaging the reconstructed face [63]. In addition, recipient skin and deep donor tissues should be collected without additional morbidity for the patient during secondary surgeries.

Protocol biopsies can identify mild histopathological changes in the dermis of the allograft skin, consistent with grade 1 rejection in the Banff classification, without any macroscopically visible signs of rejection. At present, there is no convincing evidence that subclinical rejection, defined by microscopic skin infiltration in the absence of clinical signs of rejection, should be treated to reduce the risk of long-term CR, but these patients should be carefully monitored [63]. The risk of reducing immunosuppression in patients presenting subclinical rejection in protocol biopsies is also unknown. Our reluctance to rely on the mere presence of infiltrates for treating patients can be justified by the nature of focal and mild perivascular mononuclear infiltrates that may be composed of active allogenic T cells or alternatively $FoxP3^+$ Treg cells.

This observation underlines the need for subsequent evaluation by immunohistochemistry that would allow for further characterization of the function of the cells infiltrating the graft. For immunophenotyping the infiltrates, labeling for CD3 (T lymphocytes), CD4 (T helper cells), CD8 (T suppressor/cytotoxic cells), FoxP3 (T regulatory lymphocytes), CD19 or CD20 (B lymphocytes) and CD68 (macrophages), and C4d (surrogate marker of AMR) should be performed [33]. Assessment of adhesion molecules such as LFA-1 (CD11a, on lymphocytes), ICAM-1 (CD54), E-selectin (CD62E), P-selectin (CD62P), and VE-cadherin (CD144), which are all found on the vascular endothelium, can be used to evaluate the role of lymphocyte trafficking and adhesion molecules in VCA [64].

Protocol biopsies can also identify morphological changes associated with chronic skin rejection, such as skin atrophy, fibrosis, and vasculopathy of dermal vessels.

Finally, protocol biopsy is an interesting tool for the scientific community because it provides insights into the natural history of graft rejection, as described in renal transplantation [65].

Noninvasive Immune Tests

Among the noninvasive immune tests described above, none has been developed or even validated in the setting of VCA. Nowadays, monitoring of the immunosuppressive therapy received by VCA recipients is based on measuring drug blood levels to achieve concentrations within the established therapeutic range previously defined in organ transplantation. This strategy largely prevents the toxicity of immunosuppressive drugs but is insufficient in determining the individual alloimmune response in each patient.

Based on our organ transplantation experience, we recommend performing the following tests in each episode of AR and at least at 3 and 6 months posttransplantation, and then at each transplant anniversary: (i) circulating DSA to assess humoral immune response, and (ii) peripheral blood lymphocyte subset analysis by multiparameter flow cytometry: $CD3^+$, $CD4^+$, $CD8^+$T cells, $CD20^+$B cells, CD4+CD-25^{hi}FoxP3$^+$CD127–T regulatory cells, memory and naive T cells.

Quantification of T lymphocyte subsets by flow cytometry is useful in monitoring not only the short-term effects (depletion) but also the long-term effects (reconstitution) of T depleting therapies such as Thymoglobulin or Campath-1H. Monitoring of blood T cell depletion provides guidance for adjustment of the daily dose of Thymoglobulin in the early posttransplant period. The aim is to keep the lymphocyte and/or the $CD3^+$T cell count below 200/mm^3 and 20/mm^3, respectively [66]. In the long term, Thymoglobulin-induced T cell depletion is followed by immune reconstitution, with both new thymic emigration of naive T cells and homeostatic proliferation of depletion-resistant memory T cells that may be influenced by the maintenance immunosuppression regimen. A recent preliminary study suggested that during the first year after kidney transplantation, homeostatic reconstitution following Thymoglobulin induction showed disproportionately high recovery of memory T lymphocyte subsets in patients receiving sirolimus compared to cyclosporine [67].

Immune reconstitution occurs slowly and in some individuals may be prolonged over several years. Risk factors for impaired reconstitution are not well defined, except for increasing age [68]. The blood subsets of T cells must be monitored because this secondary immunodeficiency may favor opportunistic infections and malignancies. The association between lymphocyte-depleting therapies and viral infections is well established. After treatment with Thymoglobulin, an increased incidence and severity of CMV infection [69] and pneumonia due to *Pneumocystis* [70] have been reported, especially in patients not receiving prophylactic therapy. The effect of $CD4^+$T cell lymphopenia on the risk of malignancy is still controversial; it appears to be associated with an increased incidence of some, but not all types of malignancy [71, 72].

Besides quantification of lymphocyte subsets by flow cytometry, the other biomarkers of T cell activation described above (such as soluble CD30, polyclonal lymphocyte stimulation assays, and IFN-γ ELISPOT to quantify donor-specific memory T cell reactivity) still need to be assessed in the setting of VCA in order to improve our evaluation of the patient's general immune status and T cell reactivity toward the skin allograft and to adapt immunosuppression accordingly.

So far no VCA recipient proved to be spontaneously tolerant with any immunosuppressive protocol; indeed, until now all recipients who discontinued the immunosuppressive therapy rejected their graft. On the other hand, VCA recipients have shown a low incidence of CR despite the high incidence of AR episodes, suggesting that they may have developed an operational tolerance state. It would be of utmost importance to identify a signature specific for operational tolerance in recipients of hand or face allotransplantation, as already identified in operational tolerant kidney transplant recipients [56–58].

Posttransplant Chimerism

Sixty years of research in tolerance induction in preclinical models of organ transplantation, as well as in renal transplantation in humans, have confirmed the results of the pioneer work of Medawar and colleagues who demonstrated that infusion of hematopoietic stem cells (HSC) combined with a nonmyeloablative regimen is the best way to achieve a durable tolerance state toward an allogenic organ via the development of posttransplant chimerism [73].

Interestingly, VCA (particularly upper extremities and face when the mandible is grafted) function as a vascularized bone marrow transplant (VBMT). The graft itself contains HSC within donor bones. Experimental studies in rats have clearly demonstrated that limb composite tissue allografts function as VBMT with the development of a stable mixed chimerism and donor-specific tolerance [59]. However, peripheral blood microchimerism in humans has been detected for a very short period only in two hand transplant recipients in Louisville [74], in two face transplant recipients, and in one hand transplant recipient in Lyon [63]. Peripheral chimerism was not detectable even in hand transplant patients who had received HSC infusion with a lymphocyte-depleting induction therapy with alemtuzumab but no further conditioning prior to HSC transfusion [14]. The absence of stable microchimerism might be explained by the insufficient number of HSC contained in the adult upper extremities and in the mandible, or by the rejection of the donor bone marrow by the recipient's immune system. Thus, the very promising results obtained in rodent VBMT model cannot be transposed to the clinical practice without recipient conditioning.

Based on the findings from these preliminary data, it is not possible to rely on posttransplant chimerism to assess a tolerance state in VCA. It is, however, recommended to check for blood chimerism by PCR at least during the first posttransplant year as new patients might yield different results.

In conclusion, progress in immune monitoring in VCA should come from experience in organ transplantation. Correlation of immune monitoring tests with protocol and for-cause biopsies in patients who received different combinations of immunosuppressive drugs should help to discriminate between patients in whom immunosuppression could be reduced and high immunological risk patients who should remain on high-dose immunosuppressive protocols.

Acknowledgments The authors wish to acknowledge Céline Dagot for her precious organization of CTA clinical research, all the physicians involved in the care of patients in Amiens and Lyon, and Martha Melter for the editorial review of the manuscript.

References

1. Dubernard JM, Owen E, Herzberg G, Lanzetta M, Martin X, Kapila H, et al. Human hand allograft: report on first 6 months. Lancet. 1999;353(9161):1315–20.
2. Thaunat O, Badet L, El-Jaafari A, Kanitakis J, Dubernard JM, Morelon E. Composite tissue allograft extends a helping hand to transplant immunologists. Am J Transplant. 2006;6(10):2238–42.
3. Murray JE. Organ transplantation (skin, kidney, heart) and the plastic surgeon. Plast Reconstr Surg. 1971;47(5):425–31.
4. Kanitakis J, Jullien D, Petruzzo P, Hakim N, Claudy A, Revillard JP, et al. Clinicopathologic features of graft rejection of the first human hand allograft. Transplantation. 2003;76(4):688–93.
5. Lee WP, Yaremchuk MJ, Pan YC, Randolph MA, Tan CM, Weiland AJ. Relative antigenicity of components of a vascularized limb allograft. Plast Reconstr Surg. 1991;87(3):401–11.
6. Kaufman CL, Ouseph R, Blair B, Kutz JE, Tsai TM, Scheker LR, et al. Graft vasculopathy in clinical hand transplantation. Am J Transplant. 2012;12(4):1004–16.
7. Gaspari AA, Katz SI. Induction and functional characterization of class II MHC (Ia) antigens on murine keratinocytes. J Immunol. 1988;140(9):2956–63.
8. Kanitakis J, Petruzzo P, Jullien D, Badet L, Dezza MC, Claudy A, et al. Pathological score for the evaluation of allograft rejection in human hand (composite tissue) allotransplantation. Eur J Dermatol. 2005;15(4):235–8.
9. Dubernard JM, Lengele B, Morelon E, Testelin S, Badet L, Moure C, et al. Outcomes 18 months after the first human partial face transplantation. N Engl J Med. 2007;357(24):2451–60.
10. Hautz T, Zelger B, Grahammer J, Krapf C, Amberger A, Brandacher G, et al. Molecular markers and targeted therapy of skin rejection in composite tissue allotransplantation. Am J Transplant. 2010;10(5):1200–9.
11. Eljaafari A, Badet L, Kanitakis J, Ferrand C, Farre A, Petruzzo P, et al. Isolation of regulatory T cells in the skin of a human hand-allograft, up to six years posttransplantation. Transplantation. 2006;82(12):1764–8.
12. Ravindra KV, Wu S, Bozulic L, Xu H, Breidenbach WC, Ildstad ST. Composite tissue transplantation: a rapidly advancing field. Transplant Proc. 2008;40(5):1237–48.
13. Zuber J, Grimbert P, Blancho G, Thaunat O, Durrbach A, Baron C, et al. Prognostic significance of graft Foxp3 expression in renal transplant recipients: a critical review and attempt to reconcile discrepancies. Nephrol Dial Transplant. 2013;28(5):1100–11.
14. Schneeberger S, Gorantla VS, Brandacher G, Zeevi A, Demetris AJ, Lunz JG, et al. Upper-extremity transplantation using a cell-based protocol to minimize immunosuppression. Ann Surg. 2013;257(2):345–51.
15. Petruzzo P, Kanitakis J, Badet L, Pialat JB, Boutroy S, Charpulat R, et al. Long-term follow-up in composite tissue allotransplantation: in-depth study of five (hand and face) recipients. Am J Transplant. 2011;11(4):808–16.
16. Doxiadis, II, Claas FH. Donor-specific antibodies arising after kidney transplantectomy: priming or memory? Transplantation. 2006;81(4):509–10.
17. Billen EV, Christiaans MH, Lee J, van den Berg-Loonen EM. Donor-directed HLA antibodies before and after transplantectomy detected by the luminex single antigen assay. Transplantation. 2009;87(4):563–9.

18. Mauiyyedi S, Crespo M, Collins AB, Schneeberger EE, Pascual MA, Saidman SL, et al. Acute humoral rejection in kidney transplantation: II. Morphology, immunopathology, and pathologic classification. J Am Soc Nephrol. 2002;13(3):779–87.
19. Kanitakis J, McGregor B, Badet L, Petruzzo P, Morelon E, Devauchelle B, et al. Absence of c4d deposition in human composite tissue (hands and face) allograft biopsies: an immunoperoxidase study. Transplantation. 2007;84(2):265–7.
20. Landin L, Cavadas PC, Ibanez J, Roger I, Vera-Sempere F. CD3+-mediated rejection and C4d deposition in two composite tissue (bilateral hand) allograft recipients after induction with alemtuzumab. Transplantation. 2009;87(5):776–81.
21. Hautz T, Zelger B, Brandacher G, Mueller H, Grahammer J, Zelger B, et al. Histopathologic characterization of mild rejection (grade I) in skin biopsies of human hand allografts. Transpl Int. 2012;25(1):56–63.
22. Cendales LC, Kirk AD. Nonhuman primates for the study of composite tissue allotransplantation. Transplantation. 2007;83(2):241.
23. Thaunat O. Humoral immunity in chronic allograft rejection: puzzle pieces come together. Transpl Immunol. 2012;26(2–3):101–6.
24. Schon MP, Ludwig RJ. Lymphocyte trafficking to inflamed skin–molecular mechanisms and implications for therapeutic target molecules. Expert Opin Ther Targets. 2005;9(2):225–43.
25. Robert C, Kupper TS. Inflammatory skin diseases, T cells, and immune surveillance. N Engl J Med. 1999;341(24):1817–28.
26. Pradka SP, Ong YS, Zhang Y, Davis SJ, Baccarani A, Messmer C, et al. Increased signs of acute rejection with ischemic time in a rat musculocutaneous allotransplant model. Transplant Proc. 2009;41(2):531–6.
27. Kim IK, Bedi DS, Denecke C, Ge X, Tullius SG. Impact of innate and adaptive immunity on rejection and tolerance. Transplantation. 2008;86(7):889–94.
28. Schneeberger S, Gorantla VS, van Riet RP, Lanzetta M, Vereecken P, van Holder C, et al. Atypical acute rejection after hand transplantation. Am J Transplant. 2008;8(3):688–96.
29. Schneeberger S, Lucchina S, Lanzetta M, Brandacher G, Bosmuller C, Steurer W, et al. Cytomegalovirus-related complications in human hand transplantation. Transplantation. 2005;80(4):441–7.
30. Petruzzo P, Lanzetta M, Dubernard JM, Landin L, Cavadas P, Margreiter R, et al. The international registry on hand and composite tissue transplantation. Transplantation. 2010;90(12):1590–4.
31. Hautz T, Engelhardt TO, Weissenbacher A, Kumnig M, Zelger B, Rieger M, et al. World experience after more than a decade of clinical hand transplantation: update on the Innsbruck program. Hand Clin. 2011;27(4):423–31, viii.
32. Cendales LC, Kirk AD, Moresi JM, Ruiz P, Kleiner DE. Composite tissue allotransplantation: classification of clinical acute skin rejection. Transplantation. 2006;81(3):418–22.
33. Cendales LC, Kanitakis J, Schneeberger S, Burns C, Ruiz P, Landin L, et al. The Banff 2007 working classification of skin-containing composite tissue allograft pathology. Am J Transplant. 2008;8(7):1396–400.
34. Hautz T, Zelger BG, Weissenbacher A, Zelger B, Brandacher G, Landin L, et al. Standardizing skin biopsy sampling to assess rejection in vascularized composite allotransplantation. Clin Transplant. 2013;27(2):E81–90.
35. Pomahac B, Pribaz J, Eriksson E, Bueno EM, Diaz-Siso JR, Rybicki FJ, et al. Three patients with full facial transplantation. N Engl J Med. 2012;366(8):715–22.
36. Devauchelle B, Badet L, Lengele B, Morelon E, Testelin S, Michallet M, et al. First human face allograft: early report. Lancet. 2006;368(9531):203–9.
37. Kanitakis J, Badet L, Petruzzo P, Beziat JL, Morelon E, Lefrancois N, et al. Clinicopathologic monitoring of the skin and oral mucosa of the first human face allograft: Report on the first eight months. Transplantation. 2006;82(12):1610–5.
38. Kanitakis J. The challenge of dermatopathological diagnosis of composite tissue allograft rejection: a review. J Cutan Pathol. 2008;35(8):738–44.

39. Bestard O, Cruzado JM, la Franquesa M, Grinyo JM. Biomarkers in renal transplantation. Curr Opin Organ Transplant. 2010;15(4):467–73.
40. Suthanthiran M, Schwartz JE, Ding R, Abecassis M, Dadhania D, Samstein B, et al. Urinary-cell mRNA profile and acute cellular rejection in kidney allografts. N Engl J Med. 2013;369(1):20–31.
41. Terasaki PI, Cai J. Humoral theory of transplantation: further evidence. Curr Opin Immunol. 2005;17(5):541–5.
42. Gebel HM, Liwski RS, Bray RA. Technical aspects of HLA antibody testing. Curr Opin Organ Transplant. 2013;18(4):455–62.
43. Martin L, Guignier F, Mousson C, Rageot D, Justrabo E, Rifle G. Detection of donor-specific anti-HLA antibodies with flow cytometry in eluates and sera from renal transplant recipients with chronic allograft nephropathy. Transplantation. 2003;76(2):395–400.
44. Chen G, Sequeira F, Tyan DB. Novel C1q assay reveals a clinically relevant subset of human leukocyte antigen antibodies independent of immunoglobulin G strength on single antigen beads. Hum Immunol. 2011;72(10):849–58.
45. Heidt S, Roelen DL, de Vaal YJ, Kester MG, Eijsink C, Thomas S, et al. A NOVel ELISPOT assay to quantify HLA-specific B cells in HLA-immunized individuals. Am J Transplant. 2012;12(6):1469–78.
46. Thaunat O, Morelon E, Defrance T. Am"B"valent: anti-CD20 antibodies unravel the dual role of B cells in immunopathogenesis. Blood. 2010;116(4):515–21.
47. Susal C, Pelzl S, Dohler B, Opelz G. Identification of highly responsive kidney transplant recipients using pretransplant soluble CD30. J Am Soc Nephrol. 2002;13(6):1650–6.
48. Kowalski RJ, Post DR, Mannon RB, Sebastian A, Wright HI, Sigle G, et al. Assessing relative risks of infection and rejection: a meta-analysis using an immune function assay. Transplantation. 2006;82(5):663–8.
49. Heeger PS, Greenspan NS, Kuhlenschmidt S, Dejelo C, Hricik DE, Schulak JA, et al. Pretransplant frequency of donor-specific, IFN-gamma-producing lymphocytes is a manifestation of immunologic memory and correlates with the risk of posttransplant rejection episodes. J Immunol. 1999;163(4):2267–75.
50. Heidt S, San Segundo D, Shankar S, Mittal S, Muthusamy AS, Friend PJ, et al. Peripheral blood sampling for the detection of allograft rejection: biomarker identification and validation. Transplantation. 2011;92(1):1–9.
51. Bestard O, Crespo E, Stein M, Lucia M, Roelen DL, de Vaal YJ, et al. Cross-validation of IFN-gamma Elispot assay for measuring alloreactive memory/effector T cell responses in renal transplant recipients. Am J Transplant. 2013;13(7):1880–90.
52. Braudeau C, Racape M, Giral M, Louis S, Moreau A, Berthelot L, et al. Variation in numbers of CD4 + CD25highFOXP3 + T cells with normal immuno-regulatory properties in long-term graft outcome. Transpl Int. 2007;20(10):845–55.
53. Posselt AM, Vincenti F, Bedolli M, Lantz M, Roberts JP, Hirose R. CD69 expression on peripheral CD8 T cells correlates with acute rejection in renal transplant recipients. Transplantation. 2003;76(1):190–5.
54. Maecker HT, McCoy JP, Nussenblatt R. Standardizing immunophenotyping for the Human Immunology Project. Nat Rev Immunol. 2012;12(3):191–200.
55. Vasconcellos LM, Schachter AD, Zheng XX, Vasconcellos LH, Shapiro M, Harmon WE, et al. Cytotoxic lymphocyte gene expression in peripheral blood leukocytes correlates with rejecting renal allografts. Transplantation. 1998;66(5):562–6.
56. Sagoo P, Perucha E, Sawitzki B, Tomiuk S, Stephens DA, Miqueu P, et al. Development of a cross-platform biomarker signature to detect renal transplant tolerance in humans. J Clin Invest. 2010;120(6):1848–61.
57. Brouard S, Mansfield E, Braud C, Li L, Giral M, Hsieh SC, et al. Identification of a peripheral blood transcriptional biomarker panel associated with operational renal allograft tolerance. Proc Natl Acad Sci U S A. 2007;104(39):15448–53.

58. Newell KA, Asare A, Kirk AD, Gisler TD, Bourcier K, Suthanthiran M, et al. Identification of a B cell signature associated with renal transplant tolerance in humans. J Clin Invest. 2010;120(6):1836–47.
59. Leonard DA, McGrouther DA, Kurtz JM, Cetrulo CL, Jr. Tolerance induction strategies in vascularized composite allotransplantation: mixed chimerism and novel developments. Clin Dev Immunol. 2012;2012:863264.
60. Nickel P, Presber F, Bold G, Biti D, Schonemann C, Tullius SG, et al. Enzyme-linked immunosorbent spot assay for donor-reactive interferon-gamma-producing cells identifies T-cell presensitization and correlates with graft function at 6 and 12 months in renal-transplant recipients. Transplantation. 2004;78(11):1640–6.
61. Rush D. Can protocol biopsy better inform our choices in renal transplantation? Transpl Proc. 2009;41(6 Suppl):S6–8.
62. Thaunat O, Legendre C, Morelon E, Kreis H, Mamzer-Bruneel MF. To biopsy or not to biopsy? Should we screen the histology of stable renal grafts? Transplantation. 2007;84(6):671–6.
63. Morelon E, Kanitakis J, Petruzzo P. Immunological issues in clinical composite tissue allotransplantation: where do we stand today? Transplantation. 2012;93(9):855–9.
64. Hautz T, Brandacher G, Engelhardt TO, Pierer G, Lee WP, Pratschke J, et al. How reconstructive transplantation is different from organ transplantation–and how it is not. Transpl Proc. 2011;43(9):3504–11.
65. Nankivell BJ, Borrows RJ, Fung CL, O'Connell PJ, Allen RD, Chapman JR. The natural history of chronic allograft nephropathy. N Engl J Med. 2003;349(24):2326–33.
66. Mourad G, Morelon E, Noel C, Glotz D, Lebranchu Y. The role of Thymoglobulin induction in kidney transplantation: an update. Clin Transplant. 2012;26(5):E450–64.
67. Morelon E, Lefrancois N, Besson C, Prevautel J, Brunet M, Touraine JL, et al. Preferential increase in memory and regulatory subsets during T-lymphocyte immune reconstitution after Thymoglobulin induction therapy with maintenance sirolimus vs cyclosporine. Transpl Immunol. 2010;23(1–2):53–8.
68. Ducloux D, Carron PL, Racadot E, Rebibou JM, Bresson-Vautrin C, Saint-Hillier Y, et al. CD4 lymphocytopenia in long-term renal transplant recipients. Transpl Proc. 1998;30(6):2859–60.
69. Lebranchu Y, Bridoux F, Buchler M, Le Meur Y, Etienne I, Toupance O, et al. Immunoprophylaxis with basiliximab compared with antithymocyte globulin in renal transplant patients receiving MMF-containing triple therapy. Am J Transplant. 2002;2(1):48–56.
70. Issa NC, Fishman JA. Infectious complications of antilymphocyte therapies in solid organ transplantation. Clin Infect Dis. 2009;48(6):772–86.
71. Glowacki F, Al Morabiti M, Lionet A, Labalette M, Provot F, Noel C, et al. Long-term kinetics of a T-lymphocytes subset in kidney transplant recipients: relationship with posttransplant malignancies. Transpl Proc. 2009;41(8):3323–5.
72. Ducloux D, Carron PL, Motte G, Ab A, Rebibou JM, Bresson-Vautrin C, et al. Lymphocyte subsets and assessment of cancer risk in renal transplant recipients. Transpl Int. 2002;15(8):393–6.
73. Sachs DH, Sykes M, Kawai T, Cosimi AB. Immuno-intervention for the induction of transplantation tolerance through mixed chimerism. Semin Immunol. 2011;23(3):165–73.
74. Granger DK, Briedenbach WC, Pidwell DJ, Jones JW, Baxter-Lowe LA, Kaufman CL. Lack of donor hyporesponsiveness and donor chimerism after clinical transplantation of the hand. Transplantation. 2002;74(11):1624–30.

Chapter 10
Bayesian Classifier and Molecular Marker Platforms for Immune Monitoring

Rahul M Jindal, Kristin A Stevens, Jonathan A. Forsberg and Eric A. Elster

Abbreviations

BBNs	Bayesian belief networks
CDS	Clinical decision support
USRDS	(United States Renal Data System) database
OPO	Organ Procurement Organization
TG	Transplant glomerulopathy
PPV	Positive predictive value
WRTC, www.wrtc.org	Washington Regional Transplant Community
PSAs	Public service announcements
AUC	Under the curve
VCAs	Vascularized composite allografts

Introduction

Solid organ transplantation is the definitive treatment for end-stage organ failure, and vascularized composite allografts (VCAs) are an increasingly important restorative option. However, clinical decisions regarding allocation or postopera-

E. A. Elster (✉) · R. M. Jindal · K. A. Stevens · J. A. Forsberg
Walter Reed National Military Medical Center, Bethesda, MD, USA
e-mail: eric.elster@usuhs.edu

R. M. Jindal · J. A. Forsberg · E. A. Elster
Uniformed Services University of Health Sciences, 4301 Jones Bridge Road, Bethesda, MD, USA

R. M. Jindal
George Washington University, Washington, DC, USA

J. A. Forsberg · E. A. Elster
Naval Medical Research Center, Silver Spring, MD, USA

G. Brandacher (ed.), *The Science of Reconstructive Transplantation*,
Stem Cell Biology and Regenerative Medicine, DOI 10.1007/978-1-4939-2071-6_10

tive treatment can be difficult to make. This has stimulated research investigating applications designed to analyze large amounts of data, show how that data affects outcomes, and assist in the decision-making process. As has been done in cardiology and oncology, our research uses machine learning techniques to produce a clinically relevant prognostic tool [1–3].

Machine learning uses logical algorithms that allow computers to recognize complex patterns within datasets. These patterns can then be represented mathematically into clinical decision support (CDS) tools. In the area of transplantation, machine learning has been used to estimate the likelihood of delayed graft function, transplant glomerulopathy (TG), and allograft survival.

Many types of statistical techniques can yield CDS tools, including neural networks, nomograms, and tree modeling [4–7]. Nomograms tend to use multivariate analysis in their formation, but can be limited by the number of variables they may incorporate. Neural networks and tree modeling use machine learning, but variables are selected based on univariate analysis. While both techniques identify variables significantly associated with outcome, each may leave out variables that are not predictive by themselves but can yield prognostic information when considered in combination with additional variables. In addition, both neural networks and tree modeling require the input data to be complete, and as such, are intolerant of missing data common in the clinical setting. Bayesian belief networks (BBNs), on the other hand, are designed to accommodate missing information by generating an array of conditional interdependencies. In this fashion, a joint probability distribution function can be generated which allows all variables (or features) to be displayed graphically within the same model.

Bayesian analysis has been used to examine kidney transplant outcomes with standard criteria, extended criteria, and donation after cardiac death. BBNs have been used in medicine for over a decade and are ideal for analysis of datasets containing missing information and rare events, such as elucidating disease processes, selecting ideal alternative therapies, and estimating the likelihood of certain outcomes [8] including 90-day outcomes following liver transplantation [9].

This chapter reviews their use as applied to the diagnosis of TG using biomarkers, estimating the likelihood of renal allograft survival when using deceased donor grafts, and developing a better understanding of how beliefs within ethnic populations influence organ donation within those communities. Additionally, we describe the future expanded role of machine learning in determining immune status across transplantation.

Bayesian Modeling Applied to Molecular Markers

There is an extensive area of research surrounding biomarkers and their application in wound healing, oncology, and transplantation [10–12]. In oncology, particularly for breast cancer, biomarkers such as human epidermal growth factor receptor 2 (HER2) and multigene assays are already being used to personalize therapy [13, 14]. There is also considerable interest in developing biomarkers for the early detection of breast cancer before it is evident on mammography [11].

In transplantation, biomarkers have been studied to diagnose TG. TG is a disease of the kidney allograft that, despite having a low prevalence (<10%), has a high morbidity with approximately 50% of grafts that develop TG failing within 10 years [15, 16]. While the etiology has been researched extensively, no causative factor has been found for which there is specific treatment aside from increasing immunosuppression. As such, Elster et al. [12] used a BBN approach to develop a learning model which compares gene transcripts from both the immune and fibrosis pathways in patients who developed TG and those who had stable function as defined by 6 months posttransplant with no change in renal function and no significant histological or clinical abnormalities. Their model confirmed several mechanisms which are already known, but also uncovered novel relationships such as the interdependence of ICAM-1, IL-10, CCL-3, and the development of TG, as well as the association between C4d grade and increased expression of VCAM-1, MMP-9, MMP-7, and LAMC2 in TG development. Furthermore, the models developed for gene panel 1(immune pathway) and gene panel 2 (fibrosis) were able to estimate the probability of developing TG based on available evidence. When internal cross validations of the models were performed, the predictive accuracy as measured by area under the curve (AUC) was 0.875 for GP1 and 0.859 for GP2. While these models and biomarker profiles do not fully elucidate the mechanisms involved in TG development, as in oncology, they offer the potential to direct therapy specific to a patient's particular pathology.

Bayesian Models as Applied to Kidney Transplantation

Population-based research is common in transplantation, as is the use of machine learning to develop predictive models for renal allograft function and allocation. The tools currently used to make final allocation decisions are inadequate and subjective, which can result in suboptimal graft survival. As more extended criteria are available from donors and donation following cardiac graft failure, it becomes increasingly important to develop relevant CDS tools designed to identify those donor/recipient graft pairings that are likely to fail. Machine learning has enabled the development of a prognostic model that incorporates multiple variables for a systems approach to organ allocation and made the process more objective.

This study used a BBN to create a prognostic model for deceased donor renal allograft transplantation. This probabilistic approach is useful when dealing with large databases, can tolerate missing information, and can graphically describe the probability distributions of outcomes based on the conditional interdependence of known information [17]. In other words, this type of statistical analysis allows for the use of an unlimited number of variables and identifies not only the relationship between each variable and the targeted outcome but also the relative contribution and inter-variable relationships to the probability of each outcome [9].

To create this model, first time renal transplant patients over the age of 18 and only receiving a kidney were selected from the United States Renal Data System

(USRDS) database [18]. Five thousand records were then randomly selected from 2000 to 2001, which were used to create the model. A network of 46 preoperative variables was constructed and externally validated using an additional 2000 patients from the same time period with matching demographic characteristics. This model was able to predict graft failure within the first year with a 99 % positive predictive value (PPV) and AUC of 0.81, and 3-year failure with a 94 % PPV and 0.72 AUC. A tenfold internal cross validation continued to perform well at 1 year (99 % PPV and AUC 0.81) and 3 years (98 % PPV and 0.72 AUC). Using this model as a method to optimize donor/recipient pairing equates to a 3 % increase in graft survival at 1 year and 3 years. While this number is small, given the 17,000 renal transplants performed annually, this would translate to 510 more patients with a functioning graft at 1 year than occurs with current allocation practices.

The use of a machine-learned model enabled investigators to examine a vast number of variables simultaneously from a large database and develop a robust prognostic model. This has been done for liver transplantation, but has not been used in the setting of renal transplants to date [9]. Many authors have examined donor and recipient characteristics that are associated with decreased survival such as donor age, renal insufficiency, cerebrovascular accident, and hypertension [19, 20]. None of these variables are able to estimate the likelihood of graft survival, however, but rather simply estimate how that graft will likely function. Therefore, this model may allow transplant surgeons to optimize donor–recipient pairs for high-risk grafts. It also will allow surgeons to exclude grafts with the highest risk of failure more objectively than current practices. This CDS tool may limit overall center-specific graft-loss rates and improve transplant outcomes. External validation and comparative studies must be performed to compare this model to current allocation practices.

Using Bayesian Models to Understand How Beliefs and Attitudes of Minorities Influence Organ Donation

Organ donation rates in minority populations are much lower than in the Caucasian population, and there is a need to increase those rates within minority communities. It is thought that the minority belief system and attitudes toward organ donation is the reason for decreased rates. Understanding what deters minorities from becoming donors and raising awareness about organ transplantation could contribute to increased organ donation rates in those populations.

Some organizations have already made progress toward improving minority participation in organ donation. The Task Force on Organ Transplantation (DHHS 1986) [21] issued a report recommending that "educational efforts aimed at increasing organ donation among minority populations be developed and implemented, so that the donor population will more closely reflect the ethnicity of potential transplant recipients, in order to gain the advantage of improved donor and recipient immunologic matching." After years of intensive public education to raise awareness, donation rates for minority populations increased. Also, the National Minority Organ Tissue Transplant Education Program (National MOTTEP) has done a great deal of work promoting awareness of organ donation and kidney failure in the African American population [22, 23]. They showed that culturally appropriate health education programs aimed toward minority populations can effect positive change in knowledge, attitudes, and behavior.

However, there is still a need for more research and understanding of the issues surrounding minority organ donation. In the Washington Regional Transplant Community (WRTC, www.wrtc.org) service area in 2008, 71 % of organ donations were authorized for African Americans, 50 % for Asians, and 73 % for Hispanics. In comparison, the authorization rate was 85 % for Caucasians, clearly indicating that the current effort directed toward the minority communities has not yielded a comparable number of donations.

The way beliefs and attitudes play a role in minority decision making for organ donation is currently being explored in an ongoing study. This will be accomplished by developing a BBN, which will represent the joint relationships between parameters affecting each participant's willingness to participate in organ donation. Bayesian models developed in this process may predict attitude toward donation based on the themes, personality traits, and other attributes of the participants. Ultimately, the validated model could first identify candidates for themes or beliefs that might be addressed by teaching relevant facts, and also whether changing one's view toward the theme would make him or her more likely to support organ donation. These candidate themes will be used to improve public service announcements (PSAs) already in use by the Organ Procurement Organization (OPO).

Future Directions and Summary

Machine learning, such as BBN, has long been used in health care, and its popularity has increased as its ability to handle uncertain knowledge is directly applicable to solving diagnostic dilemmas and predicting outcomes [8]. Several systems have been used, including neural networks and tree-based diagrams. We chose BBN as the graphical representation of the data as it is easily interpreted and allows the researcher to more clearly understand underlying mechanisms displayed. This has been shown by Elster et al. [12] who used BBN to analyze gene panels in renal transplant patients with and without TG. The authors identified novel gene products that when expressed together in certain combinations, could be used to estimate the likelihood of developing TG. This knowledge allows clinicians the opportunity to diagnose TG earlier, as well as direct focus toward the identified pathways in an effort to develop novel means by which to treat TG.

Bayesian analysis has also been used to develop a CDS tool for deceased donor renal allograft allocation. This tool will allow surgeons to optimize donor/recipient pairings based on pretransplant variables, which may ultimately improve graft survival. Both internal and external validations show the model to be robust with high predictive accuracy and AUC of 0.81. While this is promising, the model must still be tested in an external dataset in order to compare the model with current allocation practices.

Probabilistic approaches have also been used in an effort to better understand why historically minority populations, specifically Hispanic and African American, have had lower organ donation rates than Caucasians. For this application, a BBN is being used to model beliefs and attitudes in these patient populations. It is hoped that this approach will allow for a better understanding of the belief hierarchies of each minority population and elucidate the effects of those beliefs and attitude(s) toward organ donation. Targeted educational programs could be initiated within minority populations using this information in an effort to increase organ donation and potentially improve immunologic matching among minorities.

While machine learning approaches have been introduced in solid organ transplantation, their use as a methodology to assimilate data and assess the current immune state of an allograft of any type has not yet moved beyond these preliminary studies. With the introduction of high throughput, multiplex molecular assays, the ability to determine the state of immune engagement has grown significantly. However, there has not been a reciprocal growth in tools to analyze these data along with patient parameters in a clinically useful manner. While the aforementioned study using BBNs to interpret transcripts for diagnosing TG is a step in the right direction, it is illustrative at best. In the context of severe trauma, recent efforts have determined that machine learning can develop CDS tools that function in a manner similar to high-end clinical decision making significantly enhanced by the addition of molecular markers. As more OMIC-based assays assessing immune function (genomics, proteomics, metabolomics) become available, CDS tools will need to be developed to clinically employ such data across all aspects of transplantation [24]. In particular, VCA, with its relatively small numbers when compared to solid organ transplantation, will benefit from machine learning as robust tools can be developed from small clinical data sets [25].

Machine learning applied to transplantation has demonstrated the potential to improve the current allocation practice, give insight into mechanisms involved in TG and chronic rejection, and give insight into barriers to organ donation in minorities to allow for more tailored education to increase donation rates. Furthermore, the development of machine learning-based CDS tools offers the ability to assess immune status across transplantation and tailor therapy based on the patient's biology. This offers the potential to improve long-term survival of grafts, develop customized treatments for TG and other transplant-related outcomes, and increase donation.

Acknowledgment The views expressed in this manuscript are those of the authors and do not reflect the official policy of the Department of the Army, Department of the Navy, the Department of Defense, or the US government.

This effort was supported (in part) by the US Navy Bureau of Medicine and Surgery under the Medical Development Program and Office of Naval Research.

this title is not available for any work of the United States Government." Title 17 U.S.C. 101 defines a US government work as a work prepared by a military service member or employee of the US government as part of that person's official duties.

I/We certify that all individuals who qualify as authors have been listed; each has participated in the conception and design of this work, the analysis of data (when applicable), the writing of the document, and the approval of the submission of this version; that the document represents valid work; that if we used information derived from another source, we obtained all necessary approvals to use it and made appropriate acknowledgements in the document; and that each takes public responsibility for it.

References

1. Maskery SM, Hu H, Hooke J, et al. A Bayesian derived network of breast pathology co-occurrence. J Biomed Inform. 2008;41:242–50.
2. Diamond G, Forrester J. Analysis of probability as an aid in the clinical diagnosis of coronary artery disease. N Engl J Med. 1979;300:1350–8.
3. Staniloff H, Diamond G, Freeman M, et al. Simplified application of Bayesian analysis to multiple cardiologic tests. Clin Cardiol. 1982;5:630–6.
4. Tiong HY, Goldfarb DA, Kattan MW, et al. Nomograms for predicting graft function and survival in living donor kidney transplantation based on the UNOS registry. J Urol. 2009;181:1248–55.
5. Brier ME, Ray PC, Klein JB. Prediction of delayed renal allograft function using an artificial neural network. Nephrol Dial Transplant 2003;18:2655–9.
6. Krikov S, Khan A, Baird BC, et al. Predicting kidney transplant survival using tree-based modeling. SO—ASAIO J. 2007;53(5):592–600.
7. Goldfarb-Rumyantzev AS, Scandling JD, Pappas L, et al. Prediction of 3-yr cadaveric graft survival based on pre-transplant variables in a large national dataset. Clin Transplant. 2003;17:485–97.
8. Lucas PJF, van der Gaag LC, Abu-Hanna A. Bayesian networks in biomedicine and health-care. Artif Intel Med. 2004;30:201–14.
9. Hoot N, Aronsky D. Using Bayesian networks to predict survival of liver transplant patients. AMIA Annu Symp Proc. 2005;2005:345–9.
10. Hawksworth JS, Stojadinovic A, Gage FA, et al. Inflammatory biomarkers in combat wound healing. Annals Surg. 2009;250:1002–7.
11. Fan Y, Wang J, Yang Y, et al. Detection and identification of potential biomarkers of breast cancer. J Cancer Res Clin Oncol. 2010;136(8):1243–54.
12. Elster EA, Hawksworth JS, Cheng O, Leeser DB, Ring M, Tadaki DK, Kleiner DE, Eberhardt JS 3rd, Brown TS, Mannon RB. Probabilistic (Bayesian) modeling of gene expression in transplant glomerulopathy. Mol Diagn. 2010;12(5):653–63.
13. Albain KS, Paik S, van't Veer L. Prediction of adjuvant chemotherapy benefit in endocrine responsive, early breast cancer using multigene assays. Breast. 2009;18:S141–5.
14. Carlson RW, Moench SJ, Hammond ME, et al. HER2 testing in breast cancer: NCCN Task Force report and recommendations. J Natl Compr Canc Netw. 2006;4(Suppl 3):S1–22; quiz S23–24.
15. Banfi G, Villa M, Cresseri D, Ponticelli C. The clinical impact of chronic transplant glomerulopathy in cyclosporine era. Transplantation. 2005;80:1392–7.
16. Perkowska-Ptasinska A, Ciszek M, Chmura A, et al. Transplant glomerulopathy: clinical and pathological correlations. Transplant Proc. 2009;41:141–9.
17. Grunkemeier GL, Payne N. Bayesian analysis: a new statistical paradigm for new technology. Ann Thorac Surg. 2002;74:1901–8.

18. Brown TS, Elster EA, Stevens K, Graybill JC, Gillern S, Phinney S, Salifu MO, Jindal RM. Bayesian modeling of pre-transplant variables accurately predicts kidney graft survival. Am J Nephrol. 2012;36(6):561–9.
19. Port FK, Bragg-Gresham JL, Metzger RA, et al. Donor characteristics associated with reduced graft survival: an approach to expanding the pool of kidney donors. Transplantation. 2002;74:1281–6.
20. Morris PJ, Johnson RJ, Fuggle SV, et al. Analysis of factors that affect outcome of primary cadaveric renal transplantation in the UK. Lancet. 1999;354:1147–52.
21. The 1984 National Organ Transplant Act (42 U.S.C.A. 273 et seq) initiated a national policy regarding *Organ Transplantation*. The act provided funds to help establish "qualifies organ procurement organizations," banned the interstate sale of organs, and created a task force to study organ transplantation policy issues. The 1986 task force report was an exhaustive examination of the medical, legal, social, and economic implications of organ procurement and transplantation.
22. Callender CO, Hall MB, Miles PV. Increasing living donations: expanding the National MOTTEP community grassroots model. Minority Organ Tissue Transplant Education Program. Transplant Proc. 2002;34:2563–4.
23. Callender CO, Miles PV, Hall MB. National MOTTEP: educating to prevent the need for transplantation. Minority Organ Tissue Transplant Education Program. Ethn Dis. 2002;12:S1–34–7.
24. Anam K, Lazdun Y, Davis PM, Banas RA, Elster EA, Davis TA. Amnion-derived multipotent progenitor cells support allograft tolerance induction. Am J Transplant. 2013;13(6):1416–28.
25. Rubin DB, Schenker N. Multiple imputation in health-care databases: an overview and some applications. Stat Med. 1991;10:585–98.

Chapter 11
Migration and Communication Patterns in Skin Rejection

Johanna Grahammer, Theresa Hautz, Johann Pratschke and Stefan Schneeberger

The Skin as an Immunologic Organ

The skin is an immunologic defense organ containing various immunocompetent cells. At the same time, the skin accommodates a commensal bacterial flora and helps to differentiate between dangerous and harmless subjects [1]. The skin as the body's outermost layer is the first contact surface for many pathogens, and it is also constantly exposed to physical and chemical stressors. In the avascular epidermis, keratinocytes proliferate at the basal layer and move as maturing cells towards the stratum corneum. Cytokines are produced by keratinocytes or invading immune cells and serve as important communication signals within the epidermis [2, 3]. Epidermal Langerhans cells (LCs) are of dendritic shape and derived from bone marrow precursors. Their contribution to immunologic skin reactions is currently much debated [4]. The underlying dermis harbors a dense network of blood vessels with a constitutively pronounced level of leukocyte rolling compared to other organs [5]. This indicates a more rapid and frequent immigration of immune cells into the skin [6]. A battery of memory T cells is constantly perambulating the skin for harmful intruders [7]. At the same time, tolerogenic cells populate the skin to maintain tolerance to self and foreign antigens [8].

In reconstructive transplantation (RT), a number of experimental studies have shown that rapid infiltration of allografted skin by host T cells results in graft loss.

J. Grahammer (✉) · T. Hautz · J. Pratschke · S. Schneeberger
Center of Operative Medicine, Department of Visceral, Transplant and Thoracic Surgery, Innsbruck Medical University, Anichstrasse 35, 6020 Innsbruck, Austria
e-mail: johanna.grahammer@uki.at

J. Pratschke
Department of General, Visceral and Transplant Surgery, Charité, Berlin, Germany

S. Schneeberger
Department of Plastic and Reconstructive Surgery, The Johns Hopkins University School of Medicine, 720 Rutland Avenue, Baltimore, MD, USA

G. Brandacher (ed.), *The Science of Reconstructive Transplantation*,
Stem Cell Biology and Regenerative Medicine, DOI 10.1007/978-1-4939-2071-6_11

Under certain conditions, tolerance towards other tissues of the vascularized composite graft can be achieved while the skin is selectively rejected (*split tolerance*) [9, 10, 11]. In the clinical experience, however, survival of hand and face allografts was achieved and skin rejection episodes were found to be treatable and reversible [12]. While potential pro-tolerogenic elements such as bone marrow are part of the allograft and phenomena such as chimerism have been extensively studied in experimental and clinical trials [13, 14], pro-tolerogenic elements of the skin itself in the field of RT are poorly understood.

Intrinsic tolerogenic mechanisms of the skin have been investigated in the field of inflammatory and immune dermatoses. Data from RT is still sparse, and mainly acquired retrospectively from skin biopsies.

One study has evaluated a hand recipient 6 years after transplantation, aiming to detect evidence for regulatory responses. T cell isolation from punch biopsies revealed the presence of Foxp3 expressing cells, together with a cytokine profile supporting a tolerogenic environment that included increased levels of transforming growth factor beta (TGF-β) and interleukin-10 (IL-10) messenger RNA (mRNA) [15].

In a follow-up study of five hand transplanted patients, Hautz et al. reported similar findings indicating a pro-tolerogenic counterresponse: T cells staining positive for Foxp3 were found among the cellular infiltrate, and indoleamine 2,3-dioxygenase (IDO) staining was detected in infiltrating antigen-presenting cells (APCs). Both those markers increased with intensity of rejection and towards later time points after transplantation [16].

These findings from hand transplantation are in line with observations from renal transplantation, where Foxp3 expression in the urine was associated with acute rejection. Foxp3 expression predicted a favorable outcome and was thus considered a mechanism for "damage control" [17].

Histopathology of Acute Rejection

Episodes of acute skin rejection have been observed in the majority of hand transplant recipients [18]. T cells are the primary actors among the graft-infiltrating cells [16]. Several histologic analyses have shown skin rejection to be a dynamic process, which starts with a perivascular infiltrate that tends to spread in the dermis and eventually involves the epidermis and adnexae. Epidermal involvement starts with vacuolization and necrosis of single keratinocytes, spongiosis, and ultimately leads to dermal–epidermal separation and necrosis [19]. To allow for comparability among RT-performing centers, a classification system for acute skin rejection in RT has been proposed [20].

Macroscopic appearance of a rejection episode in hand/upper extremity transplant patients has been described as maculopapular lesions of various size and/or location that tend to spread over the allograft, with a sharp delineation at the border between host and graft tissue, along with edema and erythema of the graft, and eventually leading to erosive areas [21, 22]. However, atypical macroscopic appearance along with a difficult clinical course has been reported [23].

It was shown in the first longer-term surviving human hand allograft that was reamputated at 29 months posttransplant, that the skin of the amputated graft was most strongly affected by rejection, while other tissues such as muscles, tendons, and joints were mostly intact. Within the skin component, the epidermis and the eccrine sweat glands showed the strongest affection [24]. Also, deep tissue biopsies taken upon secondary surgeries from various patients did not show tissues other than the skin to be significantly affected [25]. These findings stand in contrast to a recently published study in nonhuman primates, where deep tissue biopsies from five long-term (200 days) surviving animals showed signs of chronic rejection. The pathologic changes included vasculopathy and neointimal proliferation, narrowing of the lumen, vessel fibrosis, and the appearance of tertiary lymphoid follicles [26]. Findings like this advocate for closer attention to deep tissues, instead of focusing only on superficial punch biopsies that are more easily obtained.

The cellular composition of an acute rejection episode has been investigated in hand allografts. Hautz et al. reported the majority of the infiltrating cells to be of $CD3^+$ T cell origin. Among those, $CD8^+$ cells were more frequent than $CD4^+$ cells during mild rejection, which, however, changed during severe rejection. Additional cells comprised B cells and macrophages, which were not identified in all samples [16]. Kanitakis reported similar findings, with the major difference that he detected a majority of $CD4^+$ cells over $CD8^+$ cells [19] in mild cases of rejection.

Leukocyte Trafficking and T Cell Epidermotropism

As it appears to be the case in acute skin rejection, T cells are also the driving force for many inflammatory dermatoses [27, 28]. While environmental factors and deterioration of the skin barrier share important contributions for the development and aggravation of many inflammatory skin conditions, it has been shown that psoriasis vulgaris is most likely a T-cell-mediated autoimmune disease [29].

The process of leukocyte trafficking into the skin is a multistep cascade, which involves various adhesion molecules and is stimulated by local inflammatory mediators (Fig. 11.1). Briefly, the leukocyte adhesion cascade is as follows: Selectins, expressed on the endothelial surface, establish loose bindings with Sialyl-Lewis^X carbohydrates on the leukocyte surface. This allows the rolling leukocyte to come closer to the endothelium with its integrin adhesion molecules; for example, lymphocyte function-associated antigen 1 (LFA-1; CD11a/CD18) and macrophage-1 (Mac-1; CD11b/CD18), which bind to endothelial intercellular adhesion molecule 1 (ICAM-1; CD54). This firm adhesion allows the leukocyte to arrest and transmigrate through the endothelium and basal membrane, thus exiting from the circulation [30]. Leukocytes and endothelial cells can upregulate adhesion molecules in response to inflammatory molecules, such as cytokines and chemokines [31].

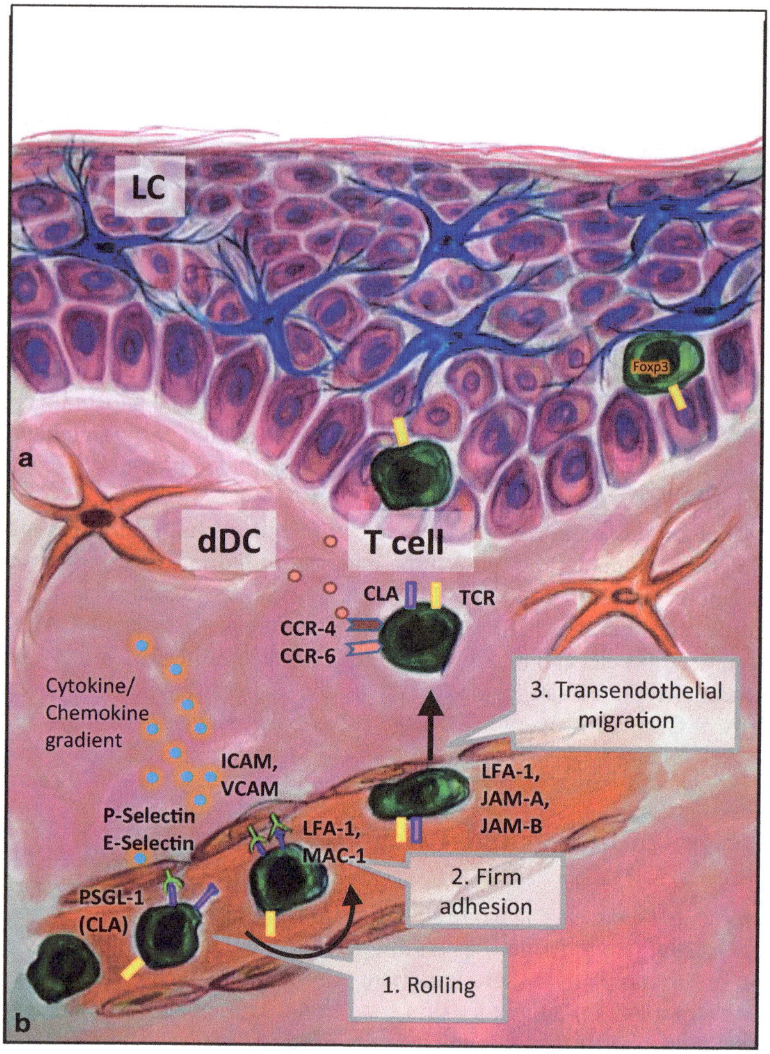

Fig. 11.1 **a** Immunocompetent cells within the skin: Langerhans cells (LC) form a network within the epidermis. Dermal dendritic cells (DC) reside below the basal membrane. T lymphocytes migrate through the skin. **b** The leukocyte adhesion cascade: Loose bindings via selectin-PSGL-1 interactions lead to rolling of the leukocyte. Further adhesion results in activation and firm attachment to the endothelial wall, and, lastly, extravasation. *PSGL-1* P-selectin glycoprotein ligand-1, *LFA-1* lymphocyte function-associated antigen 1, *MAC-1* macrophage-1, *TCR* T cell receptor, *CLA* cutaneous lymphocyte-associated antigen, *ICAM* intercellular adhesion molecule, *VCAM* vascular cell adhesion molecule, *JAM* junction adhesion molecule

Additionally, many adhesion molecules obtain better affinity and avidity following cytokine stimulation [32]. Secreted chemokines form gradients. They are of highly basic pH, which causes them to interact with the negatively charged extracellular matrix proteins and become immobilized [31].

The leukocyte adhesion cascade is not specific for leukocyte recruitment to the skin. The process of leukocytes homing to solid organs follows the same steps, with minor exceptions: No ICAM-1 is necessary for adhesion to neutrophils on hepatic sinusoids [33], and in the heart, neutrophils leave the circulation from larger coronary veins instead of postcapillary venules [34].

One molecule that distinguishes skin-homing lymphocytes from others is the surface receptor cutaneous lymphocyte-associated antigen (CLA). It contains the glycoprotein Sialyl-Lewis^{X}, and readily binds to selectins [35]. CLA^{+} T cells are found in the skin of inflammatory dermatoses in high numbers [35]. Apart from their contribution to skin autoimmunity and defense, Clark et al. have found a high number of CLA^{+} T cells in normal steady-state human skin, which even exceeded the number of circulating cells [7]. It is proposed that skin-homing T cells, apart from their detrimental effects in inflammatory skin diseases, are an important factor of immune surveillance in the steady state. Most of these isolated CLA+ T cells from noninflamed skin showed high levels of CCR4 and CCR6, and 50 % stained positive for CCR8 and CXCR6. The isolated cells contained a very diverse T cell receptor repertoire, suggesting a broad array of pathogen defense. Another recent finding was that almost all peripheral blood CD4+CD25^{hi}Foxp3 regulatory T cells (Tregs) expressed high levels of the skin-homing receptor CCR4; 80 % were found to express CLA and 73 % CCR6 [8].

The high number of skin-resident T cells under normal conditions also indicates that these skin-homing cells are already sufficiently arrested by baseline endothelial adhesion molecule expression and do not depend on inflammation-induced upregulation. When skin from psoriatic patients was transplanted into immunodeficient mice, the number of skin-resident T cells sufficed to cause psoriatic lesions in the host [36].

Taken together, these findings indicate that migration of T cells into the skin is a physiologic and frequent process necessary for defense as well as tolerance towards pathogens and injury. Skin-homing T cells express a homing marker signature that distinguishes them from T cells that are affiliated with other organs [37].

Leukocyte Adhesion Molecules in Skin Rejection

Evaluation of the expression of adhesion molecules in skin biopsies from five hand/upper extremity transplant patients indicated that a specific pattern of adhesion molecules was upregulated upon rejection. The markers LFA-1, ICAM-1, and E-selectin were highly upregulated during rejection, and expression of those markers also correlated well with the severity of rejection in immunohistochemical stains. In samples of nonrejecting skin, none of these markers was upregulated [16].

Blockade of adhesion molecules seems an attractive approach to treat skin inflammatory conditions of all kind. While some approaches have proven ineffective due to the redundancy of these pathways [38], others have revealed substantial effects. For example, the small molecule inhibitor for E- and P-selectin, efomycine

M, was effective in the treatment of plaque psoriasis [39]. In a rat hind limb transplantation study by our group, efomycine M was able to significantly prolong allograft survival when administered subcutaneously into the transplanted limb [16]. In the same experimental model, antibodies against ICAM-1 and LFA-1 were able to delay or even prevent graft rejection (unpublished results).

Other blockers of adhesion molecules, which have been applied clinically for the treatment of various inflammatory diseases, have shown ambivalent success: Natalizumab, a monoclonal antibody against the alpha-4 integrin, is approved for the treatment of severe multiple sclerosis. While causing substantial symptom relief in many patients, the side effect of progressive multifocal leukoencephalopathy was reported [40]. The same was true for efalizumab (anti-LFA-1), a substance for severe psoriasis treatment [41], which led to its withdrawal from the market.

In summary, a pattern of adhesion molecules has been shown to be upregulated during skin rejection, representing novel and interesting targets for treatment and prevention of skin rejection in RT.

In Vivo Assessment of Immune Cell Trafficking

Recently, *in vivo* imaging techniques have enhanced the ability to investigate immunological mechanisms in transplantation. Because of its accessible location, the skin has become a commonly studied subject in two-photon and single-photon microscopy [42, 43].

Horner et al. have imaged vascularized versus conventional, nonvascularized skin grafts with intravital confocal microscopy [44]. They report a fundamental difference in the kinetics and distribution of infiltrating host cells comparing vascularized versus nonvascularized skin: In nonvascularized skin grafts, the infiltrating cells are localized at the edges of the graft [45]. In primarily vascularized flaps, the cells tended to distribute more densely at the center, and were twice as many as in nonvascularized grafts. In both transplant types, the cellular infiltrate was localized in the upper dermis, clustered around hair follicles in both graft types, and additionally was cuffed around blood vessels in the vascularized skin grafts. A positive major histocompatibility complex II (MHC-II) staining could be detected in dermal endothelia of primarily vascularized grafts, but was absent in nonvascularized skin grafts. Along with their observation of the infiltrate localized primarily in the upper dermis and around hair follicles, the authors argue that not the epidermis, but different antigens might be the main target of the alloimmune response [44]. The observation period of the trial was only 4 days posttransplant and the infiltrate might have been in an early stage of rejection; any comparison with the Banff classification system [20] is therefore not meaningful.

Another *in vivo* imaging study using two-photon microscopy has characterized the contributions of both donor and host cells to acute rejection in a mouse model of nonvascularized ear skin transplantation [45]. Soon after transplantation, a burst of host CD11c positive cells infiltrated the graft; among those were mainly neutrophils (at earlier time points) and monocytes (the majority after day 9). This early influx of cells was the same in allogeneic and syngeneic transplants, indicating it to be antigen unspecific. Dermal dendritic cells (dDCs) from the graft were shown to rapidly migrate from the graft to the draining lymph nodes. In the allogeneic setting, however, the morphology of the dendritic cells resembled those of dead cells when they had reached the lymph nodes. LCs did not seem to contribute at all to the immunologic processes, but rested immobile in the epidermis until rejection was complete. Additionally, the study revealed that graft-infiltrating recipient cells were able to reach the draining lymph node and cross-prime $CD8^{+}$T cells.

These findings strongly point to the importance of the indirect pathway even at very early phases of the alloimmune response. Although the study was performed in nonvascularized skin grafts and it might be delicate to draw the same conclusions also for vascularized skin, there are other studies that advocate the importance of the indirect presentation pathway in the allogeneic response. Natural killer cells were shown to rapidly kill graft dendritic cells once they had reached the lymph node [46]. In solid organ transplantation and bone marrow transplantation, similar findings have been reported: Immunosuppressive dendritic cells (ISDC) have been cultured as a tool to prolong allograft survival, which was actually demonstrated in some animal models [47–50]. While it was long assumed that these dendritic cells would perform a "tolerogenic" direct antigen presentation to host T cells, this paradigm was recently challenged by a study, which showed that these ISDCs quickly vanished from the circulation and were uptaken by recipient dendritic cells. Depletion of recipient dendritic cells at the time of ISDC application abrogated their previously permissive effect on cardiac graft survival [51]. Such findings need further confirmation in the setting of RT, but might have an impact on the way future immunosuppressive strategies are designed.

Antigen-Presenting Cells in Skin Rejection

APCs form the bridge between innate and adaptive immunity. Knowledge about different subpopulations of APCs within the skin has only recently accumulated. Our overall understanding of APC function and their characteristic maturation and migration cycle was postulated following *in vitro* experiments. In brief, antigen uptake by the immature APC leads to maturation of the cell, downregulation of adhesion molecules, and detachment. The maturating cell travels via the lymphatic vessels to a draining lymph node, enters, and primes T cells with the help of newly upregulated molecules such as CCR7 [52–54].

In general, two major populations of resident dendritic cells have to be distinguished within the skin: the dDCs and epidermal LCs. Currently, a major debate concerning their immunogenic versus tolerogenic functions is ongoing—while some investigators use dendritic cells to induce tolerance, there are other studies that indicate their potent immunostimulatory functions. A critical confounding factor for any APC study is the environment—whether those cells are investigated in or *ex vivo*, under inflammatory or tolerogenic conditions, and if they are immature or mature [55]. A substantial part of APC research is done in the context of contact hypersensitivity (CHS). Recently, an array of transgenic animal models for skin APCs has been developed, in most studies leading to inconsistent results [56]. However, a comprehensive analysis of the current data on skin APCs is beyond the scope of this chapter, which will focus only on the aspects concerning RT.

In both hand and face transplantation, there is little but some information on the contribution of skin APCs. LCs were found to have surprising longevity in transplanted skin—in a hand transplant patient, they were isolated 10 years post transplant, where they seemed to form a stable population and renew from epidermal precursor cells [57]. Upon major destructive stimuli, such as UVC irradiation, LCs have been shown to be replenished from bone marrow precursors [58]. From an academic point of view, it would be interesting to evaluate how this replacement happens in a bone marrow containing allograft—from host monocytes, host LC precursors, or actually viable donor LC precursors from the bone marrow niche?.

Skin APCs have been exploited for their abilities to induce tolerance within the skin. In a recent study, LCs showed only a weak capacity to uptake, process, and present bacterial antigen on MHC-II molecules. Compared to dDCs, immature LCs possess a restricted repertoire of toll-like receptor (TLR) for bacterial uptake. They only poorly activated bacteria-specific T cells, as compared to dDCs. On the contrary, these bacteria-primed LCs induced Foxp3^{+} cells [59]. Studies like this indicate that the epidermis, although representing the "first barrier" for invaders, has a tolerogenic function that allows commensal, noninvasive bacteria and fungi to reside. Parallel hypothesis exist for the gut flora, where a specialized CD103^{+} dendritic cell type has been reported to induce bacteria-specific Tregs [60].

Conclusive Remarks

Specific ways to overcome skin rejection require more in-depth knowledge about mechanisms of rejection within the skin. Many of our theories to date are extrapolated from dermatology research, in particular from chronic inflammatory dermatoses. Inflammation in such conditions, however, is always a result of an orchestrated communication between skin-resident and skin-invading cells. In the setting of allotransplantation, the skin-resident cells are donor cells, whereas skin-invading cells are recipient cells. It was noted in a recent *in vivo* imaging study that dDCs from allogeneic skin grafts did not prime host T cells in the lymph node (i.e., via the direct pathway), but were dead by the time they had reached the draining node [45], indicating that the mechanisms triggering the immune response may differ in this setting.

The relevance and contribution of APCs such as LCs in skin transplantation remain unclear at this point. The long-lived, very resistant cells seem to interact with the commensal bacterial flora [59], but their contribution to skin rejection remains unclear.

T regulatory cells have been identified in hand allografts [12, 16], and like for solid organ transplantation, a role for peripheral tolerance has been implemented. Still, neither in solid nor vascularized composite allotransplantation strategies to locally enhance and enrich those cells are available.

Blocking leukocyte extravasation into the skin seems a promising approach. Leukocyte adhesion molecules are upregulated in hand allograft rejection [16], but the side effects of adhesion molecule blockers in psoriasis or multiple sclerosis limit their application in skin rejection.

Even though the immunology of transplanted skin is starting to be better explored, we do not have means to specifically target the alloimmune response towards the skin. However, the skin offers the advantage that treatment can be applied locally. We believe that local concepts for diagnosis and therapy of skin rejection have to be further exploited.

References

1. Nestle F, Di Meglio P, Qin J, et al. Skin immune sentinels in health and disease. Nat Rev Immunol. 2009;9:679–91.
2. Uchi H, Terao H, Koga T, et al. Cytokines and chemokines in the epidermis. J Dermatol Sci. 2000;24:29.
3. Gaspari A, Katz S. Induction and functional characterization of class II MHC (Ia) antigens on murine keratinocytes. J Immunol. 1988;140:2956.
4. Wang L, Bursch L, Kissenpfennig A, et al. Langerin expressing cells promote skin immune responses under defined conditions. J Immunol. 2008;180:4722–7.
5. James W, Bullard D, Hickery M. Critical role of the alpha 4 integrin/VCAM-1 pathway in cerebral leukocyte trafficking in lupus-prone MRL/fas(lpr) mice. J Immunol 2003;170:520–7.
6. Deane J, Hickery M. Molecular mechanisms of leukocyte trafficking in T-cell-mediated skin inflammation: insights from intravital imaging. Exp Rev Mol Med 2009;11:25.
7. Clark R, Chong B, Mirchandani N, et al. The vast majority of CLA + T cells are resident in normal skin. J Immunol. 2006;176:4431–9.
8. Hirahara K, Lui L, Clark R, et al. The majority of human peripheral blood CD4+ CD25 high Foxp3+ regulatory T cells bear functional skin-homing receptors. J Immunol. 2006;177:4488–94.
9. Billingham R, Brent L. Quantitative studies on tissue transplantation immunity. IV. Induction of tolerance in newborn mice and studies on the phenomenon of runt disease. Philos Trans R Soc Lond B Biol Sci 1956;242:357–414.
10. Pescovitz M, Auchincloss H, Thistlethwaite R, et al. Transplantation in miniature swine: acceptance of class I antigen mismatched renal allografts. Transpl Proc 1983;15:1124.
11. Mathes D, Randolph M, Solari M, et al. Split tolerance to a composite tissue allograft in a swine model. Transplantation. 2003;75:25–31.
12. Petruzzo P, Badet L, Gazarian A, et al. Bilateral hand transplantation: six years after the first case. Am J Transplant. 2006;6:1718–24.
13. Schneeberger S, Gorantla V, Brandacher G, et al. Upper-extremity transplantation using a cell-based protocol to minimize immunosuppression. Ann Surg 2013;257:345–51.
14. Gorantla V, Brandacher G, Schneeberger S, et al. Favoring the risk-benefit balance for upper extremity transplantation—the Pittsburgh protocol. Hand Clin. 2011;27(4):511–20.
15. Eljaafari A, Badet L, Kanitakis J, et al. Isolation of regulatory T cells in the skin of a human hand-allograft, up to six years posttransplantation. Transplantation. 2006;82:1764–8.

16. Hautz T, Zelger B, Grahammer J, et al. Molecular markers and targeted therapy of skin rejection in composite tissue allotransplantation. Am J Transplant. 2010;10:1200–9.
17. Muthukumar T, Dadhania D, Ding R, et al. Messenger RNA for Foxp3 in the urine of renal-allograft recipients. N Engl J Med. 2005;353:2342–51.
18. Petruzzo P, Lanzetta M, Dubernard J, et al. The International Registry on Hand and Composite Tissue Transplantation. Transplantation. 2010;90:1590–4.
19. Kanitakis J. The challenge of dermatopathological diagnosis of composite tissue allograft rejection: a review. J Cutan Pathol. 2008;35:738–44.
20. Cendales L, Kanitakis J, Schneeberger S, et.al. The Banff 2007 working classification of skin-containing composite tissue allograft pathology. Am J Transplant. 2008;8:1396–400.
21. Schneeberger S, Kreczy A, Brandacher G, et al. Steroid- and ATG-resistant rejection after double forearm transplantation responds to Campath-1H. Am J Transplant. 2004;4:1372–4.
22. Kanitakis J, Jullien D, Nicolas J, et al. Sequential histological and immunohistochemical study of the skin of the first human hand allograft. Transplantation. 2000;69:1380–5.
23. Schneeberger S, Gorantla V, van Riet R, et al. Atypical acute rejection after hand transplantation. Am J Transplant. 2008;8:688–96.
24. Kanitakis J, Jullien D, Petruzzo P, et al. Clinicopathologic features of graft rejection of the first human hand allograft. Transplantation. 2003;76:688–93.
25. Hautz T, Zelger B, Weissenbacher A, et al. Standardizing skin biopsy sampling to assess rejection in vascularized composite allotransplantation. Clin Transpl. 2013;27:81–90.
26. Mundinger G, Munivenkatappa R, Drachenberg C, et al. Histopathology of chronic rejection in a nonhuman primate model of vascularized composite allotransplantation. Transplantation. 2013;95:1204–10.
27. Robert C, Kupper T. Inflammatory skin diseases, T cells and immune surveillance. N Engl J Med. 1999;24:1817–28.
28. Schön M, Boehncke W. Psoriasis. N Engl J Med. 2005;18:1899–912.
29. Kaffenberger B, Wong H, Jarjour W, et al. Remission of psoriasis after allogeneic, but no not autologous, hematopoietic stem-cell transplantation. J Am Acad Dermatol. 2013;68:489–92.
30. Alon R, Feigelson S. From rolling to arrest on blood vessels: leukocyte tap dancing on endothelial integrin ligands and chemokines at sub-second contacts. Sem Immunol. 2002;14:93–104.
31. Schön M. The molecular basis for lymphocyte recruitment to the skin: clues for pathogenesis and selective therapies of inflammatory disorders. J Inv Dermatol. 2003;121:951–62.
32. Shamri R, Grabovsky V, Gauguet J, et al. Lymphocyte arrest requires instantaneous induction of an extended LFA-1 conformation mediated by endothelium-bound chemokines. Nat Immunol. 2005;6:497–506.
33. Patrick A, Rullo J, Bveaudin S, et al. Hepatic leukocyte recruitment in response to time-limited expression of TNF-alpha and IL-1beta. Am J Physiol Gastrointest Liver Physiol. 2007;293:663–72.
34. Li W, Nava R, Bribriesco A, et al. Intravital 2-photon imaging of leukocyte trafficking in beating heart. J Clin Invest. 2012;122:2499–508.
35. Fuhlbrigge R, Kiefer J, Armerding D, et al. Cutaneous lymphocyte antigen is a specialized form of PSGL-1 expressed on skin-homing T cells. Nature 1997;389:978–81.
36. Boyman O, Hefti C, Conrad B, et al. Spontaneous development of psoriasis in a new animal model shows an essential role for resident T cells and tumor necrosis factor-alpha. J Exp Med. 2004;199:731–36.
37. Butcher E, Williams M, Youngman K, et al. Lymphocyte trafficking and regional immunity. Adv Immunol 1999;72:209–53.
38. Bhushan M, Bleiker T, Ballsdon A, et al. Anti-E-selectin is ineffective in the treatment of psoriasis: a randomized trial. Br J Dermatol. 2002;146:824–31.
39. Schön M, Krahn T, Schön M, et al. Efomycine M, a new specific inhibitor of selectin, impairs leukocyte adhesion and alleviates cutaneous inflammation. Nat Med. 2002;4:366–72.

40. Boster A, Nicholas J, Topalli I, et al. Lessons learned from fatal progressive multifocal leukoencephalopathy in a patient with multiple sclerosis treated with natalizumab. JAMA Neurol. 2013;70:398–402.
41. Lebwohl M, Tyring S, Hamilton T, et al. A novel targeted T-cell modulator, efalizumab, for plaque psoriasis. N Engl J Med. 2003;349:2004–13.
42. Camirad G. New perspectives in transplantation through intravital microscopy imaging. Curr Op Org Transpl. 2013;18:6–12.
43. Germain R, Robei E, Cahalan M. A decade of imaging cellular motility and interaction dynamics in the immune system. Science. 2012;336:1676–81.
44. Horner B, Ferguson KK, Randolph M, et al. *In vivo* observations of cell trafficking in allotransplanted vascularized skin flaps and conventional skin grafts. J Plast Reconstr Aesth Surg. 2010;63:711–9.
45. Celli S, Albert M, Bousso P. Visualizing the innate and adaptive immune responses underlying allograft rejection by two-photon microscopy. Nat Med. 2011;17:744–49.
46. Garrod K, Liu F, Forrest L, et al. NK cell patrolling and elimination of donor-derived dendritic cells favor indicrect alloreactivity. J Immunol. 2010;71:418–21.
47. Min W, Gorczynski R, Huang X, et al. Dendritic cells genetically engineered to express Fas ligand induce donor-specific hyporesponsiveness and prolong allograft survival. J Immunol. 2000;164:161–7.
48. Emmer P, van der Vlag J, Adema G, et al. Dendritic cells activated by lipopolysaccharide after dexamethasone treatment induce donor-specific allograft hyporesponsiveness. Transplantation. 2006;81:1451–9.
49. Taner T, Hackstein H, Wang Z, et al. Rapamycin-treated, alloantigen-pulsed host dendritic cells induce Ag-specific T cell regulation and prolong graft survival. Am J Transplant. 2005;5:228–36.
50. Bonham C, Peng L, Liang X, et al. Marked prolongation of cardiac allograft survival by dendritic cells genetically engineered with NF-kB oligodeoxyribonucleotide decoys and adenoviral vectors encoding CTLA-Ig. J Immunol. 2002;169:3382–91.
51. Wang Z, Divito S, Shufesky W, et al. Dendritic cell therapies in transplantation revisited: deletion of recipient DCs deters the effect of therapeutic DCs. Am J Transplant. 2010;12:1398–408.
52. Schuler G, Steinman R. Murine epidermal Langerhans cells mature into potent immunostimulatory dendritic cells *in vitro*. J Exp Med. 1985;161:526–46.
53. Romani N, Clausen B, Stoitzner P. Langerhans cells and more: langerin-expressing dendritic cell subsets in the skin. Immunol Rev. 2010;234:120–41.
54. Romani N, Koide S, Crowley M, et al. Presentation of exogenous protein antigens by dendritic cells to T cell clones: intact protein is presented best by immature, epidermal Langerhans cells. J Exp Med. 1989;169:1169–78.
55. Steinman R. Decisions about dendritic cells: past, present, and future. Ann Rev Immunol. 2012;30:1–22.
56. Wang L, Bursch L, Kssenpfennig A, et al. Langerin expressing cells promote skin immune responses under defined conditions. J Immunol. 2008;180:4722–7.
57. Kanitakis J, Morelon E, Petruzzo P, et al. Self-renewal capacity of human epidermal Langerhans cells: observations made on a composite tissue allograft. Exp Dermatol. 2011;20:145–6.
58. Romani N, Tripp C, Stoitzner P. Langerhans cells come in waves. Immunity. 2012;37:766–8.
59. Van der Aar A, Picavet D, Müller F, et al. Langerhans cells favor skin flora tolerance through limited presentation of bacterial antigens and induction of regulatory T cells. J Invest Dermatol. 2013;133:1240–9.
60. Coombes J, Siddiqui K, Arancibia-Carcamo C, et al. A functionally specialized population of mucosal CD103+ DCs induces Foxp3+ regulatory T cells via TGF-beta and retinoic acid-dependent mechanism. J Exp Med. 2007;204:1757–64.

Chapter 12
Antibody-Mediated Rejection in Reconstructive Transplantation

Luis Landin, Pedro Bolado and Cesar Casado-Sanchez

Antibody-Mediated Rejection in Solid Organ Transplantation

The development of potent immunosuppressive (IS) drugs, the most popular of which are calcineurin inhibitors (CNI), has driven the improvement in rates of allograft survival in recent decades. These drugs mainly influence the mechanisms of cell-mediated rejection (CMR). However, another form of rejection—which involves antibodies (Abs) against donor-specific class-I and class-II human leukocyte antigen (HLA) and is usually referred to as humoral rejection or antibody-mediated rejection (AMR)—is increasingly recognized as a cause of allograft loss [23]. A common attribute of patients suffering from AMR is that they barely respond to conventional therapy for CMR [24, 44, 95]. High doses of IS drugs usually control CMR but do not necessarily prevent or treat AMR. Although AMR and CMR may occur simultaneously, they are independent phenomena [17, 50].

Diagnosis of AMR following kidney, heart, and liver transplantation is usually based on four criteria: (1) clinical evidence of graft dysfunction, (2) histological evidence of tissue injury, (3) immunopathological evidence of Ab action, (4) and serological evidence of anti-donor Abs at the time of biopsy.

A spectrum of clinical situations relates to these four criteria, in addition to reported exceptions (Fig. 12.1) [56].

AMR can cause two different forms of graft dysfunction: classic arterial hyperacute AMR and vascular AMR. Classic arterial hyperacute AMR is characterized by necrotizing arteritis, with mural fibrinoid necrosis and inflammation in artery walls. Endothelial cells are damaged and luminal thrombosis is common, resulting in organ infarction. Hyperacute rejection occurs minutes after transplantation and is caused by preexisting Abs against the donor. These Abs activate the classical

L. Landin (✉) · P. Bolado · C. Casado-Sanchez
Division of Plastic and Reconstructive Surgery, La Paz University Hospital,
Paseo de la Castellana, 261, 28046 Madrid, Spain
e-mail: landinsurgery@gmail.com

G. Brandacher (ed.), *The Science of Reconstructive Transplantation*,
Stem Cell Biology and Regenerative Medicine, DOI 10.1007/978-1-4939-2071-6_12

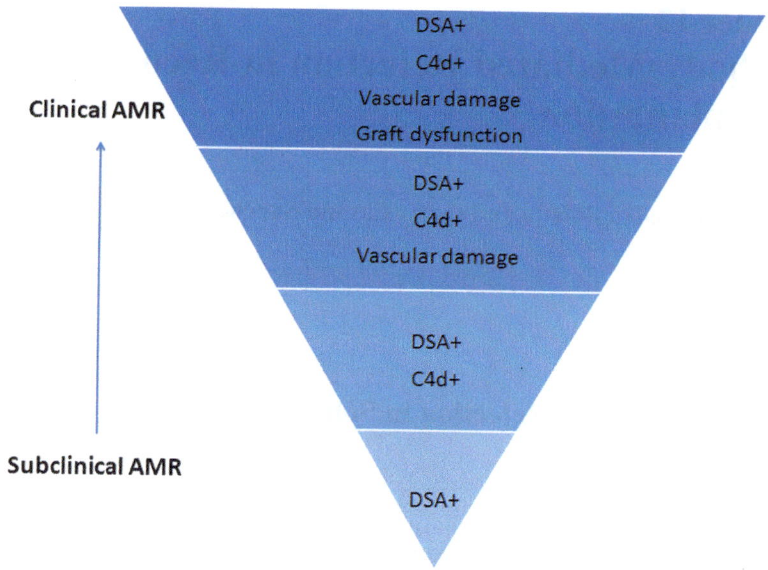

Fig. 12.1 The spectrum of clinical presentations of antibody-mediated rejection (*AMR*). AMR can only be diagnosed when the four criteria are met. The presence of donor-specific antibodies (*DSAs*) poses a risk of graft injury. By itself, the deposition of C4d is not a criterion for AMR. If DSAs and C4d are identified simultaneously to vascular histological injury, there is a high risk of graft dysfunction

pathway of the complement system and initiate the blood clotting cascade. Vessels become obstructed and neutrophils are rapidly recruited to the allograft. Hyperacute AMR differs from Ab-mediated vascular rejection in that it does not feature an inflammatory or fibrinoid component in vessel walls as its outset. Histological evidence of tissue injury in the context of vascular AMR usually includes the presence of neutrophils and macrophages in capillaries, endothelial damage such as swelling and denudation in cardiac allografts, fibrinoid necrosis, thrombi, and acute tubular injury in kidney allografts. Vascular AMR is generally characterized by abundant deposition of the complement activation product C4d in peritubular capillaries (Fig. 12.2) [44, 23, 24, 6, 26, 54].

A significant number of episodes of AMR occur in sensitized patients, in whom donor-specific antibodies (DSAs) are already present in the bloodstream. DSAs may form after implantation of the allograft, or may be present before transplantation. They may arise in patients who have been pregnant or have undergone transfusions or prior transplants. Mismatched blood type (ABO-incompatible) transplants may be subject to DSA-mediated AMR and display features of the phenomenon termed accommodation [56, 49].

Hyperacute rejection occurred more frequently before the introduction of cross-match (CXM) testing. CXM testing is used to detect the presence of Abs against the lymphocytes of an individual donor. A positive T-cell CXM test is usually considered an absolute contraindication to kidney transplantation because of the high

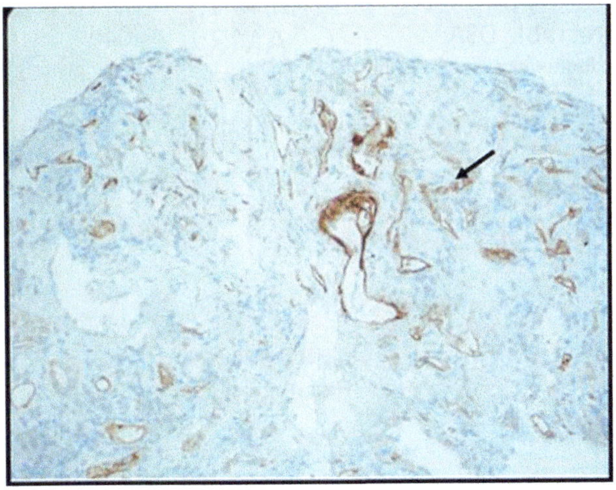

Fig. 12.2 Renal allograft biopsy from a child with chronic allograft nephropathy showing C4d deposition (*brown*) in peritubular capillaries consistent with humoral-mediated rejection (reprinted with permission, Fletcher JT)

risk of hyperacute rejection. Although CXM testing initially presented low sensitivity, the addition of antihuman globulin (AHG), flow cytometric crossmatch (FXM) testing and solid-phase beads have since improved its sensitivity: it is now capable of detecting anti-class-I and class-II Abs. The panel-reactive Ab (PRA) assay is a screening test used to measure breadth of sensitization as determined by the presence of anti-HLA Abs [27]. Since the enhancement of the PRA assay by the addition of AHG and flow cytometry, it is now able to detect very low levels of Abs and noncytotoxic Abs. The PRA tests the recipient's serum for its ability to lyse a panel of T lymphocytes that is a surrogate of potential donors. Despite continuous improvements in the technique, the panel does not reflect all potential donors and provides limited information about the specificities of the Abs.

Recipients with a positive FXM test after a negative CXM test are at very low risk of hyperacute rejection. However, they are at an increased risk of AMR, CMR, or both. The introduction of solid-phase assays permits identification of the presence of anti-HLA Abs against a wide range of HLA types and determination of their HLA specificity. The detection of DSAs in the CXM or solid-phase assays is not considered an absolute contraindication to kidney transplantation, but it represents an immunological risk of Ab-mediated injury. Whereas a positive CXM test and high DSA levels pose a notable risk of hyperacute rejection, a positive CXM test coincident with low levels of DSAs may increase the risk of AMR [25].

The presence of preformed Abs at the time of kidney transplantation was found to be associated with chronic allograft failure (CAF), which is the second leading cause of kidney allograft loss after death [91]. Up to 60 % of patients diagnosed with CAF show evidence of Ab-mediated injury. Although anti-HLA Abs are usually detectable several weeks before graft dysfunction, they are not the sole cause of AMR [91]. A study that followed 70 kidney transplants demonstrated that AMR

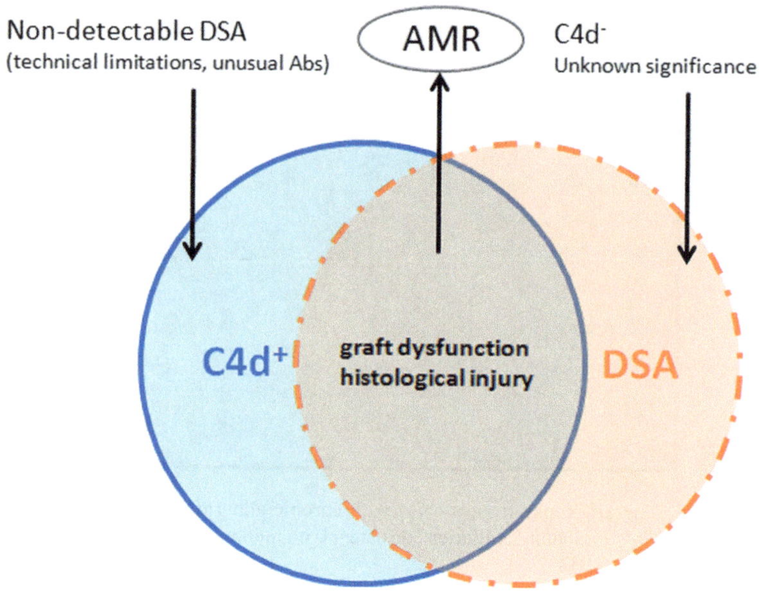

Fig. 12.3 Humoral rejection needs be diagnosed in the presence of antibodies and C4d. Other possible presentations may occur. *AMR* antibody-mediated rejection; *DSAs* donor specific antibodies

occurs across a wide spectrum of baseline DSA levels, even after a negative T-cell AHG CXM test. The risk of AMR increased with increasing baseline levels of DSA, but its occurrence remained unpredictable. Interestingly, prior transplantation did not increase the incidence or severity of AMR compared with other methods of sensitization. The authors concluded that anti-class-II DSAs, alone or in combination with anti-class-I DSAs, played an important role in AMR and could be the sole cause of AMR in as many as 16% of cases [11].

The detection of C4d deposits is considered a vital tool for the diagnosis of AMR. Before identification of the C4d component, many Ab-mediated rejection episodes went undiagnosed. Attempts to diagnose humoral rejection were unsuccessful, probably because of the rapid endocytosis of Abs from the endothelial surface [23]. Current immunopathological evidence of AMR includes positive staining for complement products such as C1q, C3d, or C4d, which may also be accompanied by Igs G, M, and A [80, 23]). C4d in transplanted kidneys is characteristically located in the peritubular capillaries, [95] and should be considered the 'fingerprint' of an Ab response [7]. However, a variety of situations can accompany the localization of C4d in peritubular capillaries. Whereas up to 90% of $C4d^{+}$ cases exhibit anti-HLA Abs, in 10% of $C4d^{+}$ recipients, the search for DSA yields no results (Fig. 12.3) [56]. This may be due to the technical limitations of assays or the presence of unusual Abs. Conversely, there is emerging evidence of episodes of AMR that are C4d− in the kidney and in other allografts [72]. Indeed, a recent study showed that a high percentage of recipients presented circulating DSAs that were not associated with C4d deposits [85, 56]). An explanation for this phenomenon might be that the pattern of C4d deposition is dynamic and can disappear after a few days. Deposition of C4d can also persist, despite anti-rejection therapy [56].

C4d is not an exclusive marker of AMR. It has also been found in protocol biopsies of stable grafts and is not even a molecule limited to allografted organs; it has been found in biopsies of normal kidneys, located at every glomerular mesangium, and at the vascular pole of the arterioles. The accumulation of C4d in glomerular capillaries has also been implicated in kidney diseases caused by the formation of immune complexes [56].

Liver allografts are less susceptible to AMR than other solid organ allografts. This has been attributed to phagocytosis by Kupffer cells of immune complexes generated after Ab binding and the absence of a basal membrane in the sinusoidal hepatic microvasculature [22]. Following liver transplants, anti-HLA Abs do not always cause clinically relevant allograft damage, whereas isoagglutinins usually cause more damage. Hyperacute rejection is followed by liver dysfunction over a period of hours to days. This can be recognized by swelling, a dusky appearance of the liver, cessation of bile flow, difficulty in achieving hemostasis, and an exaggerated need for platelets and blood replacement. Arterial thrombosis is evident on post-transplant angiograms. Areas of necrosis are visible in gross examinations of failed liver grafts. Histological examination reveals findings common to other organs, such as vessel congestion, thrombi, and necrotizing arteritis. For "vascular" humoral rejection caused by preformed anti-HLA Abs, the changes are usually less florid and necrotizing arteritis is rare. Detection of immune reactants relies on the localization of C4d deposits. However, C4d is also evident during CMR [22, 20].

In cardiac transplantation, AMR has only recently been recognized as a real and distinct clinicopathological entity [42]. Hemodynamic dysfunction has been attributed to AMR more frequently in women than in men. The presence of DSAs is strongly correlated with AMR and decreased allograft and patient survival [62]. Macroscopically, the heart is swollen and discolored, showing areas of focal necrosis. Histological examinations frequently reveal intravascular thrombi and endothelial swellings. There are abundant accumulations of macrophages in the capillaries. Mixed AMR and CMR rejection is evident in 15% of biopsies of hemodynamically abnormal cardiac allografts [47]. Similarly to the kidney, C4d deposition is a footprint of AMR in the heart. It has been associated with a poor prognosis but deposition of C4d alone should not be equated with AMR [42]. In lung transplants, histopathological criteria have not yet been established for the diagnosis of AMR.

Cellular and Molecular Basis of Antibody-Mediated Rejection

Only B-lymphocytes are able to generate humoral immunity. They originate in bone marrow and are divided into two subsets; whereas the B1 lineage contributes to innate immunity and usually produces Abs against microbes, the B2 lineage forms part of the adaptive immune system that resides in the bloodstream and peripheral lymphoid tissues. Each B cell produces only one specific type of Ab, placing a transmembrane molecule on its surface to act as receptor for the antigen (Ag) specific to that Ab. The B cell places approximately 10^5 identical Ab molecules on its surface [21]. If an Ag binds one of these molecules while the B cell resides in the secondary lymphoid tissues, it transforms into a plasma cell (PC). PCs are able to secrete a soluble version of the specific Ab with the same specificity as the membrane-bound Ab characteristic of the B-lymphocyte.

Only B lymphocytes that bind their specific Ag proliferate in successive waves; this phenomenon is known as clonal selection. Other cells, such as dendritic and T-helper follicular cells, aid the formation of a germinal center of activated B cells. It takes several days before Abs are detectable in the serum following primary contact with an Ag. This type of response to antigens is termed the acquired immune response. The maturation of B cells changes the expression of cell-surface markers. For example, the expression of CD20 and CD19 is high on unstimulated B cells, but low-to-absent on PCs. Additional markers of the later stages of B-cell development include B cell activating factor from the tumor necrosis factor family (BAFF), A proliferation inducing ligand (APRIL) and Blimp-1 [76]. Identification of B cells requires more than one marker because of the variability of B-cell phenotypes; for example, not all memory B cells express CD20. The final step in the development of persistent Ab production involves the migration of PCs back to the bone marrow. Most PCs express CD138 and CD38; these markers are not expressed on unstimulated B cells. PCs that produce Abs against specific HLA are rare, with an average frequency of a single HLA-specific PC per 2×10^6 bone marrow cells [61].

The *complement system* comprises 20 plasma proteins that are activated in a cascade-like manner (Fig. 12.4). There are different ways of activating this cascade, including the perception of microbial polysaccharides, Ab binding, and the mannose–lecithin pathway. Complement activation results from the deposition of molecules that can produce direct lysis, facilitate opsonization, and enhance phagocyte function; byproducts of the complement system act as chemotactic factors and activate mast cells. The complement components are named using the letter 'C' followed by a number related to the chronology of its discovery.

AMR damage is caused by classical complement activation when donor-reactive HLA Abs bind to allograft endothelial cells. The Abs bind C1qrs, which in turn binds C4; C4 cleaves to C4b, then recruits C2 and C3 to form C4b2a, a convertase of C3. Then, it forms C4b2a3b, which cleaves C5 into C5b, leading to the formation of the annular membrane-attack complex (MAC) C5b-9. The MAC is a transmembrane channel that is fully permeable to electrolytes and water. Owing to the high internal colloid osmotic pressure of cells, there is a net influx of $Na^+$and water into cells that contain the MAC; this causes cell lysis. The C5-9 MAC also mediates neutrophil influx and synthesis of pro-inflammatory cytokines, causing cell injury, apoptosis, and necrosis. The C5a receptor on endothelial cells seems to function in the production of adhesion molecules and may regulate apoptosis [25].

In contrast, C4d is an inert degradation product of C4 that binds the cell surface. Complement component C4d was linked to the presence of DSAs and AMR in the 1990s and has since been considered a footprint of the presence of humoral

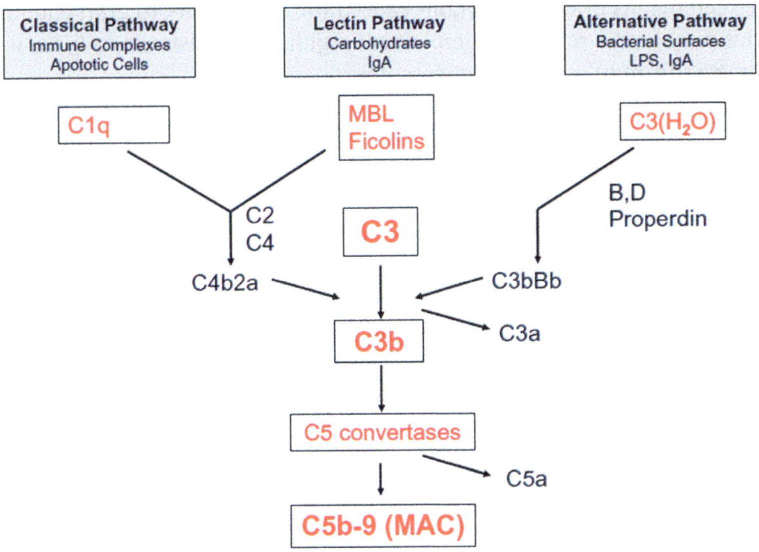

Fig. 12.4 Schematic representation of the complement system. *Ig* immunoglobulin; *LPS* lipopolysaccharide; *MAC* membrane-attack complex; *MBL* mannose-binding lectin. (Reprinted with permission, Berger ST)

rejection. C4d is a marker of complement cascade activation that is evident in the endothelium of peritubular capillaries and basal membrane of kidneys (Fig. 12.2) and in the portal capillaries of livers undergoing humoral rejection. Histological tests are best performed on frozen section specimens [91, 4]. Using immunohistochemical techniques, donor-specific IgG and IgM are undetectable on the vascular endothelium of renal allografts, even in the context of AMR [25].

Antibody-Mediated Rejection and Accommodation

Clinical and experimental research has demonstrated that allografts can develop resistance to anti-graft Abs. This is best established in ABO-incompatible kidney allografts and is usually referred to as accommodation [19, 49]. However, the mechanisms involved are not well understood.

It has been suggested that accommodation might result from a change in the expression of Abs from a complement-fixing form to a non-fixing form, or from the lack of subsets of T-helper cells [52, 81]. However, no conclusive data explain accommodation as a result of the modification of allograft antigens. Accommodated allografts seemed to change their phenotypic profile compared with ABO-compatible grafts. The evidence was stronger for ABO-incompatible grafts than for renal allograft recipients with positive-CXM tests [59]. In contrast, others have suggested a resistance to complement that might reflect modification of the complement cascade such that its effector functions cause little or no injury to graft endothelium [49]. This may be the result of complement regulation, resistance to lysis, and improved metabolism of complement complexes.

Accommodation may be more frequent than expected in functioning allografts; there is a possibility that Abs bind to the allograft that will subsequently be removed. This is consistent with the clinical finding of increased serum Abs after failed allograft removal [49].

There is a paucity of reported strategies to facilitate accommodation. Using an experimental model, Wang and colleagues reported that administration of anti-C5 Ab in presensitized allograft recipients, in conjunction with immunosuppression, prevented rejection and permitted accommodation. However, double-transplant experiments demonstrated that immunological alterations in both the graft and recipient were required for successful graft accommodation [86].

Prevention and Treatment of Antibody-Mediated Rejection

A complete patient sensitization history, which includes PRA results, CXM test results, prior transfusions, pregnancies, and previous transplants is required to assess the risk of AMR for renal, cardiac, and lung transplants. An individual cannot be reliably defined as unsensitized without a complete sensitization history [80]. Sensitive techniques should be used to determine the presence or absence of DSAs, and the specificity of DSAs should be determined before transplantation. The presence or absence of auto-Ab must be established to facilitate the interpretation of pretransplant CXM test results. In addition, both T- and B-cell CXM tests should be performed before transplantation (unless clinically contraindicated, i.e., after a long period of cold ischemia—and only then if a complete and reliable sensitization history is negative) [80]. The risk of AMR and early graft loss in patients showing a positive complement-dependent cytotoxicity (CDC) on CXM testing is high. Transplantation should be avoided unless a desensitization protocol can be used. A positive CXM test using the flow cytometry technique poses medium risk and requires high doses of immunosuppression. Recipients with negative flow cytometry results or CDC on CXM testing are at low risk of AMR.

AMR can be managed in different ways, depending on the treatment target, and some therapies can tackle several of these targets. Plasmapheresis (PP) and immunoadsorption (IA) can control B-cell responses by removing the Abs against donor antigens, whereas administration of intravenous Ig (IVIG) dilutes alloantibodies, which mediate the effector arm of AMR. Alternative approaches are to neutralize B-cell division with mycophenolate mofetil (MMF) or inhibit B-cell surface molecules with rituximab. On the other hand, the use of MMF and steroids to inhibit T-cell division, CNI to prevent the signaling of interleukin (IL)-2 to T cells, or antithymocyte globulin (ATG) to ablate T cells can restrain T cells from activating B cells. Additional drugs, such as bortezomib, act against PCs, inhibiting what is considered the basis of AMR. There is also an arsenal of drugs that prevent the fixation of complement, including Abs such as eculizumab, which decreases the spread of the complement cascade. Finally, several of these steps can be circumvented by a nonspecific treatment, such as splenectomy, which limits the development of B cells [45, 78, 5].

ATG is a polyclonal Ab preparation generated from the immunization of rabbits with human thymus. ATG has different effects on cells involved in rejection, such as anti-T-cell effects and inhibition of the interaction between $CD4^+$ helper T cells and B cells; these reduce rates of B-cell activation. However, ATG may also have direct cytotoxic effects on B cells, both by modulating the production of allo-Abs and by causing the apoptosis of B cells. These effects are due to different specificities against T cells (class-I and class-II HLA molecules, CD3, CD4, CD8, co-stimulatory pathways and cell adhesion molecules), B cells (CD20) and PCs (CD38 and CD138) [8, 9, 94]. This therapeutic option is included in different treatment algorithms, particularly when there is histological evidence of combined cellular rejection and AMR [74, 65, 71, 87, 3, 40].

IVIG is derived from the plasma of thousands of healthy blood donors and is primarily composed of IgG (90 %), a few dimers, Ag-binding fragments and traces of IgA and IgM [48]. In the 1990s, IVIG was discovered to inhibit the cytotoxic effects of allo-Abs [30, 83]. Nowadays, IVIG is used in desensitization protocols and for the treatment of AMR. IVIGs have different specificities against class-I and class-II HLA molecules, CD 40 costimulatory molecule, IL-1, IL-4, IL-6, and cytokines such as tumor necrosis factor-α, T cells and interferon-γ receptor among others [28]. IVIG has several immunomodulatory effects on activation of the complement system, including binding and blockading Fc receptors, regulating the mechanism of immune responses by diminishing allograft rejection by dendritic cells, inhibiting T-cell activation by dendritic cells, B-cell apoptosis, and downregulating the B-cell receptor [2, 82, 93, 41, 73, 89, 67, 37, 88]. Common adverse effects include headache, fever, myalgia, arthralgia, chills, hypotension, and hypertension. Slowing the infusion rate and using iso-osmolar preparations can reduce these adverse effects. Serious adverse effects include aseptic meningitis, acute renal failure related to high osmotic load, anaphylaxis associated with IgA sensitization in recipients with IgA deficiency, and thrombotic events related to rapid infusion rates [41, 18, 63, 36].)

PP acts to remove allo-Abs from the circulation and is the fastest, most effective therapeutic option for reducing the levels of DSAs. It is used in both desensitization protocols and in AMR treatment [17, 65, 60, 53]. Some of the currently available variants of PP are plasma exchange, double filtration and IA PP. These treatments are usually applied in combination with either IVIG, rituximab, MMF or CNIs, or bortezomib. Adverse effects related to PP, which occur in 5–12 % of all patients, are considered to be mild or moderate. They include volume concentration, bleeding diathesis, allergic reactions, and blood-borne pathogen transmission [92, 77, 58].

Rituximab is a chimeric monoclonal Ab against anti-CD20, an Ag expressed in the early stages of the B-cell cycle, and is absent on mature PCs. CD20 regulates the early steps involved in initiation of the cell cycle and cell differentiation. Rituximab comprises a human IgG1 heavy chain and κ-light chain constant region fused with mouse variable regions. The mechanisms of rituximab-induced depletion are

Ab-dependent T-cell-mediated cytotoxicity, complement-mediated cell killing, and induction of apoptotic cell death via CD20 [90, 31, 84, 13, 1, 16]. Rituximab has been successfully used as induction therapy for sensitized patients; however, few protocols have used it to treat AMR [68, 87]. This agent produces a marked depletion of circulating B cells and a less marked reduction of B cells present in the spleen and lymph nodes [3, 64]. Rituximab has no effect on PCs, which are the source of DSAs; accordingly, its effect on the production of allo-Abs has no relationship with the depletion of Ab-producing cells. The therapeutic benefit of rituximab may be related to depletion of B memory cells and the modification of cellular immunity, rather than a reduction in the abundances of Abs [31, 12, 75, 66, 32]. A single dose in renal transplant patients can result in prolonged B-cell depletion, with populations that are suppressed for 1–2 years [29].

Bortezomib is a proteasome inhibitor that causes apoptosis of PCs, reducing allo-Ab production in sensitized patients [46]. Bortezomib binds to the 26S proteasome, which is part of an enzyme complex that is the primary proteolytic mechanism of the eukaryotic cell [10]. The inhibition effected by bortezomib induces apoptosis by activating the terminal unfolded protein response, which is related to the high rate of Ig synthesis [57, 55, 51]. Modification of signal transduction through nuclear factor-κB (NF-κB) and inhibitors of NF-κB are alternative mechanisms of action of this drug. Bortezomib can produce digestive-related adverse effects such as nausea, diarrhea and vomiting, neurological adverse effects such as peripheral neuropathy, blood-related adverse effects such as thrombocytopenia and neutropenia, and constitutional adverse effects such as fatigue, malaise, and weakness. Adverse effects caused by bortezomib can be treated by dose reduction and supportive care [35].

Eculizumab is a humanized monoclonal Ab that blocks the activation of the complement system via its high affinity to the C5 complement protein, preventing formation of the C5b-C9 MAC. A recent clinical trial described use of eculizumab for desensitization before transplantation and for AMR treatment after transplantation. On 1-year protocol biopsy, transplant glomerulopathy was found to be present in 6.7% (1/15) eculizumab-treated recipients and in 35.7% (15/42) of control patients ($p=0.044$). The authors concluded that inhibition of terminal complement activation with eculizumab decreased the incidence of early AMR in sensitized renal transplant recipients [79].

Antibody-Mediated Rejection in Reconstructive Transplantation

AMR has not been demonstrated so far in the recipients of vascularized composite allografts. A major difficulty associated with diagnosing AMR in vascularized composite allografts is the definition of graft dysfunction. Whereas graft dysfunction can be defined by physiological parameters and their impairment in solid organ allografts, it is unusual that, for instance, allografted hands exhibit diminished function. Recipients have reported that allografted hands are less pliable and flexible during rejection episodes. However, it is impossible to quantify this perception. No

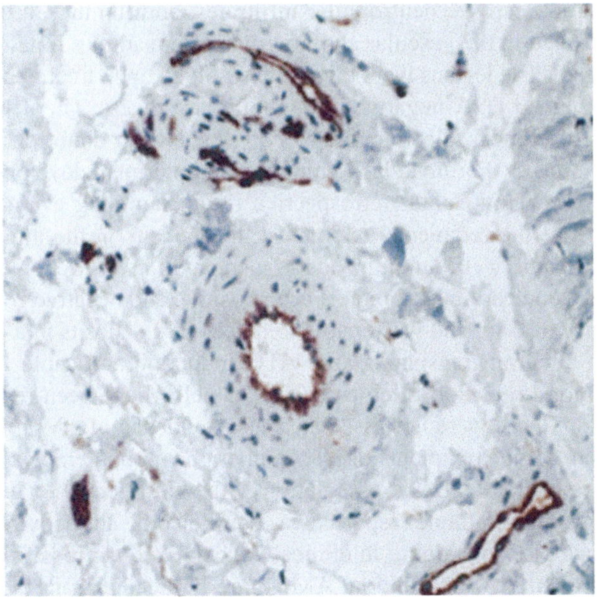

Fig. 12.5 Diffuse C4d staining was positive in the endothelium of an arteriole and its two venae comitantes. The capillaries in the surrounding tissues and subcutaneous fat were also positive for C4d (reprinted with permission, Landin L)

report has investigated the alleged changes in these patients before, during and after rejection. One potential source of data could be changes in electrodiagnostic tests.

The first evidence of C4d deposition in vascularized composite allotransplantation was seen in a nonhuman primate model of allogenic vascularized radial osteocutaneous flaps transplantation [14]. Seven cynomolgus monkeys were kept under triple standard immunosuppression using tacrolimus (Tac), MMF, and steroids. Biopsy samples showed C4d deposits in 100 % of the allografts, in addition to its deposition in native skin. Interestingly, 85 % developed anti-donor Abs. The authors concluded that the C4d marker had limited use in the diagnosis of rejection, given that it could also be found in normal skin. Unfortunately, this study did not investigate DSA levels after allograft removal.

Because the significance of C4d remained unknown, C4d staining was later recommended in the working classification scoring for rejection of skin in vascularized composite allografts for research purposes [15]. Several groups found positivity for C4d during the rejection of human hand allografts (HHA) in the absence of circulating DSAs, after administration of either ATG or alemtuzumab (Fig. 12.5) [43, 34, 69]. Specifically, C4d was found in the skin samples of three recipients of HHA that underwent rejection and also in allografts free from rejection. These patients had received alemtuzumab induction and were receiving a combination of Tac, MMF and steroids. Despite C4d deposition, the patients did not meet the aforementioned criteria for AMR. Additional reasons to avoid the treatment of alleged AMR were that arterial mural necrosis, neutrophilic infiltrates, and luminal thrombosis could not be demonstrated in the vascular and microvascular samples obtained during the secondary surgery. The lesions were clinically resolved upon treatment of CMR and the patients remained DSA-negative throughout the process. The detection of C4d in the absence of clinical rejection and in native skin cast doubt on its role in the acute rejection process. Fortunately, no functional impairment developed in the hand allografts [43]. In a multicenter retrospective tissue sampling evaluation of HHA recipients, 42% of the recipients exhibited C4d deposits in the allografted hands between the third and twelfth months post-transplantation. However, its clinical relevance remained unclear [34]. Kanitakis used positive and negative controls to evaluate the presence of C4d, and was unable to detect C4d deposits in a group of hand or face human allograft transplants after ATG induction and triple IS [38]. Other markers of humoral rejection, such as $CD20^+$ B cells, have been reported by several teams; again, these were not associated with evidence of AMR [33, 43]. More recently, four out of five recipients of HHA under Tac monotherapy after CD34 cell infusion showed positive DSA in association with skin rejection. The authors suggested that the cellular immune response was paralleled by antibody formation [70].

The relevance of C4d staining in clinical vascularized composite allotransplantation remains unknown. In addition, AMR may be occurring in the 'absence' of DSAs because Abs may be fixed to the allograft. Whether vascularized composite allografts already performed are accommodated also remains unknown. Currently, the development of intimal hyperplasia is under investigation as a form of chronic rejection of vascularized composite allografts [39]. In their report of six recipients of HHAs, one recipient lost his allograft through chronic rejection. This patient presented circulating DSAs only after HHA removal. Another patient treated by the same group presented $C4d^+$ DSA-vasculopathy and was treated with PP, IVIG, and MMF. Arterial thickening was relatively aborted. Whether this situation represents AMR leading to chronic rejection must be elucidated in future reports.

Conclusions and Future Directions

Only a few patients who have undergone reconstructive transplantation have shown circulating DSAs, whereas tissue samples from others have demonstrated the presence of C4d deposits. The presence of C4d has not been reliably associated with any induction or rejection prevention regimen. Several reports have documented the presence of CD20 in the skin of hand allografts. However, their significance remains unknown. Graft dysfunction is yet to be defined in reconstructive transplantation, while graft loss has been mostly the result of IS cessation or patient death from other complications. The diagnosis of AMR in vascularized composite allografts remains elusive and its treatment must be based on protocols developed for other kinds of solid organ transplant.

Acknowledgment There are no sources of funding that require acknowledgment. The authors have no financial or personal relationships with other people or organizations that could influence (bias) this work inappropriately. The authors have no financial interest in any of the drugs mentioned in this work.

References

1. Anolik JH, Campbell D, Felgar RE, Young F, Sanz I, Rosenblatt J, Looney RJ. The relationship of FcgammaRIIIa genotype to degree of B cell depletion by rituximab in the treatment of systemic lupus erythematosus. Arthritis Rheum. 2003;48(2):455–9. doi:10.1002/art.10764.
2. Bayry J, Lacroix-Desmazes S, Carbonneil C, Misra N, Donkova V, Pashov A, Chevailler A, Mouthon L, Weill B, Bruneval P, Kazatchkine MD, Kaveri SV. Inhibition of maturation and function of dendritic cells by intravenous immunoglobulin. Blood. 2003;101(2):758–65. doi:10.1182/blood-2002–05-1447.
3. Becker YT, Becker BN, Pirsch JD, Sollinger HW. Rituximab as treatment for refractory kidney transplant rejection. Am J Transplant: Off J Am Soc Transplant Am Soc Transplant Surg. 2004;4(6):996–1001. doi:10.1111/j.1600–6143.2004.00454.x.
4. Berger S, Daha M. Complement in glomerular injury. Semin Immunopathol. 2007;9(4):375–84. doi:10.1007/s00281-007-0090-3.
5. Blume OR, Yost SE, Kaplan B. Antibody-mediated rejection: pathogenesis, prevention, treatment, and outcomes. J Transplant. 2012;2012:201754. doi:10.1155/2012/201754.
6. Böhmig G, Regele H. Diagnosis and treatment of antibody-mediated kidney allograft rejection. Transpl Int. 2003;16(11):773–87. doi:10.1007/s00147-003-0658-3.
7. Bohmig GA, Exner M, Habicht A, Schillinger M, Lang U, Kletzmayr J, Saemann MD, Horl WH, Watschinger B, Regele H. Capillary C4d deposition in kidney allografts: a specific marker of alloantibody-dependent graft injury. J Am Soc Nephrol. 2002;13(4):1091–99.
8. Bonnefoy-Berard N, Revillard JP. Mechanisms of immunosuppression induced by antithymocyte globulins and OKT3. J Heart Lung Transplant: Off Publ Int Soc Heart Transplant. 1996;15(5):435–42.
9. Bonnefoy-Berard N, Vincent C, Revillard JP. Antibodies against functional leukocyte surface molecules in polyclonal antilymphocyte and antithymocyte globulins. Transplantation. 1991;51(3):669–73.
10. Brooks P, Fuertes G, Murray RZ, Bose S, Knecht E, Rechsteiner MC, Hendil KB, Tanaka K, Dyson J, Rivett J. Subcellular localization of proteasomes and their regulatory complexes in mammalian cells. Biochem J. 2000;346(Pt 1):155–61.
11. Burns JM, Cornell LD, Perry DK, Pollinger HS, Gloor JM, Kremers WK, Gandhi MJ, Dean PG, Stegall MD. Alloantibody levels and acute humoral rejection early after positive crossmatch kidney transplantation. Am J Transplant. 2008;8(12):2684–94. doi:10.1111/j.1600-6143.2008.02441.x.
12. Calame KL. Plasma cells: finding new light at the end of B cell development. Nat Immunol. 2001;2(12):1103–8. doi:10.1038/ni1201-1103.
13. Cartron G, Dacheux L, Salles G, Solal-Celigny P, Bardos P, Colombat P, Watier H. Therapeutic activity of humanized anti-CD20 monoclonal antibody and polymorphism in IgG Fc receptor FcgammaRIIIa gene. Blood. 2002;99(3):754–8.
14. Cendales LC, Xu H, Bacher J, Eckhaus MA, Kleiner DE, Kirk AD. Composite tissue allotransplantation: development of a preclinical model in nonhuman primates. Transplantation. 2005;80(10):1447–54.
15. Cendales LC, Kanitakis J, Schneeberger S, Burns C, Ruiz P, Landin L, Remmelink M, Hewitt CW, Landgren T, Lyons B, Drachenberg CB, Solez K, Kirk AD, Kleiner DE, Racusen L. The Banff 2007 working classification of skin-containing composite tissue allograft pathology. Am J Transplant: Off J Am Soc Transplant Am Soc Transplant Surg. 2008;8(7):1396–400. doi:10.1111/j.1600–6143.2008.02243.x.

16. Cerny T, Borisch B, Introna M, Johnson P, Rose AL. Mechanism of action of rituximab. Anti-Cancer Drugs. 2002;13(Suppl 2):S3–10.
17. Crespo M, Pascual M, Tolkoff-Rubin N, Mauiyyedi S, Collins AB, Fitzpatrick D, Farrell ML, Williams WW, Delmonico FL, Cosimi AB, Colvin RB, Saidman SL. Acute humoral rejection in renal allograft recipients: I. Incidence, serology and clinical characteristics. Transplantation. 2001;71(5):652–8.
18. Cunningham-Rundles C, Zhou Z, Mankarious S, Courter S. Long-term use of IgA-depleted intravenous immunoglobulin in immunodeficient subjects with anti-IgA antibodies. J Clin Immunol. 1993;13 (4):272–8.
19. Chopek MW, Simmons RL, Platt JL. ABO-incompatible kidney transplantation: initial immunopathologic evaluation. Transplant Proc. 1987;19(6):4553–7.
20. Dankof A, Schmeding M, Morawietz L, Günther R, Krukemeyer M, Rudolph B, Koch M, Krenn V, Neumann U. Portal capillary C4d deposits and increased infiltration by macrophages indicate humorally mediated mechanisms in acute cellular liver allograft rejection. Virchows Arch. 2005;447(1):87–93. doi:10.1007/s00428-005-1245-z.
21. Delves PJ, Martin SM, Burton DR, Roitt IM. Rott's Essential Immunology. 12th edn. Singapore: Blackwell Publishing Ltd.; 2011.
22. Demetris AJ, Minervini M, Nalesnik M, Ochoa E, Randhawa P, Sasatomi E, Wu T. Histopathology of liver transplantation. In: Ruiz P, editor. Transplantation pathology. New York: Cambridge University Press; 2009. pp. 111–84.
23. Feucht HE, Schneeberger H, Hillebrand G, Burkhardt K, Weiss M, Riethmuller G, Land W, Albert E. Capillary deposition of C4d complement fragment and early renal graft loss. Kidney Int. 1993;43(6):1333–8.
24. Feucht HE, Lederer SR, Kluth B. Humoral alloreactivity in recipients of renal allografts as a risk factor for the development of delayed graft function. Transplantation. 1998;65(5):757–8.
25. Fidler ME, Gloor JM, Lager DJ, Larson TS, Griffin MD, Textor SC, Schwab TR, Prieto M, Nyberg SL, Ishitani MB, Grande JP, Kay PA, Stegall MD. Histologic findings of antibody-mediated rejection in ABO blood-group-incompatible living-donor kidney transplantation. Am J Transplant: Off J Am Soc Transplant Am Soc Transplant Surg. 2004;4(1):101–7.
26. Fletcher J, Nankivell B, Alexander S. Chronic allograft nephropathy. Pediatr Nephrol. 2009;24(8):1465–71. doi:10.1007/s00467-008-0869-z.
27. Gebel HM, Bray RA, Nickerson P. Pre-transplant assessment of donor-reactive, HLA-specific antibodies in renal transplantation: contraindication vs. risk. Am J Transplant: Off J Am Soc Transplant Am Soc Transplant Surg. 2003;3(12):1488–500.
28. Gelfand EW. Antibody-directed therapy: past, present, and future. J Allergy Clin Immunol. 2001;108(4 Suppl):S111–6.
29. Genberg H, Hansson A, Wernerson A, Wennberg L, Tyden G. Pharmacodynamics of rituximab in kidney transplantation. Transplantation. 2007;84(12 Suppl):S33–6. doi:10.1097/01.tp.0000296122.19026.0f.
30. Glotz D, Haymann JP, Sansonetti N, Francois A, Menoyo-Calonge V, Bariety J, Druet P. Suppression of HLA-specific alloantibodies by high-dose intravenous immunoglobulins (IVIg). A potential tool for transplantation of immunized patients. Transplantation. 1993;56(2):335–7.
31. Gong Q, Ou Q, Ye S, Lee WP, Cornelius J, Diehl L, Lin WY, Hu Z, Lu Y, Chen Y, Wu Y, Meng YG, Gribling P, Lin Z, Nguyen K, Tran T, Zhang Y, Rosen H, Martin F, Chan AC. Importance of cellular microenvironment and circulatory dynamics in B cell immunotherapy. J Immunol (Baltim, Md: 1950). 2005;174(2):817–26.
32. Gonzalez-Stawinski GV, Yu PB, Love SD, Parker W, Davis RD Jr. Hapten-induced primary and memory humoral responses are inhibited by the infusion of anti-CD20 monoclonal antibody (IDEC-C2B8, Rituximab). Clin Immunol (Orlando, Fla). 2001;98(2):175–9. doi:10.1006/clim.2000.4980.
33. Hautz T, Engelhardt TO, Weissenbacher A, Kumnig M, Zelger B, Rieger M, Rumpold G, Pierer G, Ninkovic M, Gabl M, Piza-Katzer H, Pratschke J, Margreiter R, Brandacher G, Schneeberger S. World experience after more than a decade of clinical hand transplantation: update on the Innsbruck program. Hand Clin. 2011;27(4):423–31.

34. Hautz T, Zelger B, Brandacher G, Mueller H, Grahammer J, Zelger B, Andrew Lee WP, Cavadas P, Margreiter R, Pratschke J, Schneeberger S. Histopathologic characterization of mild rejection (grade I) in skin biopsies of human hand allografts. Transpl Int. 2012;25(1):56–63. doi:10.1111/j.1432-2277.2011.01369.x.
35. Jackson G, Einsele H, Moreau P, Miguel JS. Bortezomib, a novel proteasome inhibitor, in the treatment of hematologic malignancies. Cancer Treat Rev. 2005;31(8):591–602. doi:10.1016/j.ctrv.2005.10.001.
36. Jordan S, Cunningham-Rundles C, McEwan R. Utility of intravenous immune globulin in kidney transplantation: efficacy, safety, and cost implications. Am J Transplant: Off J Am Soc Transplant Am Soc Transplant Surg. 2003;3(6):653–64.
37. Jordan SC, Vo AA, Peng A, Toyoda M, Tyan D. Intravenous gammaglobulin (IVIG): a novel approach to improve transplant rates and outcomes in highly HLA-sensitized patients. Am J Transplant: Off J Am Soc Transplant Am Soc Transplant Surg. 2006;6(3):459–66. doi:10.1111/j.1600–6143.2005.01214.x.
38. Kanitakis J, McGregor B, Badet L, Petruzzo P, Morelon E, Devauchelle B, Dubernard JM. Absence of c4d deposition in human composite tissue (hands and face) allograft biopsies: an immunoperoxidase study. Transplantation. 2007;84(2):265–7. doi:10.1097/01.tp.0000266899.93315.52.
39. Kaufman CL, Ouseph R, Blair B, Kutz JE, Tsai TM, Scheker LR, Tien HY, Moreno R, Ozyurekoglu T, Banegas R, Murphy E, Burns CB, Zaring R, Cook DF, Marvin MR. Graft vasculopathy in clinical hand transplantation. Am J Transplant. 2012;12 (4):1004–16. doi:10.1111/j.1600–6143.2011.03915.x.
40. Kayler LK, Farber JL, Colombe B, LaCava D, Friedewald JJ, Ratner LE. Characterization of rejection episodes in patients following positive crossmatch and ABO-incompatible live donor renal transplantation. Transpl Int. 2006;19(2):128–39. doi:10.1111/j.1432–2277.2005.00249.x.
41. Kazatchkine MD, Kaveri SV. Immunomodulation of autoimmune and inflammatory diseases with intravenous immune globulin. N Engl J Med. 2001;345(10):747–55. doi:10.1056/NEJMra993360.
42. Lai CK, Yang H-M, Ra S, Fishbein MC. Pathology of cardiac transplantation. In: Ruiz P, editor. Transplantation pathology. New York: Cambridge University Press; 2009. pp. 185–216.
43. Landin L, Cavadas PC, Ibanez J, Roger I, Vera-Sempere F. CD3+-mediated rejection and C4d deposition in two composite tissue (bilateral hand) allograft recipients after induction with alemtuzumab. Transplantation. 2009;87(5):776–81. doi:10.1097/TP.0b013e318198dbc7, 00007890-200903150-00024 [pii].
44. Lederer SR, Kluth-Pepper B, Schneeberger H, Albert E, Land W, Feucht HE. Impact of humoral alloreactivity early after transplantation on the long-term survival of renal allografts. Kidney Int. 2001;59(1):334–41. doi:10.1046/j.1523-1755.2001.00495.x.
45. Levine MH, Abt PL. Treatment options and strategies for antibody mediated rejection after renal transplantation. Semin Immunol. 2012;24(2):136–42. doi:10.1016/j.smim.2011.08.015.
46. Locke JE, Magro CM, Singer AL, Segev DL, Haas M, Hillel AT, King KE, Kraus E, Lees LM, Melancon JK, Stewart ZA, Warren DS, Zachary AA, Montgomery RA. The use of antibody to complement protein C5 for salvage treatment of severe antibody-mediated rejection. Am J Transplant: Off J Am Soc Transplant Am Soc Transplant Surg. 2009;9(1):231–5. doi:10.1111/j.1600-6143.2008.02451.x.
47. Lones MA, Czer LS, Trento A, Harasty D, Miller JM, Fishbein MC. Clinical-pathologic features of humoral rejection in cardiac allografts: a study in 81 consecutive patients. J Heart Lung Transplant: Off Int Soc Heart Transplant. 1995; 14(1 Pt 1):151–62.
48. Lucas JG, Co JP, Nwaogwugwu UT, Dosani I, Sureshkumar KK. Antibody-mediated rejection in kidney transplantation: an update. Expert Opin Pharmacother. 2011;12(4):579–92. doi:10.1517/14656566.2011.525219.
49. Lynch RJ, Platt JL. Accommodation in organ transplantation. Curr Opin Organ Transplant. 2008;13(2):165–70. doi:110.1097/MOT.1090b1013e3282f6391e.
50. Mauiyyedi S, Crespo M, Collins AB, Schneeberger EE, Pascual MA, Saidman SL, Tolkoff-Rubin NE, Williams WW, Delmonico FL, Cosimi AB, Colvin RB. Acute humoral rejection

in kidney transplantation: II. Morphology, immunopathology, and pathologic classification. J Am Soc Nephrol. 2002;13(3):779–87.
51. Meister S, Schubert U, Neubert K, Herrmann K, Burger R, Gramatzki M, Hahn S, Schreiber S, Wilhelm S, Herrmann M, Jack HM, Voll RE. Extensive immunoglobulin production sensitizes myeloma cells for proteasome inhibition. Cancer Res. 2007;67(4):1783–92. doi:10.1158/0008-5472.can-06-2258.
52. Mohiuddin MM, Ogawa H, Yin D-P, Shen J, Galili U. Antibody-mediated accommodation of heart grafts expressing an incompatible carbohydrate antigen. Transplantation. 2003;75(3):258–62.
53. Montgomery RA, Zachary AA, Racusen LC, Leffell MS, King KE, Burdick J, Maley WR, Ratner LE. Plasmapheresis and intravenous immune globulin provides effective rescue therapy for refractory humoral rejection and allows kidneys to be successfully transplanted into cross-match-positive recipients. Transplantation. 2000;70(6):887–95.
54. Nast CC, Cohen AH. Pathology of kidney transplantation. In: Danovitch GM, editor. Handbook of kidney transplantation. 2nd edn. Philadelphia: LWW; 2010. pp. 311–30.
55. Neubert K, Meister S, Moser K, Weisel F, Maseda D, Amann K, Wiethe C, Winkler TH, Kalden JR, Manz RA, Voll RE. The proteasome inhibitor bortezomib depletes plasma cells and protects mice with lupus-like disease from nephritis. Nat Med. 2008;14(7):748–55. doi:10.1038/nm1763.
56. Nickeleit V. The pathology of kidney transplantation. In: Ruiz P, editor. Transplantation pathology. New York: Cambridge University Press; 2009. pp. 45–110.
57. Obeng EA, Carlson LM, Gutman DM, Harrington WJ Jr., Lee KP, Boise LH. Proteasome inhibitors induce a terminal unfolded protein response in multiple myeloma cells. Blood. 2006;107(12):4907–16. doi:10.1182/blood-2005-08-3531.
58. Okafor C, Ward DM, Mokrzycki MH, Weinstein R, Clark P, Balogun RA. Introduction and overview of therapeutic apheresis. J Apheresis. 2010;25(5):240–9. doi:10.1002/jca.20247.
59. Park WD, Grande JP, Ninova D, Nath KA, Platt JL, Gloor JM, Stegall MD. Accommodation in ABO-incompatible kidney allografts, a novel mechanism of self-protection against antibody-mediated injury. Am J Transplant: Off J Am Soc Transplant Am Soc Transplant Surg. 2003;3(8):952–60.
60. Pascual M, Saidman S, Tolkoff-Rubin N, Williams WW, Mauiyyedi S, Duan JM, Farrell ML, Colvin RB, Cosimi AB, Delmonico FL. Plasma exchange and tacrolimus-mycophenolate rescue for acute humoral rejection in kidney transplantation. Transplantation. 1998;66(11):1460–4.
61. Perry DK, Pollinger HS, Burns JM, Rea D, Ramos E, Platt JL, Gloor JM, Stegall MD. Two novel assays of alloantibody-secreting cells demonstrating resistance to desensitization with IVIG and rATG. Am J Transplant. 2008;8(1):133–43. doi:10.1111/j.1600-6143.2007.02039.x.
62. Petrossian GA, Nichols AB, Marboe CC, Sciacca R, Rose EA, Smith CR, Cannon PJ, Reemtsma K, Powers ER. Relation between survival and development of coronary artery disease and anti-HLA antibodies after cardiac transplantation. Circulation. 1989;80(5 Pt 2):III122–5.
63. Rault R, Piraino B, Johnston JR, Oral A. Pulmonary and renal toxicity of intravenous immunoglobulin. Clin Nephrol. 1991;36(2):83–6.
64. Reff ME, Carner K, Chambers KS, Chinn PC, Leonard JE, Raab R, Newman RA, Hanna N, Anderson DR. Depletion of B cells *in vivo* by a chimeric mouse human monoclonal antibody to CD20. Blood. 1994;83(2):435–45.
65. Rocha PN, Butterly DW, Greenberg A, Reddan DN, Tuttle-Newhall J, Collins BH, Kuo PC, Reinsmoen N, Fields T, Howell DN, Smith SR. Beneficial effect of plasmapheresis and intravenous immunoglobulin on renal allograft survival of patients with acute humoral rejection. Transplantation. 2003;75(9):1490–5. doi:10.1097/01.tp.0000060252.57111.ac.
66. Roll P, Palanichamy A, Kneitz C, Dorner T, Tony HP. Regeneration of B cell subsets after transient B cell depletion using anti-CD20 antibodies in rheumatoid arthritis. Arthritis Rheum. 2006;54(8):2377–86. doi:10.1002/art.22019.
67. Sapir T, Shoenfeld Y. Facing the enigma of immunomodulatory effects of intravenous immunoglobulin. Clin Rev Allergy Immunol. 2005;29(3):185–99. doi:10.1385/criai:29:3:185.

68. Sawada T, Fuchinoue S, Teraoka S. Successful A1-to-O ABO-incompatible kidney transplantation after a preconditioning regimen consisting of anti-CD20 monoclonal antibody infusions, splenectomy, and double-filtration plasmapheresis. Transplantation. 2002;74(9):1207–10. doi:10.1097/01.tp.0000040990.15896.76.
69. Schneeberger S, Ninkovic M, Gabl M, Hussl H, Rieger M, Loescher W, Zelger B, Brandacher G, Bonatti H, Hautz T, Boesmueller C, Piza-Katzer H, Margreiter R. First forearm transplantation: outcome at 3 years. Am J Transplant. 2007;7(7):1753–62. doi:10.1111/j.1600–6143.2007.01837.x, AJT1837 [pii].
70. Schneeberger S, Gorantla VS, Brandacher G, Zeevi A, Demetris AJ, Lunz JG, Metes DM, Donnenberg AD, Shores JT, Dimartini AF, Kiss JE, Imbriglia JE, Azari K, Goitz RJ, Manders EK, Nguyen VT, Cooney DS, Wachtman GS, Keith JD, Fletcher DR, Macedo C, Planinsic R, Losee JE, Shapiro R, Starzl TE, Lee WPA. Upper-extremity transplantation using a cell-based protocol to minimize immunosuppression. Ann Surg. 2013;257(2):345–51. doi:310.1097/SLA.1090b1013e31826d31890bb.
71. Schweitzer EJ, Wilson JS, Fernandez-Vina M, Fox M, Gutierrez M, Wiland A, Hunter J, Farney A, Philosophe B, Colonna J, Jarrell BE, Bartlett ST. A high panel-reactive antibody rescue protocol for cross-match-positive live donor kidney transplants. Transplantation. 2000;70(10):1531–6.
72. Seemayer CA, Gaspert A, Nickeleit V, Mihatsch MJ. C4d staining of renal allograft biopsies: a comparative analysis of different staining techniques. Nephrol, Dial, Transplant: Off Publ Eur Dial Transplant Assoc—Eur Ren Assoc. 2007;22(2):568–76. doi:10.1093/ndt/gfl594.
73. Semple JW, Kim M, Lazarus AH, Freedman J. Gamma-globulins prepared from sera of multiparous women bind anti-HLA antibodies and inhibit an established *in vivo* human alloimmune response. Blood. 2002;100(3):1055–9.
74. Shah A, Nadasdy T, Arend L, Brennan J, Leong N, Coppage M, Orloff M, Demme R, Zand MS. Treatment of C4d-positive acute humoral rejection with plasmapheresis and rabbit polyclonal antithymocyte globulin. Transplantation. 2004;77(9):1399–405.
75. Shapiro-Shelef M, Calame K. Plasma cell differentiation and multiple myeloma. Curr Opin Immunol. 2004;16(2):226–34. doi:10.1016/j.coi.2004.02.001.
76. Shapiro-Shelef M, Lin K-I, Savitsky D, Liao J, Calame K. Blimp-1 is required for maintenance of long-lived plasma cells in the bone marrow. J Exp Med. 2005;202(11):1471–6. doi:10.1084/jem.20051611.
77. Shepherd JC, Subramanian A, Montgomery RA, Samaniego MD, Gong G, Bergmann A, Blythe D, Dropulic L. West Nile virus encephalitis in a kidney transplant recipient. Am J Transplant: Off J Am Soc Transplant Am Soc Transplant Surg. 2004;4(5):830–3. doi:10.1111/j.1600–6143.2004.00410.x.
78. Singh N, Pirsch J, Samaniego M. Antibody-mediated rejection: treatment alternatives and outcomes. Transplant Rev (Orlando, Fla). 2009;23(1):34–46. doi:10.1016/j.trre.2008.08.004.
79. Stegall MD, Diwan T, Raghavaiah S, Cornell LD, Burns J, Dean PG, Cosio FG, Gandhi MJ, Kremers W, Gloor JM. Terminal complement inhibition decreases antibody-mediated rejection in sensitized renal transplant recipients. Am J Transplant: Off J Am Soc Transplant Am Soc Transplant Surg. 2011;11(11):2405–13. doi:10.1111/j.1600–6143.2011.03757.x.
80. Takemoto SK, Zeevi A, Feng S, Colvin RB, Jordan S, Kobashigawa J, Kupiec-Weglinski J, Matas A, Montgomery RA, Nickerson P, Platt JL, Rabb H, Thistlethwaite R, Tyan D, Delmonico FL. National conference to assess antibody-mediated rejection in solid organ transplantation. Am J Transplant. 2004;4(7):1033–41. doi:10.1111/j.1600–6143.2004.00500.x.
81. Tanemura M, Yin D, Chong AS, Galili U. Differential immune responses to α-gal epitopes on xenografts and allografts: implications for accommodation in xenotransplantation. J Clin Investig. 2000;105(3):301–10. doi:10.1172/jci7358.
82. Tha-In T, Metselaar HJ, Tilanus HW, Boor PP, Mancham S, Kuipers EJ, de Man RA, Kwekkeboom J. Superior immunomodulatory effects of intravenous immunoglobulins on human T-cells and dendritic cells: comparison to calcineurin inhibitors. Transplantation. 2006;81(12):1725–34. doi:10.1097/01.tp.0000226073.20185.b1.

83. Tyan DB, Li VA, Czer L, Trento A, Jordan SC. Intravenous immunoglobulin suppression of HLA alloantibody in highly sensitized transplant candidates and transplantation with a histo-incompatible organ. Transplantation. 1994;57(4):553–62.
84. Voso MT, Pantel G, Rutella S, Weis M, D'Alo F, Urbano R, Leone G, Haas R, Hohaus S. Rituximab reduces the number of peripheral blood B-cells *in vitro* mainly by effector cell-mediated mechanisms. Haematologica. 2002;87(9):918–25.
85. Wahrmann M, Exner M, Schillinger M, Haidbauer B, Regele H, Kormoczi GF, Horl WH, Bohmig GA. Pivotal role of complement-fixing HLA alloantibodies in presensitized kidney allograft recipients. Am J Transplant: Off J Am Soc Transplant Am Soc Transplant Surg. 2006;6 (5 Pt 1):1033–1041. doi:10.1111/j.1600–6143.2006.01285.x.
86. Wang H, Arp J, Liu W, Faas SJ, Jiang J, Gies DR, Ramcharran S, Garcia B, Zhong R, Rother RP. Inhibition of terminal complement components in presensitized transplant recipients prevents antibody-mediated rejection leading to long-term graft survival and accommodation. J Immunol. 2007;179(7):4451–63.
87. Warren DS, Zachary AA, Sonnenday CJ, King KE, Cooper M, Ratner LE, Shirey RS, Haas M, Leffell MS, Montgomery RA. Successful renal transplantation across simultaneous ABO incompatible and positive crossmatch barriers. Am J Transplant: Off J Am Soc Transplant Am Soc Transplant Surg. 2004;4(4):561–8. doi:10.1111/j.1600–6143.2004.00364.x.
88. Wassmuth R, Hauser IA, Schuler K, Erxleben H, Arnold ML, Koelman CA, Claas FH, Kalden JR. Differential inhibitory effects of intravenous immunoglobulin preparations on HLA-alloantibodies *in vitro*. Transplantation. 2001;71(10):1436–42.
89. Watanabe J, Scornik JC. IVIG and HLA antibodies. Evidence for inhibition of complement activation but not for anti-idiotypic activity. Am J Transplant: Off J Am Soc Transplant Am Soc Transplant Surg. 2005;5(11):2786–90. doi:10.1111/j.1600–6143.2005.01056.x.
90. Weng WK, Levy R. Two immunoglobulin G fragment C receptor polymorphisms independently predict response to rituximab in patients with follicular lymphoma. J Clin Oncol: Off J Am Soc Clin Oncol. 2003;21(21):3940–7. doi:10.1200/jco.2003.05.013.
91. Wilkinson A, Kasiske BL. Long-term post-transplantration managment and complications. In: Danovitch GM. editor. Handbook of kidney transplantation. Philadelphia: Lippincott WIlliams & Wilkins; 2010. pp. 217–50.
92. Yang KS, Kenpe K, Yamaji K, Tsuda H, Hashimoto H. Plasma adsorption in critical care. Ther Apheresis: Off J Int Soc Apheresis Jpn Soc Apheresis. 2002;6(3):184–8.
93. Yu Z, Lennon VA. Mechanism of intravenous immune globulin therapy in antibody-mediated autoimmune diseases. N Engl J Med. 1999;340(3):227–8. doi:10.1056/nejm199901213400311.
94. Zand MS, Vo T, Huggins J, Felgar R, Liesveld J, Pellegrin T, Bozorgzadeh A, Sanz I, Briggs BJ. Polyclonal rabbit antithymocyte globulin triggers B-cell and plasma cell apoptosis by multiple pathways. Transplantation. 2005;79(11):1507–15.
95. Zwirner J, Felber E, Herzog V, Riethmuller G, Feucht HE. Classical pathway of complement activation in normal and diseased human glomeruli. Kidney Int. 1989;36(6):1069–77.

Chapter 13
Chronic Rejection in Reconstructive Transplantation

Christina L. Kaufman, Rosemary Ouseph, Joseph E. Kutz, Yorell Manon-Matos, Huey Y. Tien, Brenda Blair and Michael R. Marvin

Introduction

Chronic rejection is a major cause of late graft loss in solid organ transplantation [24, 63]. To date, sequelae that look like chronic rejection have not been a major clinical problem for the majority of hand and face transplant recipients. Whether vascularized composite allotransplantation (VCA) may be less sensitive to chronic rejection remains to be seen. One reason most programs have not reported evidence of chronic rejection may be the lack of long-term follow-up. Of approximately 90 documented cases, only 10% of VCA recipients are out more than 10 years and 14% are out 5 or more years (www.handregistry.com, [40]. With only a quarter of the clinical cases five or more years posttransplant, long-term sequelae may be more evident as time progresses. Graft losses in VCA patients have been fairly acute, most within the first year, and all within 3 years. Clinical experience in VCA transplantation is accumulating. The time has come for this new field to deal with a specter that haunts every other type of allograft, namely chronic rejection.

The field of VCA has benefited greatly from the experience of solid organ transplantation. Losses in this new field are much lower than the graft loss encountered in the early stages of other types of allografts. After the initial kidney homograft in 1954, reports 10 years later showed mortality (not graft loss, *mortality*) rates of 35% at 1 year [52]. Especially in the past two decades, 1 year survival rates have increased in all types of solid organ transplantations, and are over 90% for renal

C. L. Kaufman (✉) · J. E. Kutz · Y. Manon-Matos · H. Tien · B. Blair
Christine M. Kleinert Institute for Hand and Microsurgery, Louisville, KY, USA
e-mail: ckaufman@cmki.org

R. Ouseph · M. R. Marvin
Department of Medicine, University of Louisville School of Medicine, Louisville, KY, USA

M. R. Marvin
Department of Surgery, Jewish Hospital Transplant Center, University of Louisville, Louisville, KY, USA

G. Brandacher (ed.), *The Science of Reconstructive Transplantation*,
Stem Cell Biology and Regenerative Medicine, DOI 10.1007/978-1-4939-2071-6_13

transplantation [30]. One year survival rates were 87, 84, 83, and 76% for heart, liver, lung, and intestine allografts respectively for 2008 recipients [34]. However, the long-term outcomes for solid organs (annualized attrition rates) have not significantly changed over the past two decades [34]. Even considering trends such as transplanting more high-risk patients, the analysis by Lodhi et al. suggests most of the recent improvements in half lives of solid organ transplants occur during the first year [34]. Chronic rejection, far less understood, and far less studied, is probably the factor most responsible for the lack of improvement in long-term outcomes.

What is Chronic Rejection?

Gather ten transplant specialists in a room, ask each to describe chronic rejection, and prepare for diverse answers. An underlying theme seems to be that as a group we are not exactly sure. A PubMed search using the term "chronic rejection" revealed 12,768 manuscripts on the subject. Two thousand four hundred of these manuscripts are review papers, with changing focus and terminology over the years. Historically, chronic rejection first revealed itself in the kidney as loss of nephrons, followed by interstitial fibrosis, and atrophy of the tubules. The common perception at the time was that this pathology was initiated by alloantibodies. It was subsequently determined that chronic rejection could also occur in the absence of antibodies [17]. A recent review by Heemann and Lutz (Heemann and Lutz 2013) noted that chronic graft loss occurs in transplants between identical twins even now. Additionally the first transplants between twins in the 1950s by Murray et al were eventually lost over the long term and had biopsies that would have been read as interstitial fibrosis today [55]. This data in twins suggests there is more to chronic rejection than an alloimmune response. Additionally, as we briefly describe below, each type of solid organ transplant seems to have a different target or cause of long-term chronic graft loss, or chronic rejection. We also present evidence for what the targets of chronic rejection may be in VCA. This evidence suggests that VCA grafts may bear the dubious distinction of having more than one primary target of chronic rejection, or sequelae that resemble chronic rejection.

Potential Causes of Chronic Rejection

The early definitions of allograft rejection were divided into three categories: hyperacute, acute, and chronic, based on the timing of the rejection episode. Hyperacute rejection occurs when preformed antibodies are present, and in renal transplantation resulted in a black kidney within minutes or hours of reestablishing blood flow [29, 54]. Advances in tissue typing and flow crossmatching have made hyperacute rejection as a result of occult preformed donor-specific antibodies something most clinicians will never see. Acute rejection was defined as occurring in the early posttransplant period (first 3–6 months) with the highest risk in the first year.

Chronic rejection was thought to occur later and was characterized as a "slow burn" phenomenon. This chronic indolent rejection is likely due to a number of different factors. Antibodies can and are involved in both types of rejections. The presence of donor-specific antibodies, especially of a de novo nature is clearly correlated with reduced graft survival [19], perhaps directly related to chronic rejection. Recently, B cells directed at donor antigens have been implicated in acute rejection [8] and have been shown to increase quantitatively shortly after renal transplantation, even in the absence of circulating donor-specific antibodies [35]. T-cell-mediated immunity is implicated more in acute rejection, but lesions in chronically rejected grafts may also have a cellular infiltrate containing T cells. Other nonimmune factors include ischemia-reperfusion injury, drug toxicity, and even infections. It has been clearly established that infection can initiate an acute rejection episode in lung [64], kidney [48], and liver [1] allografts. It is probable that any of these non-allogeneic insults will also set up an environment that leads to fibrosis, hypertrophy, and parenchymal damage in solid organ transplants. Unfortunately, based on this evidence there is no clear reason to believe that long-term VCA grafts would be spared the same fate.

Differences in Chronic Rejection Between Solid Organ Allograft Types

While acute rejection manifests as a cellular infiltrate of the parenchymal tissue in all solid organ allografts, chronic rejection seems to target different areas in different organs. A diagnosis of chronic rejection is based on biopsy and functional impairment in renal and liver transplants [10, 51]. The definitive diagnosis of chronic rejection is again generally made by biopsy of the organ in question. The heart is an exception to this generalization: Chronic rejection in heart grafts manifests as accelerated graft atherosclerosis [21]. The vasculature of the new heart undergoes a progressive but focal disease, resulting in intimal thickening and occlusion of the grafted coronary vessels [21]. Kidneys with chronic rejection have fibrosis (scarring) and damage to the microscopic blood vessels in the substance of the kidney [51]. Livers with chronic rejection have a decreased number of bile ducts on biopsy. This is referred to as the "vanishing bile duct syndrome" [22]. Transplanted lungs with chronic rejection are said to have "bronchiolitis obliterans syndrome" (BOS)—a fibro-proliferative process that gradually reduces the lumen of the bronchioles, and can result in complete occlusion [56]. Like cardiac vasculopathy, bronchiolitis obliterans is focal, making diagnosis by biopsy difficult [56]. Interestingly, this syndrome is not restricted to lung transplantation and does occur in hematopoietic stem cell transplantation as well [20].

Despite relatively different target structures, all solid organ transplants share a pattern of heterogeneous factors associated with the development of chronic rejection. Development of anti-donor human leukocyte antigen (HLA) antibodies has been associated with increased risk of chronic rejection for all organs. Numerous episodes of acute cellular rejection have been associated with an increased risk of chronic rejection in cardiac [62] and renal [41] clinical transplantation. Recently, the development of autoantibodies has been implicated in the development of chronic rejection in all types of solid organ transplantations [61]. In addition to generating alloantibodies, the presence of an allograft can induce an antibody response against antigens shared by the donor and the recipient. Sumpter and Wilkes first developed the hypothesis that rejection may be made of two phases: the first is injury to the graft, followed by development of an autoantibody [53]. They suggested that the resulting autoreactive T cell and antibody response could sustain the rejection process in the absence of alloimmunity [53]. In a recent review of BOS, Todd and Palmer cite studies showing that type 5 collagen (col(V)), and other epithelial cell surfaces such as K-a1 tubulin may be important autoantibody targets in lung transplant recipients [56]. Wilkes et al. have demonstrated that patients with severe BOS have a five- to tenfold increase in cell-mediated immunity to col(V) [5]. This theory of development of autoantibody may help to explain why different organs have different targets of chronic rejection.

In addition to comparing the targets of different kinds of rejection, multiorgan transplantation offers insight into how chronic rejection may manifest in VCA. Co-transplantation of heart and lungs, kidneys and pancreas, livers and kidneys, and multi-visceral organ transplantation are commonly performed. A protective effect of multiorgan transplantation has been noted, especially when the liver is involved [46]. However, like many issues in transplantation, exceptions abound [16, 47]. Nonetheless, a detailed examination of United Network for Organ Sharing (UNOS) data on over 133,000 transplant recipients from 1994 to 2005 revealed that multiorgan transplants had a consistently higher rate of rejection-free survival compared to singly transplanted organ recipients [46]. Interestingly, these authors also compared single and double lung transplants and single versus en bloc (double kidney) transplants. In both cases, the transplantation of larger amounts of tissue was associated with a significant decrease in the percentage of patients with acute rejection at 1 year [46]. Similar results were demonstrated for heart/lung multiorgan transplants [45], although the effect did not hold in a study of pediatric heart/lung transplantation [28].

Are VCA Transplants Protected from Chronic Rejection?

If multiple organs or larger amounts of grafted tissue are associated with protection from acute and perhaps chronic rejection, would VCA grafts be protected? The answer seems to be no, at least with respect to acute rejection. At 1 year posttransplant, at least 85 % of hand transplant recipients have had a rejection episode [38, 43]. At our own center, 100 % of patients have had multiple episodes of at least grade 1 histological rejection in the first year posttransplant. An argument can be made that rejection is much easier to detect and diagnose in hand and face transplantation, and much early rejection in solid organ transplantation goes undiagnosed. However, it is clear that VCA grafts are at minimum, certainly not protected from acute allograft rejection, at least of the skin.

The early experience in VCA suggests that there might be a protective effect of VCA for chronic rejection. As of 2011, there had been no clinical reports of chronic

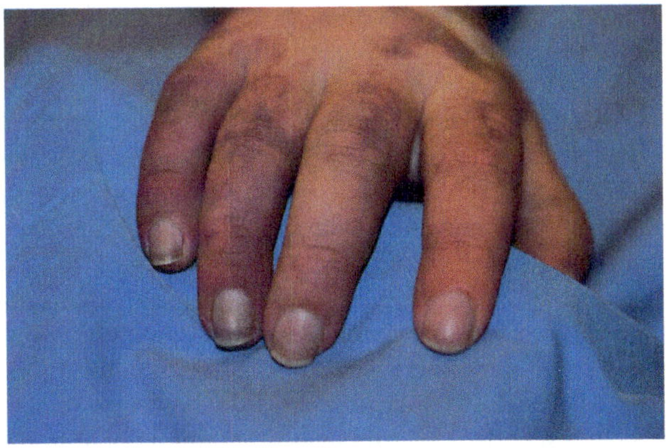

Fig 13.1 Appearance of hand allograft prior to required amputation in patient 4

rejection in a compliant hand or face transplant recipient. Criteria for diagnosing chronic rejection in hand transplantation were proposed [6]. These criteria suggested histologic and clinical features indicative of chronic injury in a VCA include vascular narrowing, loss of adnexa, skin and muscle atrophy, fibrosis of deep tissue, myointimal proliferation, and nail changes [6]. Initially, a report of chronic rejection in a hand transplant was described in the first French hand transplant recipient, who independently stopped his immunosuppression and requested that the graft be removed [32]. In hindsight, these changes were attributed to acute rejection and noncompliance, rather than chronic rejection [44]. Our group and others expected chronic rejection to manifest as fibrosis in the skin. To date there have been surprisingly few reports of atrophy and fibrosis of the skin and adnexal structures in VCA patients. It had been predicted that chronic rejection would manifest as severe dermal fibrosis or dyskeratosis [4, 23], and that hand transplant recipients might be considered as a model of scleroderma [33]. Target organs were thought to be the skin and adnexal units as has been seen in chronic graft versus host disease in recipients of bone marrow [14, 13].

It is human nature to see what you look for. Although other solid organs such as the heart demonstrated that the vasculature can be a primary target of chronic rejection, our monitoring protocols were focused on the skin. Conventional vascular monitoring of our hand transplant patients including digital brachial indices, computed tomography (CT), and magnetic resonance (MR) angiography were performed with no indication of significant issues. Nevertheless, in April of 2009, at just 9 months after his transplant, our fourth hand transplant recipient lost his graft to an aggressive confluent graft vasculopathy [27]. In his case, by the time the patient was brought back to Louisville for evaluation, the radial and ulnar arteries were almost completely occluded. Diefenbeck et al. previously reported vasculopathy in an allogeneic vascularized knee transplant [11], but this was the first finding of aggressive confluent graft vasculopathy in a hand transplant recipient. While histology of vessels taken from the amputated graft revealed almost complete obstruction in many vessels, there was at least partial blood flow up to the point of amputation (Fig. 13.1). Nonetheless, the ischemia in the graft as a result of the intimal hyperplasia necessitated amputation of the graft. This event triggered a careful evaluation of all patients previously transplanted at our center. Deep tissue biopsies and histological evaluation of excised arteries revealed measureable and in some cases significant intimal hyperplasia in all four of our other patients, despite no obvious clinical signs of graft vasculopathy, or evidence of vessel thickening on MR or CT angiography. We subsequently obtained a Vevo 2100 ultrasound biomicroscopy (UBM) unit. This device allows noninvasive imaging of the intima and media of vessels at a high resolution (up to 30 uM) in vessels near the surface of the skin. We obtained Institutional Review Board's (IRB) approval to use this research device on our patients and now routinely image the brachial, radial, ulnar, palmar arch, and digital arteries as part of the clinical trial protocol. As reported recently [27], this monitoring did allow us to image an aggressive confluent vasculopathy in patient 6 at about 6 months after transplant, which responded to intravenous immunoglobulin (IVIG), plasmapheresis, and switching from a Prograf/mycophenolate mofetil (MMF)/prednisone regimen to a Prograf/Rapamycin/Prednisone treatment protocol. The patient is now more than 2 years posttransplant.

A Rose Is Not a Rose

Unlike the experience in solid organ transplantation, it appears that chronic rejection may manifest in several different ways in VCA recipients. We have discussed the aggressive, confluent, concentric graft vasculopathy that was seen in patients 4 and 6 at our center. All arteries in the graft were involved from the ulnar and radial arteries to the digital arteries, and there was thickening of venous walls as well. In contrast, we have also seen a focal, sometimes acentric vasculopathy that is found in some but not all arteries in the graft, and progresses very slowly. We have mapped specific lesions in some of our patients that have not changed significantly in 3 years of follow-up. In some cases these thickened areas are in the digital arteries, and in some cases are in the ulnar arteries. While these focal lesions do not appear to change significantly over time, there is a clear distinction in most of our unilateral recipients between the transplanted hand and the native hand. The vessel walls have sharper images and are easier to view than the transplanted hand which has slightly thicker vessel walls in general compared to the native hand. This is not true for everyone. In patient 7, who has a unilateral transplant, but who has a native hand that endured significant damage in the original accident, it is easier to image by UBM in his transplanted hand than his native hand. In the eight hand transplant recipients transplanted at our center, two recipients have had multiple significant acute rejection episodes in the skin component of the graft, both in the first-year posttransplant and many years posttransplant. Most would predict that these two patients might have the greatest level of graft vasculopathy, and might suffer from the more confluent variety. That has not been our experience. Our second patient, who is now more than 12 years posttransplant has had eight episodes of biopsy proven skin rejection of grade 2 or higher. Our third patient has had at least 12 episodes of grade 2 or

Table 13.1 Lack of association of skin allograft rejection and vasculopathy

Patient	Numerous Acute Skin Rejection Episodes (Grade 2 or more)	Vasculopathy	Follow up
1	No (3)	Minimal	14 years
2	**Yes (8)**	**Minimal**	**12 years**
3	**Yes (12)**	**Minimal**	**6 years**
4	**No (2)**	**Obliterative**	**9 months**
5	No (3)	Minimal	4 years
6	**No (2)**	**Mod-Severe**	**2 years**
7	No (2)	Minimal	2 years
8	No (2)	Minimal	1 year

higher rejection. Both of these patients have shown minimal changes in their vessels over the past 3 years, with very good blood flow, excellent digital temperatures, and no obvious changes in vessel wall thickness by UBM imaging.

In contrast, patients 4 and 6 had relatively quiet courses with respect to skin rejection (Table 13.1). In these patients, the vasculopathy quickly escalated to the point the graft was lost to ischemia in patient 4, and without intervention, patient 6 may have lost his graft as well. The other patients to date have areas of vascular thickening, but it has not progressed noticeably, and in none of these cases does the vasculopathy seem to be affecting either blood flow or graft function in the slightest. Both forms are a type of vasculopathy, but they do not appear to be similar in any respect other than both are thickening of the vessels restricted to the graft. A similar variance in presentation is seen in other transplant vasculopathies. In cardiac transplantation, 50 % of all patients develop graft vascular disease (GVD) within 12 months, one third of which is rapidly progressive [36, 57]. The lack of correlation between skin rejection and vasculopathy is supported by recent observations in a nonhuman primate model of face transplantation [39]. In this model, animals that lost their graft upon cessation of immunosuppression also developed a near occlusive intimal thickening, and the grafts were edematous and pale, supporting a restriction of blood flow. Superficial punch skin biopsies in these animals also failed to show evidence of immunologic rejection despite a restriction of the vasculopathy to the graft vessels. Interestingly, biopsy of deeper tissues revealed an active immune response, with development of tertiary lymphoid follicles [39]. In our clinical experience, both patients with aggressive vasculopathy had surface skin biopsies that were negative for cellular infiltrates. The histology of deep tissue in the amputated graft did reveal significant infiltrates in the deeper tissues. Whether that was due to a rejection process, or severe ischemia, or both is unknown. Like our clinical experience, in the primate face transplant model antibodies did not appear to play a role in the vasculopathy. There was no immunoglobulin (Ig)G or IgM alloantibody production that correlated with intimal hyperplasia, and staining for C4d deposition was negative in both the thickened vessels and the tertiary lymphoid follicles. In our clinical trial, we saw nonspecific staining of C4d in both skin punch and deep tissue biopsies. Of note, with the exception of patient 2 who developed donor-specific antibodies (C1q negative) at year 6, none of our patients have evidence of donor-directed HLA-specific alloantibodies (DSAs). In patient 4, DSAs were negative up to and at the time of amputation, but DSAs were detected 2 days after amputation and 4 days after immunosuppression was stopped. Similar conversions to a positive DSA after graft removal have been shown in renal transplants [9]. Landin et al. have also reported that development of donor-specific antibodies does not necessarily correlate with staining of C4d on skin biopsies [31]. Kanitakis et al. examined C4d expression in four VCA recipients and found no evidence of C4d staining, despite development of DSA antibodies [25]. Dr. Anthony J. Demetris at the University of Pittsburgh has suggested that staining of vessels within adipose tissue might reduce the nonspecific staining and allow us to detect an active humoral response to the donor. Mundinger et al. reported that in their nonhuman primate model, Notch pathway receptors 1, 3, and 4 and Notch pathway ligand Jagged-1 were upregulated specifically in the areas of large vessel intimal hyperplasia compared to unaffected control vessels, suggesting this may be an important pathway in the development of graft vasculopathy [39].

Nonvascular Targets of Chronic Rejection in VCA Recipients

The graft loss from ischemia and evidence for at least some vascular thickening in comparison to the native hand in 100 % of our recipients is strong evidence that the vasculature may be a target of chronic rejection in hand transplant recipients. We and others hypothesized that the skin would be a primary target of chronic rejection, and data from experimental models and our own patients support this. Unadkat et al. have shown in a model of rat hind limb transplantation with multiple episodes of acute rejection that both the vasculature and other tissue show signs of chronic rejection (skin atrophy and fibrosis, as well as muscle atrophy and infiltration [58]). In this model, the experimental group received cyclosporine A in an irregular manner to simulate noncompliance, and rejection episodes were repeatedly treated, and animals were allowed to reject again. The investigators then followed the animals to determine the effect of multiple rejection episodes on the development of chronic rejection as defined by vasculopathy and skin and tissue changes. Animals with multiple episodes of acute rejection showed patchy hair loss with dermal atrophy and apoptotic bodies in the sebaceous glands and hair follicles demonstrating adnexal structure involvement. The dermis was thinner, however, fibrosis at the epidermo-

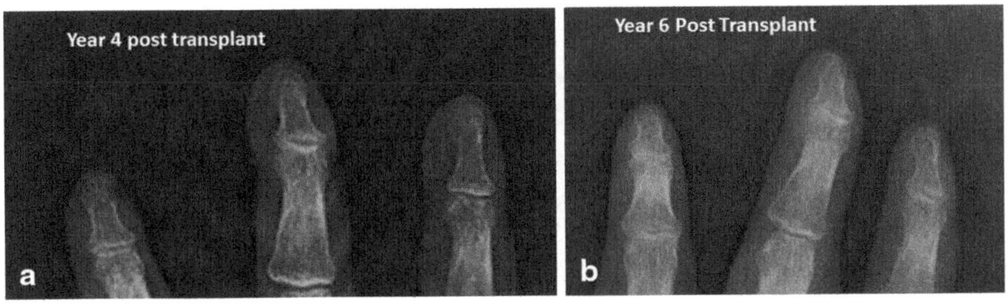

Fig 13.2 (**a**) Radiograph of fingertips of patient 3 at year 4, (**b**) and at year 6 posttransplant. Note the reduction of soft tissue at the tips of the fingers at year 6

dermal juncture resulted in significantly thicker skin in the multiple Acute Rejection (AR) animals [58]. Changes were also seen in muscle and there was an increase in osseous malunion in the multiple AR group. This association of multiple AR episodes with increased incidence of chronic rejection makes immunologic sense. Correlation of acute rejection and increased risk of chronic rejection is well documented in solid organ transplantation [41, 59, 60]. As previously stated, reports of chronic rejection directed at skin or tissue other than the vasculature have been rare in clinical VCA recipients. Careful assessment of five hand and face recipients from 10 to 2 years posttransplant failed to produce any evidence of changes in the skin or vessels, or decreases in function or sensorimotor recovery [44]. The authors suggest that strict adherence to triple-drug immunosuppression without attempts to wean or minimize may have contributed to freedom from evidence of chronic rejection [44]. A trend toward higher immunosuppression may contribute, however, other groups who have implemented immunosuppression minimization have not reported issues with chronic rejection in compliant patients to date either [3]. In fact, no center has reported loss of a VCA graft due to skin rejection in a compliant patient.

Recently, our center reported what appears to be evidence of chronic changes in the skin and adnexal structures of our third hand transplant patient who is now 6 years posttransplant [26]. Our third recipient received his unilateral graft in 2006. At year 4, the patient presented with thinning of the digits and partial loss of fingernails on the transplanted hand. At presentation, overexposure to topical steroids was suspected. Fungal scrapings were negative and skin biopsies were negative for fibrosis. At year 5, he presented with complete loss of the nails and thinning of the skin as well as a noticeable rash on the skin. Radiographs of the hand at years 4 (Fig. 13.2a) and 6 (Fig. 13.2b) were unremarkable for loss of bone in the digits, but loss of soft tissue especially at the tips was noted in year 6 (Fig. 13.2). Electromyography (EMG) conductivity is reduced at year 6. Figure 13.3 shows the changes in his skin and nails between year 2 and year 6. Significant changes in the nails were noted in year 4 (Fig. 13.3b), and the nails disappeared by year 5 (Fig. 13.3c). This subject is notable for a preexisting and stable marginal zone lymphoma (MZL) diagnosed 18 months posttransplant. The patient is also notable for multiple episodes of skin rejection over the previous 6 years, with biopsies of histologic grade 2 or higher. Note that these rejections are in the face of good compliance with

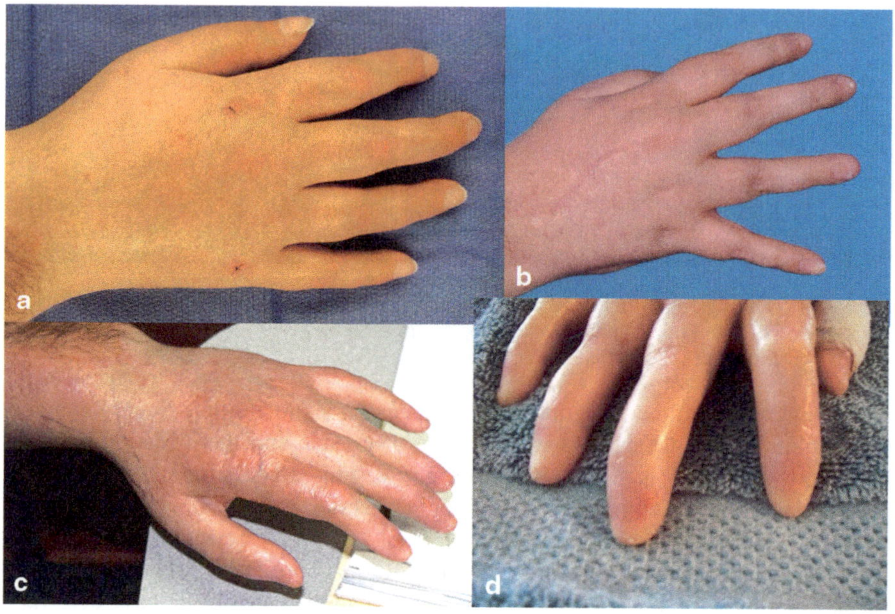

Fig 13.3 Changes in appearance, skin thinning, and loss of adnexal structures over time in patient 3. **a** Year 2 posttransplant; **b** year 4; **c** year 5; **d** year 6. Note the complete loss of fingernails by year 5 and 6 posttransplant

immunosuppressive medications. This patient is also remarkable in that while skin atrophy is present, skin biopsies to date have been negative for significant fibrosis, in contrast to the histology found in the nonhuman primate model of face transplantation [39]. Scleroderma patients have been proposed as a good model to study for chronic rejection in VCA [50]. To date, our center has not seen a VCA recipient that resembles a scleroderma patient. Antibodies do not seem to be a major initiating factor. Fibrosis and/or collagen deposition has not been a factor in either acute or chronic rejection in our patients or those of our colleagues. The field of VCA is young, and there may be a subpopulation of VCA patients with chronic rejection that resembles scleroderma. But to date, this has not been our experience.

Potential Causes of Chronic Rejection in VCA Transplantation

It is clear that nonimmune components as well as allogeneic responses contribute to vasculopathy and chronic rejection in solid organ transplantation [2, 12, 37]. The nonimmune aspects which are unique to VCA include mechanical and traumatic stress to the graft and possibly surgical techniques such as the long donor brachial artery harvest. However, mechanical stress alone and/or long artery harvest cannot be sufficient for development of vasculopathy. Patients transplanted previously and subsequently at our center who also routinely stress their grafts mechanically in manual labor and who have received a graft with a long brachial artery did not develop aggressive vasculopathy [52]. We believe that the aggressive vasculopathy we have seen in our patients is due to a "perfect storm" of both alloimmune and nonimmune factors. Our data and the data from other centers and from experimental models suggest that there may be multiple types or targets of chronic rejection in VCA grafts. We hypothesize that the four major types of chronic rejection in VCA recipients are: (1) aggressive confluent graft vasculopathy, (2) focal slowly progressing graft vasculopathy, (3) chronic rejection of the skin, with atrophy of the skin and underlying muscle, loss of adnexal structures such as fingernails, hair follicles, and sebeaucous glands, and (4) a classic chronic rejection of the skin with dermal atrophy, loss of adnexal structures, and thickening of the skin due to fibrosis. As clinical and experimental evidence accumulates, the first type of graft vasculopathy may turn out to be more of an acute type of rejection rather than a chronic rejection. We were encouraged by the fact that this type of vasculopathy was receptive to treatment. The more indolent type of vasculopathy may not be amenable to treatment. However, we would predict that few grafts would be lost to this much more focal and slowly progressing form.

There are multiple factors that can induce intimal hyperplasia and vasculopathy. Trauma alone can induce intimal hyperplasia [7, 36]. Any mechanical, cytotoxic, immunologic, and thermal injury that might result in endothelial damage can in turn initiate and propagate intimal hyperplasia [15, 36], as well as remodeling [37]. Recently, Christensen et al. reported that repeated rubbing of cage wire induced episodes of rejection in a swine model of hind limb transplantation [7]. In a rat model of sustained thrombocytopenia after injury, restoration of platelets even 2 weeks after injury can trigger smooth muscle cell proliferation [49]. Factors such as cytomegalovirus (CMV) have also been shown to induce vasculopathy [42]. In the face of these stimuli, and when allogeneic immune responses are also occurring, a situation that is conducive to the development of chronic rejection is created. Patient 4 who lost his graft received less immunosuppression than the first two patients who have done well clinically and with respect to vasculopathy. This patient also prided himself on the aggressive use of his allograft. It is not possible to know for sure, but one can hypothesize both of these scenarios contributed to the aggressive vasculopathy and loss of the graft. This hypothesis was the impetus for starting patient 6 on standard triple-drug immunosuppression. Nonetheless, patient 6 also developed severe vasculopathy despite conventional levels of immunosuppression (induction, followed by triple-drug therapy with maintenance steroids). This patient had wound coverage issues and infectious complications requiring multiple surgical debridements. This level of mechanical trauma could also have contributed to his aggressive vasculopathy.

Diagnosis/Monitoring of Chronic Rejection in VCA

Most VCA programs have excellent monitoring protocols which involve testing blood levels of immunosuppressive drugs, frequent interactions with nurse coordinators to discuss clinical changes and problems, protocol skin biopsies, and annual monitoring of vascular changes by digital brachial indices, MR or CT angiography, and CT scans. Sensorimotor function and EMG changes are also monitored. Based on our experience, we highly recommend that programs implement vascular monitoring using high-resolution ultrasound imaging. Ideally, centers should have access to an ultrasound unit with probes of at least 20–40 mHz. If smaller vessels such as the digital arteries will be monitored, probes of 50–70 mHz are recommended. Note that the higher resolution probes are not effective for vessels more than a centimeter or so under the skin. Our established recipients who are doing well are monitored on an annual basis, and recent transplant recipients are monitored monthly or as clinical course indicates.

Treatment of Chronic Rejection

In the case of the second patient with aggressive vasculopathy, treatment with IVIG, plasmapheresis, and conversion from MMF to Rapamycin was associated with cessation of progression of the vasculopathy, and this patient still has his bilateral grafts at 34 months posttransplant, over 2 years after the event that occurred 6 months after the transplant. Unfortunately in the case of patient 3, who has what appears to be more conventional chronic rejection with skin and muscle atrophy and loss of adnexal structures, we have only maintained his immunosuppression levels as high as possible, balancing graft projection with protection of his kidneys, blood sugar, and lipid levels. We continue to follow and report the progression of chronic rejection-like sequelae in this recipient.

Research Directions

Our center and others are focusing their research efforts in animal models to determine what immune and nonimmune factors affect the development of chronic rejection. We have the same obstacles found in solid organ transplantation with the additional challenges of mechanical, traumatic, and toxic stress which can initiate or exacerbate rejection in VCA recipients. Evidence suggests these nonimmune stressors can induce acute rejection, and associated evidence that they may play a role in chronic rejection as well. It will be important to determine which patients are more susceptible to these stressors, and also if the sensitivity of our VCA recipients to these challenges changes over time. Ideally, patients would become less sensitive to mechanical or traumatic stressors as time posttransplant increases.

Perhaps the most promising research areas with respect to managing or preventing chronic rejection are immunomodulatory and/or tolerance-inducing strategies which should control the alloimmune portions of the initiating events and thus significantly reduce the rate of chronic rejection as well as the amount of immunosuppressive agents needed to control both acute and chronic rejection. The recent findings of developments of autoimmune antibodies in transplant recipients and how these antibodies may participate in accommodation of the VCA graft will also be a key area of research.

Summary

While the presence of donor bone microenvironment could minimize or even mitigate sequelae of chronic rejection in VCA recipients, early evidence from our center on hand transplants at the mid-forearm or more distal suggests that protection will be incomplete. All of our transplants are at the mid-forearm or more distal. The outcome may be different in recipients of larger amounts or more hematopoietically active bone. It appears that more than one structure will be targeted by long-term chronic rejection-like sequelae. Unlike solid organ transplantation, at least some VCA recipients will have the option of removing the graft, should chronic rejection prove to be as debilitating as it has been for solid organ transplantation. Like solid organ transplantation, there is a clear need in VCA for tolerance-inducing protocols that should prevent the onset of both acute and chronic rejection.

Acknowledgments The authors would like to acknowledge the Department of Defense, Office of Army Research, and the Office of Naval Research as well as the Jewish Hospital Foundation, the Kleinert Kutz Hand Care Center, and the Christine M. Kleinert Institute for Hand and Microsurgery for support of these studies.

Disclosure: The authors do not have any conflicts to disclose regarding this manuscript.

References

1. Abdel Massih RC, Razonable RR. Human herpesvirus 6 infections after liver transplantation. World J Gastroenterol. 2009;15(21):2561–9.
2. Azuma H, Tullius SG, Heemann UW, Tilney NL. Nonimmune factors may contribute to chronic rejection of kidney transplants. Transplant Proc. 1994;26(4):2109–10. (Available from: PM:8066686)
3. Brandacher G, Lee WP, Schneeberger S. Minimizing immunosuppression in hand transplantation. Expert Rev Clin Immunol. 2012;8(7):673–83.
4. Brenner MJ, Tung TH, Jensen JN, Mackinnon SE. The spectrum of complications of immunosuppression: is the time right for hand transplantation? J Bone Joint Surg Am. 2002;84-A(10):1861–70. (Available from: PM:12377920)
5. Burlingham WJ, Love RB, Jankowska-Gan E, Haynes LD, Xu Q, Bobadilla JL, Meyer KC, Hayney MS, Braun RK, Greenspan DS, Gopalakrishnan B, Cai J, Brand DD, Yoshida S, Cummings OW, Wilkes DS. IL-17-dependent cellular immunity to collagen type V predisposes to obliterative bronchiolitis in human lung transplants. J Clin Invest. 2007;117(11):3498–506.
6. Cendales LC, Kanitakis J, Schneeberger S, Burns C, Ruiz P, Landin L, Remmelink M, Hewitt CW, Landgren T, Lyons B, Drachenberg CB, Solez K, Kirk AD, Kleiner DE, Racusen L. The Banff 2007 working classification of skin-containing composite tissue allograft pathology. Am J Transplant. 2008;8(7):1396–400.

7. Christensen J, Ibrahim Z, Wu L, Leto Barone A, Furtmuller M, Sarhane K, Al-Rakan M, Schneeberger S, Shores JT, Bonawitz S, Cooney D, Sacks J, Lee WP, Brandacher G. Mechanical trauman can reactivate an alloimmune response against the skin component in a large animal vascularized composite allotransplantation (VCA) model of operational tolerance. Am Transpl Congr 2013;13[S5]:203.
8. Clatworthy MR. B Cell responses to allograft—more common than we thought? Am J Transplant. 2013;13(7):1629–30.
9. Clatworthy MR, Espeli M, Torpey N, Smith KG. The generation and maintenance of serum alloantibody. Curr Opin Immunol. 2010;22(5):669–81. (Available from: PM:20932734)
10. Demetris AJ, Adeyi O, Bellamy CO, Clouston A, Charlotte F, Czaja A, Daskal I, El-Monayeri MS, Fontes P, Fung J, Gridelli B, Guido M, Haga H, Hart J, Honsova E, Hubscher S, Itoh T, Jhala N, Jungmann P, Khettry U, Lassman C, Ligato S, Lunz JG, III, Marcos A, Minervini MI, Molne J, Nalesnik M, Nasser I, Neil D, Ochoa E, Pappo O, Randhawa P, Reinholt FP, Ruiz P, Sebagh M, Spada M, Sonzogni A, Tsamandas AC, Wernerson A, Wu T, Yilmaz F. Liver biopsy interpretation for causes of late liver allograft dysfunction. Hepatology 2006;44(2):489–501.
11. Diefenbeck M, Nerlich A, Schneeberger S, Wagner F, Hofmann GO. Allograft vasculopathy after allogeneic vascularized knee transplantation. Transpl Int. 2011;24(1):e1–5. (Available from: PM:20958796)
12. Edelman ER, Danenberg HD. Rapamycin for cardiac transplant rejection and vasculopathy: one stone, two birds? Circulation 2003;108(1):6–8. (Available from: PM:12847052)
13. Ferrara JL, Deeg HJ. Graft-versus-host disease. N Engl J Med. 1991;324(10):667–74. (Available from: PM:1994250)
14. Ferrara JL, Levine JE, Reddy P, Holler E. Graft-versus-host disease. Lancet 2009;373(9674):1550–61.
15. Fingerle J, Johnson R, Clowes AW, Majesky MW, Reidy MA. Role of platelets in smooth muscle cell proliferation and migration after vascular injury in rat carotid artery. Proc Natl Acad Sci U S A. 1989;86(21):8412–8416. (Available from: PM:2813399)
16. Giovanelli M, Gupte GL, Sharif K, Mayer DA, Mirza DF. Chronic rejection after combined liver and small bowel transplantation in a child with chronic intestinal pseudo-obstruction: a case report. Transpl Proc. 2008;40(5):1763–67.
17. Hancock WH, Whitley WD, Tullius SG, Heemann UW, Wasowska B, Baldwin WM III, Tilney NL. Cytokines, adhesion molecules, and the pathogenesis of chronic rejection of rat renal allografts. Transplantation. 1993;56(3):643–50.
18. Heemann U, Lutz J. Pathophysiology and treatment options of chronic renal allograft damage. Nephrol Dial Transpl. 2013;28(10):2438–46.
19. Hidalgo LG, Campbell PM, Sis B, Einecke G, Mengel M, Chang J, Sellares J, Reeve J, Halloran PF. De novo donor-specific antibody at the time of kidney transplant biopsy associates with microvascular pathology and late graft failure. Am J Transplant. 2009;9(11):2532–41.
20. Holbro A, Lehmann T, Girsberger S, Stern M, Gambazzi F, Lardinois D, Heim D, Passweg JR, Tichelli A, Bubendorf L, Savic S, Hostettler K, Grendelmeier P, Halter JP, Tamm M. Lung histology predicts outcome of bronchiolitis obliterans syndrome after hematopoietic stem cell transplantation. Biol Blood Marrow Transplant. 2013;19(6):973–80.
21. Hornick P, Rose M. Chronic rejection in the heart. Methods Mol Biol. 2006;333:131–44.
22. Hubscher SG, Buckels JA, Elias E, McMaster P, Neuberger J. Vanishing bile-duct syndrome following liver transplantation–is it reversible? Transplantation. 1991;51(5):1004–10.
23. Jensen JN, Mackinnon SE. Composite tissue allotransplantation: a comprehensive review of the literature–Part II. J Reconstr Microsurg. 2000;16(2):141–57.
24. Joosten SA, van KC, Paul LC. Pathogenesis of chronic allograft rejection. Transpl Int. 2003;16(3):137–45.
25. Kanitakis J, McGregor B, Badet L, Petruzzo P, Morelon E, Devauchelle B, Dubernard JM. Absence of c4d deposition in human composite tissue (hands and face) allograft biopsies: an immunoperoxidase study. Transplantation 2007;84(2):265–7. (Available from: PM:17667820)

26. Kaufman CL, Marvin MR, Ouseph R, Manon-Matos Y, Blair B, Kutz JE. Evidence for chronic rejection of the skin and adnexal structures in a hand transplant recipient at six years post transplant without significant graft vasculopathy. Am J Transplant. 2013;13[S5]:286.
27. Kaufman CL, Ouseph R, Blair BW, Kutz JE, Tsai TM, Scheker LR, Tien HY, Moreno R, Ozyrekoglu T, Banegas R, Murphy EE, Burns C, Zavan B, Cook DF, Marvin MR. Graft vasculopathy in clinical hand transplantation. Am J Transplant. 2012;12(4):1004–16.
28. Khan MS, Heinle JS. Pediatric thoracic multiorgan transplantation. Tex Heart Inst J. 2012;39(6):850–1.
29. Kissmeyer-Nielsen F, Olsen S, Petersen VP, Fjeldborg O. Hyperacute rejection of kidney allografts, associated with pre-existing humoral antibodies against donor cells. Lancet. 1966;2(7465):662–5.
30. Lamb KE, Lodhi S, Meier-Kriesche HU. Long-term renal allograft survival in the United States: a critical reappraisal. Am J Transplant. 2011;11(3):450–62.
31. Landin L, Cavadas PC, Ibanez J, Roger I, Vera-Sempere F. CD3 + -mediated rejection and C4d deposition in two composite tissue (bilateral hand) allograft recipients after induction with alemtuzumab. Transplantation 2009;87(5):776–81. (Available from: PM:19295326)
32. Lanzetta M, Petruzzo P, Margreiter R, Dubernard JM, Schuind F, Breidenbach W, Lucchina S, Schneeberger S, van HC, Granger D, Pei G, Zhao J, Zhang X. The International Registry on Hand and Composite Tissue Transplantation. Transplantation 2005;79(9):1210–4. (Available from: PM:15880072)
33. Lineaweaver WC. Chronic rejection, hand transplantation, and the monkey's paw. Microsurgery. 2006;26(6):419–20.
34. Lodhi SA, Lamb KE, Meier-Kriesche HU. Solid organ allograft survival improvement in the United States: the long-term does not mirror the dramatic short-term success. Am J Transplant. 2011;11(6):1226–35.
35. Lynch RJ, Silva I, Chen BJ, Punch JD, Cascalho M, Platt JL. Cryptic B cell response to renal transplantation. Am J Transplant. 2013;13(7):1713–23.
36. Mitchell RN. Graft vascular disease: immune response meets the vessel wall. Annu Rev Pathol. 2009;4:19–47. (Available from: PM:18717641)
37. Mitchell RN, Libby P. Vascular remodeling in transplant vasculopathy. Circ Res. 2007;100(7):967–78. (Available from: PM:17431198)
38. Morelon E, Kanitakis J, Petruzzo P. Immunological issues in clinical composite tissue allotransplantation: where do we stand today? Transplantation. 2012;93(9):855–9.
39. Mundinger GS, Munivenkatappa R, Drachenberg CB, Ha JS, Vaca EE, Shipley ST, Papadimitriou JC, Bartlett ST, Rodriguez ED, Barth RN. Histopathology of chronic rejection in a nonhuman primate model of vascularized composite allotransplantation. Transplantation 2013;95(10):1204–10.
40. Murphy BD, Zuker RM, Borschel GH. Vascularized composite allotransplantation: an update on medical and surgical progress and remaining challenges. J Plast Reconstr Aesthet Surg. 2013;(13):10.
41. Nankivell BJ, Kuypers DR. Diagnosis and prevention of chronic kidney allograft loss. Lancet 2011;378(9800):1428–37.
42. Orloff SL, Hwee YK, Kreklywich C, Andoh TF, Hart E, Smith PA, Messaoudi I, Streblow DN. Cytomegalovirus latency promotes cardiac lymphoid neogenesis and accelerated allograft rejection in CMV naive recipients. Am J Transplant. 2011;11(1):45–55. (Available from: PM:21199347)
43. Petruzzo P, Lanzetta M, Dubernard JM, Landin L, Cavadas P, Margreiter R, Schneeberger S, Breidenbach W, Kaufman C, Jablecki J, Schuind F, Dumontier C. The international registry on hand and composite tissue transplantation. Transplantation. 2010;90(12):1590–4. (Available from: PM:21052038)
44. Petruzzo P, Kanitakis J, Badet L, Pialat JB, Boutroy S, Charpulat R, Mouly J, Gazarian A, Lanzetta M, Brunet M, Devauchelle B, Testelin S, Martin X, Dubernard JM, Morelon E. Long-term follow-up in composite tissue allotransplantation: in-depth study of five (hand and face) recipients. Am J Transplant. 2011;11(4):808–16. (Available from: PM:21446980)

45. Pinderski LJ, Kirklin JK, McGiffin D, Brown R, Naftel DC, Young KR Jr., Smith K, Bourge RC, Tallaj JA, Rayburn BK, Benza R, Zorn G, Leon K, Wille K, Deierhoi M, George JF. Multi-organ transplantation: is there a protective effect against acute and chronic rejection? J Heart Lung Transpl. 2005;24(11):1828–33.
46. Rana A, Robles S, Russo MJ, Halazun KJ, Woodland DC, Witkowski P, Ratner LE, Hardy MA. The combined organ effect: protection against rejection? Ann Surg. 2008;248(5):871–9.
47. Reichman TW, Marino SR, Milner J, Harland RC, Cochrane A, Millis JM, Testa G. Acute humoral rejection in an ABO compatible combined liver-kidney transplant–the kidney is not always protected. Am J Transplant. 2009;9(8):1957–60.
48. Reischig T. Cytomegalovirus-associated renal allograft rejection: new challenges for antiviral preventive strategies. Expert Rev Anti Infect Ther. 2010;8(8):903–10.
49. Sirois MG, Simons M, Kuter DJ, Rosenberg RD, Edelman ER. Rat arterial wall retains myointimal hyperplastic potential long after arterial injury. Circulation 1997;96(4):1291–8. (Available from: PM:9286961)
50. Sivakumar B, Haloob N, Puri A, Latif A, Ghani S, Brough V, Molloy J, Clarke A, Denton CP, Butler PE. Systemic sclerosis as a model of chronic rejection in facial composite tissue transplantation. J Plast Reconstr Aesthet Surg. 2010;63(10):1669–76.
51. Solez K, Colvin RB, Racusen LC, Haas M, Sis B, Mengel M, Halloran PF, Baldwin W, Banfi G, Collins AB, Cosio F, David DS, Drachenberg C, Einecke G, Fogo AB, Gibson IW, Glotz D, Iskandar SS, Kraus E, Lerut E, Mannon RB, Mihatsch M, Nankivell BJ, Nickeleit V, Papadimitriou JC, Randhawa P, Regele H, Renaudin K, Roberts I, Seron D, Smith RN, Valente M. Banff 07 classification of renal allograft pathology: updates and future directions. Am J Transplant. 2008;8(4):753–60.
52. Starzl TE, Marchioro TL, Rifkind D, Rowlands DT Jr, Waddell WR. Clinical experience with organ transplantation. South Med J. 1965;58:131–47.
53. Sumpter TL, Wilkes DS. Role of autoimmunity in organ allograft rejection: a focus on immunity to type V collagen in the pathogenesis of lung transplant rejection. Am J Physiol Lung Cell Mol Physiol. 2004;286(6):L1129–39.
54. Terasaki PI, Marchioro TL, Starzl TE. Sero-typing of human lymphocyte antigens: preliminary trials on long term kidney homograft survivors. First Histocompatibility Workshop, 83. 1964.
55. Tilney NL. Renal transplantation between identical twins: a review. World J Surg. 1986;10(3):381–8.
56. Todd JL, Palmer SM. Bronchiolitis obliterans syndrome: the final frontier for lung transplantation. Chest. 2011;140(2):502–8.
57. Tuzcu EM, Kapadia SR, Sachar R, Ziada KM, Crowe TD, Feng J, Magyar WA, Hobbs RE, Starling RC, Young JB, McCarthy P, Nissen SE. Intravascular ultrasound evidence of angiographically silent progression in coronary atherosclerosis predicts long-term morbidity and mortality after cardiac transplantation. J Am Coll Cardiol. 2005;45(9):1538–42. (Available from: PM:15862431)
58. Unadkat JV, Schneeberger S, Horibe EH, Goldbach C, Solari MG, Washington KM, Gorantla VS, Cooper GM, Thomson AW, Lee WP. Composite tissue vasculopathy and degeneration following multiple episodes of acute rejection in reconstructive transplantation. Am J Transplant. 2010;10(2):251–61. (Available from: PM:20041866)
59. Weigt SS, Wallace WD, Derhovanessian A, Saggar R, Saggar R, Lynch JP, Belperio JA. Chronic allograft rejection: epidemiology, diagnosis, pathogenesis, and treatment. Semin Respir Crit Care Med. 2010;31(2):189–207.
60. Weiss MJ, Madsen JC, Rosengard BR, Allan JS. Mechanisms of chronic rejection in cardiothoracic transplantation. Front Biosci. 2008;13:2980–8. (Available from: PM:17981771)
61. Win TS, Pettigrew GJ. Humoral autoimmunity and transplant vasculopathy: when allo is not enough. Transplantation. 2010;90(2):113–20.
62. Yamani MH, Yousufuddin M, Starling RC, Tuzcu M, Ratliff NB, Cook DJ, Abdo A, Crowe T, Hobbs R, Rincon G, Bott-Silverman C, McCarthy PM, Young JB. Does acute cellular rejection correlate with cardiac allograft vasculopathy? J Heart Lung Transplant. 2004;23(3):272–6.

63. Yates PJ, Nicholson ML. The aetiology and pathogenesis of chronic allograft nephropathy. Transpl Immunol. 2006;16(3–4):148–57.
64. Zeevi A, Morel P, Spichty K, Dauber J, Yousem S, Williams P, Grgurich W, Pham S, Iacono A, Keenan R, Duquesnoy R, Griffith B. Clinical significance of CMV-specific T helper responses in lung transplant recipients. Hum Immunol. 1998;59(12):768–75.

Chapter 14
Cell-Based Immunomodulatory Concepts and Tolerance Protocols for Reconstructive Transplantation

Angelo A. Leto Barone and Victor W. Wong

Introduction

The investigation of cell-based protocols to induce clinical tolerance in VCAs has permeated the past decade of research in the field of reconstructive transplantation. The major limitations of immunosuppressive drugs currently in clinical use include increased propensity to develop infections, organ toxicity (i.e., nephrotoxicity and neurotoxicity) and risk of carcinogenesis.

The establishment of tolerance, defined as the lack of a destructive immune response against donor tissues in the absence of immunosuppression, would free vascularized composite allotransplantation (VCA) recipients from the burden of long-term immunosuppressive medication and potentially expand the indication of such procedures to post-oncological reconstruction or surgical correction of congenital malformations in pediatric recipients. Currently, the main strategy to induce tolerance in basic science, translational, and clinical studies is via central or peripheral immunoregulatory mechanisms. Central tolerance entails the deletion of leukocytes directed against the donor (*donor-reactive leukocytes*) and is mainly achieved by establishment of cellular chimerism. In contrast, peripheral tolerance is most commonly achieved via the induction of T cell anergy or expansion of regulatory T cells (Tregs) [105].

The experience in solid-organ transplantation has provided key insight into the mechanisms underlying donor-specific nonresponsiveness following VCA. However, in contrast to solid organs, tolerance induction in VCA is limited by the presence of skin, the most immunogenic of tissues [56], making tolerance induction to this component particularly challenging. Furthermore, limited data are available on the existence or prevention of chronic rejection in VCA [43, 72] and whether VCA recipients rendered tolerant by cellular approaches are protected from chronic

A. A. Leto Barone (✉) · V. W. Wong
Department of Plastic and Reconstructive Surgery,
The Johns Hopkins University School of Medicine, Baltimore, MD, USA
e-mail: aletobarone@jhmi.edu

G. Brandacher (ed.), *The Science of Reconstructive Transplantation*,
Stem Cell Biology and Regenerative Medicine, DOI 10.1007/978-1-4939-2071-6_14

injury responsible for late graft loss. Immunomodulation and tolerance induction in VCA have been attempted using both cell-based therapies and biologic agents. In this chapter, we focus on cell-based therapies and tolerance protocols that combine the two approaches.

Cell-Based Immunomodulation

Actively acquired transplantation tolerance was shown to be possible for the first time in 1953 by Billingham, Brent, and Medawar, who reported indefinite acceptance of donor hematopoietic cells in mice following intravenous injection within 24 h of birth [8]. The mechanism that was deemed responsible for the development of tolerance was the coexistence of host and donor-derived cells, a phenomenon termed *chimerism.*

The pioneering studies investigating tolerance induction through establishment of mixed chimerism were conducted in mice, and entailed the use of myeloablative regimens including lethal irradiation of the recipients and bone marrow reconstitution using a combination of host and donor T-cell-depleted bone marrow [36]. Donor-specific unresponsiveness *in vitro*, and acceptance of donor skin grafts in these animals proved that mixed chimerism was capable of inducing tolerance in mice without incurring graft-versus-host disease (GvHD). However, the myoablative irradiation used in this protocol produced severe side effects due to the ablation of the recipient immune system and therefore would not be acceptable for clinical use [77].

The need for alternatives of such toxic irradiation/conditioning regimens prompted the investigation and development of T-cell-depleting agents and immunoregulatory drugs. Such regimens would severely decrease T cell number or promote their tolerance toward the donor antigens encountered, modulating T cells both in the thymus and the peripheral blood [77, 96]. T cell depletion in the host along with irradiation allowed donor-specific tolerance of organs and bone marrow engraftment across a major histocompatibility complex (MHC) mismatch. Addition of thymic irradiation (7 Gy) was capable of reducing the high-dose total-body irradiation (TBI) from 6 to 3 Gy, while enabling the elimination of mature alloreactive thymocytes [70, 85]. The addition of monoclonal antibodies against T-cell-specific targets has been shown to be comparable to thymic irradiation for induction of mixed chimerism and tolerance [96, 97]. However, both thymic irradiation and monoclonal antibodies may be replaced by the use of drugs that provide co-stimulatory blockade, such as cytotoxic T-lymphocyte antigen 4-immunoglobulin (CTLA4-Ig), which interferes with the second signal pathway of T cell activation [102] (Fig. 14.1).

In the attempt to translate these protocols to the clinic, multiple large animal studies have been conducted in the past decade. In fact, protocols that are feasible in mice and small animals often do not yield the same results in large animal models and humans due to several biological differences that affect both metabolic responses to drugs and the intrinsic–innate and adaptive immunologic responses [75, 77].

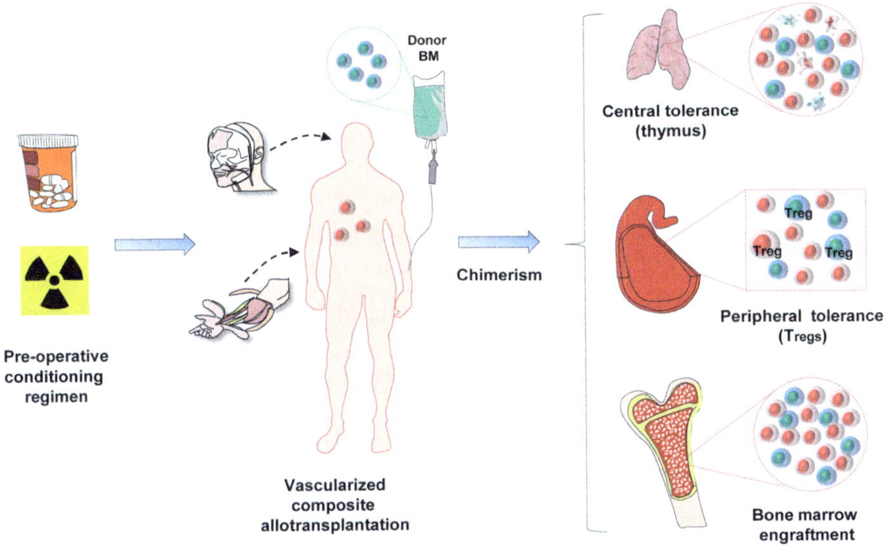

Fig. 14.1 Schematic representation of possible mechanisms of tolerance in reconstructive transplantation. *BM* bone marrow, *Tregs* T regulatory cell

The first clinical evidence that cellular strategies may promote tolerance induction was prompted by the observation that patients undergoing myeloablative therapy combined with both solid-organ transplantation and allogeneic bone marrow transplantation (BMT) became tolerant of their kidney allograft [18]. The repopulation of host myeloid lineages with donor hematopoietic cells induced tolerance toward the donor kidney graft, highlighting the potential role of cell-based approaches for establishing tolerance in VCA. Since then, numerous studies have investigated the use of different cell types as possible immunomodulators. These cell populations include hematopoietic stem cells (HSCs), Tregs, dendritic cells (DCs), and mesenchymal stem cells (MSCs) among others.

Mixed Chimerism

Chimerism is defined as the coexistence in the host of donor and recipient hematopoietic cells. The use of myeloablative conditioning regimens leads to the establishment of a condition termed *full chimerism,* in which the complete replacement of the recipient's bone marrow with donor cells occurs. Severe morbidity is associated with the establishment of full chimerism, largely due to the toxicity of myeloablative regimens that expose the patient at risk for severe infections and GvHD. For this reason, this procedure was historically reserved to patients receiving organs from human leukocyte antigen (HLA)-matched siblings. More recently,

Sykes et al. have shown that mixed chimerism can be induced in adult recipients of HLA-mismatched BMT by a nonmyeloablative conditioning regimen, allowing for wider application of such protocols in patients receiving an organ from unrelated donors [94].

The preliminary studies to achieve these goals were possible due to the development of the Massachusetts General Hospital (MGH) miniature swine as a model for translational research in transplantation [76]. These animals were bred to fix the MHC antigens (while maintaining variability for minor antigens). This allowed selection of a variety of MHC mismatches including but not limited to:

1. Syngeneic (complete match of both alleles for MHC class I and II in both donor and recipient swine leukocyte antigens (SLA), i.e. $SLA^{aa} \rightarrow SLA^{aa}$)
2. Single-haplotype mismatch (sharing of only one allele of class I and II, i.e., $SLA^{ac} \rightarrow SLA^{ad}$)
3. Full mismatch (no sharing of any haplotype of class I and II between donor and recipient, i.e., $SLA^{cc} \rightarrow SLA^{dd}$) clinical scenarios

For this reason, the MGH miniature swine model allowed the investigation of different conditioning regimens for tolerance induction of organs and testing of such protocols, while simulating different genetic disparity, such as those present in HLA-identical siblings (full MHC match, minor antigens mismatch only), parents to offsprings, and single haplotype-mismatched siblings (single haplotype MHC mismatch) or nonmatched living-related or cadaveric-donor transplants (full MHC mismatch) [73, 76].

Most of the preliminary studies aimed at the treatment of hematologic malignancies by means of mixed chimerism were conducted in the MGH miniature swine and utilized leukapheresis to mobilize large quantities of HSCs from peripheral blood [14, 34]. Such protocols, however, are not readily translatable to the scenario of clinical reconstructive transplantation, as drug-induced mobilization of HSCs and leukapheresis is not feasible in the deceased donor. Clinically, translational protocols for VCA, indeed, require the use of induction therapies that can be administered in the immediate pre/perioperative time period.

Currently, mixed chimerism has been the only approach to successfully achieve tolerance in solid-organ transplantation in nonhuman primates (NHPs), despite numerous protocols capable of inducing tolerance in other animal models including rodents and swine [16, 93]. Based on promising experimental results, clinical trials have been developed to study combined BMT and kidney transplants first in MHC-matched patients affected by multiple myeloma-induced renal failure and, subsequently, in kidney transplants in MHC-mismatch recipients with no malignancy [47, 90]. Importantly, recipients of these cell-based tolerance protocols achieved long-term tolerance to their allografts, with the longest immunosuppression-free survival exceeding 10 years. Furthermore, no evidence of chronic rejection has been observed and a lower-than-expected incidence of GvHD has been reported following these protocols. (We will defer an in-depth discussion of mixed chimerism-based approaches to Chap. 16.)

T Regulatory Cells

Tregs, previously known as T suppressor cells, play a crucial role in maintaining tolerance to self-antigens and balancing immune responses that otherwise would be harmful to the host. In fact, Treg dysfunction has been implicated in a variety of autoimmune and inflammatory diseases in humans. Tregs can be classified as either thymus-derived naturally occurring ($CD4^{+}$/CD25hi/$FOXP3^{+}$) or peripherally induced (FOXP3) and upregulated in response to antigen exposure [81]. FOXP3 is a master transcription factor that regulates Treg activation and IL-2, IL-10, and transforming growth factor β (TGF-β) pathways appear essential in controlling Treg interactions with other cells.

In transplantation, Tregs are thought to be highly important in maintaining immune suppression and preventing acute and chronic graft rejection. Experimental transplantation models that target Treg function have been highly promising and often utilize either *in vivo* or *ex vivo* techniques to activate this cell population. *In vivo* strategies involve antigen exposure to induce expansion of Treg populations, resulting in prevention of allograft rejection in multiple animal models (Issa and Wood 2012). Co-stimulatory blockade—e.g., targeting cluster of differentiation (CD)28, CTLA4, or CD40L)—or lymphocyte depletion protocols (monoclonal antibodies, cytotoxic agents, radiation) have also shown success in activating Tregs and promoting graft tolerance [71, 100].

Ex vivo strategies that employ Tregs have utilized blood obtained from the peripheral circulation or umbilical cord. However, isolation of these cells is predicated on being able to efficiently identify characteristic surface markers. Given the nonspecific nature of CD25 and FOXP3 expression, other markers have been described to isolate Tregs, including CD44, CD45RA, CD49b, CD69, CD127, CD152, and latency-associated peptide (LAP). For example, Tregs with low expression of CD127 have been demonstrated to suppress vessel allograft and skin graft rejection in humanized mice models [39]. Additionally, Tregs expressing LAP may have superior *in vivo* immunosuppressive capabilities mediated via TGF-β signaling. This subset of Tregs was shown to successfully block experimental autoimmune encephalomyelitis in mice [13].

Ex vivo expansion of Tregs has been described using αCD3/αCD28 microbeads and recombinant human interleukin (IL)-2 [78]. Selection of human Tregs based on CD69 and CD71 expression was utilized to reduce immune-mediated injury to human skin grafts in a mouse model, suggesting that anti-donor Tregs can be "customized" for clinical transplantation protocols [80]. Recently, scalable techniques have been described to harvest and efficiently expand Treg populations from peripheral blood using artificial antigen-presenting cells (APCs) (loaded with anti-CD3 antibody) expressing the high-affinity fragment crystallizable (Fc) receptor and CD86 [32]. These populations were infused into mice and markedly reduced GvHD lethality. Collectively, these studies demonstrate the potential of Treg modulation to maintain transplant graft tolerance.

Dendritic Cells

DCs are bone marrow-derived APCs that play a critical role in modulating the immune response. Although comprising less than 1 % of circulating blood mononuclear cells, they are found throughout the body and appear to regulate T cell physiology. Multiple subsets of DCs have been characterized based on distinct phenotypic and functional properties. Although initially described as key initiators of the adaptive immune response, they are now recognized as sentinel regulators of both innate and adaptive immune responses [62].

The two broadest categories of DCs include classical DCs and plasmacytoid DCs. Classical DCs are largely involved in antigen presentation and T cell activation, whereas plasmacytoid DCs function to secrete interferons and other cytokines [91]. Importantly, both classes of DCs do not directly engage in effector activities. Basic science studies on DC function in mice have established surface marker profiles (e.g., $CD8^+/CD103^+$ and $CD11c^+$) to characterize key functional properties but likely underestimate the significant heterogeneity in DC function [21, 28]. Much less is known about human DCs but genomic analyses demonstrate substantial conservation between mice and human DCs [17]. Recently, heterogeneous DC subpopulations that express Blood Dendritic Cell Antigen-3 (BDCA-3) have been identified in human blood, lymph nodes, and spleen and may correlate with established murine markers. However, much work remains in identifying and characterizing DC subsets in humans.

Given the primary role of DCs in regulating the immune response, recent transplant immunology research has focused on targeting subpopulations of DCs that may be responsible for promoting and maintaining tolerance [69]. In a mouse cardiac allograft model, immature bone marrow-derived DCs were able to significantly prolong graft survival compared to mature DCs [66]. In a mouse study on allogeneic BMT, a subset of DCs displaying high levels of MHC and low levels of co-stimulatory markers induced protective effects against GVHD, demonstrating the therapeutic potential of DCs [83].

Researchers have also demonstrated the ability to augment the tolerogenic effects of DCs via manipulation of co-stimulation pathways. For example, modulation of the CD40 pathway regulates DC–T cell interactions and promotes cardiac allograft survival in mice [65]. Additionally, blockade of the CD80/CD86–CD28 co-stimulatory pathway has been shown to prolong survival of hepatic allografts in a mouse model 63]. In a different mouse cardiac allograft model, preoperatively infused DC precursors treated with immunomodulatory signals have been demonstrated to extend allograft survival [1]. Most recently, DCs treated with the co-stimulation blocking agent CTLA4Ig were infused preoperatively in a NHP kidney allograft model, resulting in significantly prolonged graft survival [22]. These studies and others highlight the therapeutic promise of DC-based approaches for composite tissue transplantation.

Donor Hematopoietic Cells

Multiple approaches for tolerance induction have been developed that employ donor BMT [87]. Donor hematopoietic cells reach the recipient thymus, resulting in the negative selection of donor-reactive T cells and theoretically limiting the host response to donor antigens. Several groups have demonstrated improved chimerism using intraosseous delivery of bone marrow cells compared to standard intravenous delivery in a rat model [49]. Additionally, greater numbers of donor cells were detected in thymic and lymphoid tissues after intraosseous transplantation compared to intravenous techniques [49]. Enhanced chimerism and hind-limb transplant survival was also demonstrated in rats when intraosseous delivery of HSC was employed compared to no HSC therapy [88].

The combination of nonmyeloablative regimens with donor bone marrow cell delivery may be a viable strategy for tolerance induction in composite tissue allotransplantation. In a rat osteomyocutaneous flap model, recipient rats were pretreated with TBI, transplantation of selectively depleted T cells (lacking αβ-T cells), anti-lymphocyte serum, and short-course tacrolimus to generate mixed allogeneic chimeras [74]. Although peripheral blood chimerism was lost by 6 months, the donor bone compartment maintained long-term chimerism that appeared to sustain tolerance to the allograft.

Successful combined bone marrow and kidney transplantation from HLA single-haplotype-mismatched living-related donors has demonstrated the clinical potential of nonmyeloablative conditioning protocols for tolerance induction [46]. Importantly, these clinical trials highlight the relevance of transplanted hematopoietic cells in promoting immune tolerance. In addition to pretransplantation strategies, posttransplantation infusion of HSCs has also been used successfully, most notably in the first partial human face transplant [20]. Ongoing studies are examining the role of donor HSC infusions as maintenance immunosuppression following hand transplantation. Taken together, these studies suggest that hematopoietic-based strategies may be a key component of future tolerance induction protocols.

Facilitating Cells

Facilitating cells are a rare population of bone marrow-derived cells that enable allogeneic stem cell engraftment. Extensive studies have demonstrated that facilitating cells are neither mature T, B, or natural killer (NK) cells [15]. They exhibit lower levels of CD8 expression and lack the T cell antigen receptor (TCR) phenotype characteristic of T cells [42]. Studies have suggested that these cells are related to precursor plasmacytoid DCs and may have a similar cytokine activation profile [3, 25].

Experimental models of BMT have demonstrated that facilitating cells promote stem cell engraftment, inhibit GvHD, and induce tolerance [15]. Researchers have even shown that facilitating cells can sustain the engraftment of both adult and fetal stem cells in animal models of BMT [26]. Although the utility of these cells in solid-organ transplantation has not been extensively explored, facilitating cells represent a promising approach to establish allograft tolerance, while avoiding myeloablation and minimizing the risk of GvHD [40]. More recently, Leventhal et al. used mobilized cells enriched for HSCs and graft-facilitating cells combined with a nonmyeloablative conditioning in kidney transplant recipients across an MHC mismatch (nonrelated donor/recipient pair) [61]. In this phase 2 clinical trial, five of the eight kidney recipients displayed durable chimerism and were rendered immunosuppression-free 1 year after transplantation. No signs of GvHD or engraftment syndrome (a constellation of symptoms and signs following HSC transplantation and that include fever, erythrodermatous skin rash, and noncardiogenic pulmonary edema) [89] were observed in these patients. Testing this encouraging protocol in a larger patient cohort would provide further data to potentially allow its widespread use in the clinic.

Mesenchymal/Adipose-Derived Stem Cells

MSCs are a multipotent population of progenitor cells that have been described in many human tissues. Although originally identified in bone marrow, mesenchymal-like stem cells have also been found in adipose tissue (the adipose-derived stem cell—ASC) but their true identity and physiologic relevance remain unclear [59, 60]. *In vitro*, these cells can be directed to differentiate into several mesenchymal lineage cells, including fat, bone, and cartilage. They have also been shown to exhibit potent immunoregulatory properties [103] and have been increasingly explored as therapeutic cells for a wide variety of diseases [55].

The immunosuppressive effects of MSCs have been studied in various allotransplantation models in mice. For example, allogeneic islet graft take was markedly improved by therapeutic MSCs via mechanisms linked to matrix metalloproteinases and T cell expression of CD25 [19]. MSCs have also been used with rapamycin to augment cardiac allograft survival [27]. Interestingly, labeled MSCs could be tracked to lymphoid organs and cardiac grafts in tolerant recipients. In a kidney-allotransplant model, the delivery of MSCs pretransplantation promoted Treg pathways and graft survival, whereas posttransplantation delivery of MSCs resulted in early graft rejection [11]. The role of MSCs in VCA remains unclear, but the potent paracrine mechanisms observed in wound regeneration models suggest that MSCs may be a valuable tool to induce tolerance [2].

In this regard, MSCs have also been studied in large animal models of vascularized composite tissue transplantation. In a swine hind-limb allotransplantation model, MSC delivery was shown to prolong allograft survival and the combined use of MSCs, BMT, and cyclosporine A demonstrated the greatest degree of survival with no signs of GvHD and the lowest levels of rejection [52]. Similarly, a swine hemi-face allotransplantation model demonstrated improved graft survival with combined MSC infusion and cyclosporine A [54]. These findings were associated with increased markers for regulatory T cells and suppressed inflammatory signaling.

Researchers have also studied the immunomodulatory effects of intravenous ASC delivery following hind-limb transplantation in a rat model [53]. Allografts were shown to persist longer with ASC infusion via mechanisms linked to blockade of inflammatory cytokines and regulation of T cell function. ASCs have even been shown to augment engraftment of stem cells and to prevent GvHD *in vivo* [107]. Given their ease of harvest, relative abundance in Westernized societies, and the growing obesity epidemic worldwide, ASCs have tremendous translational potential as a therapeutic cell type [64].

Other Regulatory Cell Populations

Regulatory immune cells, including the previously discussed regulatory T cells, play a critical role in modulating the host immune response and can dictate long-term outcomes after allotransplantation. The broad classification of immune cells as T cells, B cells, macrophages, etc., grossly oversimplifies the complexity of these important cell populations. Recent immunology research has shed light on several cell subpopulations that may facilitate cell-based approaches to induce allograft tolerance [37, 38]. These cell subsets and others have been utilized in clinical studies and shown to improve graft tolerance. We refer the reader to an excellent recent review on this topic [2012].

Studies in renal transplant recipients that no longer require immunosuppression have demonstrated greater numbers of B cells and increased expression of B cell-related genes [79]. Regulatory B cells are a rare, immature subpopulation of B cells that have been described in both mice and humans ($CD19^{+}/CD24^{hi}/CD38^{hi}$). A major property of these cells is their propensity to secrete IL-10, a potent immunomodulatory cytokine that controls T cell differentiation [23]. Another important signaling pathway appears to be CD40-associated activation of IL-10 and induction of Treg cells. Although B-cell-activated T cells appear more potent than plasmacytoid DC-activated T cells, the relevance to human allotransplantation remains unclear [109]. Studies in kidney transplant recipients suggest that regulatory B cells may facilitate the appearance of Treg cells later after induction therapy, suggesting that therapeutic expansion of regulatory B cells may be a viable strategy in the clinical setting [30].

Regulatory macrophages play a diverse role in tissue repair and immune regulation. Although this cell population remains ill-defined, these cells secrete IL-10 and can develop a suppressive phenotype upon interaction with Treg and B cells [95, 104]. In mice, blockade of macrophage populations in recipient animals worsened GvHD and resulted in higher numbers of donor T cells in a HSC transplantation model, indicating that macrophages may have a protective role in transplantation [29]. In promising early studies, kidney transplant recipients that received regulatory macrophage infusions required reduced levels of tacrolimus and maintained good early graft function [35]. These studies and others suggest that cell-based approaches to regulate the immunologic response after allotransplantation may be an effective strategy to promote graft tolerance.

Tolerance Protocols in VCA

Most of the research on allograft tolerance over the past few decades has focused on solid organs. However, recent research has increasingly focused on VCA. The presence of skin, the most immunogenic of tissues, in most VCA cases makes the translation of these protocols from solid-organ transplantation to VCA extremely challenging. In these protocols, heavy immunosuppressive therapy has been necessary to avoid rejection of these tissues. In this section, we discuss the regimens utilized in large animal models (swine and NHPs) and humans.

Swine

Several groups have utilized the MGH miniature swine models to study tolerance induction in VCA using different protocols based on mixed chimerism and various surgical models [51, 59, 98]. Furthermore, swine, canine, and NHP models have been used to investigate diverse strategies to reduce the number and dose of immunosuppressive drugs [5, 6, 12, 68].

The first attempts to achieve tolerance in VCA were performed using heterotopic limb transplants in the swine model. These experiments entailed the use of an osteomyocutaneous composite graft that included the distal part of the femur, the entire knee joint, and parts of tibia and fibula together with the surrounding muscles. The donor hind-limb allograft was transplanted to a subcutaneous abdominal wall pocket in the recipient animals. In contrast with the controls, which rejected the allografts in approximately 12 weeks, experimental animals receiving a calcineurin inhibitor (cyclosporine A) displayed long-term graft survival for up to 47 weeks following transplantations [57]. The presence of a bone marrow component in the donor graft-induced transient mixed microchimerism in the peripheral blood, assessed by the presence of pig allelic antigen positive (PAA^{+}) donor cells in recipients lacking such antigens (PAA-). Interestingly, despite loss of chimerism in this model, the experimental animals remained tolerant of their allografts [9]. However, a skin component was not present in this model and additional immunomodulation, by means of BMT, to assess the validity of this model across disparate mismatch barriers was not attempted in these early studies. When a skin paddle was added to the design of the graft, further studies on MHC-matched/minor antigens mismatch models elucidated the concept of "split tolerance," according to which the simultaneous acceptance of some of the graft tissues (i.e., muscle and bone) and the rejection of a more immunogenic tissue, such as skin, was possible due to the high immunogenicity of the skin component [67].

In the scenario of clinical reconstructive transplantation, however, it seems unlikely to find a fully matched donor/recipient couple to which such regimens would apply. For this reason, subsequent studies have explored clinically relevant models involving transplantation of vascularized composite allografts across partial and full MHC mismatches. In one of these studies, following T cell depletion, cytokine-mobilized peripheral blood mononuclear cells (CM-PBMCs) or bone marrow cells were transplanted at the time of VCA together with a short course of a calcineurin inhibitor. Donor-specific unresponsiveness and tolerance to the musculoskeletal components of the grafts were observed in both the CM-PBMC and the BMT recipients. Similar to the previous studies, the skin component was not accepted in both models, and peripheral chimerism was observed only in the group receiving CM-PBMCs. However, skin GvHD was observed in all the animals that displayed macrochimerism, making this protocol suboptimal for translation in the clinic [31].

Tolerance to skin in VCA across a single-haplotype MHC mismatch in MGH miniature swine was first reported by Horner et al. Donors were mobilized using stem cell factor and IL-3 and then leukapheresed to collect the donor hematopoietic product, which was then infused to recipients conditioned with 100 cGy TBI, T cell depletion and a 45-day course of cyclosporine A. Using this protocol, one animal achieved long-term multilineage mixed chimerism in peripheral blood and tissue chimerism in the thymus and in the bone marrow without developing GvHD [33]. More recently, the establishment of multilineage mixed chimerism across MHC haploidentical donor/recipient combinations was obtained using a conditioning regimen with CD3 immunotoxin (pCD3-DT390), TBI, hematopoietic cell transplantation and a 45-day course of cyclosporine A, leading to long-term survival of VCAs in this model [58]. As these studies further support the evidence that stable mixed chimerism can induce indefinite survival of a VCA, migration toward more clinically applicable protocols that do not require cell mobilization and donor treatment before transplant is paramount, since the possibility of conditioning a patient in the setting of clinical VCA is unlikely. The multilineage chimerism and bone marrow engraftment is thought to provide constant antigenic stimulation, thereby inducing *central tolerance* by prompting negative selection of donor-reactive thymocytes.

Furthermore, it has been hypothesized that peripheral regulatory mechanisms by T regulatory cells may allow for *peripheral tolerance* induction by deletion of donor-reactive clones escaped from the thymus [106].

More recently, the consideration that vascularized bone marrow may not be present in every type of VCA has led to the design of a reliable VCA model lacking a bone segment. This myocutaneous graft is harvested from the pig groin (based on the saphenous and femoral vessels) and reanastomosed to the common carotid artery and internal jugular vein in the cervical region [59]. This model allows the assessment of tolerance induction protocols in absence of a vascularized source of donor bone marrow cells. Further studies, however, are needed for achievement of stable tolerance induction of skin-containing allografts. The use of biologic agents, such as co-stimulatory blockade, may help to overcome this limitation and is discussed ahead in this chapter.

Nonhuman Primates

Tolerance induction regimens for solid-organ transplantation or VCA require testing in large animal models before clinical applicability, with NHPs representing an ideal preclinical model for the assessment of such protocols. The initial protocols tested in NHP included nonmyeloablative TBI, thymic irradiation, and T cell depletion. The severe T cell depletion obtained with these regimens, however, carried a high rate of mortality due to infections and development of lymphoma. In the attempt to reduce the toxicity of such regimens and obtain an effective T cell depletion, horse anti-thymocyte globulin (ATG), splenectomy, and a short course of cyclosporine A was added to these protocols, allowing for detection of high-level multilineage chimerism [44]. Unlike the long-term chimerism observed in mice studies, this protocol only allowed for *transient chimerism*. Surprisingly, despite cessation of cyclosporine and loss of mixed chimerism, the NHP did not reject its renal allograft, which was deemed viable for several years without any signs of rejection [44]. To reduce the toxicity of this protocol, the TBI was fractioned over 2 days. In this study, 11 of the 14 recipients developed transient mixed chimerism and eight of them survived long-term without need for any immunosuppressive drugs, with the longest kidney recipient surviving over 14 years with a functioning graft. Stable tolerance was confirmed using skin grafting, by evidence of acceptance of donor skin and prompt rejection of skin from third-party animals.

Similar protocols have been used in NHP to translate these encouraging findings from solid organ to reconstructive transplantation. NHP models offer relevant anatomical and immunological similarity to humans for preclinical studies. However, transplantation of full hand and face allografts is extremely challenging both surgically and from a postoperative management standpoint. Aiming to simulate a clinical VCA (e.g., a hand transplant) while reducing surgical level of complexity, costs, and discomfort to the animals, Cendales et al. have described a sensate osteomyocutaneous radial forearm flap. This flap was performed in 14 allogeneic transplants and entailed the use of an immunosuppressive regimen similar to those used in clinical transplantation, such as a calcineurin inhibitor (tacrolimus), an antiproliferative agent (mycophenolate mofetil), and methylprednisolone allowing for evaluation of VCA histology. Other VCA models in NHP were also investigated by Barth et al., who described a partial face transplant model by isolation of an osteomyocutaneous flap based on the common carotid artery and the internal and external jugular vein [5]. In this protocol, long-term survival of the grafts on tacrolimus monotherapy was associated with development of posttransplant lymphoproliferative disorder (PTLD). Later studies from the same group have shown that the addition of mycophenolate mofetil to the regimen prevented PTLD [6]. Furthermore, in this study, the authors investigated the role of vascularized bone marrow in this model, demonstrating that the lack of the vascularized bone segment was associated with acute rejection by 2 weeks [7]. The role of mixed chimerism or other cell-based approaches to induce tolerance in these models remains largely unknown.

Immunomodulatory Approaches

Alternative approaches to modulate the immune response have entailed the use of different biologic agents directed toward specific target molecules. Some of these drugs are intended to avoid acute rejection and, simultaneously, decrease sensitization to donor antigens, usually due to donor-specific antibody formation. The addition of these compounds to the conditioning regimens aimed to replace splenectomy as a way to improve graft survival. A high number of residual T cells remains in lymph nodes and in the spleen following the common conditioning regimens used in these studies, making sensitization likely. The addition of splenectomy in some protocols, such as the delayed kidney transplant model, has been shown to lead to long-term graft survival [45]. Replacing splenectomy with co-stimulatory blockade molecules such as anti-CD154 (also known as anti-CD40L) or anti-CD152 (also known as *CTLA4*) has been associated with high-level and long-lasting chimerism, leading to long-term graft survival [77].

The mechanism of action of co-stimulatory blockade molecules is linked to the modulation of T cell activation. In order for alloantigens to robustly activate naïve T cells, two signals are required. Following processing of donor antigen by the APCs in the context of the MHC, the first signal is delivered upon interaction of the TCR with the MHC/Ag complex. The second signal is a co-stimulatory signal that can be delivered by another surface molecule. Failure to produce co-stimulation via a second signal induces a state of T cell anergy [108] (Fig. 14.2a). The two co-stimulatory pathways best known to allow normal development and maintenance of immunity involve the CD80/CD86–CD28 and the CD40–CD40L pathway. In the CD80/CD86–CD28 second signal pathway, the CD80 (also known as B7.1) and CD86 (alternatively known as B7.2) molecules on the surface of APCs deliver a co-stimulatory signal by interacting with the CD28 cell surface proteins expressed on naïve T cells [24, 41] (Fig. 14.2b). Furthermore, following TCR interaction with the MHC/Ag complex present on the APC, CD40L expression is induced in T cells, allowing interaction with its receptor CD40, which results in upregulation of CD80 and CD86 by the APCs. Moreover, the CD80/CD86–CD28 interaction triggers upregulation of CD40L surface molecules when the TCR is occupied [24]. The redundancy of these mechanisms, and the fact that these two pathways regulate T-cell-dependent immune responses critically, hints at the importance of such interrelated pathways.

The presence of this co-stimulatory signal acts as a silencing signal, leading to T cell anergy and has been, therefore, used to promote tolerance induction. CTLA4 is similar to CD28 and therefore binds CD80/CD86, acting as a "switch off" signal. For this reason, much interest has developed around the use of biologic agents, such as monoclonal antibodies (mAb) targeting CD80/CD86 (named CTLA4-Ig) or targeting CD154, for tolerance induction in solid-organ transplantation. However, anti-CD40L mAb have displayed severe thrombogenic effects, limiting their clinical applicability [50]. Recently, our group and others have investigated co-stimulatory blockage by means of perioperative CTLA4-Ig infusions for reconstructive transplantation with promising results in mice, pig, and NHP models [48, 92, 99, 101].

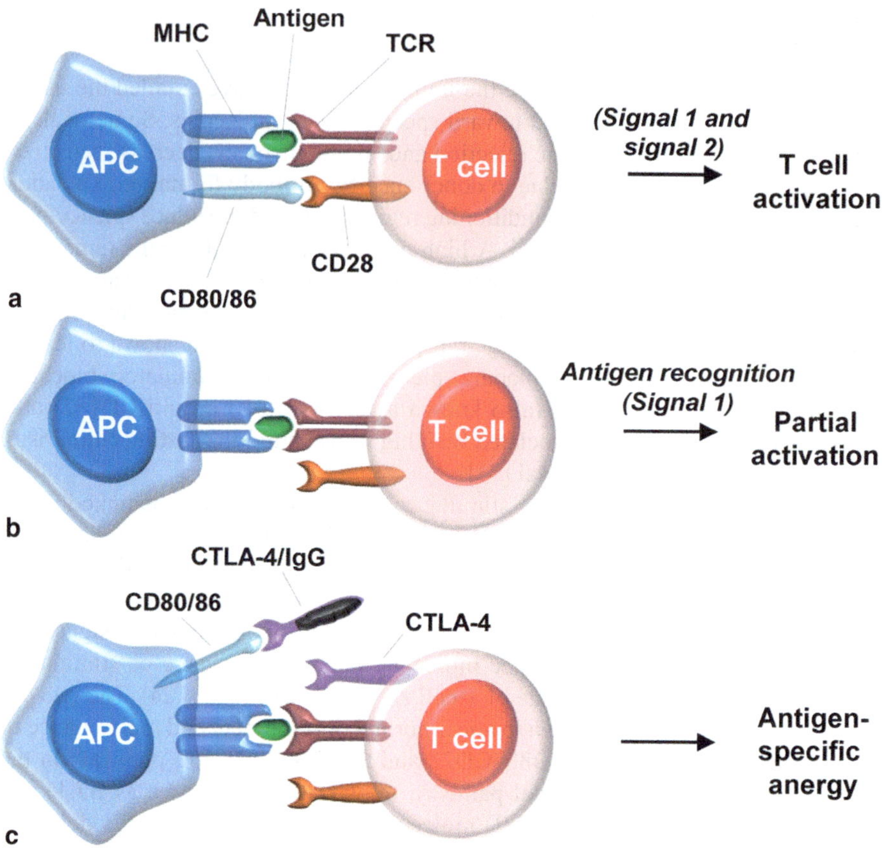

Fig. 14.2 T cell activation and co-stimulatory signals; a) T-cell activation via signal 1 and 2; b) Partial T cell activation by lack of signal 2; c) Antigen-specific anergy following co-stimulatory blockade. MHC: Major histocompatibility complex, CD: cluster of differentiation, APC: antigen-presenting cells, TCR: T cell antigen receptor; CTLA4-IgG: cytotoxic T-lymphocyte antigen 4-immunoglobulin G.

Clinical Protocols

Most of the reconstructive transplantation programs performing limb or face transplantation currently still use conventional immunosuppressive regimens similar to those used for solid-organ transplantation. These regimens entail conditioning with ATG and long-term immunosuppression with triple therapy (tacrolimus, mycophenolate mofetil, and corticosteroids) [86]. The use of these protocols in VCA has allowed long-term survival of skin-bearing allografts and successfully prevented acute rejection. Scarce data are available to date on the rate of chronic rejection in VCA, although it seems that, if compared to solid-organ transplantation, chronic rejection occurs less frequently in VCA. Despite the encouraging results obtained so far, these immunosuppressive regimens show a high rate of acute rejection episodes. It has been observed that 85 % of patients experience at least one episode of

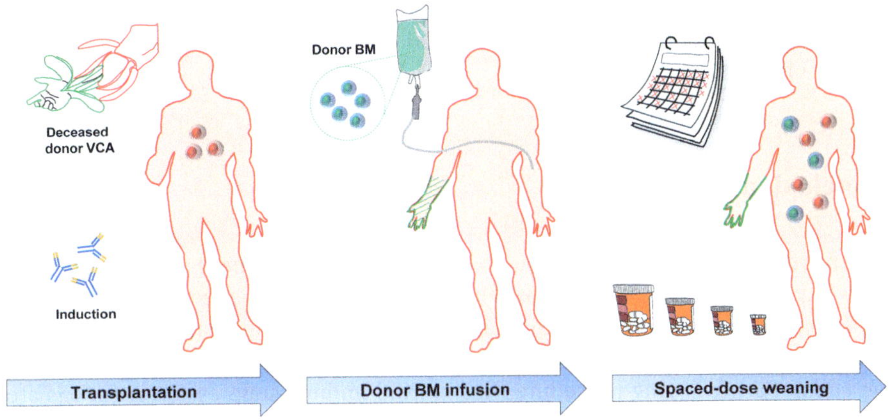

Fig. 14.3 Schematic representation of immunomodulation using minimization protocols in clinical reconstructive transplantation. *VCA* vascularized composite allotransplantation, *BM* bone marrow

acute rejection in the first year following transplantation and require an increase in the dose of immunosuppression to avoid graft loss [72].

The need to treat transplant recipients with lifelong immunosuppression has fueled the attempt to develop alternate protocols to minimize and ultimately obviate immunosuppression. At the University of Pittsburgh, and subsequently at Johns Hopkins University, the team led by W. P. Andrew Lee and colleagues has developed the first protocol for minimization of immunosuppression in reconstructive transplantation. Based on encouraging results in the swine model, this group used an induction regimen using monoclonal antibody (Campath 1 H) at the time of transplant, followed by treatment with a single immunosuppressive agent (tacrolimus) and donor bone marrow infusion 2 weeks post transplant (Fig. 14.3). To date, this novel protocol has been successfully used in a total of ten upper-extremity transplants performed in six recipients, including the first double-hand transplant and first above-elbow transplant performed in the USA. In compliant patients, this protocol allowed also reduction of the dose tacrolimus dose (2–4 mg/day), was well tolerated and, despite lack of detectable chimerism, enabled long-term survival of the allografts [82]. Episodes of acute rejection were infrequent using this protocol and were easily manageable using topical immunosuppression. No chronic rejection has been observed to date in any of the recipients [10, 84, 99]. Of note, the Louisville group has reported the development of intimal hyperplasia in one patient under similar immunosuppression minimization protocols. The vascular damage resulted in allograft ischemia and reamputation was performed 275 days post transplant [43]. Long-term follow-up in more patients will be necessary to determine whether reduced immunosuppression protocols are a viable option for long-term allograft maintenance. Recently, intra-thymic HSC transplantation has been described in the macaque as a potential way to induce longer and more robust T cell reconstitution. Such techniques may require less HSCs and maintain long-term thymopoiesis [4]. The future direction for clinical reconstructive transplantation may entail the translation of successful preclinical experimental protocols on swine and NHP to clinical scenarios.

Conclusions

Reconstructive transplantation is an exciting field of medicine that could dramatically improve the lives of many patients. The clinical experience with solid-organ transplantation has allowed researchers to make tremendous gains in developing tolerance protocols for VCA. However, because VCA is currently a life-enhancing and not a life-saving procedure, less morbid immunosuppression regimens are needed to justify performing these surgeries. Ongoing research in small and large animal models has allowed researchers to develop innovative tolerance induction protocols that may have clinical applicability. Similar to preclinical studies in animals, human recipients of concomitant BMT and kidney transplant have developed allograft tolerance and maintain a functioning graft long term without immunosuppressive medications. This observation highlights the potential of cell-based approaches to modulate the immune system and promote tolerance in VCA. It is possible that the use of conditioning regimens combining both cell-based therapies and biologic agents may provide a synergistic effect to enable long-term composite allograft function without immunosuppression.

References

1. Abe M, et al. Plasmacytoid dendritic cell precursors induce allogeneic T-cell hyporesponsiveness and prolong heart graft survival. Am J Transplant. 2005;5(8):1808–19.
2. Antony AK, et al. Composite tissue allotransplantation and dysregulation in tissue repair and regeneration: a role for mesenchymal stem cells. Front Immunol. 2013;4:188.
3. Asselin-Paturel C, et al. Mouse type I IFN-producing cells are immature APCs with plasmacytoid morphology. Nat Immunol. 2001;2(12):1144–50.
4. De Barros SC, Zimmermann VS, Taylor N. Concise review: hematopoietic stem cell transplantation: targeting the thymus. Stem Cell (Dayton, Ohio). 2013;31(7):1245–51.
5. Barth RN, Bluebond-Langner R, et al. Facial subunit composite tissue allografts in nonhuman primates: I. Technical and immunosuppressive requirements for prolonged graft survival. Plast reconstruct Surg. 2009;123(2):493–501.
6. Barth RN, Nam AJ, et al. Prolonged survival of composite facial allografts in non-human primates associated with posttransplant lymphoproliferative disorder. Transplantation 2009;88(11):1242–50.
7. Barth RN, et al. Vascularized bone marrow-based immunosuppression inhibits rejection of vascularized composite allografts in nonhuman primates. Am J Transplant. 2011;11(7):1407–16. http://www.ncbi.nlm.nih.gov/pubmed/21668624. Accessed: 4 Aug 2013.
8. Billingham RE, Brent L, Medawar PB. Actively acquired tolerance of foreign cells. Nature 1953;172(4379):603–6. http://www.ncbi.nlm.nih.gov/pubmed/13099277. Accessed: 24 May 2013.

9. Bourget JL, et al. Tolerance to musculoskeletal allografts with transient lymphocyte chimerism in miniature swine. Transplantation 2001;71(7):851–6. http://www.ncbi.nlm.nih.gov/pubmed/11349715. Accessed: 31 July 2013.
10. Brandacher G, Lee WPA, Schneeberger S. Minimizing immunosuppression in hand transplantation. Expert Rev Clin Immunol. 2012;8(7):673–84.
11. Casiraghi F, et al. Localization of mesenchymal stromal cells dictates their immune or proinflammatory effects in kidney transplantation. Am J Transplant. 2012;12(9):2373–83. http://www.ncbi.nlm.nih.gov/pubmed/22642544.
12. Cendales LC, et al. Composite tissue allotransplantation: development of a preclinical model in nonhuman primates. Transplantation 2005;80(10):1447–54. http://www.ncbi.nlm.nih.gov/pubmed/16340790. Accessed: 14 June 2013.
13. Chen ML, et al. Latency-associated peptide identifies a novel CD4 + CD25 + regulatory T cell subset with TGFbeta-mediated function and enhanced suppression of experimental autoimmune encephalomyelitis. J Immunol. 2008;180(11):7327–37. http://www.ncbi.nlm.nih.gov/pubmed/18490732.
14. Colby C, et al. Cytokine-mobilized peripheral blood progenitor cells for allogeneic reconstitution of miniature swine. Transplantation 2000;69(1):135–40. http://www.ncbi.nlm.nih.gov/pubmed/10653392. Accessed: 27 July 2013.
15. Colson YL, Shinde Patil VR, Ildstad ST. Facilitating cells: novel promoters of stem cell alloengraftment and donor-specific transplantation tolerance in the absence of GVHD. Crit Rev Oncol Hematol. 2007;61(1):26–43. http://www.ncbi.nlm.nih.gov/pubmed/17150368.
16. Cosimi AB, Sachs DH. Mixed chimerism and transplantation tolerance. Transplantation 2004;77(6):943–6. http://www.ncbi.nlm.nih.gov/pubmed/15077044. Accessed: 28 July 2013.
17. Crozat K, et al. Comparative genomics as a tool to reveal functional equivalences between human and mouse dendritic cell subsets. Immunol Rev. 2010;234(1):177–98. http://www.ncbi.nlm.nih.gov/pubmed/20193019.
18. Dey B, Sykes M, Spitzer TR. Outcomes of recipients of both bone marrow and solid organ transplants. A review. Medicine 1998;77(5):355–69. http://www.ncbi.nlm.nih.gov/pubmed/9772924. Accessed: 23 July 2013.
19. Ding Y, et al. Mesenchymal stem cells prevent the rejection of fully allogenic islet grafts by the immunosuppressive activity of matrix metalloproteinase-2 and -9. Diabetes 2009;58(8):1797–806. http://www.ncbi.nlm.nih.gov/pubmed/19509016.
20. Dubernard JM et al. Outcomes 18 months after the first human partial face transplantation. N Engl J Med. 2007;357(24):2451–60. http://www.ncbi.nlm.nih.gov/pubmed/18077810.
21. Dudziak D, et al. Differential antigen processing by dendritic cell subsets *in vivo*. Science 2007;315(5808):107–11. http://www.ncbi.nlm.nih.gov/pubmed/17204652.
22. Ezzelarab MB, et al. Regulatory dendritic cell infusion prolongs kidney allograft survival in nonhuman primates. Am J Transplant. 2013. http://www.ncbi.nlm.nih.gov/pubmed/23758811.
23. Fillatreau S, et al. B cells regulate autoimmunity by provision of IL-10. Nat Immunol. 2002;3(10):944–50. http://www.ncbi.nlm.nih.gov/pubmed/12244307.
24. Foy TM, et al. Immune regulation by CD40 and its ligand GP39. Ann Rev Immunol. 1996;14:591–617. http://www.ncbi.nlm.nih.gov/pubmed/8717526. Accessed: 3 Aug 2013.
25. Fugier-Vivier IJ, et al. Plasmacytoid precursor dendritic cells facilitate allogeneic hematopoietic stem cell engraftment. J Exp Med. 2005;201(3):373–83. http://www.ncbi.nlm.nih.gov/pubmed/15699072.
26. Gaines BA, et al. Facilitating cells enable engraftment of purified fetal liver stem cells in allogeneic recipients. Exp Hematol. 1996;24(8):902–13. http://www.ncbi.nlm.nih.gov/pubmed/8690049.
27. Ge W, et al. Infusion of mesenchymal stem cells and rapamycin synergize to attenuate alloimmune responses and promote cardiac allograft tolerance. Am J Transplant. 2009;9(8):1760–72. http://www.ncbi.nlm.nih.gov/pubmed/19563344.
28. Den Haan JM, Lehar SM, Bevan MJ. CD8(+) but not CD8(–) dendritic cells cross-prime cytotoxic T cells *in vivo*. J Exp Med. 2000;192(12):1685–96. http://www.ncbi.nlm.nih.gov/pubmed/11120766.

29. Hashimoto D, et al. Pretransplant CSF-1 therapy expands recipient macrophages and ameliorates GVHD after allogeneic hematopoietic cell transplantation. J Exp Med. 2011;208(5):1069–82. http://www.ncbi.nlm.nih.gov/pubmed/21536742.
30. Hester J, et al. Th17 cells in alemtuzumab-treated patients: the effect of long-term maintenance immunosuppressive therapy. Transplantation 2011;91(7):744–50. http://www.ncbi.nlm.nih.gov/pubmed/21412187.
31. Hettiaratchy S, et al. Tolerance to composite tissue allografts across a major histocompatibility barrier in miniature swine. Transplantation 2004;77(4):514–21. http://www.ncbi.nlm.nih.gov/pubmed/15084927.
32. Hippen KL, et al. Massive *ex vivo* expansion of human natural regulatory T cells (T(regs)) with minimal loss of *in vivo* functional activity. Sci Transl Med. 2011;3(83):83ra41. http://www.ncbi.nlm.nih.gov/pubmed/21593401.
33. Horner BM, et al. Induction of tolerance to an allogeneic skin flap transplant in a preclinical large animal model. Transplant Proc. 2009;41(2):539–41. http://www.pubmedcentral.nih.gov/articlerender.fcgi?artid=2796481&tool=pmcentrez&rendertype=abstract. Accessed: 2 Aug 2013.
34. Huang CA, et al. Stable mixed chimerism and tolerance using a nonmyeloablative preparative regimen in a large-animal model. J Clin Invest. 2000;105(2):173–81. http://www.pubmedcentral.nih.gov/articlerender.fcgi?artid=377429&tool=pmcentrez&rendertype=abstract. Accessed: 27 July 2013.
35. Hutchinson JA, et al. Cutting edge: immunological consequences and trafficking of human regulatory macrophages administered to renal transplant recipients. J Immunol. 2011;187(5):2072–8. http://www.ncbi.nlm.nih.gov/pubmed/21804023.
36. Ildstad ST, Sachs DH. Reconstitution with syngeneic plus allogeneic or xenogeneic bone marrow leads to specific acceptance of allografts or xenografts. Nature 1984;307(5947):168–70. http://www.ncbi.nlm.nih.gov/pubmed/6361574. Accessed: 13 June 2013.
37. Issa F, Wood KJ. Translating tolerogenic therapies to the clinic – where do we stand? Front Immunol. 2012;3:254. http://www.ncbi.nlm.nih.gov/pubmed/22934094.
38. Issa, F, Wood KJ. Translating tolerogenic therapies to the clinic – where do we stand? Front Immunol. 2012;3:254. http://www.pubmedcentral.nih.gov/articlerender.fcgi?artid=3422982&tool=pmcentrez&rendertype=abstract. Accessed: 1 July 2013.
39. Issa F, et al. *Ex vivo*-expanded human regulatory T cells prevent the rejection of skin allografts in a humanized mouse model. Transplantation 2010;90(12):1321–7. http://www.ncbi.nlm.nih.gov/pubmed/21048528.
40. Jacquet EG, et al. Facilitating cells as a venue to establish mixed chimerism and tolerance. Pediatr Transplant. 2003;7(5):348–57. http://www.ncbi.nlm.nih.gov/pubmed/14738294.
41. June CH, et al. The B7 and CD28 receptor families. Immunol Today. 1994;15(7):321–31. http://www.ncbi.nlm.nih.gov/pubmed/7522010. Accessed: 3 Aug 2013.
42. Kaufman CL, et al. Phenotypic characterization of a novel bone marrow-derived cell that facilitates engraftment of allogeneic bone marrow stem cells. Blood 1994;84(8):2436–46. http://www.ncbi.nlm.nih.gov/pubmed/7919363.
43. Kaufman CL, et al. Graft vasculopathy in clinical hand transplantation. Am J Transplant. 2012;12(4):1004–16. http://www.ncbi.nlm.nih.gov/pubmed/22325051.
44. Kawai T, et al. Mixed allogeneic chimerism and renal allograft tolerance in cynomolgus monkeys. Transplantation 1995;59(2):256–62. http://www.ncbi.nlm.nih.gov/pubmed/7839449. Accessed: 2 Aug 2013.
45. Kawai T, et al. Long-term outcome and alloantibody production in a non-myeloablative regimen for induction of renal allograft tolerance. Transplantation 1999;68(11):1767–75. http://www.ncbi.nlm.nih.gov/pubmed/10609955. Accessed: 3 Aug 2013.
46. Kawai T, et al. HLA-mismatched renal transplantation without maintenance immunosuppression. N Engl J Med. 2008;358(4):353–61. http://www.ncbi.nlm.nih.gov/pubmed/18216355.
47. Kawai T, et al. HLA-mismatched renal transplantation without maintenance immunosuppression. N Engl J Med. 2013;368(19):1850–2. http://www.ncbi.nlm.nih.gov/pubmed/23656665. Accessed: 28 July 2013.

48. Kirk AD, et al. CTLA4-Ig and anti-CD40 ligand prevent renal allograft rejection in primates. Proc Natl Acad Sci U S A. 1997;94(16):8789–94. http://www.pubmedcentral.nih.gov/articlerender.fcgi?artid=23132&tool=pmcentrez&rendertype=abstract. Accessed: 3 Aug 2013.
49. Klimczak A, et al. Donor-origin cell engraftment after intraosseous or intravenous bone marrow transplantation in a rat model. Bone Marrow Transplant. 2007;40(4):373–80. http://www.ncbi.nlm.nih.gov/pubmed/17572707.
50. Koyama I, et al. Thrombophilia associated with anti-CD154 monoclonal antibody treatment and its prophylaxis in nonhuman primates. Transplantation 2004;77(3):460–2. http://www.ncbi.nlm.nih.gov/pubmed/14966427. Accessed: 3 Aug 2013.
51. Kuo Y-R, et al. Swine hemi-facial composite tissue allotransplantation: a model to study immune rejection. J Surg Res. 2009;153(2):268–73. http://www.ncbi.nlm.nih.gov/pubmed/18708190. Accessed: 10 June 2013.
52. Kuo YR, et al. Mesenchymal stem cells prolong composite tissue allotransplant survival in a swine model. Transplantation 2009;87(12):1769–77. http://www.ncbi.nlm.nih.gov/pubmed/19543052.
53. Kuo YR, et al. Modulation of immune response and T-cell regulation by donor adipose-derived stem cells in a rodent hind-limb allotransplant model. Plast Reconstr Surg, 2011;128(6):661e–72e. http://www.ncbi.nlm.nih.gov/pubmed/22094768.
54. Kuo YR, Chen CC, Goto S, Huang YT, et al. Immunomodulatory effects of bone marrow-derived mesenchymal stem cells in a swine hemi-facial allotransplantation model. PLoS One. 2012;7(4):e35459. http://www.ncbi.nlm.nih.gov/pubmed/22558153.
55. Kuo YR, Chen CC, Goto S, Lin PY, et al. Mesenchymal stem cells as immunomodulators in a vascularized composite allotransplantation. Clin Dev Immunol. 2012:854846. http://www.ncbi.nlm.nih.gov/pubmed/23227090.
56. Lee WP, et al. Relative antigenicity of components of a vascularized limb allograft. Plast Reconstr Surg. 1991;87(3):401–11. http://www.ncbi.nlm.nih.gov/pubmed/1998012.
57. Lee WP, et al. Tolerance to limb tissue allografts between swine matched for major histocompatibility complex antigens. Plast Reconstr Surg. 2001;107(6):1482–90; discussion 1491–2. http://www.ncbi.nlm.nih.gov/pubmed/11335822. Accessed: 31 July 2013.
58. Leonard DA, et al. Vascularized composite allograft tolerance across MHC barriers in a large animal model. Am J Transplant. 2014. http://www.ncbi.nlm.nih.gov/pubmed/24405666.
59. Leto Barone AA, et al. Adipose-derived stem cells: true or false? A different point of view. J Craniofac Surg. 2013;24(4):1072. http://www.ncbi.nlm.nih.gov/pubmed/23851741. Accessed: 9 Aug 2013.
60. Leto Barone AA, et al. The gracilis myocutaneous free flap in swine: an advantageous preclinical model for vascularized composite allograft transplantation research. Microsurgery 2013;33(1):51–5. http://www.ncbi.nlm.nih.gov/pubmed/22707437. Accessed: 28 July 2013.
61. Leventhal J, et al. Chimerism and tolerance without GVHD or engraftment syndrome in HLA-mismatched combined kidney and hematopoietic stem cell transplantation. Sci Transl Med. 2012;4(124):124ra28. http://www.pubmedcentral.nih.gov/articlerender.fcgi?artid=3610325&tool=pmcentrez&rendertype=abstract. Accessed: 24 July 2013.
62. Lewis KL, Reizis B. Dendritic cells: arbiters of immunity and immunological tolerance. Cold Spring Harb Perspect Biol. 2012;4(8):a007401. http://www.ncbi.nlm.nih.gov/pubmed/22855722.
63. Li W, et al. Costimulation blockade promotes the apoptotic death of graft-infiltrating T cells and prolongs survival of hepatic allografts from FLT3 L-treated donors. Transplantation 2001;72(8):1423–32. http://www.ncbi.nlm.nih.gov/pubmed/11685115.
64. Lindroos B, Suuronen R, Miettinen S. The potential of adipose stem cells in regenerative medicine. Stem Cell Rev. 2011;7(2):269–91. http://www.ncbi.nlm.nih.gov/pubmed/20853072.
65. Lu L, et al. Blockade of the CD40-CD40 ligand pathway potentiates the capacity of donor-derived dendritic cell progenitors to induce long-term cardiac allograft survival. Transplantation 1997;64(12):1808–15. http://www.ncbi.nlm.nih.gov/pubmed/9422424.
66. Lutz MB, et al. Immature dendritic cells generated with low doses of GM-CSF in the absence of IL-4 are maturation resistant and prolong allograft survival *in vivo*. Eur J Immunol. 2000;30(7):1813–22. http://www.ncbi.nlm.nih.gov/pubmed/10940870.

67. Mathes DW, et al. Split tolerance to a composite tissue allograft in a swine model. Transplantation 2003;75(1):25–31. http://www.ncbi.nlm.nih.gov/pubmed/12544866. Accessed: 31 July 2013.
68. Mathes DW, et al. Tolerance to vascularized composite allografts in canine mixed hematopoietic chimeras. Transplantation 2011;92(12):1301–8. http://www.pubmedcentral.nih.gov/articlerender.fcgi?artid=3310209&tool=pmcentrez&rendertype=abstract. Accessed: 28 July 2013.
69. Morelli AE, Thomson AW. Tolerogenic dendritic cells and the quest for transplant tolerance. Nat Rev Immunol. 2007;7(8):610–21. http://www.ncbi.nlm.nih.gov/pubmed/17627284.
70. Nikolic B, Khan A, Sykes M. Induction of tolerance by mixed chimerism with nonmyeloblative host conditioning: the importance of overcoming intrathymic alloresistance. Biol Blood Marrow Transplant. 2001;7(3):144–53. http://www.ncbi.nlm.nih.gov/pubmed/11302548. Accessed: 25 July 2013.
71. Page E, et al. Lymphodepletional strategies in transplantation. Cold Spring Harb Perspect Med. 2013;3(7). http://www.ncbi.nlm.nih.gov/pubmed/23818516.
72. Petruzzo P, et al. Long-term follow-up in composite tissue allotransplantation: in-depth study of five (hand and face) recipients. Am J Transplant. 2011;11(4):808–16. http://onlinelibrary.wiley.com/doi/10.1111/j.1600-6143.2011.03469.x/abstract.
73. Popitz-Bergez FA, et al. Bone marrow transplantation in miniature swine. II. Effect of selective genetic differences on marrow engraftment and recipient survival. Transplantation 1988;45(1):27–31. http://www.ncbi.nlm.nih.gov/pubmed/3122386. Accessed: 27 July 2013.
74. Rahhal DN, et al. Dissociation between peripheral blood chimerism and tolerance to hindlimb composite tissue transplants: preferential localization of chimerism in donor bone. Transplantation 2009;88(6):773–81. http://www.ncbi.nlm.nih.gov/pubmed/19920776.
75. Sachs DH. Tolerance: of mice and men. J Clin Invest. 2003;111(12):1819–21. http://www.ncbi.nlm.nih.gov/pubmed/12813017.
76. Sachs DH, et al. Transplantation in miniature swine. I. Fixation of the major histocompatibility complex. Transplantation 1976;22(6):559–67. http://www.ncbi.nlm.nih.gov/pubmed/137560. Accessed: 27 July 2013.
77. Sachs DH, et al. Immuno-intervention for the induction of transplantation tolerance through mixed chimerism. Semin Immunol. 2011;23(3):165–73. http://www.pubmedcentral.nih.gov/articlerender.fcgi?artid=3178004&tool=pmcentrez&rendertype=abstract. Accessed: 28 May 2013.
78. Sagoo P, Lombardi G, Lechler RI. Regulatory T cells as therapeutic cells. Curr Opin Organ Transplant. 2008;13(6):645–53. http://www.ncbi.nlm.nih.gov/pubmed/19060557.
79. Sagoo P, et al. Development of a cross-platform biomarker signature to detect renal transplant tolerance in humans. J Clin Invest. 2010;120(6):1848–61. http://www.ncbi.nlm.nih.gov/pubmed/20501943.
80. Sagoo P, et al. Human regulatory T cells with alloantigen specificity are more potent inhibitors of alloimmune skin graft damage than polyclonal regulatory T cells. Sci Transl Med. 2011;3(83):83ra42. http://www.ncbi.nlm.nih.gov/pubmed/21593402.
81. Sakaguchi S, et al. Regulatory T cells and immune tolerance. Cell 2008;133(5):775–87. http://www.ncbi.nlm.nih.gov/pubmed/18510923.
82. Sarhane KA, et al. Minimization of immunosuppression and tolerance induction in reconstructive transplantation. Curr Surg Rep. 2012;1(1):40–6. http://link.springer.com/10.1007/s40137-012-0008-4. Accessed: 28 June 2013.
83. Sato K, et al. Regulatory dendritic cells protect mice from murine acute graft-versus-host disease and leukemia relapse. Immunity 2003;18(3):367–79. http://www.ncbi.nlm.nih.gov/pubmed/12648454.
84. Schneeberger S, et al. Upper-extremity transplantation using a cell-based protocol to minimize immunosuppression. Ann Surg. 2013;257(2):345–51. http://www.ncbi.nlm.nih.gov/pubmed/23001085. Accessed: 2 July 2013.
85. Sharabi Y, Sachs DH. Mixed chimerism and permanent specific transplantation tolerance induced by a nonlethal preparative regimen. J Exp Med. 1989;169(2):493–502. http://www.

pubmedcentral.nih.gov/articlerender.fcgi?artid=2189213&tool=pmcentrez&rendertype=abstract. Accessed: 9 June 2013.
86. Siemionow M, Ozturk C. An update on facial transplantation cases performed between 2005 and 2010. Plast Reconstr Surg. 2011;128(6):707e–20e. http://www.ncbi.nlm.nih.gov/pubmed/22094772.
87. Siemionow M, Klimczak A. Chimerism-based experimental models for tolerance induction in vascularized composite allografts: Cleveland clinic research experience. Clin Dev Immunol. 2013:831410. http://www.ncbi.nlm.nih.gov/pubmed/23573114.
88. Siemionow M, et al. Intraosseus transplantation of donor-derived hematopoietic stem and progenitor cells induces donor-specific chimerism and extends composite tissue allograft survival. Transplant Proc. 2005;37(5):2303–8. http://www.ncbi.nlm.nih.gov/pubmed/15964405.
89. Spitzer TR. Engraftment syndrome following hematopoietic stem cell transplantation. Bone Marrow Transplant. 2001;27:893–8.
90. Spitzer TR, et al. Combined histocompatibility leukocyte antigen-matched donor bone marrow and renal transplantation for multiple myeloma with end stage renal disease: the induction of allograft tolerance through mixed lymphohematopoietic chimerism. Transplantation 1999;68(4):480–4. http://www.ncbi.nlm.nih.gov/pubmed/10480403. Accessed: 28 July 2013.
91. Steinman RM. Decisions about dendritic cells: past, present, and future. Annu Rev Immunol. 2012;30:1–22. http://www.ncbi.nlm.nih.gov/pubmed/22136168.
92. Steurer W, et al. Prolonged allograft survival following *ex vivo* blockade of costimulatory signals B7-1/2 and CD40. Transplant Proc. 2001;33(1–2):274–5. http://www.ncbi.nlm.nih.gov/pubmed/11266816. Accessed: 3 Aug 2013.
93. Sykes M, Sachs DH. Mixed chimerism. Philos Trans R Soc Lond B. 2001;356(1409):707–26. http://www.pubmedcentral.nih.gov/articlerender.fcgi?artid=1088458&tool=pmcentrez&rendertype=abstract. Accessed: 9 June 2013.
94. Sykes M, et al. Mixed lymphohaemopoietic chimerism and graft-versus-lymphoma effects after non-myeloablative therapy and HLA-mismatched bone-marrow transplantation. Lancet 1999;353(9166):1755–9. http://www.ncbi.nlm.nih.gov/pubmed/10347989. Accessed: 24 July 2013.
95. Tiemessen MM, et al. CD4 + CD25 + Foxp3 + regulatory T cells induce alternative activation of human monocytes/macrophages. Proc Natl Acad Sci U S A. 2007;104(49):19446–51. http://www.ncbi.nlm.nih.gov/pubmed/18042719.
96. Tomita Y, Sachs DH, et al. Additional monoclonal antibody (mAB) injections can replace thymic irradiation to allow induction of mixed chimerism and tolerance in mice receiving bone marrow transplantation after conditioning with anti-T cell mABs and 3-Gy whole body irradiation. Transplantation 1996;61(3):469–77. http://www.ncbi.nlm.nih.gov/pubmed/8610363. Accessed: 25 July 2013.
97. Tomita Y, Khan A, Sykes M. Mechanism by which additional monoclonal antibody (mAB) injections overcome the requirement for thymic irradiation to achieve mixed chimerism in mice receiving bone marrow transplantation after conditioning with anti-T cell mABs and 3-Gy whole body irradi. Transplantation 1996;61(3):477–85. http://www.ncbi.nlm.nih.gov/pubmed/8610364. Accessed: 25 July 2013.
98. Ustüner ET, et al. Swine composite tissue allotransplant model for preclinical hand transplant studies. Microsurgery 2000;20(8):400–6. http://www.ncbi.nlm.nih.gov/pubmed/11150991. Accessed: 28 July 2013.
99. Wachtman GS, et al. Biologics and donor bone marrow cells for targeted immunomodulation in vascularized composite allotransplantation: a translational trial in swine. Transplant Proc. 2011;43(9):3541–4. http://www.ncbi.nlm.nih.gov/pubmed/22099837. Accessed: 3 Aug 2013.
100. Waldmann H, et al. Infectious tolerance and the long-term acceptance of transplanted tissue. Immunol Rev. 2006;212:301–13. http://www.ncbi.nlm.nih.gov/pubmed/16903922. Accessed: 5 Aug 2013.

101. Weaver TA, et al. Alefacept promotes co-stimulation blockade based allograft survival in nonhuman primates. Nat Med. 2009;15(7):746–9. http://www.pubmedcentral.nih.gov/articlerender.fcgi?artid=2772128&tool=pmcentrez&rendertype=abstract. Accessed: 3 Aug 2013.
102. Wekerle T, et al. Extrathymic T cell deletion and allogeneic stem cell engraftment induced with costimulatory blockade is followed by central T cell tolerance. J Exp Med. 1998;187(12):2037–44. http://www.pubmedcentral.nih.gov/articlerender.fcgi?artid=2212372&tool=pmcentrez&rendertype=abstract. Accessed: 19 July 2013.
103. Wolbank S, et al. Dose-dependent immunomodulatory effect of human stem cells from amniotic membrane: a comparison with human mesenchymal stem cells from adipose tissue. Tissue Eng. 2007;13(6):1173–83. http://www.ncbi.nlm.nih.gov/pubmed/17518752. Accessed: 12 Oct 2012.
104. Wong SC, et al. Macrophage polarization to a unique phenotype driven by B cells. Eur J Immunol. 2010;40(8):2296–307. http://www.ncbi.nlm.nih.gov/pubmed/20468007.
105. Wood KJ, Bushell A, Hester J. Regulatory immune cells in transplantation. Nature Rev Immunol. 2012;12(6):417–30. http://www.ncbi.nlm.nih.gov/pubmed/22627860. Accessed: 31 May 2013.
106. Wu A, et al. Regulatory mechanism of peripheral tolerance: *in vitro* evidence for dominant suppression of host responses during the maintenance phase of tolerance to renal allografts in miniature swine. Transplant Immunol. 2003;11(3–4):367–74. http://www.ncbi.nlm.nih.gov/pubmed/12967789. Accessed: 2 Aug 2013.
107. Yanez R, et al. Adipose tissue-derived mesenchymal stem cells have *in vivo* immunosuppressive properties applicable for the control of the graft-versus-host disease. Stem Cell. 2006;24(11):2582–91. http://www.ncbi.nlm.nih.gov/pubmed/16873762.
108. Zheng XX, et al. CTLA4 signals are required to optimally induce allograft tolerance with combined donor-specific transfusion and anti-CD154 monoclonal antibody treatment. J Immunol (Baltimore, MD: 1950). 1999;162(8):4983–90. http://www.ncbi.nlm.nih.gov/pubmed/10202046. Accessed: 3 Aug 2013.
109. Zheng J, et al. CD40-activated B cells are more potent than immature dendritic cells to induce and expand CD4(+) regulatory T cells. Cell Mol Immunol. 2010;7(1):44–50. http://www.ncbi.nlm.nih.gov/pubmed/20081875.

Chapter 15
Mixed Chimerism for Tolerance Induction of Vascularized Composite Allografts

David A. Leonard, Josef M. Kurtz and Curtis L. Cetrulo

Introduction

Vascularized composite allograft (VCA) transplantation was introduced to the modern clinical arena in 1998, when the first successful hand transplant was performed in Lyon, France [1]. The first partial face transplant followed in 2005 [2], and since then more than 80 upper extremity and 20 face transplant procedures have been performed worldwide [3]. VCA transplantation can now be considered an established option in the management of disfiguring injury, amputation, and massive tissue loss, particularly when the specialized anatomic and functional units of the craniofacial region and upper extremity are involved. Although most frequently associated with these regions, any somatic unit composed of multiple tissues, transplanted in a primarily vascularized manner, may be considered a VCA, and abdominal wall, larynx, and lower extremity transplants have also all been performed. Follow up of some of these patients now extends more than 10 years, and results are encouraging, with both impressive objective functional outcomes [3–6] and patient satisfaction and quality of life [3, 7–10]. Indeed, some of these cases, performed many years following injury on patients who had already undergone autologous reconstruction

C. L. Cetrulo (✉)
Vascularized Composite Allotransplantation Laboratory, Transplantation Biology Research Center & Division of Plastic & Reconstructive Surgery, Massachusetts General Hospital/Harvard Medical School, 15 Parkman Street, Boston, MA 02114, USA
e-mail: ccetrulo@mgh.harvard.edu

D. A. Leonard
Transplantation Biology Research Center & Division of Plastic & Reconstructive Surgery, Massachusetts General Hospital/Harvard Medical School, MGH-East, Building 149-9019, 13th Street, Boston, MA 02129, USA
e-mail: david.leonard@tbrc.mgh.harvard.edu

J. M. Kurtz
Transplantation Biology Research Center, Massachusetts General Hospital/Harvard Medical School and Emmanuel College, 400 The Fenway, Boston, MA 02115, USA
e-mail: josef.kurtz@tbrc.mgh.harvard.edu

G. Brandacher (ed.), *The Science of Reconstructive Transplantation*,
Stem Cell Biology and Regenerative Medicine, DOI 10.1007/978-1-4939-2071-6_15

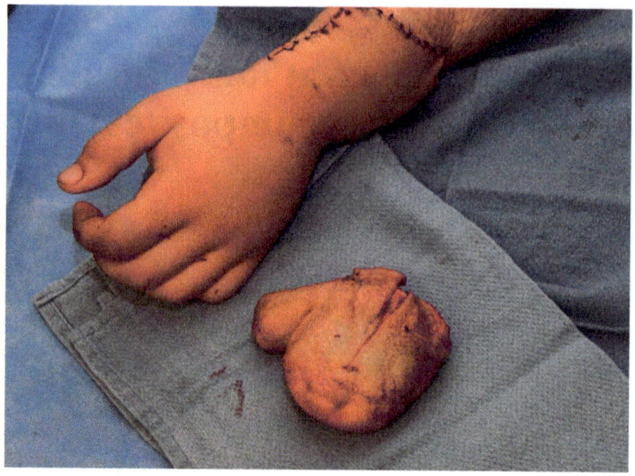

Fig. 15.1 Vascularized composite allotransplantation achieves high-quality restoration of function and form following complex injuries to specialized anatomic subunits

procedures, clearly demonstrate the challenges posed by reconstruction of complex functional units, and illustrate the paradigm shift represented by VCA Fig. 15.1.

However, despite these considerable benefits and the positive clinical outcomes to date, VCA transplantation is not, in contrast to solid organ transplantation, an acutely life-preserving procedure. Therefore, the risk-to-benefit ratio of the surgery, the necessary lifelong immunosuppressive therapy, and the sequelae of immunologic rejection, whether acute or chronic, must be carefully considered. Indeed, the ethical aspects of VCA transplantation, particularly with regards to face transplants, remain a topic of considerable debate [11–13]. While various immunosuppression minimization approaches have shown considerable promise, and are undoubtedly closer to clinical application for VCA if not already in clinical trial [14], we believe that transplant tolerance, defined as a state of specific unresponsiveness to donor antigen, permitting long-term rejection-free acceptance of transplanted tissues without immunosuppression, represents the ultimate goal of research in this field. The potential significance of transplant tolerance for VCA was recognized prior to the first face transplant, by the 2004 working party of the Royal College of Surgeons who stated that "clearly, if it did prove feasible to induce transplant tolerance clinically, this would overcome all the immunological disadvantages of facial transplantation. Indeed, many of the ethical objections to proceeding with clinical evaluation of the procedure would be eliminated" [12]. While clinical evaluation of face transplantation has clearly proceeded, and grafts have been designed to restore a range of injuries from partial loss of the skin and muscles of facial expression [15] to complex three-dimensional craniomaxillofacial defects [16–18] with impressive results, recipients of face transplants, and other VCAs, have experienced a burden of immunosuppressive side effects comparable to that observed in solid organ transplantation, including infection, metabolic and renal impairment, and malignancy [3]. Given the necessity for lifelong immunosuppression to prevent VCA rejection, it can be expected that the impact of these side effects will accrue over time, further highlighting the benefit of a tolerance protocol when one considers that the traumatic extremity and craniofacial injury suitable for reconstructive transplantation are predominantly experienced by a relatively young cohort of patients.

While numerous VCAs have now been performed, the incidence and presentation of chronic rejection in these grafts have not yet become clear. To date, allograft vasculopathy has been reported in one knee transplant [19] and one hand, which presented with progressive ischemia necessitating amputation 9 months post transplant, with widespread intimal hyperplasia of donor arteries demonstrated on histology [20]. While these cases may be attributed to chronic rejection, nonimmune factors may also have contributed, particularly in the hand where procurement of a long segment of isolated donor brachial artery and over use of the hand post transplant were described [20]. In contrast to the small number of confirmed cases of chronic rejection to date, the incidence of acute rejection episodes has been reported to exceed 85 % at 1 year [3], and this has been associated with an increased risk of chronic rejection in animal models of VCA [21]. The apparently low incidence of chronic rejection in VCAs may include some degree of artifact due to the relatively small number of cases available for analysis, but none the less stands in contrast to kidney transplantation, for example, where 10-year survival for living donor transplants is 59 % and only 43 % for cadaveric grafts [22]. Chronic vasculopathy has been associated with humoral immunity and the development of donor-specific antibody in various experimental models of organ transplantation [23], and unfortunately has proved resistant to conventional immunosuppression, despite the considerable improvements observed in control of acute rejection over the past 20 years. Until such time as multicenter cohorts of VCA recipients are available for long-term follow-up and analysis, it is likely that experience in organ transplantation will continue to offer important insights for development of VCA protocols, and further investigation of chronic rejection in the context of VCA would be welcome. Induction of VCA tolerance not only avoids the risks associated with long-term treatment with conventional immunosuppressive regimens but also holds the potential to prevent chronic rejection.

Tolerance and Mixed Chimerism

Tolerance of transplanted organs or tissues, including skin grafts, may be achieved in small animal models by numerous protocols; however, nearly all have failed in large animal studies, which remain an important step in the translation from bench to bedside [24]. Hematopoietic mixed chimerism, defined as the coexistence of hematopoietic cells of donor and recipient origin within an individual has been successfully applied for induction of transplant tolerance in small animals [25], in large animal models [26], and recently in clinical trials in renal transplantation [27–29].

The initial association of hematopoietic mixed chimerism with immune tolerance stemmed from an observation in naturally occurring bovine chimeras (termed free-martin cattle) almost 70 years ago [30]. Medawar and colleagues confirmed this observation in a mouse model by intrauterine injection of fetuses with a preparation of cells from a mismatched adult donor, with subsequent skin grafts from the donor strain accepted indefinitely following birth and maturation [31]. Later experiments by Ildstad and Sachs demonstrated the induction of chimerism and tolerance of transplanted skin grafts in adult mice reconstituted with a mixture of donor and recipient bone marrow following myeloablative conditioning [32]. Ablation of the recipient immune system and hematopoiesis, combined with depletion of mature T cells from the donor bone marrow prior to transplant, permitted engraftment of donor hematopoietic cells while avoiding graft versus host disease (GvHD) and permitted development of a new immune system tolerant of donor and recipient antigens. However, these early protocols consisting of lethal total body irradiation (TBI) and chemotherapeutic agents would be unacceptable for use in patients for the sole purpose of inducing VCA tolerance. Furthermore, rodents have also demonstrated considerable resistance to GvHD following hematopoietic stem cell (HSC) transplantation across major histocompatibility complex (MHC) barriers in comparison to large animal models and humans. Paired with a considerable incidence of engraftment failure when the donor hematopoietic cell graft is T-cell-depleted prior to transplant to reduce the risk of GvHD [33–36], the use of myeloablative protocols becomes unjustifiable.

Thus, subsequent studies focused on development of progressively less toxic, nonmyeloablative conditioning regimens based on T-cell-depleting antibodies, low-dose total body, or targeted thymic irradiation and/or costimulatory blockade [37]. Protocols of this sort have resulted in engraftment of donor HSCs and stable multilineage mixed chimerism and induction of donor-specific unresponsiveness with considerably reduced morbidity and mortality. Indeed, this approach was successfully extended to induction of chimerism and organ transplant tolerance [38–40], and more recently VCA tolerance [41, 42] across MHC barriers in the Massachusetts General Hospital (MGH) miniature swine model.

Mixed chimerism has also been successfully applied for the induction of kidney allograft tolerance across MHC barriers in nonhuman primates [26], and in both human leukocyte antigen (HLA)-matched and HLA-mismatched settings in humans [27–29, 43]. Interestingly, both stable [28] and transient [27] mixed chimerism have been demonstrated to contribute to tolerance of renal allografts in clinical trials. The successful induction of kidney tolerance by transient mixed chimerism is consistent with previous data in both small and large animal models achieving induction of renal allograft tolerance by very mild regimens, in which the kidney has been demonstrated to contribute to its own tolerance [44]. This suggests that in the transient chimerism model in nonhuman primates and clinical protocols, the transplanted kidneys may be contributing to the maintenance of tolerance following loss of mixed chimerism, and the mechanisms may or may not be similar for VCA tolerance. The recent report of a protocol achieving engraftment of HSCs, durable high-level chimerism, and acceptance of kidney transplants without long-term immunosuppression from HLA-mismatched donors [28] is an interesting development while may be of significance for more immunogenic transplants, including VCAs, which lack the tolerogenic capacity of renal or liver grafts. All in all, there is currently no one protocol which will reliably and safely induce stable multilineage mixed chimerism across MHC barriers in humans without GvHD, and further work in this area is undoubtedly required.

Mechanisms of Tolerance in Mixed Chimerism

Definitions of immune tolerance vary, but throughout this discussion we will define tolerance as a state of specific unresponsiveness to donor antigens associated with long-term rejection-free acceptance of transplanted tissues without immunosuppression. Currently, mechanisms for induction and maintenance of immune tolerance are considered in three main groups: deletion, anergy, and regulation/suppression. These mechanisms may be further classified as occurring centrally, referring to intrathymic events such as the clonal deletion of alloreactive thymocytes, or generation of T regulatory cells (Tregs), or peripherally within either secondary lymphoid tissues or at the graft site itself. Similarly, the relative importance of each mechanism may vary during induction and maintenance phases of tolerance, and in protocols achieving transient or limited, rather than stable, multi-lineage mixed chimerism. We will consider each of these mechanisms and scenarios in turn, with a specific focus on tolerance mediated by mixed chimerism, and its relevance to VCA Fig. 15.2.

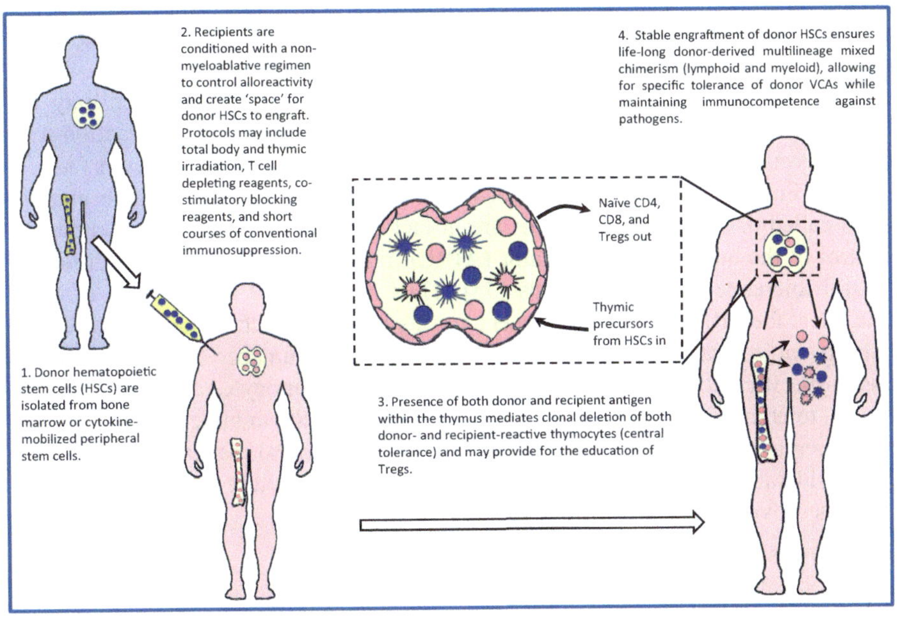

Fig. 15.2 Mechanisms of tolerance in mixed chimerism. *VCA* vascularized composite allograft, *CD* cluster of differentiation

Central and Peripheral Mechanisms of Tolerance

The establishment of immune tolerance is a natural process during immune development, which is essential for discrimination of "self" and "nonself," recognition of pathogens, and avoidance of autoimmunity. If one considers HSC transplantation and the establishment of mixed chimerism as the redefinition of self, it is coherent that the mechanisms by which transplant tolerance is achieved will overlap with those establishing self-tolerance during natural immune system development. Induction of tolerance of donor tissues or organs requires specific unresponsiveness of those immune cells which would otherwise be capable of exerting a destructive effect on the graft. Although cells of the innate immune system, particularly n*atural killer* (NK) cells, have been implicated in some situations [45, 46], allograft rejection is principally mediated by the adaptive immune system. Furthermore, while it has been demonstrated that B cells may be directly tolerized by mechanisms including deletion [47, 48] and anergy [49, 50], evidence for humoral rejection in VCA is lacking [20, 51], and induction of T cell tolerance may be sufficient for tolerance of donor antigens at a functional level [52]. Therefore, we will focus our discussion on the mechanisms of T cell tolerance, as these appear most relevant to VCA.

The thymus is the primary site of T cell development, and also the first step in establishing tolerance at the "population" level. Thus, T cell tolerance established by intrathymic mechanisms is classified as *central* tolerance. In contrast, *peripheral* tolerance is established in mature T cells following encounter with antigen in the peripheral tissues. The underlying mechanisms by which both central and peripheral T cell tolerance can be established include deletion, anergy, and regulation. In the context of self-tolerance, peripheral mechanisms were historically viewed as critical in tolerizing T cells, recognizing tissue-specific antigens which may not have been encountered in the thymus during development through interaction with either thymic epithelium or hematopoietic-derived antigen-presenting cells (APCs) [53,54]. Furthermore, it has now been demonstrated that a considerable number of so-called tissue-specific antigens are in fact expressed in the thymus, under the control of the autoimmune regulator (AIRE) transcription factor and can mediate central tolerance of these antigens [55]. Thus, with regard to induction of VCA tolerance, the efficiency of these mechanisms may reflect the challenge of establishing skin tolerance, even in mixed chimeras, as presumably the skin expresses tissue-specific antigens not presented to developing thymocytes during the time of central tolerance induction. Nonetheless, the role of AIRE in mixed chimerism and transplant tolerance has not been established and peripheral mechanisms are likely of considerable importance, particularly in controlling mature alloreactive T cells which survive pre-transplant conditioning.

Clonal Deletion

Clonal deletion is the core mechanism of central tolerance, mediated in the thymus by recognition of autoantigens presented by medullary and cortical thymic epithelial cells and dendritic cells of hematopoietic origin at the cortico-medullary junction [56, 57]. During T cell development, lymphoid progenitors arising in the bone marrow traffic to the thymus and enter the subcapsular region of the cortex where they proliferate and sequentially express cluster of differentiation (CD)2, CD44, and CD25 and undergo rearrangement of the T cell receptor (TCR) beta chain [58]. Successful rearrangement and cell-surface expression of the beta chain with CD3 and pTa, forming the pre-TCR, is followed by expression of CD4 and CD8. These double-positive thymocytes commence alpha chain rearrangement and migrate toward the cortico-medullary junction. Positive selection, mediated by binding of the rearranged TCR to complexes of MHC and self-peptide on cortical epithelial cells with sufficient affinity to provide a survival signal, and ensure cognate T cell/APC interaction in the periphery is essential within 3–4 days to avoid "death by neglect." But those thymocytes with TCR with high affinity for self-peptide/MHC expressed by hematopoietic-derived dendritic cells at the cortico-medullary junction undergo deletion, also termed negative selection, or clonal diversion to become Tregs [59]. Successfully selected thymocytes differentiate into CD4 and CD8 single-positive populations and migrate to the thymic medulla, expressing high levels of rearranged $\alpha\beta$ TCR and addressins to permit homing to secondary lymphoid organs as naïve T cells.

In mixed chimeras, while thymic epithelial cells remain of solely recipient origin, hematopoietic cells of both donor and recipient origin migrate to the thymus and mediate deletion of both donor-reactive and host-reactive T cell precursors. This results in establishment of a peripheral T cell repertoire tolerant of both donor and host, thus simultaneously controlling graft rejection and GvHD [60, 61]. Furthermore, the persistence of recipient-origin APC populations, alongside cells of donor origin, in the periphery of mixed chimeras facilitates appropriate recipient-restricted immune responses and maintenance of immunocompetence. This is in contrast to full chimeras prepared by myeloablative conditioning, in which the MHC disparity between the thymic epithelial cells responsible for positive selection, and the entirely donor-derived APC pool in the periphery against which the recipient-restricted T cells are significantly limited in mounting an effective response [25].

One of the primary mechanisms by which mixed chimerism maintains long-term, robust tolerance is through the continual presence of donor-derived cells mediating negative selection of newly developing thymocytes. This was demonstrated in a murine model, in which elimination of chimerism using donor-specific anti-MHC antibody resulted in abrogation of tolerance and emergence in the periphery of T cells recognizing donor-presented superantigen. In contrast, if recipient thymectomy was performed prior to elimination of chimerism, donor-specific tolerance was preserved and cells with donor-reactive TCR could not be detected [62]. This result is consistent with a central deletional mechanism of tolerance playing a primary role in maintenance of tolerance in this case, since alternate mechanisms such as anergy and regulation typically required persistence of antigen for maintenance of tolerance.

While central deletion of donor- and recipient-reactive T cells is a highly robust mechanism of tolerance, as cells deleted in the thymus will never be available in the periphery and therefore unable to contribute to pathology, the thymus does not express all antigens expressed by all peripheral tissues. Deletion of mature T cells in the periphery has been described for both CD4 and CD8 T cells in antigen-specific manners [63–65]. Peripheral deletion occurs in a similar fashion to central deletion, through apoptotic cell death, with evidence for both Fas- and Bim- apoptosis pathways operating in a coordinated manner to achieve elimination of chronically self-stimulated T cells *in vivo* [59].

Achieving robust tolerance is clearly of significant importance in the development of protocols for clinical application, and the strength of deletional tolerance has been well established, permitting acceptance of highly immunogenic grafts (including skin and small bowel) across stringent histocompatibility barriers in small animal models [66–68]. Given the well-established challenges of inducing tolerance of transplanted skin, and the facility with which mixed chimerism achieves clonal deletion, it is likely that protocols achieving VCA tolerance will be built on a primarily deletional mechanism. However, other mechanisms, principally anergy and regulation, may contribute, particularly in the early post-transplant phase, prior to completion of the deletion process [69].

Anergy

The classic definition of anergy is a state in which a T cell is rendered unresponsive to an antigen by binding of a peptide-MHC complex (pMHC) in the absence of appropriate costimulatory signals [70–72]. Cellular anergy has been associated with reduced interleukin (IL)-2 production, altered intracellular signaling and tyrosine phosphorylation patterns [73–75]. Recent studies have demonstrated that inhibition of the mammalian target of rapamycin (mTOR) pathway following T cell activation via CD3 and CD28 is sufficient for induction of anergy, and that nutrient-sensing pathways such as activation of the hypoxia and adenosine triphosphate (ATP) deprivation sensor adenosine monophosphate (AMP)-activated protein kinase (AMPK), may play a role in this process [76].

TCR engagement in the context of co-inhibitory molecule ligation such as programmed death 1 (PD-1) and cytotoxic T lymphocyte antigen-4 (CTLA-4) has also been shown to induce an anergic state in T cells. Binding of PD-1 by its ligands PD-L1 and PD-L2 can limit the expansion of autoreactive T cells [77]. Similarly, CTLA-4 binding of CD80/86 can transduce a negative signal that blocks the cell cycle progression [78]. Interestingly, Tregs may promote local hypoxia via expression of CD39 and CD73 which convert ATP to AMP, and AMP to adenosine, respectively, suggesting that the Tregs may in part modulate T cell activity through anergy [79, 80]. Overall, a constellation of genes has been associated with anergy, and interestingly many of which overlap with the expression profiles of cells undergoing peripheral deletion [59, 81].

Evidence for anergy in mixed chimerism was demonstrated in a murine bone marrow transplantation (BMT) model, using a costimulatory blockade-based induction regimen. Peripheral donor-reactive CD4 T cells were rapidly rendered anergic post transplant, with evidence of complete unresponsiveness to donor at week 1, prior to progressive deletion over a period of weeks [82]. Similarly, the sequential deletion of T cells following induction of anergy has been reported in several other studies [83, 84]. However, induction of anergy does not always precede deletion, as anergic T cells have been shown to survive long term [85,86]) and may in fact exhibit regulatory activity by modulating the activity of other effector T cells [87, 88].

Classically [89], but not universally [90], T cell anergy may be abrogated *in vitro* by reexposure to cognate antigen in the presence of exogenous IL-2. There is also evidence for reversal of anergy *in vivo* both by infection [91] and removal of antigen [92–95]. Clearly, these findings would raise concern regarding the reliability and robustness of any protocol in which anergy was the primary mechanism for maintenance of tolerance over the long term.

Regulation

Specialized cell populations, initially termed "suppressor cells," capable of attenuating and controlling immune responsiveness in an antigen-specific manner, were identified in the 1970s [96, 97]. Efforts to define the cellular characteristics and molecular mechanisms responsible for suppression achieved limited success until 1990 when it was demonstrated in a rodent model of cardiac allograft transplantation that tolerance could be adoptively transferred by $CD4^{+}CD25^{+}$ cells [98]. Subsequently, considerable strides have been made in characterizing this cell population, now referred to as Tregs, in both mice and humans [99, 100].

One of the major hurdles in dissecting the specific roles and contribution of Tregs to transplantation tolerance is the heterogeneity of both cell surface marker and transcription factor expression in various model systems and species. While regulatory function has been demonstrated for non-$CD4^{+}$ cell populations including $CD8^{+}$ [101–103], $CD8^{+}CD28-$ [104], $TCR^{+}CD4-CD8-$ [105], natural killer T (NKT) [106, 107], and myeloid suppressor cells [108], $CD4^{+}$ Tregs remain the most fully characterized. The cardinal markers associated with this population of Tregs in humans are considered to be cell-surface expression of CD25 (the alpha subunit of the high affinity IL-2 receptor), constitutive expression of the transcription factor forkhead box P3 (FoxP3) and both cell-surface and cytoplasmic expression of the co-inhibitory receptor cytotoxic T lymphocyte antigen 4 (CTLA-4) [109]. Each of these molecules has been demonstrated to be of functional significance in Tregs, respectively providing survival signals to the Treg (and presumably mediating suppression by IL-2 absorption), promoting transcription of factors essential for Treg development and survival (including CD25 and FOXP3), and outcompeting CD28 for binding of CD80 and CD86 on APCs [110]. Furthermore, Tregs may downregulate expression of CD80/CD86 via a CTLA-4-dependent mechanism, exerting suppressive activity through APC modification [100, 111]. Another area of active research in Tregs that has direct implication for the field of transplantation is the

Fig. 15.3 Tolerance is induced rapidly following hematopoietic stem cell transplantation, with the contribution of nondeletional mechanisms declining over time as clonal deletion progresses

origin of Treg generation. It has been shown that Tregs can arise both during thymic development through intermediate-affinity interactions with antigens expressed in the context of MHC class II on nonhematopoietic thymic epithelial cells in the cortex (nTregs) [112–114] or from naïve CD4 T cell precursors in the periphery (iTregs) [115].

While the principle role of Tregs is maintenance of self-tolerance, and resolving active immune responses to pathogens, they are also the focus of considerable research in the field of transplantation, where they have been demonstrated to play a role in prevention of organ allograft rejection [116–118] and to attenuate GvHD following BMT [53, 119]. Populations of cells enriched for Tregs have also been identified in allograft recipients, both in the peripheral lymphoid tissues [116] and at the graft-site itself [120]. Indeed, $FoxP3^+$ cells have been identified in the skin of accepted hand transplants as much as 6 years post transplant on conventional immunosuppressive protocols [121], and experimentally, in the skin of VCAs transplanted between MHC-matched canines [122].The homing and activation of Tregs are now recognized to mirror that of conventional T cells, and appear to be critical for optimal protection of transplants, including skin grafts, against anti-allo responses [123, 124]. However, as yet the functional significance of Tregs for survival or tolerance of VCAs has not been clearly demonstrated.

The evidence for Treg function in the maintenance of tolerance in mixed chimerism models is sparse [125], but during the induction phase of chimerism, or in models where only low levels of chimerism are achieved, it is conceivable that the contribution of regulatory mechanisms to tolerance may be more significant [126, 127], as in this setting the deletion of donor-reactive T cells appears to be less complete [128] Fig. 15.3.

Mixed Chimerism: Transient Versus Stable

In elucidating and determining the contribution by each of the various mechanisms of tolerance invoked through use of HSC transplantation and mixed chimerism, another major aspect that must be considered is the nature of the chimerism achieved, specifically whether it is *transient* or *stable* in nature. The majority of protocols utilized in small animal models (from extensive T cell depletion to costimulatory blockade) result in engraftment and establishment of long-term multilineage chimerism. In these situations, the continual contribution and differentiation of pluripotent HCSs into common lymphoid and myeloid progenitors ensure sustained input of donor-derived cells, facilitating both central and peripheral mechanisms of tolerance. However, in the translation of these small animal models into large animal models and the clinic, it has become evident that establishing stable chimerism with contribution from both host-derived and donor-derived HSCs is much more challenging: in most cases, recipients either fail to maintain donor chimerism for the long term (as determined by peripheral blood analysis by either flow cytometry or polymerase chain reaction (PCR), and lack of evidence for donor HSC in bone marrow samples), or they spontaneously convert to full-donor chimerism, with the attendant risk of GvHD.

With regards to the requirement of sustained donor hematopoiesis for establishment and maintenance of tolerance for organ transplantation, the current data in large animal and clinical trials do not yet provide clear insight into the necessity of the various mechanisms for optimal outcomes. While the initial studies described for combined HSC and kidney transplant described successful engraftment of donor HSCs and long-term hematopoiesis [43], these patients were being treated for both kidney failure and hematological disorders. However, further studies demonstrated that only transient chimerism was necessary for long-term kidney acceptance [27, 129], suggesting that peripheral mechanisms were sufficient and that the kidney may play an active role in promoting its acceptance. From a therapeutic perspective, this has the benefit of avoiding the risk of GvHD in patients, although recent studies have demonstrated the establishment of long-term donor chimerism in the absence of GvHD and concurrent acceptance of kidneys without immunosuppression [28].

Infusion of donor bone marrow cells has been included as part of the induction regimen for some cases from the outset of the clinical VCA era, with mixed evidence for effect. The recipient of the first face transplant received two donor bone marrow infusions, but chimerism was not detected; there was no evidence for modulation of anti-donor responses and comparable to protocols not including bone marrow, the recipient has experienced a number of rejection episodes despite conventional immunosuppression [2]. In contrast, a protocol including alemtuzumab induction and donor bone marrow infusion has been shown to have a positive effect, permitting acceptance of upper extremity transplants on significantly reduced maintenance immunosuppression (tacrolimus monotherapy) [14]. From a mechanistic perspective, as these patients have not received any conditioning that would likely facilitate engraftment of donor HSCs and the establishment of long-term chimerism, the infusion of donor marrow most likely is providing a window of donor-specific nonresponsiveness, mediated through either peripheral deletion of donor-reactive lymphocytes or induction of regulatory cells. However, in contrast to the kidney, as yet there is little evidence to suggest that VCAs will be able to contribute to their own long-term acceptance, and the short-term effects of bone marrow infusion may not be sufficient to ensure immunosuppression-free survival due to the inherent qualities of the VCA Fig. 15.4.

Fig. 15.4 Protocols achieving stable mixed chimerism result in the persisting contribution of donor cells to the thymus, and sustained clonal deletion. Protocols achieving transient or limited chimerism may require greater contribution from nondeletional mechanisms to maintain tolerance

Issues Specific to VCA Tolerance

VCA transplantation has only relatively recently entered substantial clinical application, and consequently, many well-accepted paradigms in solid organ transplantation remain to be specifically tested in VCA models. While it is likely that many of the basic principles of transplant acceptance and rejection will be shared, and protocols comparable to those sufficient for prevention of solid organ transplant rejection in clinical practice have achieved an impressively low rate of VCA loss to rejection [3], some aspects specific to VCA warrant further consideration and investigation.

The relative difficulty in achieving acceptance of skin across allogeneic barriers in comparison to kidney and other organ transplants is well established. Classically, this has been attributed to skin's unique mode of transplantation (as a secondarily vascularized split or full-thickness skin graft), or to the presence of skin-specific alloantigens [130]. Presumably, the prolonged period of relative ischemia and subsequent ischemia–reperfusion injury following secondary establishment of circulation contribute to a local inflammatory milieu and presentation of donor antigen to the immune system in a strongly immunogenic fashion. However, it has been observed in both experimental models and clinical practice that the skin component of a primarily VCA is particularly susceptible to acute rejection episodes, suggesting a unique intrinsic factor of the skin.

It is now well established that skin provides more than a simple mechanical barrier in the protection against pathogens, and is a site of significant immunologic activity. The identification of Langerhan's cells as professional APCs within the epidermis [131], and the ability of epidermal cells to stimulate T cells directly [132] suggested a model where skin can serve as a site of immune sampling and effector immune function. Elucidation of a network of multiple immunologically active cell types within skin suggests that in fact the "skin immune system" may be considered another unique and distinct component of the overall immune system [133].

The scale of the skin immune system has only recently begun to be appreciated, as recognition that the skin of a normal, healthy adult contains approximately 1×10^6 T cells/cm^2;almost twice the number circulating in blood. These T cells express the skin-homing addressins cutaneous lymphocyte-associated antigen (CLA) and chemokine (C-C motif) receptor (CCR)4, and the majority lack expression of CD62L and CCR4, a phenotype consistent with effector memory T cells. The presence of such a significant population of T effector memory (T_{EM}) cells in normal skin stands in contrast to the conventional model where T_{EM} remain primarily in the circulation until recruited to the tissues by inflammation. Furthermore, these skin-resident T_{EM} cells persist long term without recirculation, and appear to provide global cutaneous immunity which accumulates over time [134, 135].

Tregs have also been identified in human skin, where they represent between 5 and 10% of the resident T cell population [136, 137]. Interestingly, expansion of cutaneous Tregs has recently been demonstrated in response to corticosteroid treatment, through upregulation of TGF-β secretion by Langerhans' cells [138], suggesting a mechanism through which topical application of clobetasol may act in the treatment of acute rejection episodes in VCA patients.

While skin represents a well-established challenge to transplant tolerance induction, both when transplanted in isolation and more recently as a component of VCA, there is evidence to suggest that VCAs may in fact be less immunogenic than the sum of their parts [139]. One frequent outcome in large animal studies has been the induction of split tolerance, with long-term acceptance of musculoskeletal components of limb allografts but rejection of skin, across both minor histocompatibility antigen (miHA) and MHC barriers. In some cases, rejection has been targeted specifically at epidermis with maintenance of viable dermal tissue [140, 141].

The establishment of mixed chimerism by HSC transplantation, and the association of chimerism and tolerance have been extensively discussed. Clearly, the conditioning regimens necessary to achieve engraftment of donor HSCs retain some degree of toxicity and morbidity, despite considerable advances in the development of reduced intensity regimens. It has been suggested that the vascularized BMT (VBMT) may offer benefits over cellular BMT in this regard, by providing a source of donor HSCs pre-engrafted within appropriate cellular architecture.

Much of the data concerning the immunologic role of VBMT has been collected from experimental models of hind-limb transplantation in rats [142, 143]. The utility of VBMT as a source of functional hematopoiesis was demonstrated by the complete repopulation with donor cells of bone marrow and secondary lymphoid organs in lethally irradiated recipients [144, 145]. Subsequent studies demonstrated induction of mixed chimerism, long-term-immunosuppression-free survival of the allograft, and *in vitro* evidence of tolerance of donor antigen in a parental to F1 hybrid setting [146]. More recently, a number of small animal models of VBMT-containing VCA have been introduced, permitting study of both immunologic aspects of these procedures and functional responses to transplantation and various conditioning and immunosuppressive regimens [147].

VCAs containing a vascularized bone marrow compartment have also been studied in large animal systems, including heterotopic hind-limb models in both porcine [140, 141] and canine systems [122], and partial face allografts in nonhuman primates [148]. In a series of studies in MGH miniature swine, tolerance of musculoskeletal components, but not skin, was observed in untreated recipients of MHC-matched, minor antigen-mismatched limb transplants. In contrast to some small animal studies, evidence for sustained donor hematopoiesis was not observed either in peripheral blood or in the recipient bone marrow compartment and by 48 weeks post transplant donor cells could no longer be isolated from the donor marrow compartment, which had been repopulated with recipient origin cells [149]. In a subsequent study in which limb transplants were performed in conjunction with either cellular BMT or mobilized HSCs following irradiation-free conditioning, split tolerance was once again demonstrated across MHC barriers. Multilineage mixed chimerism was detected early in some, but not all, recipients and when present was observed to gradually decline over time [140]. Thus, the acceptance of some components of the VCA cannot be entirely attributed to the VBMT. The potential role of VBMT was studied more specifically using a nonhuman primate model of heterotopic partial face transplantation, in which the facial segment was transplanted with or without hemi-mandible [150]. Immunosuppression was maintained with calcineurin inhibitors and antiproliferative agents, but no conditioning or cytoreductive treatments were given. In contrast to rejection of VCAs lacking VBMT within 50 days, none of those containing VBMT rejected while on immunosuppression. However, tolerance was not achieved in this model, as withdrawal of immunosuppression between 205 and 430 days post transplant lead to rejection, suggesting that attenuation of the immune response against VCAs while under therapeutic immunosuppression may offer encouragement for development of clinical immunosuppression minimization protocols for VCAs containing functional marrow compartments Fig. 15.5.

Practical Considerations for Translating VCA Tolerance to Clinical Practice

Significant strides have been made in our understanding of the immunobiology of VCA since the introduction of these procedures to the clinical arena. Encouragingly, many of these findings have been reproducible in large animal models, which represent an important step from laboratory bench to clinical practice. However, these studies have also served to illustrate the unique aspects of achieving tolerance in VCA.

In contrast to kidneys, for which transient chimerism has proven sufficient, it appears that stable multi-lineage mixed chimerism may be a prerequisite for VCA tolerance. To date, the application of HSC transplantation for induction of stable chimerism in humans remains limited by the toxicity of conditioning regimens traditionally used to achieve engraftment, and by the risk of GvHD which until relatively recently precluded transplantation in all but MHC-matched scenarios [151]. However, progress in the field of BMT for treatment of malignant and nonmalignant diseases with increased focus on nonmyeloablative and reduced intensity conditioning regimens, including recent protocols utilizing post-transplant cyclophosphamide therapy for preferential depletion of alloreactive T cell clones, has resulted in the introduction of MHC mismatched BMT to clinical practice [152]. Thus, the use of HSC transplantation for induction of tolerance is continuing to become a more viable option.

Another major consideration which clinically applicable VCA tolerance protocols must address is the obvious necessity of procurement from deceased donors.

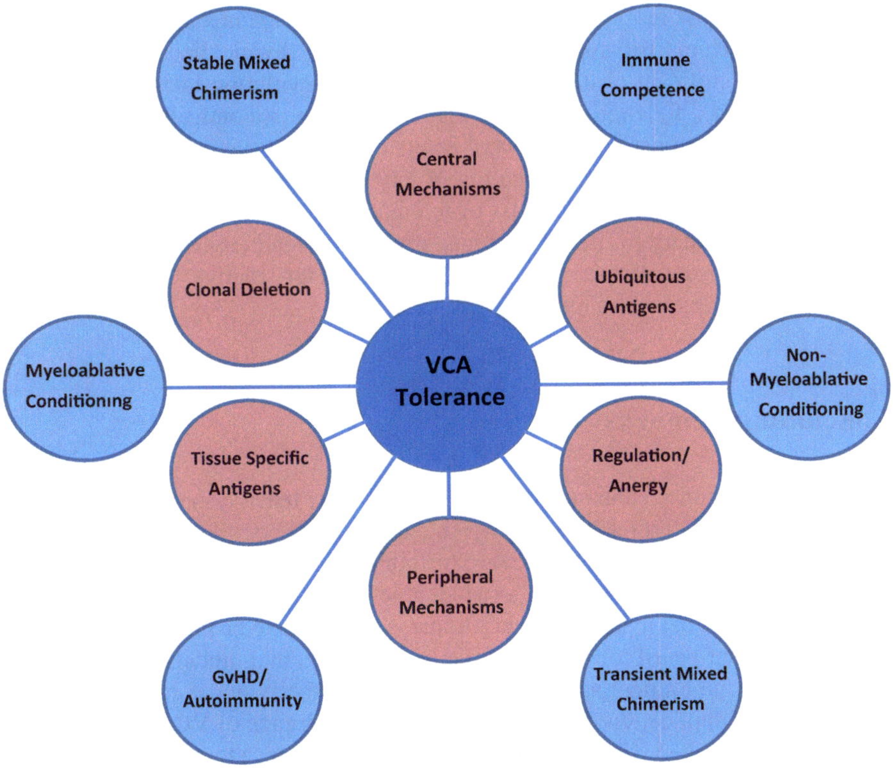

Fig. 15.5 Immune tolerance requires a dynamic balance of multiple interrelated mechanisms. Clinically applicable protocols for VCA tolerance will also have to strike a balance between the level and duration of chimerism necessary for tolerance, the intensity of conditioning necessary to achieve chimerism, and potential sequelae. *VCA* vascularized composite allograft

This is in contrast to current tolerance protocols for kidney transplantation in which there are two significant differences: First, HSCs have been collected from living donors which influence the structure of the recipient conditioning regimen, as conditioning of a VCA recipient prior to the day of transplant will be impractical. Second, the choice of HSC source used for induction of chimerism. While mechanistic studies in small animal models typically utilize bone marrow for induction of chimerism which similarly can serve as a source of HSCs in clinical practice, it has become more common to use cytokine-mobilized peripheral blood stem cells as an alternative. While cytokine-mobilized HSC have successfully been utilized for induction of chimerism across MHC barriers without GvHD [40, 153] and induction of VCA tolerance [41], the necessity for treatment of the donor with cytokines to mobilize hematopoietic stem and progenitor cells from the marrow to the peripheral blood for collection by apheresis limits applicability in the deceased donor setting.

Finally, mechanisms involved in the induction and maintenance of tolerance through mixed chimerism approaches are dependent on thymic function. While this may seem a significant obstacle in light of thymic involution in adulthood, persistence of thymic function, despite the absence of an anatomically well-defined thymus, can be demonstrated in normal adults even in old age [154]. Furthermore, in experiments in aged mice, abrogation of long-term mixed chimerism with anti-donor MHC antibody resulted in loss of tolerance and re-emergence of anti-donor T cells in the periphery [62]; thus demonstrating that, even in senescence, sufficient thymic function is maintained to generate de novo donor-reactive T cells in the absence of donor antigen within the thymus.

Concluding Remarks

Transplantation and reconstructive surgery have an extensive shared history, and researchers at the interface of these two specialties have contributed much to both fields [155]. The introduction of VCA to the clinical arena has shifted the emphasis on skin transplantation from being a stringent test of experimental protocols to becoming a clinical goal in its own right. Progress continues to be made in translating mixed chimerism-based tolerance protocols to large animal models and, for renal transplantation, to clinical application. However, further work is required before regular clinical application of protocols achieving tolerance of VCAs through the mechanisms described above becomes reality, and we believe that such an achievement will be inherently reliant on the availability of safe and effective protocols for establishment of stable mixed chimerism across HLA barriers. To this end, it should be acknowledged the state of the art with regard to surgical aspects of VCA is impressive, allowing these complex procedures to be performed with relatively high frequency and with a low incidence of surgical complications. Progress towards the implementation of tolerance protocols will require preclinical and translation research efforts, and effective collaborations between immunologist, hematologist, and surgical scientist in the field of VCA (Figs. 15.1, 15.2, 15.3, 15.4, and 15.5).

References

1. Dubernard JM, Owen E, Herzberg G, et al. Human hand allograft: report on first 6 months. Lancet 1999;353:1315–20.
2. Dubernard JM, Lengele B, Morelon E, et al. Outcomes 18 months after the first human partial face transplantation. N Engl J Med. 2007;357:2451–60.
3. Petruzzo P, Lanzetta M, Dubernard JM, et al. The international registry on hand and composite tissue transplantation. Transplantation 2010;90:1590–4.

4. Kaufman CL, Breidenbach W. World experience after more than a decade of clinical hand transplantation: update from the Louisville hand transplant program. Hand Clin. 2011;27: 417–viii.
5. Petruzzo P, Kanitakis J, Badet L, et al. Long-term follow-up in composite tissue allotransplantation: in-depth study of five (hand and face) recipients. Am J Transplant. 2011;11:808–16.
6. Siemionow M, Ozturk C. An update on facial transplantation cases performed between 2005 and 2010. Plast Reconstr Surg. 2011;128: 707e–20e.
7. Jensen SE, Butt Z, Bill A, et al. Quality of life considerations in upper limb transplantation: review and future directions. J Hand Surg Am. 2012;37:2126–35.
8. Petruzzo P, Testelin S, Kanitakis J, et al. First human face transplantation: 5 years outcomes. Transplantation 2012;93:236–40.
9. Singhal D, Pribaz JJ, Pomahac B. The Brigham and Women's Hospital face transplant program: a look back. Plast Reconstr Surg. 2012;129:81e–8e.
10. Cavadas PC, Landin L, Thione A, et al. The Spanish experience with hand, forearm, and arm transplantation. Hand Clin. 2011;27:443–53, viii.
11. Chang J, Mathes DW. Ethical, financial, and policy considerations in hand transplantation. Hand Clin. 2011;27:553–60, xi.
12. Morris PJ, Bradley JA, Doyal L, et al. Facial transplantation: a working party report from the Royal College of Surgeons of England. Transplantation 2004;77:330–8.
13. Paradis C, Siemionow M, Papay F, et al. Ethical considerations in the first American face transplant. Plast Reconstr Surg. 2010;126:896–901.
14. Schneeberger S, Gorantla VS, Brandacher G, et al. Upper-extremity transplantation using a cell-based protocol to minimize immunosuppression. Ann Surg. 2013;257:345–51.
15. Devauchelle B, Badet L, Lengele B, et al. First human face allograft: early report. Lancet 2006;368:203–9.
16. Dorafshar AH, Bojovic B, Christy MR, et al. Total face, double jaw, and tongue transplantation: an evolutionary concept. Plast Reconstr Surg. 2013;131:241–51.
17. Pomahac B, Pribaz JJ, Bueno EM, et al. Novel surgical technique for full face transplantation. Plast Reconstr Surg. 2012;130:549–55.
18. Siemionow MZ, Papay F, Djohan R, et al. First U.S. near-total human face transplantation: a paradigm shift for massive complex injuries. Plast Reconstr Surg. 2010;125:111–22.
19. Schneeberger S, Landin L, Jableki J, et al. Achievements and challenges in composite tissue allotransplantation. Transpl Int. 2011;24:760–9.
20. Kaufman CL, Ouseph R, Blair B, et al. Graft vasculopathy in clinical hand transplantation. Am J Transplant. 2012;12:1004–16.
21. Unadkat JV, Schneeberger S, Horibe EH, et al. Composite tissue vasculopathy and degeneration following multiple episodes of acute rejection in reconstructive transplantation. Am J Transplant. 2010;10:251–61.
22. UNOS. 2009 Annual Report of the U.S. Organ Procurement and Transplantation Network and the Scientific Registry of Transplant Recipients: Transplant Data 1999–2008; 2013. Rockville, MD, HHS/HRSA/HSB/DOT. 9–2–2013.
23. Bishop DK, Li W, Chan SY, Ensley RD, Shelby J, Eichwald EJ. Helper T lymphocyte unresponsiveness to cardiac allografts following transient depletion of CD4-positive cells. Implications for cellular and humoral responses. Transplantation 1994;58:576–84.
24. Sachs DH. Tolerance: of mice and men. J Clin Invest. 2003;111:1819–21.
25. Sykes M, Sachs DH. Mixed chimerism. Philos Trans R Soc Lond B. 2001;356:707–26.
26. Kawai T, Cosimi AB, Colvin RB, et al. Mixed allogeneic chimerism and renal allograft tolerance in cynomolgus monkeys. Transplantation 1995;59:256–62.
27. Kawai T, Cosimi AB, Spitzer TR, et al. HLA-mismatched renal transplantation without maintenance immunosuppression. N Engl J Med. 2008;358:353–61.
28. Leventhal J, Abecassis M, Miller J, et al. Chimerism and tolerance without GVHD or engraftment syndrome in HLA-mismatched combined kidney and hematopoietic stem cell transplantation. Sci Transl Med. 2012;4:124ra28.

29. Scandling JD, Busque S, Dejbakhsh-Jones S, et al. Tolerance and withdrawal of immunosuppressive drugs in patients given kidney and hematopoietic cell transplants. Am J Transplant. 2012;12:1133–45.
30. Owen RD. Immunogenetic consequences of vascular anastomoses between bovine twins. Science 1945;102:400–1.
31. Billingham RE, Brent L, Medawar PB. Actively acquired tolerance of foreign cells. Nature 1953;172:603–6.
32. Ildstad ST, Sachs DH. Reconstitution with syngeneic plus allogeneic or xenogeneic bone marrow leads to specific acceptance of allografts or xenografts. Nature 1984;307(5947): 168–70.
33. Bordignon C, Keever CA, Small TN, et al. Graft failure after T-cell-depleted human leukocyte antigen identical marrow transplants for leukemia: II. *In vitro* analyses of host effector mechanisms. Blood 1989;74: 2237–43.
34. Kernan NA, Bordignon C, Heller G, et al. Graft failure after T-cell-depleted human leukocyte antigen identical marrow transplants for leukemia: I. Analysis of risk factors and results of secondary transplants. Blood 1989;74:2227–36.
35. Martin PJ, Hansen JA, Torok-Storb B, et al. Graft failure in patients receiving T cell-depleted HLA-identical allogeneic marrow transplants. Bone Marrow Transplant. 1988;3:445–56.
36. Martin PJ. Prevention of allogeneic marrow graft rejection by donor T cells that do not recognize recipient alloantigens: potential role of a veto mechanism. Blood 1996;88:962–9.
37. Sykes M. Mixed chimerism and transplant tolerance. Immunity 2001;14:417–24.
38. Fuchimoto Y, Huang CA, Yamada K, et al. Mixed chimerism and tolerance without whole body irradiation in a large animal model. J Clin Invest. 2000;105:1779–89.
39. Guzzetta PC, Sundt TM, Suzuki T, Mixon A, Rosengard BR, Sachs DH. Induction of kidney transplantation tolerance across MHC barriers by bone marrow transplantation in miniature swine. Transplantation 1991;51:862–6.
40. Huang CA, Fuchimoto Y, Scheier-Dolberg R, Murphy MC, Neville DM Jr, Sachs DH. Stable mixed chimerism and tolerance using a nonmyeloablative preparative regimen in a large-animal model. J Clin Invest. 2000;105:173–81.
41. Horner BM, Randolph MA, Duran-Struuck R, et al. Induction of tolerance to an allogeneic skin flap transplant in a preclinical large animal model. Transplant Proc. 2009;41:539–41.
42. Leonard DA, Kurtz JM, Mallard C et al. Vascularized composite allograft tolerance across MHC barriers in a large animal model. Am J Transplant 2014; 14: 343–55
43. Spitzer TR, Delmonico F, Tolkoff-Rubin N, et al. Combined histocompatibility leukocyte antigen-matched donor bone marrow and renal transplantation for multiple myeloma with end stage renal disease: the induction of allograft tolerance through mixed lymphohematopoietic chimerism. Transplantation 1999;68:480–4.
44. Sachs DH, Sykes M, Kawai T, Cosimi AB. Immuno-intervention for the induction of transplantation tolerance through mixed chimerism. Semin Immunol. 2011;23:165–73.
45. Kitchens WH, Uehara S, Chase CM, Colvin RB, Russell PS, Madsen JC. The changing role of natural killer cells in solid organ rejection and tolerance. Transplantation 2006;81:811–7.
46. Kondo T, Morita K, Watarai Y, et al. Early increased chemokine expression and production in murine allogeneic skin grafts is mediated by natural killer cells. Transplantation 2000;69:969–77.
47. Li Y, Ma L, Shen J, Chong AS. Peripheral deletion of mature alloreactive B cells induced by costimulation blockade. Proc Natl Acad Sci U S A. 2007;104:12093–8.
48. Li Y, Ma L, Yin D, Shen J, Chong AS. Long-term control of alloreactive B cell responses by the suppression of T cell help. J Immunol. 2008;180:6077–84.
49. Benschop RJ, Aviszus K, Zhang X, Manser T, Cambier JC, Wysocki LJ. Activation and anergy in bone marrow B cells of a novel immunoglobulin transgenic mouse that is both hapten specific and autoreactive. Immunity 2001;14:33–43.
50. Fang W, Weintraub BC, Dunlap B, et al. Self-reactive B lymphocytes overexpressing Bcl-xL escape negative selection and are tolerized by clonal anergy and receptor editing. Immunity 1998;9:35–45.

51. Kanitakis J, McGregor B, Badet L, et al. Absence of c4d deposition in human composite tissue (hands and face) allograft biopsies: an immunoperoxidase study. Transplantation 2007;84:265–7.
52. Sykes M, Wood K, Sachs D. Transplantation immunology. In: Paul W, editor. Fundamental immunology. Philadelphia, PA: Lippincott Williams and Wilkins; 2012.
53. Hoffmann MW, Heath WR, Ruschmeyer D, Miller JF. Deletion of high-avidity T cells by thymic epithelium. Proc Natl Acad Sci U S A. 1995;92:9851–5.
54. Oukka M, Colucci-Guyon E, Tran PL, et al. CD4 T cell tolerance to nuclear proteins induced by medullary thymic epithelium. Immunity 1996;4:545–53.
55. Anderson MS, Venanzi ES, Klein L, et al. Projection of an immunological self shadow within the thymus by the aire protein. Science 2002;298:1395–401.
56. Starr TK, Jameson SC, Hogquist KA. Positive and negative selection of T cells. Annu Rev Immunol. 2003;21:139–76.
57. Griesemer AD, Sorenson EC, Hardy MA. The role of the thymus in tolerance. Transplantation 2010;90:465–74.
58. Koch U, Radtke F. Mechanisms of T cell development and transformation. Annu Rev Cell Dev Biol. 2011;27:539–62.
59. Xing Y, Hogquist KA. T-cell tolerance: central and peripheral. Cold Spring Harb Perspect Biol. 2012;4:a006957.
60. Manilay JO, Pearson DA, Sergio JJ, Swenson KG, Sykes M. Intrathymic deletion of alloreactive T cells in mixed bone marrow chimeras prepared with a nonmyeloablative conditioning regimen. Transplantation 1998;66:96–102.
61. Tomita Y, Khan A, Sykes M. Role of intrathymic clonal deletion and peripheral anergy in transplantation tolerance induced by bone marrow transplantation in mice conditioned with a nonmyeloablative regimen. J Immunol. 1994;153:1087–98.
62. Khan A, Tomita Y, Sykes M. Thymic dependence of loss of tolerance in mixed allogeneic bone marrow chimeras after depletion of donor antigen. Peripheral mechanisms do not contribute to maintenance of tolerance. Transplantation 1996;62:380–7.
63. Kurts C, Kosaka H, Carbone FR, Miller JF, Heath WR. Class I-restricted cross-presentation of exogenous self-antigens leads to deletion of autoreactive CD8(+) T cells. J Exp Med. 1997;186:239–45.
64. Rocha B, von Boehmer H. Peripheral selection of the T cell repertoire. Science 1991;251: 1225–8.
65. Webb SR, Hutchinson J, Hayden K, Sprent J. Expansion/deletion of mature T cells exposed to endogenous superantigens *in vivo*. J Immunol. 1994;152:586–97.
66. Main JM, Prehn RT. Successful skin homografts after the administration of high dosage X radiation and homologous bone marrow. J Natl Cancer Inst 1955;15:1023–9.
67. Orloff MS, Fallon MA, DeMara E, Coppage ML, Leong N, Cerilli J. Induction of specific tolerance to small-bowel allografts. Surgery 1994;116:222–8.
68. Sharabi Y, Sachs DH. Mixed chimerism and permanent specific transplantation tolerance induced by a nonlethal preparative regimen. J Exp Med. 1989;169:493–502.
69. Kurtz J, Wekerle T, Sykes M. Tolerance in mixed chimerism—a role for regulatory cells? Trends Immunol. 2004;25:518–23.
70. Ramsdell F, Fowlkes BJ. Clonal deletion versus clonal anergy: the role of the thymus in inducing self tolerance. Science 1990;248:1342–8.
71. Schwartz RH. A cell culture model for T lymphocyte clonal anergy. Science 1990;248: 1349–56.
72. Schwartz RH. T cell anergy. Annu Rev Immunol. 2003;21:305–34.
73. Jenkins MK, Pardoll DM, Mizuguchi J, Chused TM, Schwartz RH. Molecular events in the induction of a nonresponsive state in interleukin 2-producing helper T-lymphocyte clones. Proc Natl Acad Sci U S A. 1987;84:5409–13.
74. McKay DB, Irie HY, Hollander G, et al. Antigen-induced unresponsiveness results in altered T cell signaling. J Immunol. 1999;163:6455–61.

75. Migita K, Eguchi K, Kawabe Y et al. Defective TCR-mediated signaling in anergic T cells. J Immunol. 1995;155:5083–7.
76. Powell JD, Delgoffe GM. The mammalian target of rapamycin: linking T cell differentiation, function, and metabolism. Immunity 2010;33:301–11.
77. Probst HC, McCoy K, Okazaki T, Honjo T, van den Broek M. Resting dendritic cells induce peripheral CD8+ T cell tolerance through PD-1 and CTLA-4. Nat Immunol. 2005;6:280–6.
78. Greenwald RJ, Boussiotis VA, Lorsbach RB, Abbas AK, Sharpe AH. CTLA-4 regulates induction of anergy *in vivo*. Immunity 2001;14:145–55.
79. Chappert P, Schwartz RH. Induction of T cell anergy: integration of environmental cues and infectious tolerance. Curr Opin Immunol 2010;22:552–9.
80. Sitkovsky MV. T regulatory cells: hypoxia-adenosinergic suppression and re-direction of the immune response. Trends Immunol. 2009;30:102–8.
81. Parish IA, Rao S, Smyth GK, et al. The molecular signature of CD8+ T cells undergoing deletional tolerance. Blood 2009;113:4575–85.
82. Kurtz J, Shaffer J, Lie A, Anosova N, Benichou G, Sykes M. Mechanisms of early peripheral CD4 T-cell tolerance induction by anti-CD154 monoclonal antibody and allogeneic bone marrow transplantation: evidence for anergy and deletion but not regulatory cells. Blood 2004;103:4336–43.
83. Chai JG, Bartok I, Scott D, Dyson J, Lechler R. T:T antigen presentation by activated murine CD8+ T cells induces anergy and apoptosis. J Immunol. 1998;160:3655–65.
84. Fink PJ, Fang CA, Turk GL. The induction of peripheral tolerance by the chronic activation and deletion of CD4+ V beta 5+ cells. J Immunol. 1994;152:4270–81.
85. Migita K, Ochi A. The fate of anergic T cells *in vivo*. J Immunol. 1993;150:763–70.
86. Rocha B, Grandien A, Freitas AA. Anergy and exhaustion are independent mechanisms of peripheral T cell tolerance. J Exp Med. 1995;181:993–1003.
87. Chai JG, Bartok I, Chandler P et al. Anergic T cells act as suppressor cells *in vitro* and *in vivo*. Eur J Immunol. 1999;29:686–92.
88. Lombardi G, Sidhu S, Batchelor R, Lechler R. Anergic T cells as suppressor cells *in vitro*. Science 1994;264:1587–9.
89. Schwartz RH. Costimulation of T lymphocytes: the role of CD28, CTLA-4, and B7/BB1 in interleukin-2 production and immunotherapy. Cell 1992;71:1065–8.
90. Bhandoola A, Cho EA, Yui K, Saragovi HU, Greene MI, Quill H. Reduced CD3-mediated protein tyrosine phosphorylation in anergic CD4+ and CD8+ T cells. J Immunol. 1993;151:2355–67.
91. Rocken M, Urban JF, Shevach EM. Infection breaks T-cell tolerance. Nature 1992;359: 79–82.
92. Morecki S, Leshem B, Eid A, Slavin S. Alloantigen persistence in induction and maintenance of transplantation tolerance. J Exp Med. 1987;165:1468–80.
93. Pape KA, Merica R, Mondino A, Khoruts A, Jenkins MK. Direct evidence that functionally impaired CD4+ T cells persist *in vivo* following induction of peripheral tolerance. J Immunol. 1998;160:4719–29.
94. Ramsdell F, Fowlkes BJ. Maintenance of *in vivo* tolerance by persistence of antigen. Science 1992;257:1130–4.
95. Rocha B, Tanchot C, von Boehmer H. Clonal anergy blocks *in vivo* growth of mature T cells and can be reversed in the absence of antigen. J Exp Med. 1993;177:1517–21.
96. Gershon RK, Kondo K. Infectious immunological tolerance. Immunology 1971;21:903–14.
97. Kilshaw PJ, Brent L, Pinto M. Suppressor T cells in mice made unresponsive to skin allografts. Nature 1975;255:489–91.
98. Hall BM, Pearce NW, Gurley KE, Dorsch SE. Specific unresponsiveness in rats with prolonged cardiac allograft survival after treatment with cyclosporine. III. Further characterization of the CD4+ suppressor cell and its mechanisms of action. J Exp Med. 1990;171:141–57.
99. Sakaguchi S, Miyara M, Costantino CM, Hafler DA. FOXP3+ regulatory T cells in the human immune system. Nat Rev Immunol. 2010;10:490–500.

100. Sakaguchi S, Vignali DA, Rudensky AY, Niec RE, Waldmann H. The plasticity and stability of regulatory T cells. Nat Rev Immunol. 2013;13:461–7.
101. Gilliet M, Liu YJ. Generation of human CD8 T regulatory cells by CD40 ligand-activated plasmacytoid dendritic cells. J Exp Med. 2002;195:695–704.
102. Hutchinson IV. Suppressor T cells in allogeneic models. Transplantation 1986;41:547–55.
103. Zhou J, Carr RI, Liwski RS, Stadnyk AW, Lee TD. Oral exposure to alloantigen generates intragraft CD8+ regulatory cells. J Immunol. 2001;167:107–13.
104. Ciubotariu R, Colovai AI, Pennesi G et al. Specific suppression of human CD4+ Th cell responses to pig MHC antigens by CD8+ . J Immunol. 1998;161:5193–202.
105. Zhang ZX, Yang L, Young KJ, DuTemple B, Zhang L. Identification of a previously unknown antigen-specific regulatory T cell and its mechanism of suppression. Nat Med. 2000;6:782–9.
106. Seino KI, Fukao K, Muramoto K, et al. Requirement for natural killer T (NKT) cells in the induction of allograft tolerance. Proc Natl Acad Sci U S A. 2001;98:2577–81.
107. Zeng D, Lewis D, Dejbakhsh-Jones S, et al. Bone marrow NK1.1(−) and NK1.1(+) T cells reciprocally regulate acute graft versus host disease. J Exp Med. 1999;189:1073–81.
108. Wood KJ, Bushell A, Hester J. Regulatory immune cells in transplantation. Nat Rev Immunol. 2012;12:417–30.
109. Sakaguchi S, Yamaguchi T, Nomura T, Ono M. Regulatory T cells and immune tolerance. Cell 2008;133:775–87.
110. Peach RJ, Bajorath J, Brady W, et al. Complementarity determining region 1 (CDR1)- and CDR3-analogous regions in CTLA-4 and CD28 determine the binding to B7–1. J Exp Med. 1994;180: 2049–58.
111. Oderup C, Cederbom L, Makowska A, Cilio CM, Ivars F. Cytotoxic T lymphocyte antigen-4-dependent down-modulation of costimulatory molecules on dendritic cells in CD4+ CD25+ regulatory T-cell-mediated suppression. Immunology 2006;118:240–9.
112. Bensinger SJ, Bandeira A, Jordan MS, Caton AJ, Laufer TM. Major histocompatibility complex class II-positive cortical epithelium mediates the selection of CD4(+)25(+) immunoregulatory T cells. J Exp Med. 2001;194:427–38.
113. Seddon B, Mason D. The third function of the thymus. Immunol Today 2000;21:95–9.
114. Shevach EM. Regulatory T cells in autoimmmunity*. Annu Rev Immunol. 2000;18: 423–49.
115. Wood KJ, Sakaguchi S. Regulatory T cells in transplantation tolerance. Nat Rev Immunol. 2003;3:199–210.
116. Hara M, Kingsley CI, Niimi M et al. IL-10 is required for regulatory T cells to mediate tolerance to alloantigens *in vivo*. J Immunol. 2001;166:3789–96.
117. Kingsley CI, Karim M, Bushell AR, Wood KJ. CD25+ CD4+ regulatory T cells prevent graft rejection: CTLA-4- and IL-10-dependent immunoregulation of alloresponses. J Immunol. 2002;168:1080–6.
118. Sakaguchi S, Sakaguchi N, Shimizu J et al. Immunologic tolerance maintained by CD25+ CD4+ regulatory T cells: their common role in controlling autoimmunity, tumor immunity, and transplantation tolerance. Immunol Rev. 2001;182:18–32.
119. Cohen JL, Trenado A, Vasey D, Klatzmann D, Salomon BL. CD4(+)CD25(+) immunoregulatory T cells: new therapeutics for graft-versus-host disease. J Exp Med. 2002;196:401–6.
120. Graca L, Cobbold SP, Waldmann H. Identification of regulatory T cells in tolerated allografts. J Exp Med. 2002;195:1641–6.
121. Eljaafari A, Badet L, Kanitakis J, et al. Isolation of regulatory T cells in the skin of a human hand-allograft, up to six years posttransplantation. Transplantation 2006;82:1764–8.
122. Mathes DW, Hwang B, Graves SS, et al. Tolerance to vascularized composite allografts in canine mixed hematopoietic chimeras. Transplantation 2011;92:1301–8.
123. Issa F, Robb RJ, Wood KJ. The where and when of T cell regulation in transplantation. Trends Immunol. 2013;34:107–13.

124. Issa F, Hester J, Milward K, Wood KJ. Homing of regulatory T cells to human skin is important for the prevention of alloimmune-mediated pathology in an *in vivo* cellular therapy model. PLoS One. 2012;7:e53331.
125. Kurtz J, Wekerle T, Sykes M. Tolerance in mixed chimerism – a role for regulatory cells? Trends Immunol. 2004;25:518–23.
126. Bemelman F, Honey K, Adams E, Cobbold S, Waldmann H. Bone marrow transplantation induces either clonal deletion or infectious tolerance depending on the dose. J Immunol. 1998;160:2645–8.
127. Fehr T, Takeuchi Y, Kurtz J, Wekerle T, Sykes M. Early regulation of CD8 T cell alloreactivity by CD4+. Eur J Immunol. 2005;35:2679–90.
128. Wekerle T, Kurtz J, Ito H, et al. Allogeneic bone marrow transplantation with co-stimulatory blockade induces macrochimerism and tolerance without cytoreductive host treatment. Nat Med. 2000;6:464–9.
129. Kawai T, Sachs DH, Sykes M, Cosimi AB. HLA-mismatched renal transplantation without maintenance immunosuppression. N Engl J Med. 2013;368:1850–2.
130. Steinmuller D. The enigma of skin allograft rejection. Transplant Rev. 1998;12:42–57.
131. Stingl G, Katz SI, Clement L, Green I, Shevach EM. Immunologic functions of Ia-bearing epidermal Langerhans cells. J Immunol. 1978;121:2005–13.
132. Rubenfeld MR, Silverstone AE, Knowles DM, et al. Induction of lymphocyte differentiation by epidermal cultures. J Invest Dermatol. 1981;77:221–4.
133. Bos JD, Zonneveld I, Das PK, Krieg SR, van der Loos CM, Kapsenberg ML. The skin immune system (SIS): distribution and immunophenotype of lymphocyte subpopulations in normal human skin. J Invest Dermatol. 1987;88:569–73.
134. Clark RA, Watanabe R, Teague JE, et al. Skin effector memory T cells do not recirculate and provide immune protection in alemtuzumab-treated CTCL patients. Sci Transl Med. 2012;4:117ra7.
135. Jiang X, Clark RA, Liu L, Wagers AJ, Fuhlbrigge RC, Kupper TS. Skin infection generates non-migratory memory CD8+ T(RM) cells providing global skin immunity. Nature 2012;483:227–31.
136. Clark RA, Kupper TS. IL-15 and dermal fibroblasts induce proliferation of natural regulatory T cells isolated from human skin. Blood 2007;109:194–202.
137. Clark RA, Huang SJ, Murphy GF, et al. Human squamous cell carcinomas evade the immune response by down-regulation of vascular E-selectin and recruitment of regulatory T cells. J Exp Med. 2008;205:2221–34.
138. Stary G, Klein I, Bauer W, et al. Glucocorticosteroids modify Langerhans cells to produce TGF-beta and expand regulatory T cells. J Immunol. 2011;186:103–12.
139. Lee WP, Yaremchuk MJ, Pan YC, Randolph MA, Tan CM, Weiland AJ. Relative antigenicity of components of a vascularized limb allograft. Plast Reconstr Surg. 1991;87:401–11.
140. Hettiaratchy S, Melendy E, Randolph MA, et al. Tolerance to composite tissue allografts across a major histocompatibility barrier in miniature swine. Transplantation 2004;77:514–21.
141. Mathes DW, Randolph MA, Solari MG, et al. Split tolerance to a composite tissue allograft in a swine model. Transplantation 2003;75:25–31.
142. Black KS, Hewitt CW, Fraser LA, et al. Composite tissue (limb) allografts in rats. II. Indefinite survival using low-dose cyclosporine. Transplantation 1985;39:365–8.
143. Hewitt CW, Black KS, Fraser LA, et al. Composite tissue (limb) allografts in rats. I. Dose-dependent increase in survival with cyclosporine. Transplantation 1985;39:360–4.
144. Lukomska B, Durlik M, Cybulska E, Olszewski WL. Comparative analysis of immunological reconstitution induced by vascularized bone marrow versus bone marrow cell transplantation. Transpl Int. 1996;9(Suppl 1): S492–6.
145. Lukomska B, Durlik M, Cybulska E, Olszewski WL. Reconstitution of lymphoid tissue after vascularized bone marrow transplantation. Ann Transplant. 1996;1:67–9.
146. Hewitt CW, Ramsamooj R, Patel MP, Yazdi B, Achauer BM, Black KS. Development of stable mixed T cell chimerism and transplantation tolerance without immune modulation in recipients of vascularized bone marrow allografts. Transplantation 1990;50:766–72.

147. Siemionow M, Nasir S. Impact of donor bone marrow on survival of composite tissue allografts. Ann Plast Surg. 2008;60:455–62.
148. Barth RN, Bluebond-Langner R, Nam A, et al. Facial subunit composite tissue allografts in nonhuman primates: I. Technical and immunosuppressive requirements for prolonged graft survival. Plast Reconstr Surg. 2009;123:493–501.
149. Mathes DW, Randolph MA, Bourget JL, et al. Recipient bone marrow engraftment in donor tissue after long-term tolerance to a composite tissue allograft. Transplantation 2002;73:1880–5.
150. Barth RN, Rodriguez ED, Mundinger GS, et al. Vascularized bone marrow-based immunosuppression inhibits rejection of vascularized composite allografts in nonhuman primates. Am J Transplant. 2011;11:1407–16.
151. Sykes M, Preffer F, McAfee S, et al. Mixed lymphohaemopoietic chimerism and graft-versus-lymphoma effects after non-myeloablative therapy and HLA-mismatched bone-marrow transplantation. Lancet 1999;353:1755–9.
152. Luznik L, O'Donnell PV, Fuchs EJ. Post-transplantation cyclophosphamide for tolerance induction in HLA-haploidentical bone marrow transplantation. Semin Oncol. 2012;39: 683–93.
153. Cina RA, Wikiel KJ, Lee PW, et al. Stable multilineage chimerism without graft versus host disease following nonmyeloablative haploidentical hematopoietic cell transplantation. Transplantation 2006;81:1677–85.
154. Haynes BF, Markert ML, Sempowski GD, Patel DD, Hale LP. The role of the thymus in immune reconstitution in aging, bone marrow transplantation, and HIV-1 infection. Annu Rev Immunol. 2000;18:529–60.
155. Murray JE. Organ transplantation (skin, kidney, heart) and the plastic surgeon. Plast Reconstr Surg. 1971;47:425–31.

Chapter 16
Bone Marrow-Derived *Ex Vivo* Created Hematopoietic Chimeric Cells to Support Engraftment and Maintain Long-Term Graft Survival in Reconstructive Transplantation

Maria Siemionow, Joanna Cwykiel and Maria Madajka

List of abbreviations

ACI	August Copenhagen Irish (An inbred rat strain derived by breeding an August male with an Irish coat and a Copenhagen 2331 female)
Anti-αβTCR monoclonal antibody	Anti-αβ T cell receptor monoclonal antibody
BMCs	Bone marrow cells
BMT	Bone marrow transplantation
CsA	Cyclosporine A
CTL	cytotoxic T lymphocytes
GVHD	Graft-versus-host disease
Gy	Gray (radiation unit)
HLA	Human leukocyte antigen
LBN	Lewis Brown Norway (An inbred rat strain derived by breeding a Brown Norway male and a Lewis female)
mAb	monoclonal antibodies
MHC	Major histocompatibility complex
MLR	Mixed lymphocyte reaction
NK	Natural killer cells
TBI	Total body irradiation
T-reg	Regulatory T cells
VCA	Vascularized composite allotransplantation

M. Siemionow (✉) · J. Cwykiel
Department of Orthopaedics, University of Illinois at Chicago, 900 S Ashland Ave, Room 3356 MBRB, Chicago, IL 60607, USA
e-mail: siemiom@uic.edu

M. Madajka
Department of Plastic Surgery, Cleveland Clinic, Cleveland, OH, USA

G. Brandacher (ed.), *The Science of Reconstructive Transplantation*,
Stem Cell Biology and Regenerative Medicine, DOI 10.1007/978-1-4939-2071-6_16

Chimerism as an Approach to the Problem of Tolerance Induction in Transplantation

The Origins of Chimerism in Medicine

In Greek mythology, there is a fabled creature possessing the strength and body parts of many animals called chimera. One of the most famous chimera statues is "The Chimera of Arezzo" of 1553, which is made of bronze and found on display in the National Archaeological Museum of Florence, Italy. This chimera is a mythic three-headed monster. The special feature characterizing the chimera is its indestructibility; such a creature is able to compensate for its shortcomings because it has the strength of another creature in addition to its own. The term "chimera" permanently entered medical vernacular and is understood today as an experimental animal or a human that accepts another's genetic makeup, tolerates it, and becomes its permanent host. These donor-derived cells live and differentiate, in the host body and can be delivered to the host body in a variety of methods. The most familiar method is chimerism resulting from organ transplant, which was first displayed in the case of kidney and liver transplantation and originally treated more as an interesting side effect rather than a common posttransplantation interaction between donor and recipient. For almost 70 years, scientists have been debating over the interpretation of this phenomenon. We know now that donor cell engraftment in the recipient body is usually a sign of transplantation success [72]. This means the body has developed tolerance toward the foreign organ and has become a chimera. However, it must be emphasized that the word *chimerism* is understood not only as a result of solid organ and VCA transplantation, but also as a natural phenomenon during pregnancy. This aspect of chimerism is not covered in this chapter.

Several basic questions are important to understand chimerism. These questions include how long the chimeric state will last, whether these cells will induce or be induced to create a state of tolerance to support graft survival, and whether the development of chimerism will lead to chronic problems in the long-term. It is also important to know if these genetically distinct cell types will live without any side effects in the body untill the end of the recipient's life. These essential questions open the scientific debate that supports continuation of multidisciplinary projects in many scientific centers around the world.

Chimerism and Tolerance

The existence of chimerism is often interpreted as evidence of transplantation success and is applicable to solid organ, bone marrow, and stem cell transplantation. For many decades, scientists have been debating over the interpretation of this phenomenon. Chimerism can be viewed today as a form of a tolerance induction therapy to support engraftment and maintain long-term graft survival in solid organs, as well as in the relatively new and still developing field of reconstructive transplantation. The main goal of such a therapeutic approach is to understand a complex mechanism of self and nonself discrimination by the recipient's immune system and would apply to our new bone marrow-based cellular therapy of donor–recipient chimeric cell creation where the cells are carrying antigens specific for both donors. Such an advantage would open unlimited options for induction and modulation therapies in patients receiving solid organ or VCA transplants. Traditional cellular therapies in the form of bone marrow-derived cells are still the most common protocols for tolerance induction in transplantation. This therapeutic approach is changing, toward the development of more defined cellular therapies including chimeric cell protocols.

The first application of chimerism as an approach for tolerance induction was reported in the literature in the early 1950s [9, 10]. Billingham, Brent, and Medawar published a paper in 1953 describing the process of acquiring tolerance to foreign cells [10]. They reported the induction of transplantation tolerance by injection of donor hematopoietic cells into neonatal mice within the first 24 h after birth [92]. These classic experiments performed by Billingham showed that it was possible to engraft splenic leukocytes or bone marrow cells (BMCs) into the newborn mice recipients which were immunologically immature and unable to reject them [108]. Moreover, the recipient mice carrying donor leukocytes were able to accept a skin graft from the donor leukocyte strain which indicated that leukocyte chimerism was supporting the acquired tolerance [108]. Many years later, in 1984, Ildstadt and Sachs published the results of their study about tolerance induction through mixed chimerism [37]. Sachs' experimental design included a total body irradiation of the recipient mice and its reconstitution by a mixture of the T-cell-depleted bone marrow of the recipient with the bone marrow of the donor [92]. As the results showed, animals receiving a mixture of donor and recipient BMCs became and remained lymphohematopoietic chimeras for the rest of their lives without the development of graft-versus-host disease (GVHD) or complications due to T cell depletion of the host bone marrow [92]. Moreover, long-term graft acceptance was achieved after the transplantation of tail skin grafts onto the lateral thorax of the irradiated mice. Despite the success of achieving long-term tolerance following skin graft transplantation, one major disadvantage of this experimental model was the need for recipient conditioning in the form of irradiation. Irradiation can cause irreversible changes, such as depletion of the bone marrow compartment, and patients may develop toxic side effects. Due to these reasons, the irradiation protocol tested in this experimental model has significant limitations for clinical application. However, despite the negative aspects of total body irradiation, this study proved that mixed chimerism may lead to the development of long-term tolerance following skin graft transplantation [37, 92]. Another study performed by Ilstadt in 1984 showed significant limitations in the use of animal models to study the role of chimerism in tolerance induction [37]. The development of animal models closely relevant to a clinical scenario is always challenging, and there are no perfect animal models for tolerance and chimerism studies. Based on the experimental data obtained in the

Table 16.1 Milestone experimental protocols in the development of chimerism approach

Protocol	Description	References
In vivo T cell depletion	6 Gy TBI	[48]
In vivo T cell depletion	Irradiation of the thymic/mediastinal area, reduction in TBI to 3 Gy	[99]
In vivo T cell depletion	Thymic irradiation combined with T-cell-depleting antibodies	[116]
In vivo T cell depletion	High dose of nonmyeloablative TBI—6.5 Gy	[5]
Co-stimulation blockers	Anti-CD40 ligand with or without CTLA4-Ig and 3 Gy TBI	[98, 122]
Co-stimulation blockers	Anti-CD4 and anti-CD9 with anti-CD40 ligand	[26]
Administration of T-reg cells	T-reg cells administrated after myeloablative conditioning (8.5 Gy)	[40]
Administration of T-reg cells	T-reg cells combined with BMT, depleting antibodies and TBI (10 Gy)	[83]
In vivo T cell depletion and BMT	Canine model, low-dose TBI combined with BMT and immunosuppression	[50, 51]
In vivo T-cell depletion	Miniature swine model, CD3 antibody and CsA	[34]
Co-stimulation blockers	Nonhuman primates model, TBI, CsA, anti-CD154 antibody	[45]
Co-stimulation blockers	Rhezus monkey model, nonmyeloablative conditioning with busulfan, anti-CD40 ligand antibody, sirolimus, anti-CD25 antibody, BMT	[47]

TBI total body irradiation, *BMT* bone marrow transplantation

late 1970s and 1980s, which included irradiation protocols, a new approach to study chimerism and tolerance induction was proposed. A total depletion of the recipient's T cells was replaced by the introduction of T-cell-depleting monoclonal antibodies (Table 16.1). This scenario seemed to be less toxic and much more clinically acceptable.

It needs to be emphasized that the animal studies assessing the mechanism of tolerance induction via chimerism are crucial for the transition of this approach toward clinical applications.

Scientific efforts and experimental protocols of Billingham, Ilstadt, Sachs, and many other investigators studying tolerance induction in animal models opened a new era in the field of solid organ and VCA transplantation. As a natural follow-up to these experimental studies, clinical trials on tolerance induction were introduced. The first reports referring to chimerism were based on bone marrow transplantation (BMT). In the late 1970s, reports confirmed that donor-specific blood transfusion in humans often led to the improved allograft acceptance in kidney transplant recipients [75, 76, 93, 119]. Consequently, studies on mice [65], dogs [15, 61, 29], and monkeys [113, 114] showed that the infusion of the donor BMCs combined with the depletion of recipient T cells prolonged allograft survival without the need for toxic and chronic immunosuppression. The first clinical attempt to use donor BMCs in combination with polyclonal antilymphocyte globulin therapy was performed by Monaco et al. in kidney transplant patients [66]. The use of BMT was performed as a supportive cellular therapy in solid organ transplants. Initially, piloted in only a few patients, bone marrow grafts (11×10^9 BMCs) were given 21–25 days after kidney transplantation in combination with immunosuppressive protocol consisting of: azathioprine, antilymphocyte globulin, and prednisone [66]. The early results were encouraging. Decreased levels of kidney graft rejection and decreasing levels of donor responsiveness were observed [1]. In another study, post-kidney transplant administration of donor BMCs ($2–3 \times 10^8$ BMCs/kg) was performed and supported by cyclosporine A (CsA), prednisone, azathioprine, and antilymphocyte globulin immunosuppressive protocol [61]. Although not randomized, improved kidney graft survival was observed at both 12 and 18 months in the group that received donor BMCs [6]. However, no differences were observed in renal graft function or in the rejection episodes between the two groups [1, 31]. In 1997, the first randomized trial was performed in liver transplant patients and supported by perioperative BMT showing significantly better results in BMT cases, thus favoring the protocol of multiple donor bone marrow infusions. Specifically, both patient and liver graft survival were greater in patients who received multiple bone marrow infusions than the controls who did not receive bone marrow or who received a bone marrow infusion only on the day of liver transplant. In addition, this study confirmed that cytoablative conditioning was not necessary to improve allograft survival when the recipient was given multiple bone marrow infusions [1, 86]. Recently, another study on kidney transplantation was performed following total lymphoid irradiation, antithymocyte globulin, CsA, prednisone, and mycophenolate mofetil (MMF) followed by the administration of 1×10^6 $CD3^+$ T cells and 8×10^6 $CD34^+$-enriched hematopoietic cells/kg [94]. Impressively, this regimen established mixed chimerism and tolerance toward the kidney allograft to such an extent that all immunosuppressive medications were discontinued at 6 months after transplantation. Subsequently, this protocol led to stable kidney graft function and withdrawal of immunosuppression in eight out of 12 patients [1, 95].

In contrast to the augmentation of chimerism by BMT, donor microchimerism commonly refers to chimerism that is detectable after solid organ transplantation without BMT support. In these cases, microchimerism is a consequence of passenger leukocytes migrating out of transplanted tissues. Starzl reported one of the first cases indicating the existence of chimerism after liver transplantation [110, 115]. His group reported systemic microchimerism after a successful liver transplant and suggested that immunosuppression in transplant recipients promotes the microchimeric state. Examples of clinical studies confirmed that the essential condition required for the development of donor chimerism is the migration of passenger leukocytes from transplanted tissue into the host. Confirmation of leukocyte migration was reported in kidney transplants where a large leukocyte population originating from the recipient cells was found [109]. Donor-derived cells were also found in the skin and lymph nodes of the recipients [109]. The presence of donor DNA was confirmed by polymerase chain reaction (PCR) and detection of peripheral blood chimerism. Moreover, donor-specific nonreactivity was confirmed by mixed lymphocyte reaction (MLR) assays. The fact that donor cells were found in the lymphoid organs, peripheral blood, and bone marrow compartments confirms their ability to migrate and be distributed throughout the body. This suggests that a cumulative effect of donor cell engraftment must be considered as a significant phenomenon.

In 1990, Larsen et al. published their observations regarding migration and function of dendritic leukocytes after transplantation [55]. They reported that the infused donor BMCs or passenger leukocytes that migrated out of the transplanted organ in the nonimmunosuppressed murine recipients first circulated in the peripheral blood and then disappeared [63]. Starzl and Elwood observed a similar continuing pattern of BMCs circulation in the posttransplantation patients [24, 91]. The highest concentration of the infused BMCs was recorded in the peripheral blood up to three months following transplantation and then gradually decreased to the minimal levels by one year after cellular therapy infusion [24]. These observations indicate that there are preferential sites for bone marrow-derived cell engraftment as well as differences in cell concentration and localization throughout the recipient body.

Questions thus arise as to how the chimeric cells replicate or prolong their proliferation and long-term survival in the recipient body? What are the immunological mechanisms and functions that allow for the acceptance and existence of the chimeric cells? Based on animal experimental models and experiences in human organ transplantation, the phenomenon of mutual immunosuppression plays an important role in the maintenance of graft survival.

Chimerism-Based Strategies for Tolerance Induction: Cleveland Clinic Experience

Since confirmation of the crucial role of passenger leukocytes by Starzl in his pioneering work, the quest for discovering immunomodulatory properties of chimerism and its correlation with tolerance induction in solid organs and vascularized composite allograft (VCA) transplants is still being investigated [20, 110]. During the last decade, the Siemionow laboratory has been actively involved in the development of tolerance-inducing strategies by introducing new VCA models and performing basic science research on new tolerance inducing immunosuppressive protocols and cell-based therapies.

Siemionow's team designed a variety of VCA experimental models and evaluated the effects of different tolerance inducing protocols on the development of donor-specific chimerism and VCA transplant survival (Table 16.2). Using the single-tissue (skin) vascularized allotransplantation model, the Siemionow team showed that donor-derived cells present in the graft are migrating to the recipient's lymphoid compartments and leading to chimerism induction. Furthermore, these studies confirmed the importance of immunosuppressive protocol adjustments for chimerism induction and maintenance.

Table 16.2 Experimental VCA models for chimerism and tolerance induction developed and tested in the Siemionow laboratory

Model	Immunosuppressive/immunodepletive therapy	Supportive cellular therapy	Study conclusion	References
Total abdominal wall transplantation (LBN to Lewis rat)	CsA tapered from 16 mg/kg/day to 2 mg/kg/day over 4-week period	–	Single-tissue component (skin) under adequate immunosuppressive therapy is capable of long-term chimerism induction and extends allograft survival	[70]
Vascularized and nonvascularized skin allograft transplanta-tionLBN to Lewis rat)	CsA tapered from 16 mg/kg/day to 2 mg/kg/day over 4-week period	–	Development of donor-derived chimerism is affected by the allograft size and vascularization	[71]
Limb semi-allogenic transplant(LBN to Lewis rat)	Comparison between fluocinolone acetonide (50 mg/ml) applied topically, cyclosporine A 4 mg/kg/day,combined systemic cyclosporine with topical fluocinolone acetonide	Vascularized bone marrow component of the graft	Extended allograft survival was accomplished by combination of low-dose CsA and topical steroids	[38]
	Comparison between ALS only (0.4 mL/kg), CsA only (16 mg/kg), and a combination of CsA and ALS, conditioning was administered 12 h before surgery at three different intervals (7, 14, and 21 days)	Vascularized bone marrow component of the graft	Correlation of donor-specific hema-topoietic chimerism with transplant tolerance	[80]
Limb allogenic trans-plant (BN to Lewis rat)	CsA and ALS were administered 2 h before surgery and continued for 21 days. Tapered CsA,16 mg/kg/day,1st week; 8 mg/kg/day, 2nd week; 4 mg/kg/day, 3rd week. ALS—0.4 ml/day for the 1st week, every other day for the 2nd week, and twice weekly during the 3rd week	Vascularized bone marrow component of the graft	Extended survival of the allogenic limb transplant associated with tran-sient chimerism	[78]
Face/scalp semi-allogenic (LBN to Lewis rat)	CsA 16 mg/kg/day every 24 h after transplanta-tion; tapered to 2 mg/kg/day over 4-week period and maintained at that level thereafter	–	Confirmation of the feasibility of the model with > 170 days survival on maintenance immunosuppression	[117]

Table 16.2 (continued)

Model	Immunosuppressive/immunodepletive therapy	Supportive cellular therapy	Study conclusion	References
Limb semi-allogenic transplant (LBN to Lewis rat)	35-day course of 250 μg/kg/day anti-αβTCR antibody and 16 mg/kg/day CsA	Vascularized bone marrow component of the graft	Confirmation of induction of donor-specific tolerance to rat hind-limb allografts (graft survival over XX)	[100]
Skin allograft (LBN to Lewis rat)	35-day course of 250 μg/kg/day anti-αβTCR antibody and 16 mg/kg/day CsA	"crude" bone marrow transplantation	Combination of anti-αβTCR antibody and CsA extended skin allograft survival up to 65 days	[105]
Limb semi-allogenic transplant (LBN to Lewis rat)	21-, 7-, and 5-day course of 250 μg/kg/day anti-αβTCR antibody and 16 mg/kg/day CsA	Vascularized bone marrow component of the graft	Clinical tolerance and immunocompetence were confirmed by skin grafting *in vivo* and MLR *in vitro*. High level of donor chimerism in the peripheral blood of long-term survivors was detected	[79]
Limb allogenic transplant (BN to Lewis rat)	7-day protocol of 250 μg/kg/day anti-αβTCR antibody and 16 mg/kg/day CsA	Vascularized bone marrow component of the graft	Seven days combined therapy of anti-αβTCR and CsA induced tolerance in rat hind-limb allografts. Tolerance was directly associated with stable, donor-specific chimerism	[102]
Hemiface transplantation(LBN to Lewis rat and ACI to Lewis rat)	CsA 16 mg/kg/day, tapered to 2 mg/kg/day and maintained at that level thereafter	–	Survival in 100% of LBN graft up to 400 days and ACI up to 330 days Low-dose immunosuppression facilitates engraftment of donor-derived cells into the lymphoid organs (lymph nodes and spleen) and supports chimerism induction	[103]
Hemiface/calvaria transplantation(LBN to Lewis rat)	CsA 16 mg/kg/day tapered to 2 mg/kg per day over 4-week period and maintained thereafter	Vascularized bone marrow component of graft	New reconstructive graft model with survival up to 220 days. Development of donor chimerism—predominantly in the B cell population	[125]

Table 16.2 (continued)

Model	Immunosuppressive/immunodepletive therapy	Supportive cellular therapy	Study conclusion	References
Maxilla transplantation(LBN to Lewis rat)	CsA 16 mg/kg/day tapered to 2 mg/kg per day over 3-week period and maintained thereafter	Vascularized bone marrow component of graft	Maxilla graft survival up to 105 days. High level of donor-specific chimerism for T cell and B cell lineages was maintained in the peripheral blood	[126]
Hemiface/mandible/tongue transplantation (LBN to Lewis rat)	CsA 16 mg/kg/day tapered to 2 mg/kg per day over 4-week period and maintained thereafter	Vascularized bone marrow component of graft	Long-term allograft survival (up to 385 days) correlated with development and maintenance of donor-specific chimerism in lymphoid organs and BM compartment	[52, 53]
Composite osseo-musculocutaneous sternum, ribs, thymus, pectoralis muscles, and skin allotransplantation model (LBN to Lewis rat)	CsA 16 mg/kg/day tapered to 2 mg/kg/day within 4 weeks and maintained thereafter	Vascularized bone marrow component of graft	Long-term allograft survival correlated with development and maintenance of donor-specific chimerism	[12]

ACI – August Copenhagen Irish
CsA cyclosporine A, *LBN* Lewis-Brown Norway, *VCA* vascularized composite allotransplantation, *αβTCR* αβT-cell receptor, *MLR* mixed lymphocyte reaction, *ALS* antilymphocyte serum

Single-tissue component models provided experience and knowledge that allowed progression to more surgically and immunologically advanced multi-tissue models. In 2000, the Siemionow team was the first to perform a full composite face/scalp allograft transplantation in the rat. The report on the semi-allogenic full-face transplant model in 2003 discussed the role of tapered and low-dose CsA monotherapy on the long-term allograft survival [101]. The development of the less surgically challenging hemiface transplantation model allowed assessment of the immunological responses of face transplant recipients to different immunosuppressive protocols. The study, which compared allogenic and semi-allogenic hemiface transplants, confirmed engraftment of donor-derived cells into lymphoid tissues, such as lymph nodes and spleen in both experimental groups. There was a significant difference between the groups, however, in the level of donor-specific chimerism for both the T and B cell lineages. Following a fully allogenic face transplantation, a higher level of chimerism was observed for both the CD4 and CD8 T cell populations, whereas a significantly reduced chimerism level was found for the B cell lineage when compared with the semi-allogenic hemiface transplants.

The introduction of multi-tissue transplant models containing bone marrow components such as limb, calvaria, maxilla, or hemiface/mandible/tongue transplantation confirmed that the addition of the bone marrow component in combination with adjusted immunosuppressive protocol significantly increased allograft survival (400–700 days). High chimerism levels in these experimental designs encouraged Siemionow's team to further evaluate the immunomodulatory role of BMT and its potential application as the supportive therapy facilitating allograft acceptance and long-term survival.

Selection of Immunosuppressive Protocol for Chimeric Cell Therapy

According to literature reports, it is confirmed that the highest success rate for extension of allograft survival, chimerism, and tolerance induction is achieved by application of immunosuppressive protocols combining targets which are affecting different stages of the immunologic response [92]. When used as a monotherapy, immunosuppressants, such as immunodepletive agents (Table 16.1), calcineurin inhibitors (CsA, Tacrolimus), mammalian target of rapamycin (mTOR) inhibitors (sirolimus), steroids (methylprednisolone), or co-stimulatory blockade antibodies, are not effective in tolerance induction and have to be used daily in high concentrations. This increases drug-related side effects, morbidity, and mortality. On the other hand, cessation of these immunosuppressive agents leads to acute rejection. Thus, major research centers worldwide are focusing on the development of protocols inducing tolerance and/or chimerism in VCA models by taking advantage of the synergistic effect of combining induction therapy with immunodepletive antibodies, followed by maintenance therapy using calcineurin inhibitors, steroids, and MMF. The immunosuppressive protocols designed for hand transplants include

Table 16.3 Immunosuppressive protocols in the clinical cases of hand and face VCA transplantation

Protocol	Induction therapy	Maintenance immunosuppression	Cellular treatment	References
Conventional Hand, face	ATG or alemtuzumab	Triple therapy of corticosteroids, calcineurin inhibitors, and mycophenolate mofetil	N/A	[81, 96]
Louisville Hand	Alemtuzumab	Tacrolimus and mycophenolate mofetil	N/A	[44]
Innsbruck Hand	Alemtuzumab	Tacrolimus and prednisone	N/A	[13, 30]
Pittsburgh Hand	Alemtuzumab	Tacrolimus	Donor-specific bone marrow cell transfusion within 2 weeks after transplantation	[97]
Lyon Face	ATG	Tacrolimus prednisone and mycophenolate mofetil	Donor bone marrow transplantation at day 4 and 11 after transplantation	[32, 82]

VCA vascularized composite allotransplantation, *ATG* antithymocyte globulin

conventional triple therapy, Louisville protocol, Innsbruck protocol, and Pittsburgh protocol; protocols supporting face transplants include conventional triple therapy and Lyon protocol (Table 16.3).

Over the past decade, the majority of VCA clinical cases used immunosuppressive protocols originating from kidney transplant experience [17, 81, 96]. However, the dosage of immunosuppressive agents used in clinical VCA is either similar or higher when compared to kidney transplant protocols due to the histological heterogeneity of different tissue components of VCA transplants. Therefore, there is still a great need for the development of new protocols that would facilitate induction of tolerance associated with donor-specific chimerism. The introduction of supportive cellular therapies is a new and exciting option for the prevention of both acute and chronic allograft rejection and the promotion of tolerance induction. Cellular therapies may enhance the development of chimerism and help to induce tolerance secretory functions or through cell-to-cell interactions. It is crucial, however, to adjust current immunosuppressive protocols to the specific types and characteristics of supportive cellular therapies.

The Siemionow laboratory has tested different dosages and timing of the ALS induction therapy, as well as introduced a new immunodepletive agent—the anti-αβ T cell receptor (anti-αβTCR) monoclonal antibody, to the field of VCA. The study of the rat semi-allogenic limb transplantation showed that a 21-day protocol of combined ALS and CsA induced tolerance and significantly prolonged limb allograft survival (over 420 days). Extended survival of the allograft was associated with a high level of donor chimerism measured in the peripheral blood of the recipient. Tolerance was confirmed *ex vivo* by MLR assay as well as *in vivo* by the acceptance of the donor skin graft and rejection of the third-party graft (Table 16.2; 80]. Although application of the same immunosuppressive protocol of ALS and CsA in the case of allogenic limb transplantation between BN and Lewis rats extended allograft survival (up to 56 days) and was associated with a transient, low level of chimerism, it is important to note that tolerance was not induced [78].

In the process of optimizing an immunosuppressive protocol to support bone marrow-based therapies such as donor–recipient chimeric cell transplantation, the Siemionow group applied the selectively blocking anti-αβTCR monoclonal antibody induction therapy combined with CsA. The main advantage of using the αβTCR monoclonal antibody is the selective depletion of both the immature and mature alloreactive T lymphocytes responsible for the first signal of T cell activation, when preserving small populations of gamma-delta T cells known for their tolerogenic properties.

An additional advantage of this protocol is decreased expression of intragraft pro-inflammatory cytokines such as interleukin 2 (IL-2) and interferon γ (IFNγ), which are associated with allo-recognition, development of rejection, and endothelial activation. Increased expression of Th2 cytokines (IL-10 and IL-4) suggests that the mechanism of action of this antibody includes a switch from Th1 to Th2 response, allowing for long-term acceptance of the allografts [28]. Additionally, the anti-αβTCR antibody can facilitate the beneficial effects of cellular therapies due to the fact that it does not target cells of the myeloid origin (monocytes, granulocytes) or lymphoid cells such as natural killer (NK) cells, gamma delta (γδ) T cells and B cells. These cells may play a significant role in the development of chimerism and tolerance induction and protect the recipient against potential infections. Induction therapy with the anti-αβTCR antibody was combined with the use of the calcineurin inhibitor CsA. The use of CsA as a monotherapy requires higher dosages, which are toxic and cause a number of serious adverse reactions such as nephrotoxicity and hepatotoxicity. It has been shown that the application of high doses of CsA can reduce the number of interdigitating cells, the size of the thymic medulla, and can also change the morphology of epithelial cells [85]. Thus, high doses of CsA obstructs engraftment of the donor cells and prevents the effective development of a new repertoire of tolerogenic T cells that could facilitate allograft survival by changing the microenvironment in the recipient thymus [85].

Siemionow's group developed a new clinically applicable nonmyeloablative 35-day protocol of anti-αβTCR monoclonal antibody (250 μg/kg/day) in combination with tapered CsA (16 to 2 mg/kg/day) therapy based on the previous experience with the limb allograft transplantation model under the ALS and CsA immunosuppressive protocol [100]. Experimental animals, which received hind-limb transplants under the anti-αβTCR/CsA protocol, developed tolerance (survival over 720 days) as well as stable donor-specific chimerism observed in the T cell population. Further studies revealed that optimizing the duration of 21, 7, and 5 days of the anti-αβTCR/CsA protocol resulted in long-term limb allograft survival at all time points. Long-term survival of limb allografts was associated with stable chimerism maintenance. The 7-day αβTCR/CsA protocol was chosen to support the future chimeric cell therapy due to the fact that it provides fast immunodepletion of T cells and is sufficient in creating immunological unresponsiveness to the newly introduced donor-derived cells, thus allowing for cell engraftment into the bone marrow niches and lymphoid organs—specifically into the thymus where the interaction with a new repertoire of T lymphocytes may occur and lead to negative selection of the host alloreactive T cells. This protocol is nonmyeloablative and safer for patients by providing protection against bacterial and viral infections. It also provides stable chimerism for both the T and B cell lineages and does not require preconditioning which makes it clinically applicable to cadaveric VCA transplants.

Chimerism as a Base for Development of Chimeric Cell Therapy: Primary and Secondary Chimeric Animals' Creation

Multiple studies have shown that transplants containing a bone component are more efficacious in chimerism induction and maintenance compared to those without, such as vascularized skin allografts [12, 106]. Vascularized donor bone components provide a continuous source of hematopoietic stem cells that mature, differentiate, and are ultimately responsible for chimerism establishment and preservation in the recipient. In the case of solid organs transplants or VCA that do not contain a bone component, such as face or abdominal wall transplants, bone marrow infusion could be an alternative therapy that supports chimerism induction and increases allograft survival. Billingham et al. demonstrated in their study using neonatal mice that transplantation of hematopoietic cells can induce tolerance of the recipient to the skin allograft via chimerism development [10]. Several animal and clinical case studies confirmed the beneficial effect of bone marrow-derived therapies for survival of the solid organ allografts [8, 46, 49].

To assess the role of hematopoietic cells in chimerism and tolerance induction, the Siemionow group designed a series of experiments in which chimeric animals were created by an adoptive transfer of allogenic (ACI responder to transfused ACI ($RT1^a$)) or semi-allogenic ($RT1^{l+n}$) BMCs to Lewis rat recipients ($RT1^l$). Animals received a nonmyeloablative 7-day immunosuppressive protocol of anti-αβTCR/CsA on the day of transplantation, and chimerism in both allogenic and semi-allogenic models was successfully developed. Interestingly, Siemionow's group detected major histocompatibility complex (MHC) antigens specific for both the donor and recipient, ACI ($RT1^a$) donor and Lewis ($RT1^l$) recipient in the allogenic model and LBN ($RT1^{l+n}$) donor and Lewis ($RT1^l$) recipient in the semi-allogenic model, which were observed on the surface from the isolated recipient (Lewis rat, $RT1^l$) bone marrow cells. To further investigate the properties of these *in vivo* created donor–recipient ($RT1^a/RT1^l$) and ($RT1^{l+n}/RT1^l$) chimeric cells, the Siemionow group harvested BMCs from the primary chimera animals. Next, donor–recipient chimeric cells were isolated using specific monoclonal MHC $RT1^a$ and $RT1^n$ antibodies and a magnetic-activated cell sorting (MACS) method. Isolated donor–recipient

($RT1^a/RT1^l$) and ($RT1^{l+n}/RT1^l$) chimeric cells were used as a supportive therapy for VCA transplantation of allogenic ($RT1^a$) or semi-allogenic ($RT1^{l+n}$) skin graft to Lewis recipients ($RT1^l$) thereby creating secondary chimeric cells. VCA survival in these animals was up to 365 days post transplant, compared to 84 days in animals receiving no supportive cellular therapy (manuscript in preparation). These results showed that *in vivo* created donor–recipient chimeric cells carry pro-tolerogenic properties, which significantly improve VCA survival and could be a breakthrough in tolerance induction treatment.

Evaluation of *in vivo* created donor–recipient chimeric cells in a more complex hemiface transplantation model (manuscript in preparation) confirmed that chimeric cell therapy successfully induces chimerism and increases survival of fully MHC-mismatched hemiface allografts.

Donor–Recipient Chimeric Cells: Potential Mechanism of Chimeric Cell Creation *in vivo*

The mechanism of the creation of cells, which express markers characteristic of both the donor and recipient, may be explained based on the knowledge gained from stem cell research. There are only a few hypotheses describing the fate of donor-derived cells in the recipient tissues. Following bone marrow infusion, stem cells may undergo different processes, such as differentiation, transdifferentiation, and cell fusion [22], or following cell maturation, they may undergo surface antigen transfer (trogocytosis). Siemionow's group considers trogocytosis and/or *in vivo* cell fusion as the most probable mechanism of spontaneous *in vivo* creation of chimeric cells.

Trogocytosis

Trogocytosis is a widespread phenomenon of the rapid exchange of the membrane antigens between two interacting cells [42]. In 1972, Cone et al. provided the first evidence of trogocytosis (from Greek trogo-, meaning nibble or gnaw [19, 35] when they observed transfer of MHC class II proteins from B to T cells [19]. In the literature, it is described as a transfer of membrane patches containing proteins from the surface of one cell to another following "immune synapse" formation [35]. Trogocytosis is not a protein-selective process. During the transfer, some of the adjacent proteins can also be passively transferred together with the membrane patch. It is regulated by Src kinases and depends on several factors, including adenosine triphosphate (ATP), calcium mobilization, and actin cytoskeleton reorganization [107].

The mechanism leading to trogocytosis *in vivo*, types of cells involved in the process, and functional consequences are still under investigation. Trogocytosis is currently perceived as a process that can generate complex adaptations in the immune system and results in immune plasticity beyond genetics and epigenetic programming. Acquisition of membrane patches by immune cells may significantly change their phenotype and augment or diminish immune function by generation of activated or immunosuppressive and regulatory cells [25, 56, 57].

Several studies confirmed that lymphocytes, including both B and T cells, are able to acquire or lose proteins relevant for their function [14, 23, 62, 69, 90, 128]. It has also been reported in murine models that CD4- and CD8-positive T cells can acquire APC-derived MHC class I and II molecules and the co-stimulatory molecules B7-1 (CD80), B7-2 (CD86), and I-CAM1 (CD54; 33, 35, 36, 87, 111, 123]. It has been demonstrated that human NK cells can acquire MHC class I or II protein and viral receptors from the target cells [68, 69, 89]. An *in vitro* study performed by Huang et al. showed that cytotoxic T lymphocytes (CTLs) after trogocytosis of MHC molecules loaded with antigenic peptide recognized by their own T cell receptor, can be eliminated from the organism through cytolysis (called "fratricide"; [33]). Trogocytosis observed in murine models is unidirectional, which means that proteins are transferred exclusively from the recipient to donor cells [16]. In the human experimental model, however, the process of trogocytosis is bidirectional. The phenomenon of intercellular membrane transfer is currently being tested as a tool for engineering immune cells presenting preferential membrane protein patterns for therapeutic application and could be used in the future for adoptive immunotherapy.

Trogocytosis in the Field of Transplantation

Trogocytosis may affect the immune response either in a quantitative (intensity of the process provides positive or negative modulation, e.g., "fratricide") or qualitative (transfer of rarely expressed or functionally atypical molecules, e.g., human leukocyte antigen G (HLA-G)) manner. There have been several attempts to use trogocytosis as a tool to transiently modify cells without direct genetic intervention and to use them for adoptive transfer.

Using trogocytosis, Somanchi et al. engineered human NK cells presenting surface CCR7 receptors. Authors reported that *in vitro* modified NK cells were capable of migrating toward CCL19 and CCL21 chemokines in a migration assay. Furthermore, *in vivo* studies showed that after adoptive transfer of NK cells, the presence of CCR7 receptors on their surface facilitated their homing into the lymph nodes of the recipient [107]. The results obtained by Somanchi stand in line with the observations of Marcenaro et al. who suggested that NK cells presenting the CCR7 receptor will be capable of migrating to lymph nodes and playing an active role in prevention of graft-versus-host and host-versus-graft reaction [62].

Few studies observed the effect of trogocytosis on the modification of the function of regulatory T (T-reg) cells. In the study by LeMounlt et al., authors suggested that the trogocytosis of HLA-G could have a major impact on the immune response by generating T-reg cells from effector T cells. HLA-G is a nonclassical HLA class I molecule, which can inhibit the function of NK and CTLs cells. Effector T cells can inhibit immune responses toward alloantigens by temporarily displaying HLA-G acquired through membrane transfers. This type of transient regulatory cell may constitute an "emergency" immune suppression mechanism used by HLA-G-expressing tissues to protect themselves against immune aggression [57].

Another mechanism of modifying T regulatory cells' function was demonstrated by Ford McIntyre et al. Acquisition of alloantigen by murine αβ-$TCR^{+}CD3^{+}CD4^{-}CD8^{-}$NK1.1-double negative T-reg cells via trogocytosis allowed these cells to eliminate antigen-specific syngeneic CD8 positive T cells. Interestingly, T-regs presenting alloantigen were not cytotoxic toward antigen nonspecific CD8-positive T cells.

Yamanaka et al. assessed human hematopoietic stem cell engraftment following xenotransplantation of human bone marrow to nonobese diabetic/severe combined immunodeficiency (NOD/SCID) mice. Unexpectedly, the authors observed the presence of mouse MHC class I protein on the surface of almost 100 % human cells, suggesting that this mechanism was adopted by the human cells to evade host immune surveillance [124].

In a follow-up study by Chow T et al., authors showed that the transfer of MHC class I from host to the donor cells plays an important role in protecting donor cells from NK cell and macrophage-mediated rejection during hematopoietic stem cell transplantation and engraftment [18]. The authors hypothesize that by expressing MHC class I antigens from the host cells, human donor cells obtain the identity of the host, which allows for the evasion of detection by the immune cells. After the development of donor chimerism, transplanted cells do not require host MHC class I protein transfer to survive.

The Phenomenon of Cell-to-Cell Fusion

The phenomenon of cellular fusion is defined as the process of merging of entrapped contents between two membrane-enclosed aqueous compartments that involve the mixing of the membrane contents [21]. Examples of cell fusion can be found in multiple species (yeast, nematodes, arthropods, and mammals) and can involve various cell types (gametes, epithelia, myoblasts, and macrophages; [77]). It has been shown that cell fusion plays a critical role in organism development and is involved in a variety of biological processes, such as sexual reproduction, development of trophoblasts, placenta, muscles, and bones, as well as immune response and tumorogenesis. The first observation of spontaneous fusion between mammalian cells was reported by Barski in 1961 [7]. The discovery of spontaneous fusion *in vitro* between pluripotent embryonic stem cells and mouse BMCs [112] or brain progenitor cells [127] created an interest in cell fusion as a process applicable to tissue regeneration.

The work of pioneers such as Terada [112] and Ying [127] revealed that cells created by spontaneous cell fusion could express characteristics of undifferentiated cells or properties of both types of cells undergoing fusion. The possibility of the formation of stable, multinucleated heterokaryons as a result of spontaneous fusion of bone marrow derived cells with several types of cells, such as skeletal muscle, cardiac muscle, liver, monocytes, macrophages, intestinal cells, and Purkinje neurons, was confirmed by multiple *in vivo* studies [2, 41, 54, 74, 88, 118, 120, 121].

In these experiments, fused cells not only presented mixed phenotype, but also overtook the function of injured recipient cells and helped facilitate the process of tissue regeneration. Several articles suggest that spontaneous cell fusion is triggered by changes in the cytokine microenvironment that are inevitable during injury. Pro-inflammatory cytokines, such as IL-4, IL-13, IFNγ, tumor necrosis factor α (TNFα), IL-1, and IL-3, were shown as potential participants in a fusion process of monocytes, macrophages, and osteoblasts [3, 39, 64]. The interest in application of bone marrow-derived cells in various medical fields, including tissue regeneration and transplantation, is increasing due to their potential therapeutic effects.

Cell Fusion in the Field of Transplantation

The infusion of bone marrow-derived cells and presence of chimerism in peripheral blood followed by migration of donor-derived cells to lymphoid organs was associated with prolonged survival of the transplants, and in some cases even tolerance induction [4, 43, 67, 84, 103]. In the field of transplantation, however, due to inappropriate immunosuppressive protocols, a low number of fusion events or low interest in this process has resulted in a paucity of publications on the subject of spontaneous cell fusion. Bonde et al. [11] reported that spontaneous fusion occurred during coculturing of BMCs derived from two different mice strains, and an *in vivo* study confirmed this result following allogenic and syngeneic transplantation. This study reported that fused cells expressed both donor and recipient MHC antigens on their surface.

Siemionow's group also performed experiments on cell fusion of bone marrow-derived cells (PSRC supplement 2010). Results of this preliminary study were in line with those from Bonde et al. and confirmed the phenomenon of spontaneous cell fusion leading to the *in vivo* creation of donor–recipient chimeric cells that can facilitate face allograft survival (article in press). A short immunomodulatory protocol of anti-αβTCR monoclonal antibody and CsA was used to facilitate the engraftment of spontaneously created donor–recipient chimeric cells.

Ex Vivo Cell Fusion as a New Approach for Tolerance Induction

The successful establishment of a protocol creating *in vivo* donor–recipient chimeric cells either via the mechanism of trogocytosis or spontaneous cell fusion is opening the door for using bone marrow-derived cells as a tool for the development of novel therapeutic products. Although further research on donor–recipient chimeric cells is necessary in order to fully understand the underlying mechanisms of their creation and action *in vivo*, the creation of tolerance-inducting cells could be a breakthrough modality in solid organ and VCA transplantation.

Ex Vivo Creation of the Donor–Recipient Chimeric Cells: Animal Model

There are several disadvantages for creating donor–recipient chimeric cells *in vivo* in the clinical setting. The most challenging issue in the clinical execution of the primary chimera protocol (Fig. 16.1a, b) is the critical time frame which is required following bone marrow infusion for the development of a chimeric pro-tolerogenic environment in the transplant recipient. In the clinical scenario, pretreatment of the transplant recipient, which is required in order to create *in vivo* donor–recipient chimeric cells, will be possible only in cases of living organ donor transplantation. To overcome this hindrance and progress toward a more clinically applicable model, Siemionow's group adopted a new approach to create donor–recipient chimeric cells via *ex vivo* cell fusion. Donor–recipient chimeric cell therapy was created by fusion of donor and recipient bone marrow-derived cells using the polyethylene glycol (PEG) technique. The application of PEG in combination with dimethyl sulfoxide (DMSO) was confirmed to be more efficient in creating a higher number of fused cells than the use of PEG alone [27, 73]. Due to its hydrophobic properties, PEG decreases the distance between cells by removing water and causing their aggregation. Addition of DMSO facilitates formation of pores in the membrane lipid bilayer (Fig. 16.2).

A detailed donor–recipient chimeric cell fusion protocol using PEG/DMSO solution is shown in Fig. 16.1c. Briefly, BMCs were harvested from the tibia and femur bones of two fully MHC-mismatched ACI ($RT1^a$) and Lewis ($RT1^l$) rats by a flushing technique. Next, erythrocytes were removed and white blood cells from each donor were separately stained with PKH 26 (red) or PKH67 (green) cell membrane fluorescent dye. Fluorescent staining of cells was applied in order to detect and separate the double stained (green and red) donor–recipient chimeric cells, which were created during fusion. Separation of fused double-stained cells was performed using fluorescence activated cell sorting. Only fused donor–recipient chimeric cells were used for further evaluation, culturing, and therapeutic application. It is feasible to perform PEG-mediated *ex vivo* fusion protocol in the surgical unit. This protocol will yield a higher number of donor–recipient chimeric cells when compared with *in vivo* protocol of spontaneous cell fusion. The PEG protocol does not require cells of similar diameter nor specific proportions as does the electrofusion protocol [60].

Siemionow's team successfully confirmed feasibility of the *ex vivo* fusion protocol and creation of the donor–recipient chimeric cells. The assessment of donor–recipient chimeric cells was performed by flow cytometry, PCR, and immunostaining assays. The results confirmed the presence of MHC class I antigens derived from

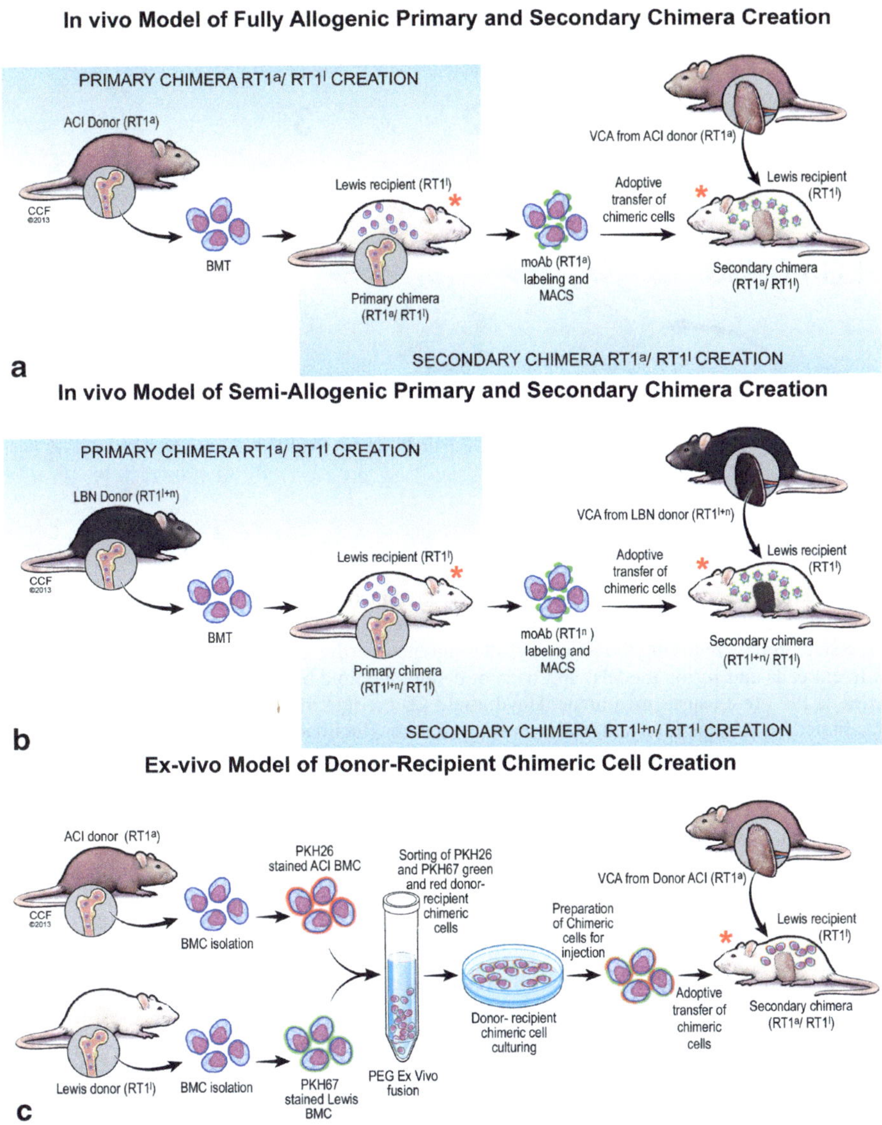

Fig. 16.1 Experimental model of *in vivo* and *ex vivo* creation of the donor–recipient chimeric cells. Panels **a** and **b**: Primary and secondary chimeras will be created *in vivo* by bone marrow transplantation across an MHC barrier between LBN ($RT1^{n+l}$; semi-allogenic model—Panel **a**) or ACI ($RT1^{a}$) donors (fully allogenic model—Panel **b**) and Lewis ($RT1^{l}$) recipients. Creation of the primary chimeric animals will be performed by transplantation of 70×10^{6} BMC harvested from the LBN ($RT1^{n+l}$) or ACI ($RT1^{a}$) rat femurs and tibias. Isolated BMC will be transplanted directly to the bone of the naïve Lewis ($RT1^{l}$) rat recipients. The primary chimeric animals will serve as a source of the donor–recipient chimeric cells of MHC-mismatched phenotypes. Chimeric cells will be harvested from the BM compartment of the primary chimeras and be purified using a mAb specific for the $RT1^{n}$ and $RT1^{a}$ MHC class I by magnetic-activated cell sorting (MACS) technique. Freshly isolated chimeric cells from the ACI or LBN donors will be delivered via intraosseous

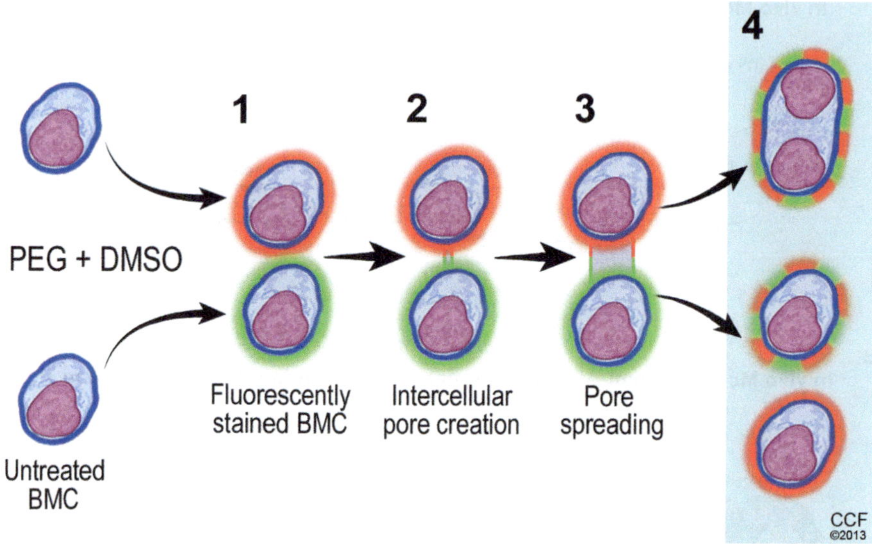

Fig. 16.2 The mechanism of polyethylene glycol/dimethyl sulfoxide (PEG/DMSO) induced donor–recipient chimeric cells creation via *in vitro* cell fusion. PEG-mediated cell fusion is a three-step process requiring the following: (1) aggregation (the intercellular distance may vary for different cells and fusion models) approach of membrane lipid bilayers due to hydrophobic properties of PEG that causes membrane dehydration; (2) the intermediate membrane destabilization (facilitated by PEG) is followed by the creation of pores (facilitated by DMSO) in the membranes of cells undergoing fusion; (3) positive osmotic pressure created by PEG improves stabilization of fusion intermediates and leads to expansion of the pores, cell swelling and cell-to-cell fusion. The products of PEG/DMSO-solution-induced cell fusion may include (4) heterokaryon and synkaryon cells, as well as cells that did not undergo the fusion process. For a more detailed description of cell fusion mechanisms, review articles by Lentz [58, 59]

both donors on the surface of chimeric cells. The presence of ACI and Lewis-specific genomic sequences was also confirmed by PCR. The phenotype evaluation showed that more than 40 % of chimeric cells were carrying hematopoietic stem cell/progenitor cell marker—CD90. Additionally, MLR assay showed immunologic unresponsiveness of chimeric cells, and the colony-forming unit assay revealed that

---

injection into the naïve Lewis recipients at the time of VCA transplant coming from the respective bone marrow donor. Both the primary and secondary chimeric animals will be additionally treated with a 7-day protocol of combined αβ-TCR mAb (250 μg/day) and CsA (16 mg/kg/day) therapy. Panel **c**: Donor–recipient chimeric cells will be created *ex vivo* by the chemical polyethylene glycol (PEG)-induced cell fusion of the bone marrow cells harvested from the ACI ($RT1^a$) and Lewis ($RT1^l$) rat donors. Isolated bone marrow cells will be separately stained with two different (red and green) fluorescent dyes. Next, the *ex vivo* fusion will be performed using PEG. Supportive therapy using the fused donor–recipient chimeric cells will be given based on the double fluorescent staining and injected into the bone of Lewis ($RT1^l$) rat recipients along with the donor matching (ACI or LBN) VCA (skin allograft) transplant. ACI August Copenhagen Irish *MHC* major histocompatibility complex, *LBN* Lewis-Brown Norway, *BMC* bone marrow cells, *VCA* vascularized composite allotransplantation, *mAb* monoclonal antibody. * 7-day protocol of combined αβ-TCR mAb (250 μg/day) and CsA (16 mg/kg/day) therapy chimeric cells are capable of creating the same types of colonies as the normal unprocessed BMCs.

Donor–recipient chimeric cells were tested *in vivo* as a supportive therapy for allogenic vascularized skin allograft under Siemionow's 7-day immunosuppressive laboratory protocol of the anti-αβTCR monoclonal antibody (250 μg/kg/day) and CsA (16 mg/kg/day). The results of this study showed an increased survival of the allograft confirming pro-tolerogenic properties of donor–recipient chimeric cells. Prolonged survival of the fully MHC-mismatched allografts was associated with the presence of donor-derived cells in the peripheral blood and lymphoid organs (lymph nodes and thymus) of the recipient rats.

One of the potential mechanisms of action of donor–recipient chimeric cells may be similar to the phenomenon observed by Chow et al. where cells presenting MHC of recipient origin acquired via trogocytosis were protected from the recipient's immune response [18]. Donor–recipient chimeric cells presenting both the donor and recipient MHC on their cell membrane may be viewed by the host as self-cells, and thus may engraft, induce chimerism, and create allograft acceptance.

Another possible mechanism of action of chimeric cells may be related to the migration of donor–recipient chimeric cells to the recipient lymphoid organs, such as the thymus, where these cells can influence the selection process for donor-reactive T cells, causing induction of chimerism and acceptance of the allograft.

Hematopoietic Donor–Recipient Chimeric Cells from the Animal Model to Humans

Ex vivo fused human donor–recipient chimeric cells are the ultimate goal of developing donor(s)-specific transferable tolerance. This novel approach will allow for the generation of custom-made cellular therapies with high specificity and sensitivity designed for the individual patient.

The promising results from the *in vivo* testing of murine donor-recipient chimeric cells led Siemionow's group to test the feasibility of *ex vivo* creation of chimeric cells using human cord blood cells as a proof of concept. In contrast to collecting bone marrow from living donors, cord blood is easily available and can be harvested in a noninvasive way without any discomfort for the donor. Moreover, cord blood is less expensive and contains a sufficient number of hematopoietic stem cells required for cord blood cell fusion. Preliminary experiments by Siemionow's group confirmed feasibility of their human chimeric cells protocol. Lymphocytotoxicity tests confirmed that the newly created chimeric cells presented on their surface the HLA class I and II characteristic for both cord blood cell donors. These results were confirmed at the DNA level by short-tandem repeat PCR (STR-PCR). Human chimeric cells were capable of creating the same colony types with number of cells, which were comparable to the normal, unprocessed cord blood controls.

As previously observed in the murine model, *ex vivo* fusion of human cord blood cells did not significantly increase the apoptotic cell number.

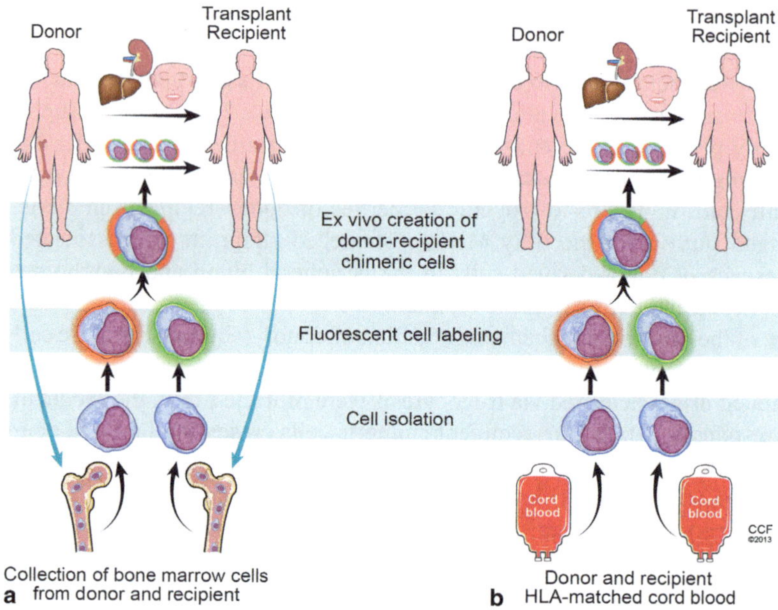

Fig. 16.3 Future applications of the *ex vivo* created donor–recipient chimeric cells used as supportive therapy in the clinical scenario. Human donor–recipient chimeric cells can be utilized as a supportive therapy for solid organ (living donor kidney, liver transplantation) and in the future for vascularized composite allotransplantation (VCA). Progenitor cells derived from sources, such as bone marrow or cord blood, will be isolated, fluorescently labeled using two different cell membrane dyes (PKH26 and PKH67), and fused *ex vivo* using the PEG technique creating the donor–recipient chimeric cells. Based on the double fluorescent staining, the *ex vivo* fused chimeric cells will be sorted out and delivered via either the intraosseous or intravenous route to the recipient at the day of solid organ or VCA transplants. *Panel A*—Patients receiving a transplant from a living donor will be supported with the bone marrow-derived donor–recipient chimeric cells collected from both the donor and the transplant recipient. *Panel B*—If access to the donor and/or recipient's bone marrow cells is not possible (i.e., recipient is suffering from severe bone marrow deficiencies due to gamma irradiation or organ donor deceased), the donor and recipient HLA-matched cord blood cells can be used to create *ex vivo* donor–recipient chimeric cells and applied as a supportive therapy. [*PEG* polyethylene glycol, *VCA* vascularized composite allotransplantation, *HLA* human leukocyte antigen]

As a future direction, Siemionow's group will focus on the creation of donor–recipient chimeric cells from human bone marrow, which serves as a tremendous reservoir of donor- and recipient-specific hematopoietic stem cells. It will be a primary source of cells for *ex vivo* cell fusion and the creation of chimeric cells for therapeutic applications. In cases where bone marrow is not available (i.e., recipient is suffering from bone marrow malignancies or severe deficiencies) or its collection results in a very low yield, an alternative option will be to use cord blood cells. The possibility of interchangeable application of either bone marrow or cord blood cells will provide assurance that a sufficient number of chimeric cells can be acquired always (Fig. 16.3). Additionally, if clinical trials determine that multiple injections of chimeric cells are required, two sources of cells will be available for chimeric cell creation.

Following the creation of human chimeric cells via *ex vivo* cell fusion, cellular therapy will be applied to transplant recipients by direct intraosseous transplantation according to Siemionow group's tolerance-inducing protocol. The ultimate goal is to use chimeric cell supportive therapy to facilitate the development of a tolerogenic microenvironment for engraftment and long-term allograft survival. Additionally, human chimeric cell therapy may have clinical applications in the treatment of diseases based on BMT; this new approach may serve as a platform for supportive cellular immunotherapy for transplants, where efficacious and stable engraftment is needed without recipient conditioning.

Finally, this innovative, chimeric cell supportive therapy represents a breakthrough modality in the field of reconstructive transplantation and may allow for the reduction or elimination of lifelong immunosuppressive therapy.

References

1. Al-Adra DP, Anderson CC. Mixed chimerism and split tolerance: mechanisms and clinical correlations. Chimerism. 2011;2(4):89–101.
2. Alvarez-Dolado M, Pardal R, et al. Fusion of bone-marrow-derived cells with Purkinje neurons, cardiomyocytes and hepatocytes. Nature. 2003;425(6961):968–73.
3. Anderson JM. Multinucleated giant cells. Curr Opin Hematol. 2000;7(1):40–7.
4. Ayala R, Grande S, et al. Long-term follow-up of donor chimerism and tolerance after human liver transplantation. Liver Transpl. 2009;15(6):581–91.
5. Bachar-Lustig E, Rachamim N, et al. Megadose of T cell-depleted bone marrow overcomes MHC barriers in sublethally irradiated mice. Nat Med. 1995;1(12):1268–73.
6. Barber WH, Mankin JA, et al. Long-term results of a controlled prospective study with transfusion of donor-specific bone marrow in 57 cadaveric renal allograft recipients. Transplantation. 1991;51(1):70–5.
7. Barski G, Sorieul S et al. Hybrid type cells in combined cultures of two different mammalian cell strains. J Natl Cancer Inst. 1961;26:1269–91.
8. Beitinjaneh A, Burns LJ, et al. Solid organ transplantation in survivors of hematopoietic cell transplantation: a single institution case series and literature review. Clin Transpl. 2010;24(4):E94–102.
9. Billingham RE, Medawar PB. Desensitization to skin homografts by injections of donor skin extracts. Ann Surg. 1953;137(4):444–9.
10. Billingham RE, Brent L, et al. Actively acquired tolerance of foreign cells. Nature. 1953;172(4379):603–6.
11. Bonde S, Pedram M, et al. Cell fusion of bone marrow cells and somatic cell reprogramming by embryonic stem cells. FASEB J. 2010;24(2):364–73.
12. Bozkurt M, Klimczak A, et al. Composite osseomusculocutaneous sternum, ribs, thymus, pectoralis muscles, and skin allotransplantation model of bone marrow transplantation. Microsurgery. 2013;33(1):43–50.
13. Brandacher G, Ninkovic M, et al. The Innsbruck hand transplant program: update at 8 years after the first transplant. Transpl Proc. 2009;41(2):491–4.
14. Brown R, Suen H, et al. Trogocytosis generates acquired regulatory T cells adding further complexity to the dysfunctional immune response in multiple myeloma. Oncoimmunology. 2012;1(9):1658–60.
15. Caridis DT, Liegeois A, et al. Enhanced survival of canine renal allografts of ALS- treated dogs given bone marrow. Transpl Proc. 1973;5(1):671–4.

16. Caumartin J, Lemaoult J, et al. Intercellular exchanges of membrane patches (trogocytosis) highlight the next level of immune plasticity. Transpl Immunol. 2006;17(1):20–2.
17. Chang J, Davis CL, et al. The impact of current immunosuppression strategies in renal transplantation on the field of reconstructive transplantation. J Reconstr Microsurg. 2012;28(1):7–19.
18. Chow T, Whiteley J, et al. The transfer of host MHC class I protein protects donor cells from NK Cell and macrophage mediated rejection during hematopoietic stem cell transplantation and engraftment in Mice. Stem Cells. 2013 Oct;31(10):2242-52.
19. Cone RE, Sprent J, et al. Antigen-binding specificity of isolated cell-surface immunoglobulin from thymus cells activated to histocompatibility antigens. Proc Natl Acad Sci U S A. 1972;69(9):2556–60.
20. Demetris AJ, Murase N, et al. The dichotomous functions of passenger leukocytes in solid-organ transplantation. Adv Nephrol Necker Hosp. 1995;24:341–54.
21. Dittmar T, Zanker KS. Cell fusion in health and disease. Volume II: cell fusion in disease. Introduction. Adv Exp Med Biol. 2011;714:1–3.
22. Eisenberg LM, Eisenberg CA. Stem cell plasticity, cell fusion, and transdifferentiation. Birth Defects Res C Embryo Today. 2003;69(3):209–18.
23. Eisenberg G, Uzana R, et al. Imprinting of lymphocytes with melanoma antigens acquired by trogocytosis facilitates identification of tumor-reactive T cells. J Immunol. 2013;190(11):5856–65.
24. Elwood ET, Larsen CP, et al. Microchimerism and rejection in clinical transplantation. Lancet. 1997;349(9062):1358–60.
25. Ford McIntyre MS, Young KJ, et al. Cutting edge: *in vivo* trogocytosis as a mechanism of double negative regulatory T cell-mediated antigen-specific suppression. J Immunol. 2008;181(4):2271–5.
26. Graca L, Daley S, et al. Co-receptor and co-stimulation blockade for mixed chimerism and tolerance without myelosuppressive conditioning. BMC Immunol. 2006;7:9.
27. Gurtovenko AA, Anwar J. Modulating the structure and properties of cell membranes: the molecular mechanism of action of dimethyl sulfoxide. J Phys Chem B. 2007;111(35):10453–60.
28. Hancock WW, Sayegh MH, et al. Blocking of mononuclear cell accumulation, cytokine production, and endothelial activation within rat cardiac allografts by CD4 monoclonal antibody therapy. Transplantation. 1992;53(6):1276–80.
29. Hartner WC, De Fazio SR, et al. Prolongation of renal allograft survival in antilymphocyte-serum-treated dogs by postoperative injection of density-gradient-fractionated donor bone marrow. Transplantation. 1986;42(6):593–7.
30. Hautz T, Engelhardt TO, et al. World experience after more than a decade of clinical hand transplantation: update on the Innsbruck program. Hand Clin. 2011;27(4):423–31, viii.
31. Heaf J. Bone marrow function during quadruple immunosuppressive therapy after renal transplantation. Clin Nephrol. 1993;40(6):332–8.
32. Hequet O, Morelon E, et al. Allogeneic donor bone marrow cells recovery and infusion after allogeneic face transplantation from the same donor. Bone Marrow Transpl. 2008;41(12):1059–61.
33. Huang JF, Yang Y, et al. TCR-Mediated internalization of peptide-MHC complexes acquired by T cells. Science. 1999;286(5441):952–4.
34. Huang CA, Fuchimoto Y, et al. Posttransplantation lymphoproliferative disease in miniature swine after allogeneic hematopoietic cell transplantation: similarity to human PTLD and association with a porcine gammaherpesvirus. Blood. 2001;97(5):1467–73.
35. Hudrisier D, Riond J, et al. Cutting edge: CTLs rapidly capture membrane fragments from target cells in a TCR signaling-dependent manner. J Immunol. 2001;166(6):3645–9.
36. Hwang I, Huang JF, et al. T cells can use either T cell receptor or CD28 receptors to absorb and internalize cell surface molecules derived from antigen-presenting cells. J Exp Med. 2000;191(7):1137–48.
37. Ildstad ST, Sachs DH. Reconstitution with syngeneic plus allogeneic or xenogeneic bone marrow leads to specific acceptance of allografts or xenografts. Nature. 1984;307(5947):168–70.

38. Inceoglu S, Siemionow M, et al. The effect of combined immunosuppression with systemic low-dose cyclosporin and topical fluocinolone acetonide on the survival of rat hind-limb allografts. Ann Plast Surg. 1994;33(1):57–65.
39. Jimi E, Akiyama S, et al. Osteoclast differentiation factor acts as a multifunctional regulator in murine osteoclast differentiation and function. J Immunol. 1999;163(1):434–42.
40. Joffre O, Gorsse N, et al. Induction of antigen-specific tolerance to bone marrow allografts with CD4 + CD25 + T lymphocytes. Blood. 2004;103(11):4216–21.
41. Johansson CB, Youssef S, et al. Extensive fusion of haematopoietic cells with Purkinje neurons in response to chronic inflammation. Nat Cell Biol. 2008;10(5):575–83.
42. Joly E, Hudrisier D. What is trogocytosis and what is its purpose? Nat Immunol. 2003;4(9):815.
43. Kanamoto A, Maki T. Chimeric donor cells play an active role in both induction and maintenance phases of transplantation tolerance induced by mixed chimerism. J Immunol. 2004;172(3):1444–8.
44. Kaufman CL, Breidenbach W. World experience after more than a decade of clinical hand transplantation: update from the Louisville hand transplant program. Hand Clin. 2011;27(4):417–21, vii–viii.
45. Kawai T, Sogawa H, et al. CD154 blockade for induction of mixed chimerism and prolonged renal allograft survival in nonhuman primates. Am J Transpl. 2004;4(9):1391–8.
46. Kawai T, Cosimi AB, et al. HLA-mismatched renal transplantation without maintenance immunosuppression. N Engl J Med. 2008;358(4):353–61.
47. Kean LS, Adams AB, et al. Induction of chimerism in rhesus macaques through stem cell transplant and costimulation blockade-based immunosuppression. Am J Transpl. 2007;7(2):320–35.
48. Ko S, Deiwick A, et al. The functional relevance of passenger leukocytes and microchimerism for heart allograft acceptance in the rat. Nat Med. 1999;5(11):1292–7.
49. Koenecke C, Hertenstein B, et al. Solid organ transplantation after allogeneic hematopoietic stem cell transplantation: a retrospective, multicenter study of the EBMT. Am J Transpl. 2010;10(8):1897–906.
50. Kuhr CS, Allen MD, et al. Tolerance to vascularized kidney grafts in canine mixed hematopoietic chimeras. Transplantation. 2002;73(9):1487–92.
51. Kuhr CS, Yunusov M, et al. Long-term tolerance to kidney allografts in a preclinical canine model. Transplantation. 2007;84(4):545–7.
52. Kulahci Y, Siemionow M. A new composite hemiface/mandible/tongue transplantation model in rats. Ann Plast Surg. 2010;64(1):114–21.
53. Kulahci Y, Klimczak A, et al. Long-term survival of composite hemiface/mandible/tongue allografts correlates with multilineage chimerism development in the lymphoid and myeloid compartments of recipients. Transplantation. 2010;90(8):843–52.
54. LaBarge MA, Blau HM. Biological progression from adult bone marrow to mononucleate muscle stem cell to multinucleate muscle fiber in response to injury. Cell. 2002;111(4):589–601.
55. Larsen CP, Austyn JM, et al. The role of graft-derived dendritic leukocytes in the rejection of vascularized organ allografts. Recent findings on the migration and function of dendritic leukocytes after transplantation. Ann Surg. 1990;212(3):308–15; discussion 316–7.
56. LeMaoult J, Caumartin J, et al. Exchanges of membrane patches (trogocytosis) split theoretical and actual functions of immune cells. Hum Immunol. 2007;68(4):240–3.
57. LeMaoult J, Caumartin J, et al. Immune regulation by pretenders: cell-to-cell transfers of HLA-G make effector T cells act as regulatory cells. Blood. 2007;109(5):2040–8.
58. Lentz BR. Polymer-induced membrane fusion: potential mechanism and relation to cell fusion events. Chem Phys Lipids. 1994;73(1–2):91–106.
59. Lentz BR. PEG as a tool to gain insight into membrane fusion. Eur Biophys J. 2007;36(4–5):315–26.
60. Li LH, Hensen ML, et al. Electrofusion between heterogeneous-sized mammalian cells in a pellet: potential applications in drug delivery and hybridoma formation. Biophys J. 1996;71(1):479–86.
61. Liegeois A, Caridis T, et al. Renal allograft in the dog. A new method for survival prolongation. Nouv Presse Med. 1973;2(20):1351–4.

62. Marcenaro E, Pesce S, et al. KIR2DS1-dependent acquisition of CCR7 and migratory properties by human NK cells interacting with allogeneic HLA-C2 + DCs or T-cell blasts. Blood. 2013;121(17):3396–401.
63. Mathew JM, Miller J. Immunoregulatory role of chimerism in clinical organ transplantation. Bone Marrow Transpl. 2001;28(2):115–9.
64. Merkel KD, Erdmann JM, et al. Tumor necrosis factor-alpha mediates orthopedic implant osteolysis. Am J Pathol. 1999;154(1):203–10.
65. Monaco AP, Wood ML. Studies on heterologous antilymphocyte serum in mice. VII. Optimal cellular antigen for induction of immunologic tolerance with antilymphocyte serum. Transpl Proc. 1970;2(4):489–96.
66. Monaco AP, Clark AW, et al. Possible active enhancement of a human cadaver renal allograft with antilymphocyte serum (ALS) and donor bone marrow: case report of an initial attempt. Surgery. 1976;79(4):384–92.
67. Muramatsu K, Kuriyama R, et al. Intragraft chimerism following composite tissue allograft. J Surg Res. 2009;157(1):129–35.
68. Nakamura K, Nakayama M, et al. Fratricide of natural killer cells dressed with tumor-derived NKG2D ligand. Proc Natl Acad Sci U S A. 2013;110(23):9421–6.
69. Nakayama M, Takeda K, et al. Natural killer (NK)-dendritic cell interactions generate MHC class II-dressed NK cells that regulate CD4 + T cells. Proc Natl Acad Sci U S A. 2011;108(45):18360–5.
70. Nasir S, Bozkurt M, et al. Large antigenic skin load in total abdominal wall transplants permits chimerism induction. Ann Plast Surg. 2008;61(5):572–9.
71. Nasir S, Bozkurt M, et al. Correlation of chimerism with graft size and revascularization in vascularized and nonvascularized skin allografts. Ann Plast Surg. 2009;62(4):430–8.
72. Nikbin B, Talebian F, et al. Chimerism: a new look. Urol J. 2007;4(1):1–9.
73. Norwood TH, Zeigler CJ, et al. Dimethyl sulfoxide enhances polyethylene glycol-mediated somatic cell fusion. Somatic Cell Genet. 1976;2(3):263–70.
74. Nygren JM, Jovinge S, et al. Bone marrow-derived hematopoietic cells generate cardiomyocytes at a low frequency through cell fusion, but not transdifferentiation. Nat Med. 2004;10(5):494–501.
75. Opelz G, Terasaki PI. Absence of immunization effect in human-kidney retransplantation. N Engl J Med. 1978;299(8):369–74.
76. Opelz G, Terasaki PI. Improvement of kidney-graft survival with increased numbers of blood transfusions. N Engl J Med. 1978;299(15):799–803.
77. Oren-Suissa M, Podbilewicz B. Evolution of programmed cell fusion: common mechanisms and distinct functions. Dev Dyn. 2010;239(5):1515–28.
78. Ozer K, Gurunluoglu R, et al. Extension of composite tissue allograft survival across major histocompatibility barrier under short course of anti-lymphocyte serum and cyclosporine a therapy. J Reconstr Microsurg. 2003;19(4):249–56.
79. Ozer K, Izycki D, et al. Development of donor-specific chimerism and tolerance in composite tissue allografts under alphabeta-T-cell receptor monoclonal antibody and cyclosporine a treatment protocols. Microsurgery. 2004;24(3):248–54.
80. Ozer K, Oke R, et al. Induction of tolerance to hind limb allografts in rats receiving cyclosporine A and antilymphocyte serum: effect of duration of the treatment. Transplantation. 2003;75(1):31–6.
81. Petruzzo P, Lanzetta M, et al. The International Registry on hand and composite tissue transplantation. Transplantation. 2010;90(12):1590–4.
82. Petruzzo P, Testelin S, et al. First human face transplantation: 5 years outcomes. Transplantation. 2012;93(2):236–40.
83. Pilat N, Wekerle T. Combining Treg therapy with mixed chimerism: Getting the best of both worlds. Chimerism. 2010;1(1):26–9.
84. Rahhal DN, Xu H, et al. Dissociation between peripheral blood chimerism and tolerance to hindlimb composite tissue transplants: preferential localization of chimerism in donor bone. Transplantation. 2009;88(6):773–81.

85. Rezzani R. Cyclosporine A and adverse effects on organs: histochemical studies. Prog Histochem Cytochem. 2004;39(2):85–128.
86. Ricordi C, Karatzas T, et al. High-dose donor bone marrow infusions to enhance allograft survival: the effect of timing. Transplantation. 1997;63(1):7–11.
87. Riond J, Elhmouzi J, et al. Capture of membrane components via trogocytosis occurs *in vivo* during both dendritic cells and target cells encounter by CD8(+) T cells. Scand J Immunol. 2007;66(4):441–50.
88. Rizvi AZ, Swain JR, et al. Bone marrow-derived cells fuse with normal and transformed intestinal stem cells. Proc Natl Acad Sci U S A. 2006;103(16):6321–5.
89. Rosenits K, Keppler SJ, et al. T cells acquire cell surface determinants of APC via *in vivo* trogocytosis during viral infections. Eur J Immunol. 2010;40(12):3450–7.
90. Rossi EA, Goldenberg DM, et al. Trogocytosis of multiple B-cell surface markers by CD22-targeting with epratuzumab. Blood. 2013;122(17):3020-9.
91. Rugeles MT, Aitouche A, et al. Evidence for the presence of multilineage chimerism and progenitors of donor dendritic cells in the peripheral blood of bone marrow-augmented organ transplant recipients. Transplantation. 1997;64(5):735–41.
92. Sachs DH, Sykes M, et al. Immuno-intervention for the induction of transplantation tolerance through mixed chimerism. Semin Immunol. 2011;23(3):165–73.
93. Salvatierra O, Jr, Vincenti F, et al. Deliberate donor-specific blood transfusions prior to living related renal transplantation. A new approach. Ann Surg. 1980;192(4):543–52.
94. Scandling JD, Busque S, et al. Tolerance and chimerism after renal and hematopoietic-cell transplantation. N Engl J Med. 2008;358(4):362–8.
95. Scandling JD, Busque S, et al. Induced immune tolerance for kidney transplantation. N Engl J Med. 2011;365(14):1359–60.
96. Schneeberger S, Landin L, et al. Achievements and challenges in composite tissue allotransplantation. Transpl Int. 2011;24(8):760–9.
97. Schneeberger S, Gorantla VS, et al. Upper-extremity transplantation using a cell-based protocol to minimize immunosuppression. Ann Surg. 2013;257(2):345–51.
98. Seung E, Iwakoshi N, et al. Allogeneic hematopoietic chimerism in mice treated with sublethal myeloablation and anti-CD154 antibody: absence of graft-versus-host disease, induction of skin allograft tolerance, and prevention of recurrent autoimmunity in islet-allografted NOD/Lt mice. Blood. 2000;95(6):2175–82.
99. Sharabi Y, Sachs DH. Mixed chimerism and permanent specific transplantation tolerance induced by a nonlethal preparative regimen. J Exp Med. 1989;169(2):493–502.
100. Siemionow M, Ortak T, et al. Induction of tolerance in composite-tissue allografts. Transplantation. 2002;74(9):1211–7.
101. Siemionow M, Gozel-Ulusal B, et al. Functional tolerance following face transplantation in the rat. Transplantation. 2003;75(9):1607–9.
102. Siemionow MZ, Izycki DM, et al. Donor-specific tolerance in fully major histocompatibility major histocompatibility complex-mismatched limb allograft transplants under an anti-alphabeta T-cell receptor monoclonal antibody and cyclosporine A protocol. Transplantation. 2003;76(12):1662–8.
103. Siemionow M, Demir Y, et al. Development and maintenance of donor-specific chimerism in semi-allogenic and fully major histocompatibility complex mismatched facial allograft transplants. Transplantation. 2005;79(5):558P.
104. Siemionow MZ, Demir Y, et al. Facial tissue allograft transplantation. Transpl Proc. 2005;37(1):201–4.
105. Siemionow M, Ozer K, et al. A new method of bone marrow transplantation leads to extension of skin allograft survival. Transpl Proc. 2005;37(5):2309–14.
106. Siemionow M, Zielinski M, et al. Intraosseus transplantation of donor-derived hematopoietic stem and progenitor cells induces donor-specific chimerism and extends composite tissue allograft survival. Transpl Proc. 2005;37(5):2303–8.

107. Somanchi SS, Somanchi A, et al. Engineering lymph node homing of *ex vivo*-expanded human natural killer cells via trogocytosis of the chemokine receptor CCR7. Blood. 2012;119(22):5164–72.
108. Starzl TE. Chimerism and tolerance in transplantation. Proc Natl Acad Sci U S A. 2004;101(Suppl 2):14607–14.
109. Starzl TE, Demetris AJ, et al. Chimerism and donor-specific nonreactivity 27 to 29 years after kidney allotransplantation. Transplantation. 1993;55(6):1272–7.
110. Starzl TE, Demetris AJ, et al. Migratory nonparenchymal cells after organ allotransplantation: with particular reference to chimerism and the liver. Prog Liver Dis. 1994;12:191–213.
111. Tatari-Calderone Z, Semnani RT, et al. Acquisition of CD80 by human T cells at early stages of activation: functional involvement of CD80 acquisition in T cell to T cell interaction. J Immunol. 2002;169(11):6162–9.
112. Terada N, Hamazaki T, et al. Bone marrow cells adopt the phenotype of other cells by spontaneous cell fusion. Nature. 2002;416(6880):542–5.
113. Thomas FT, Carver FM, et al. Long-term incompatible kidney survival in outbred higher primates without chronic immunosuppression. Ann Surg. 1983;198(3):370–8.
114. Thomas JM, Carver FM, et al. Renal allograft tolerance induced with ATG and donor bone marrow in outbred rhesus monkeys. Transplantation. 1983;36(1):104–6.
115. Thomson AW, Lu L, et al. *In vitro* propagation and homing of liver-derived dendritic cell progenitors to lymphoid tissues of allogeneic recipients. Implications for the establishment and maintenance of donor cell chimerism following liver transplantation. Transplantation. 1995;59(4):544–51.
116. Tomita Y, Sachs DH, et al. Additional monoclonal antibody (mAB) injections can replace thymic irradiation to allow induction of mixed chimerism and tolerance in mice receiving bone marrow transplantation after conditioning with anti-T cell mABs and 3-Gy whole body irradiation. Transplantation. 1996;61(3):469–77.
117. Ulusal BG, Ulusal AE, et al. A new composite facial and scalp transplantation model in rats. Plast Reconstr Surg. 2003;112(5):1302–11.
118. Vassilopoulos G, Wang PR, et al. Transplanted bone marrow regenerates liver by cell fusion. Nature. 2003;422(6934):901–4.
119. Walker J, Opelz G, et al. Correlation of MLC response with graft survival in cadaver and related donor kidney transplants. Transplant Proc. 1978;10(4):949–51.
120. Wang X, Willenbring H, et al. Cell fusion is the principal source of bone-marrow-derived hepatocytes. Nature. 2003;422(6934):897–901.
121. Weimann JM, Charlton CA, et al. Contribution of transplanted bone marrow cells to Purkinje neurons in human adult brains. Proc Natl Acad Sci U S A. 2003;100(4):2088–93.
122. Wekerle T, Sayegh MH, et al. Extrathymic T cell deletion and allogeneic stem cell engraftment induced with costimulatory blockade is followed by central T cell tolerance. J Exp Med. 1998;187(12):2037–44.
123. Xiang J, Huang H, et al. A new dynamic model of CD8 + T effector cell responses via CD4 + T helper-antigen-presenting cells. J Immunol. 2005;174(12):7497–505.
124. Yamanaka N, Wong CJ, et al. Bone marrow transplantation results in human donor blood cells acquiring and displaying mouse recipient class I MHC and CD45 antigens on their surface. PLoS One. 2009;4(12):e8489.
125. Yazici I, Unal S, et al. Composite hemiface/calvaria transplantation model in rats. Plast Reconstr Surg. 2006;118(6):1321–7.
126. Yazici I, Carnevale K, et al. A new rat model of maxilla allotransplantation. Ann Plast Surg. 2007;58(3):338–44.
127. Ying QL, Nichols J, et al. Changing potency by spontaneous fusion. Nature. 2002;416(6880):545–8.
128. Zhang Y, McClellan M, et al. Daclizumab reduces CD25 levels on T cells through monocyte-mediated trogocytosis. Mult Scler. 2014 Feb;20(2):156–64.

Chapter 17
Mesenchymal Stem Cells as Immune Modulators in VCA

Daniel J. Ceradini and Marc A. Soares

Introduction

Mesenchymal stem cells (MSCs), often also referred to as mesenchymal stromal cells, were first identified within the bone marrow more than 40 years ago [29–31]. Since then, MSCs have gained prominence and interest over the past two decades [10, 85] primarily for their ability to differentiate into cells that make up the tissue of mesodermal [10] and even nonmesodermal [46] origin. As our understanding of MSCs has increased, we now recognize several key features of this cell population that have significance to transplantation biology and regenerative medicine: (1) MSCs are participants in tissue repair through direct regeneration of precursor cells, and more importantly, (2) MSCs alter the organismal response to injury and inflammation through dynamic interactions with surrounding cell populations. Particularly in the context of vascularized composite allotransplantation (VCA), MSCs hold significant therapeutic potential to modulate the inflammatory milieu to suppress rejection and improve the host response to foreign antigens. While the clinical data are early, there is compelling scientific evidence that MSCs can modulate the allograft response to ischemia-reperfusion injury (IRI), locally suppress both the innate and adaptive immune system, and skew the immune response to promote peripheral allograft tolerance. Despite this immense potential, significant challenges remain in translating their properties into a consistent therapeutic strategy. We will discuss several new approaches to harness the MSC's immunomodulatory potential, and highlight its particular utility in the context of composite allografts.

D. J. Ceradini (✉) · M. A. Soares
Institute of Reconstructive Plastic Surgery, New York University School of Medicine New York University School of Medicine, 305 E 33rd Street, New York, NY 10016, USA
e-mail: Ceradd01@nyumc.org

G. Brandacher (ed.), *The Science of Reconstructive Transplantation*,
Stem Cell Biology and Regenerative Medicine, DOI 10.1007/978-1-4939-2071-6_17

What is a Mesenchymal Stem Cell?

There is considerable debate as to what defines an MSC. This controversy has largely evolved from lack of incontrovertible *in vivo* evidence that this cell population can self-renew, an essential property of any stem cell. However, as new techniques in molecular biology have allowed us to observe *in vivo* behavior of cell populations, there is now early evidence [77, 95] that MSCs possess this fundamental stem cell property. Unlike its capacity to self-renew, there is no debate about the MSC's ability to differentiate into cells of multiple tissue types including bone, fat, cartilage, tendon, muscle, and marrow stroma [10]. MSC multipotentiality, as in all stem-cell populations, is a highly regulated, controlled, stepwise process involving multiple cell lineages whose fate is determined by the local microenvironments. Newer research has suggested that MSCs also have pluripotential capability, able to transdifferentiate into tissues of endodermal and neuroectodermal origin (neurons, hepatocytes, and endothelia), and are not necessarily restricted to mesenchyme-derived tissues [46]. While scientifically interesting, MSC pluripotentiality is likely to have limited significance in the context of VCA.

Although MSCs were originally isolated from the bone marrow, MSCs exist in almost all tissues in both peripheral reservoirs, such as the dermis, tooth pulp, and hair follicles, and the central reservoir of the marrow [7, 96]. Though studies on MSC populations demonstrate significant heterogeneity in gene expression even from within the same tissue reservoir [115], they retain functional similarity in their ability to regulate immune tolerance, wound healing, inflammation, and fibrosis [49].

There is an evolving definition of what molecular characteristics encompass an MSC. Initially identified in the marrow, the Tissue Stem Cell Committee of the International Society of Cellular Therapy (ISCT) designated the term "mesenchymal stem cell" for marrow-derived, nonhematopoietic, and plastic-adherent cells expanded under standard culture conditions [43]. Further phenotypic refinements to this imprecise definition included surface marker characterization (positive for cluster of deviation (CD)90, CD105, CD73, (STRO-1); negative for CD14, CD11b, CD45, CD34, CD19, and human leukocyte antigen; HLA-DR), as well as the potential to differentiate into bone, fat, and cartilage. Despite significant research and study, no one has been able to identify a single marker that can definitively distinguish MSCs from other cell types. However, establishing minimum criteria for MSCs may actually hamper our understanding of their physiologic role. As characteristics of MSCs vary according to the source of tissue, generating a global definition of MSCs may be too simplistic or unnecessary. Specific definitions of particular MSC subsets may suffice, provided that they accurately and reproducibly define the cells under study.

Given the evolving phenotypic characterization, we are still developing an understanding of the MSC's physiologic role. Our understanding of MSC function primarily centered around two early-observed roles: (1) the ability to form mesenchymal-derived tissue in the context of tissue injury and (2) the ability to support hematopoiesis within the bone marrow, critically maintaining the "stem-cell niche."

The earliest studies [10, 30] of MSCs first noted their osteogenic potential, spurring its study in the tissue-engineering literature [105]. Out of those studies, it became clear that MSCs were directly involved in tissue regeneration in the context of injury. Although there is limited direct evidence of migration of MSCs to sites of injury, it is reasonable to assume that severe tissue damage mobilizes and recruits remote MSCs to injured sites [47]. These recruited and resident MSCs regulate the repair process by differentiation into several kinds of stromal and/or damaged cell types, as well as by providing a microenvironment through the interaction with many types of tissue cells including fibroblasts, endothelial, and epithelial cells. As immune cells are also directed to sites of inflammation, MSCs are able to effectively exert their immunomodulatory properties following induction by the local inflammatory milieu [87]. This stromal MSC–immune cell interaction is critical in providing a microenvironment for tissue regeneration and wound repair.

Outside of sites of injury, multiple studies detailed how marrow-derived MSCs create microenvironments that maintain hematopoietic stem cells (HSCs) and their derivatives [11, 39, 73, 77, 79, 95]. Specifically, in the bone marrow, while MSCs and MSC-derived stroma maintain HSC quiescence within the endosteal niche, they also control HSC proliferation, differentiation, and recruitment within the subendothelial or "vascular" niche of the marrow sinusoids [50, 95, 117]. It soon became clear that MSCs not only participate in direct tissue regeneration following injury but also craft a complex and dynamic network to tightly regulate different arms of hematopoiesis and ultimately alter the immunologic profile of the tissue compartments [112] (Fig. 17.1). Indeed, the powerful immunomodulatory properties of MSCs have already been used to treat a host of immune-related challenges of clinical transplantation, potentially improving outcomes in clinical transplantation, particularly in the context of VCA.

Mechanisms of MSC-Mediated Modulation of the Immune System

The biological response to allografts is a process that involves both the innate and adaptive components of the immune system. Modern biology has demonstrated multiple overlapping mechanisms in which both the innate and adaptive immune systems potentiate rejection. As a process, the cascade of allograft rejection is initiated at the time of reperfusion, initially fueled by inflammation related to IRI, and ultimately executed by immunologic attack of the graft vasculature and underlying parenchyma. Strategies to encourage graft tolerance have focused on not only reducing IRI but also reducing the immunologic sensitivity of the recipient's immune system. To this effect, there has been a growing body of literature supporting the MSC's integral role in promoting transplant tolerance [12, 27, 48, 53, 81, 109–112, 116, 118]. In both ischemia- and immunologic-related injury, MSCs

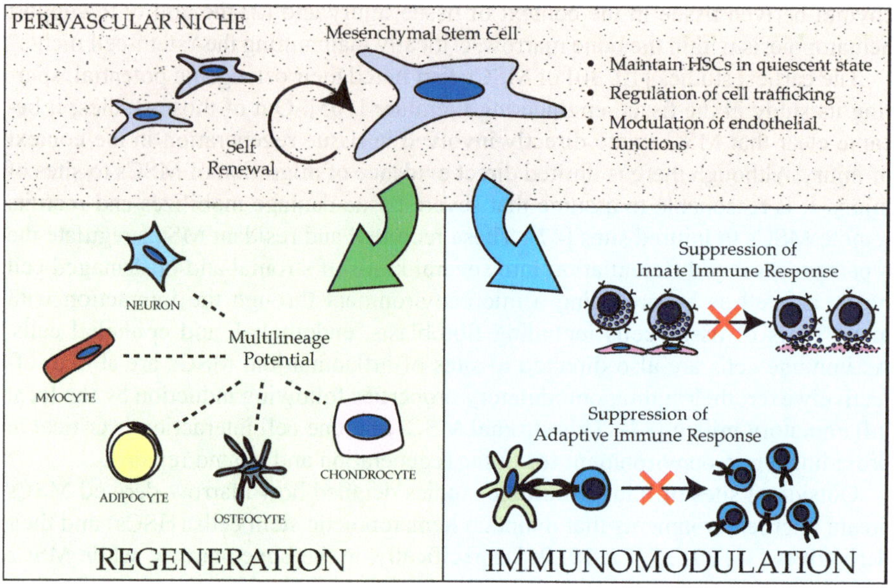

Fig. 17.1 Regenerative and immunomodulatory properties of mesenchymal stem cells (*MSCs*). MSCs help define and maintain the stem cell niche in the bone marrow where they undergo self-renewal. They are thought to function in this compartment by maintaining hematopoietic stem cells (*HSCs*) in a quiescent state, regulating the trafficking of both primitive and more differentiated leukocytes, and modulating sinusoidal endothelial cell function. Although largely unknown, they likely have similar functions in nonbone marrow stem cell niches in peripheral tissues. In addition to this supportive and regulatory function, MSCs are identified by their regenerative capacity to differentiate into different tissues of mesenchymal origin. Furthermore, there is evidence to suggest that transdifferentiation into nonmesodermal derivatives is also a possibility. In contrast to their regenerative properties, MSC have recently been identified as powerful immunomodulatory cells, with numerous functions which suppress both the innate and adaptive immune responses. This function is critical for their potential use in VCA. (*VCA* vascularized composite allotransplantation)

have significant therapeutic potential from their ability to secrete soluble factors and communicate with surrounding cells to modulate the local inflammatory microenvironment. These dynamic interactions have multipronged effects that can be exploited for use in VCA.

The Adaptive Immune System and MSCs

Although it has been almost 30 years since the hypothesis that marrow-derived stem cells could promote tolerance when transplanted into a mismatched recipient [44], only recently have MSCs been shown as a critical mediator of transplant tolerance. In the context of the adaptive immune system, MSCs have been shown to specifically inhibit T cell proliferation *in vitro* [56, 57, 93], and more importantly, have the potential to prolong allogeneic skin engraftment in nonhuman primates *in vivo* [3]. When we look at specific mediators of allograft rejection, cytotoxic T cells (CTLs) represent effector cells of the adaptive immune system and are powerful mediators of parenchymal damage and inflammation. In experimental models of allograft rejection, CTLs lose their lytic abilities in the presence of MSCs [90, 101]. While the exact molecular mechanisms are still a matter of debate and vary between experimental models, most studies agree that MSCs do not constitutively exert their immunomodulatory functions. Rather, they are induced into that role through frequent cross-talk interactions with surrounding T lymphocytes and inflammatory cytokines within the allograft environment. Several specific factors have been hypothesized to contribute to this phenomena, including transforming growth factor (TGF)-β, hepatocyte growth factor [19], prostaglandin E2 (PGE-2; [1]), as well as nitric oxide [97] and indoleamine 2,3 dioxygenase [56].

PGE-2 is a small, short-acting lipid-signaling molecule that has long been linked to inflammation. The pathways for prostaglandin synthesis are mediated by the cyclooxygenase enzyme (COX-1 and COX-2), which produces PGE-2 from arachidonic acid. MSCs have been shown to have baseline expression of COX-2 [25] that is significantly increased in the presence of inflammatory cytokines such as interferon (IFNγ), TNFα, and interleukin 6 (IL-6). MSC-elaborated PGE-2 has been linked to suppressor T-cell activation *in vitro* and *in vivo* [1, 80]. Beyond T-cell activation, MSC-derived PGE-2 has been shown to inhibit production of T helper (Th)17 T-cell populations, a cell population that impairs peripheral tolerance to allografts [40].

Nitric oxide (NO) is a rapidly diffusing gaseous and bioactive molecule, particularly in the context of vascular biology and transplant immunology. NO production is catalyzed by the nitric oxide synthases (NOS) by a variety of cell types, including endothelial cells and MSCs. NO exerts its effects locally at high concentrations, often induced by inflammatory conditions. MSC-derived NO can suppress T-cell proliferation and promote T cell apoptosis [93]. In an elegant study in a murine system [93], MSCs have been shown to chemoattract T cells using various chemokines, and once in proximity, they release NO to exert local immunosuppression. In experimental models of graft-versus-host disease (GvHD), MSCs produce nitric oxide (NO) in a dose-dependent manner in response to interactions with $CD4^+$ or $CD8^+$ T cells, facilitating their immunomodulatory effects [93, 97].

Of recent interest, the tryptophan-metabolizing enzyme, indoleamine 2,3 dioxygenase (IDO) has been implicated in the suppression of T cell proliferation and apoptosis of activated T cells [48, 81]. Tryptophan is an essential amino acid required for T cell proliferation and is depleted through expression of the tryptophan-metabolizing enzyme, indoleamine 2,3 dioxygenase. Via local tryptophan depletion and subsequent production of pro-apoptotic downstream metabolites, IDO provides the common basis for tolerance induction in a variety of physiologic conditions including pregnancy, autoimmunity, tumor immunosurveillance, and transplantation [9]. MSC expression of IDO can be induced via stimulation from IFNγ, or through toll-like-receptor (TLR) 3 and TLR4 ligands. Moreover, MSC-derived IDO has been shown to "reeducate" immune cells towards a more immunosuppressive phenotype through promotion of a Th1–Th2 switch, altering the helper T cell balance and overall adaptive immune response. MSCs also modulate Th17 differentiation to favor induction of tolerogenic T regulatory cells (Tregs; [25, 56, 71, 82, 103, 107]). As Tregs critically regulate alloreactive T cell responses as well as induce transplant tolerance [27, 107], harnessing MSCs to induce both peripheral and central tolerance to allografts is an area of particular benefit to VCA applications.

B cell behavior, the other main cellular component of the adaptive immune system, is also altered in the presence of MSCs. Though less studied, MSCs have been shown to inhibit B cell proliferation in murine studies [18]. Specifically, allogeneic MSCs have been shown to inhibit the proliferation, activation, and immunoglobulin (IgG) secretion by B cells. IFNγ-induced IDO expression by MSCs has been linked to these effects as well [56].

Collectively, the powerful effects of MSCs on multiple pathways of the adaptive immune system (Fig. 17.2) are a promising approach to facilitating tolerance and eliminating the need for lifelong immunosuppression.

The Innate Immune System, IRI, and MSCs

Traditionally, the focus of transplant immunology has been on targeting the mechanisms of adaptive immunity. However, there is an emerging consensus that the potency of rejection is strongly influenced by the activity of the innate immune system as well as external factors, particularly including IRI [51]. MSCs have an equally important role in modulating innate immune system response to allografts (Fig. 17.3).

Natural Killer (NK) cells have a multifaceted role in allograft rejection. They contribute to the innate immune system's ability to track pathogens through a surveillance role, and, within the lymph nodes, they produce significant amounts of IFNγ to activate T cell responses to potentiate allograft rejection [51, 58, 75]. NK cells cocultured with MSCs in the presence of IL-2 and IFNγ demonstrate reduced lytic capacity against traditional targets lacking MHC-1 expression [56]. In a manner analogous to its interaction with T cells, MSCs suppress not only natural killer (NK) cytotoxicity but also their ability to proliferate, largely through IL-6-, IDO-, and PGE-mediated mechanisms [103].

Like NKs, dendritic cells (DCs) are also members of the innate immune system. They are present mainly in tissues exposed to the external environment and serve as the most potent of the antigen-presenting cells (APCs) which are essential for immune system recognition of alloantigens. Both donor-derived and recipient-derived DCs have critical roles in triggering allograft rejection through direct and indirect pathways of allorecognition. MSCs interfere with the activation and maturation of DCs [1, 4, 21, 111] and tilt the immune response towards generation of tolerogenic phenotypes via IL-6-mediated mechanisms [34, 104]. Furthermore, MSCs downregulate expression of DC maturation markers including the major histocompatibility complex (MHC) class II, CD40, CD80, and CD86, and impair the ability of DCs

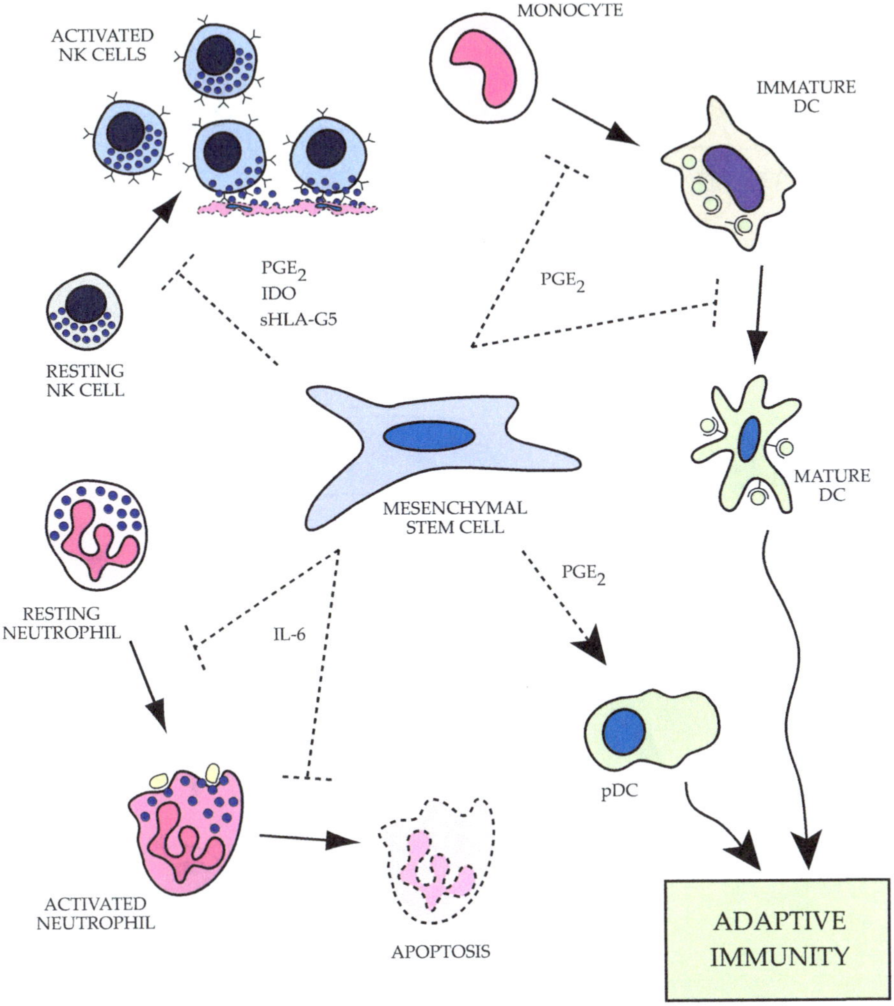

Fig. 17.2 Impact of mesenchymal stem cells (*MSCs*) on functions of the innate immune system. *Top left:* MSCs alter the phenotype of natural killer (*NK*) cells by inhibiting proliferation, cytokine expression and cell cytotoxicity via expression of a number of factors including prostaglandin E_2 (*PGE_2*), indoleamine 2,3-dioxygenase (*IDO*) and soluble HLA-G5. *Top right:* MSCs block the differentiation of monocytes into immature dendritic cells by preventing entry into the cell cycle. In addition, the maturation process of myeloid-derived dendritic cells is altered resulting in retention of more immature characteristics and an impaired ability to present antigen and activate the adaptive immune response. *Bottom right:* Plasmacytoid dendritic cells (*pDC*) significantly increase IL-10 expression in the presence of MSCs, promoting a more robust regulatory adaptive immune response. Bottom left: MSCs actively maintain neutrophils in a resting state, prevent apoptosis of both resting and activated cells, and decrease the respiratory burst associated with invading pathogens or inflammatory mediators via IL-6 secretion. PGE2 prostaglandin E_2, IDO indoleamine 2,3 dioxygenase, *sHLA-G5* secretory isoform of HLA-G, *DC* dendritic cells

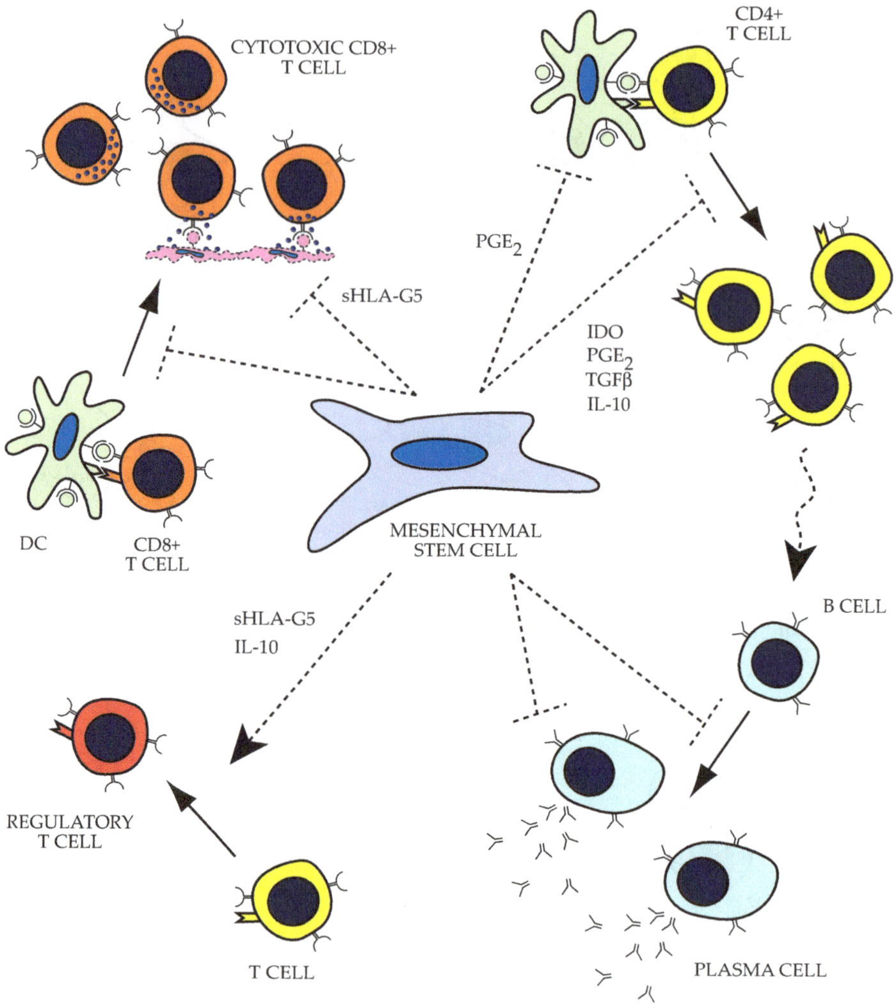

Fig. 17.3 Impact of mesenchymal stem cells (*MSCs*) on functions of the adaptive immune system. *Top left:* MSCs inhibit CD8+ T cell activation through a major histocompatibility complex -independent pathway (MHC-independent pathway). In addition, they also directly inhibit CD8+ cytotoxicity via expression of sHLA-G5. *Top right:* Several factors produced by MSCs actively contribute to direct inhibition of CD4+ activation, proliferation, and helper T cell function. *Bottom right:* MSCs inhibit B cell proliferation, differentiation, antibody production, and chemotaxis via its impact on helper T-cell function, as well as direct inhibition by MSC-expressed factors that are poorly understood. *Bottom left:* MSCs recruit and facilitate the expansion of regulatory T cells directly through expression of sHLA-G5 and indirectly through stimulation of IL-10 production by plasmacytoid dendritic cells

to home to the lymph node *in vivo* [26]. DCs generated in the presence of MSCs produce high levels of anti-inflammatory cytokines including interleukin-10 (IL-10) and lower levels of TNFα. Functionally, tolerogenic DCs generated by MSCs have failed to induce activation of $CD4^+$ T cells *in vitro* and *in vivo* [104]. The capacity of MSC-educated DCs to induce a state of peripheral tolerance would greatly improve outcomes in VCA, and early data have shown this mechanism to contribute to kidney allograft survival in the setting of low-dose immunosuppression [34].

Innate immune phagocytes including neutrophils and macrophages have been shown to promote graft rejection through tissue damage, production of pro-inflammatory cytokines, and activation of antigen-specific T cells [120]. Although it has not been directly shown in a transplant-related model, MSCs can reprogram macrophages to take on more "anti-inflammatory" (M2) phenotypes to mitigate propagation of injury to surrounding tissues [72, 74, 91, 98, 112]. Furthermore, MSCs actively maintain neutrophils in a resting state, prevent apoptosis, and decrease the respiratory burst associated with invading pathogens or inflammatory mediators via IL-6 secretion (Raffaghello 2008, p. 136).

IRI is an unavoidable consequence of VCA and has been associated with an increased incidence of both acute and chronic rejection in the solid organ literature [65]. Reactive oxygen species produced after reperfusion have been linked to the induction of adaptive immune responses through the activation of APCs. During reperfusion, allograft donor DCs are activated and recipient $CD4^+$ T cells, monocytes, and macrophages infiltrate the reperfused graft. This results in a strong cytokine and chemokine release, including damage-associated molecular patterns (DAMPs; [24, 120]). Accordingly, there has been significant interest in using MSCs to mitigate this process. MSCs have been shown to protect against ischemia-reperfusion renal injury via inhibition of apoptosis and stimulation of cell proliferation [32]. Specifically, in the context of transplantation, MSC administration reduced allograft inflammatory gene expression and recruitment of APCs into the allograft in a model of renal allotransplantation [37]. Independent of their immunomodulatory activities, MSCs have been shown to release an array of growth factors to accelerate tissue repair including epidermal growth factor (EGF), fibroblast growth factor (FGF), platelet-derived growth factor (PDGF), transforming growth factor-β (TGF-β), vascular endothelial growth factor (VEGF), insulin-like growth factor 1 (IGF-1), angiopoietin-1, and stromal cell-derived factor-1 (SDF-1), all of which influence both stromal cells and endothelial cells [54, 55].

Taken together, there is overwhelming evidence demonstrating that MSCs modulate the function of immune cells *in vitro*, particularly with regard to T cell and APC behavior. While there is limited *in vivo* evidence corroborating the molecular mechanisms of these observations, as more complex *in vivo* models are developed, we will gain a better understanding of how to develop clinically effective MSC-based therapies.

MSC Targeting Strategies in Transplantation

In contrast to the multitude of studies examining the immunomodulatory properties of MSCs, there are relatively few studies examining therapeutic targeting strategies specifically in transplantation. Within transplantation, most models focus on MSCs in systemically delivered stem-cell transplants, with few studies focusing on solid organ (heart, liver, kidney), and even fewer on VCA [60, 62]. Analogous to solid organ transplantation, accelerated arteriosclerosis and vascular injury are hallmarks of chronic composite allograft rejection [99]. Both acute rejection and chronic rejection are characterized by significant vascular damage and accumulation of inflammatory cells within the allograft. Accordingly, strategies to utilize MSC-based cytotherapy should focus on protection of allograft's endothelial barrier from injury. As MSCs have been shown to take residence in perivascular space [6, 8], they serve as ideal candidates for promotion of the endothelial barrier properties that are necessary to limit cytotoxic damage and influx of immune cells that prime the rejection response.

As with many therapies in transplantation, obtaining targeted delivery is a limiting factor in clinical efficacy. Generally, two approaches of systemic administration have been used for MSC applications. One is intravascular injection, utilizing the capabilities of MSCs to migrate to specific inflammatory tissues *in vivo*. The engraftment was demonstrated in animal models and capable of persisting as long as 13 months after transplantation [70]. However, studies observing MSC trafficking after systemic intravenous infusion have demonstrated that they largely end up accumulating in the lung, liver, and spleen rather than solely the site of injury [41, 42, 125, 36]. The other is site-directed delivery, such as direct injection [38], which can be impractical in the case of a composite allograft where there are multiple tissue compartments. In rodent transplant models [119, 124], MSCs effectively migrate to the site of allograft rejection during chronic rejection, suggesting that the inflammatory milieu of rejection can improve targeting and engraftment of MSCs. Pre-transplant infusion of MSCs appeared to be more effective as compared with peri-transplant administration. A recurring theme in these studies is that long-term graft acceptance is correlated to MSC-dependent expansion of Tregs or tolerogenic DCs in sites of immunologic importance. However, techniques to effectively traffic exogenous MSCs to those sites remains elusive [12, 13, 45, 88] and will remain a significant obstacle to clinical use of MSC immunotherapy.

As accelerated arteriosclerosis and vascular injury are hallmarks of solid/composite allograft rejection [99], strategies to utilize MSC therapy should focus on maintenance of the endothelial barrier from both ischemic and immunologic injury. As MSCs have been shown to take residence in perivascular space [6, 8], they serve as ideal candidates for promotion of the endothelial barrier properties that are necessary to limit cytotoxic damage and influx of immune cells that prime the rejection response. Novel strategies to improve MSC targeting have been discussed in the literature [17, 23, 35, 108, 114, 122] including hypoxic/pharmacologic preconditioning [122], genetic engineering of MSCs [17], and magnetic-based guidance [23,

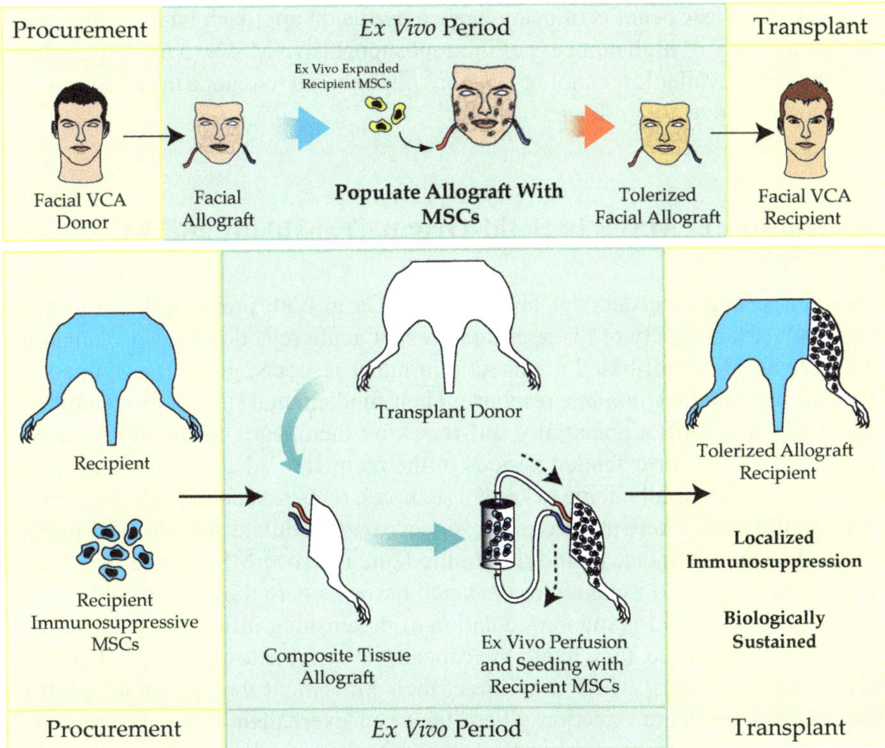

Fig. 17.4 The tolerization of composite tissue allografts. While MSCs are powerful immunomodulatory agents, effective targeting of these cells to exactly where they are needed remains a critical limitation of their use. Due to these limitations, we have recently developed a novel method of delivering immunosuppressive MSCs via the allograft vasculature during the obligate *ex vivo* period in between procurement and transplantation. Using this strategy, large numbers of immunomodulatory MSCs home to the perivascular niche within the allograft where they take up residence to intercept infiltrating leukocytes, potentially converting these effector cells to suppressive leukocytes, tolerogenic dendritic cells, and regulatory T cells

35, 108, 114] to focus delivery to the region of interest using systemic delivery after the inciting inflammatory event. Recent work has proposed a new approach that can utilize MSC-based cell therapy to modify the allograft itself at the time of transplantation to facilitate graft tolerance [16, 76, 78, 100, 101] (Fig. 17.4). *Ex vivo* MSC perfusion utilizes clinically available organ perfusion technology [113] to engineer allografts that are primed for peripheral tolerance. MSCs are seeded into the allograft during the *ex vivo* period between procurement and transplantation where they take up permanent residence in the perivascular space of the allograft. Machine perfusion allows tight biologic control of the allograft environment in an isolated circuit and is clinically available for use in kidney, lung, and liver transplantation. The efficacy of pulsatile perfusion has been tested over the past 30 years, demonstrating a reduction in delayed graft function and improved long-term graft survival

[16, 113]. The clear benefits of using *ex vivo* perfusion approach include the ability to target delivery of high numbers of immunosuppressive MSCs to the allograft using a clinically available technology that has been shown to reduce the effects of IRI injury and improve allograft survival.

Applications of MSCs in Solid-Organ Transplant and VCA

There is a strong precedent set for use of MSCs in both preclinical and clinical settings to treat a variety of diseases and prevent acute rejection in transplantation. MSCs are an ideal cell-based therapeutic in many respects, given their impact on the innate and adaptive immune response. Their fundamental stem or progenitor cell properties make them a potentially self-renewing therapeutic agent which has the potential to persist for extended periods in the recipient. When delivered systemically, MSC preferentially home to known stem-cell niches, such as the bone marrow in a hand allograft, where they take up residence and modulate donor hematopoietic cells in this compartment. In addition to the bone marrow, MSCs have been demonstrated to migrate to ischemic or damaged tissues where they participate in both tissue regeneration and immunomodulation of the ensuing inflammatory response. Within the target tissue, they retain the capacity to migrate towards an ischemic or inflammatory stimulus, which may direct their movement through an allograft to areas undergoing acute rejection where they can exert their immunosuppressive effect. They also secrete a number of cytokines to attract subsets of leukocytes that egress from circulation during inflammatory processes, such as acute rejection, effectively attracting these effector cells to modulate their immune response. Finally, allogeneic MSCs evade the host immune system despite expressing intermediate levels of MHC class I on the cell surface [66, 86, 109], and fail to stimulate recipient lymphocyte proliferation [19, 67, 109]. This relative hypoimmunogenicity permits transplantation across allogeneic barriers, making third-party sources of MSCs a practical and viable alternative for use in VCA.

Since the first demonstration of prolongation of allograft survival in skin grafts over a decade ago [3], there have been numerous studies in small animal models to establish the powerful immunosuppressive role of systemically delivered MSCs in heart [12, 15, 22, 34, 45, 88, 124], islet cell (Ding 2009, p. 120; Li 2010, p. 122; Xu 2012, p. 124; Kim 2011, p. 128; Solari 2009, p. 130), kidneys (Ge 2010, p. 131; Casiraghi 2012, p. 87), and liver transplantation (Wang 2009, p. 30). While many of these experiments used no additional immunosuppression, more practical preclinical studies used low-dose immunosuppression that more accurately reflects the likely treatment strategy in human subjects. It remains to be determined, however, what the impact of the immunosuppressive drug regimen on MSC function and downstream generation of a regulatory T cell response will be (Zeiser 2006, p. 134; Wang 2009, p. 135).

There is an early yet growing clinical use of MSCs in solid-organ transplantation. In renal transplantation, a clinical feasibility study [83] demonstrated that autologous bone marrow-derived MSCs can be safely administered 7 days post transplant with concomitant increase in allograft Treg populations and stable serum creatinine at 1-year post transplant. Similarly, recent clinical trials in living-related kidney transplantation revealed that induction therapy with MSCs before allograft revascularization and then again 2 weeks following transplant resulted in lower incidence of acute rejection, decreased risk for opportunistic infections, and better estimated renal function after 1 year [106]. However, there is some evidence that post-transplant administration can induce a transient inflammatory allograft injury, not unlike "engraftment syndrome," although there appeared to be no clinical consequences at 1-year posttransplant [83, 84]. Phase I studies are currently being conducted in liver transplantation [89] as well as lung transplantation [94].

Probably the most impressive *in vivo* evidence and use of MSC-based immunomodulation in humans is in the treatment of GvHD after allogeneic hematopoietic stem-cell transplantation. In phase II clinical studies, MSC infusions were shown to be safe and effective in treating steroid-resistant acute GvHD [68] and have shown to reduce the incidence of severe GvHD when used prophylactically [5, 64]. While extremely promising, further long-term studies are needed to conclusively demonstrate safety and efficacy.

Much like solid-organ transplantation, vascularized composite allotransplantation requires systemic lifelong immunosuppression. However, unlike the solid-organ transplant population, the recipients of reconstructive composite allografts are relatively healthy patients without evidence of end-organ failure or chronic debilitating disease. Considering the potential detrimental effects of systemic lifelong immunosuppression including shortened life span, organ failure, cancer, and even death, the need for more innovative therapeutic approaches towards achieving tolerance without the need for such harsh immunosuppressive drugs is even more evident in VCA.

There are important intrinsic differences in the composite allograft biological environment compared to solid organs that may represent an opportunity to use MSCs as potent immunomodulators. Unlike many solid organs, composite allografts frequently contain skin, fat, lymph nodes, and bone marrow, which are known stem-cell niches where MSCs may engraft and proliferate to exert their immunosuppressive effect. Thus, the allograft itself becomes a reservoir of immunomodulatory MSCs exactly where they are needed to exert their effect. The skin-containing allografts are a rich source of dendritic-antigen-presenting cells which frequently occupy similar tissue compartments that MSCs would engraft, potentially increasing the likelihood of generating large numbers of tolerogenic cell types. Furthermore, the very nature of a skin-bearing allograft being visible to the patient allows for continuous monitoring for signs of rejection, which would likely be a benefit when undertaking clinical trials of MSC therapy or minimization protocols. While there are no active clinical trials using exclusively MSCs for VCA, several preclinical studies have demonstrated a marked immunosuppressive effect of systemically delivered MSCs following VCA on the incidence and severity of acute rejection, the peripheral blood T regulatory response, inflammatory cytokine expression profile, and ultimately allograft survival [2, 59, 61, 63]. These promising findings serve as a foundation for eventual translation of MSC therapy to clinical cases.

Future Directions and Considerations

As promising as MSCs appear as therapeutic agents, some consideration must be made to their safety profile. One of the current disadvantages of systemic immunosuppression regimens is their inability to distinguish between pathologic and protective immune responses; it is important to critically examine the effects of MSC-based immunomodulation. There is little existing knowledge regarding the *in vivo* survival of MSCs, their potential to contribute to systemic immune suppression, ectopic tissue formation, or malignancy [14, 28, 94]. As MSCs have been shown to confer tumor immunity in several experimental studies, it is unclear whether MSC therapy can initiate malignant transformation of benign processes or shield indolent malignancies from immune surveillance in humans. While existing clinical studies have not reported adverse effects from MSC infusions in the context of stem-cell transplants or in GvHD, our experience is still relatively young [68] and warrants careful attention.

Route of administration, homing, and persistence of MSCs in the allograft are critical factors that require further investigation before translation to clinical care. Many of the published work on MSCs utilize systemic intravenous administration, which is likely inefficient to effectively target these cells to the graft. Furthermore, it is unclear whether the MSC immunomodulatory effect will persist over time, as the allograft microenvironment changes markedly from a significant inflammatory state immediately following reperfusion and resultant IRI (favoring more immunomodulation), to a more stable and less inflammatory state once healing is complete (favoring less immunomodulation). Finally, although there is some initial evidence that MSCs undergo self-renewal *in vivo*, it is unclear whether this will occur in the perivascular niche within that allograft tissue, or whether this is even required to make a significant impact on the need for systemic immunosuppression. Nonetheless, while there are many questions yet to be answered, the use of MSCs in clinical transplantation has the potential to be a paradigm-changing therapeutic approach to immunosuppression.

References

1. Aggarwal S, Pittenger MF. Human mesenchymal stem cells modulate allogeneic immune cell responses. Blood. 2005;105:1815–22.
2. Aksu AE, Horibe E, Sacks J, Ikeguchi R, Breitinger J, Scozio M, Unadkat J, Feili-Hariri M. Co-infusion of donor bone marrow with host mesenchymal stem cells treats GVHD and promotes vascularized skin allograft survival in rats. Clin Immunol. 2008;127:348–58.

3. Bartholomew A, Sturgeon C, Siatskas M, Ferrer K, Mcintosh K, Patil S, Hardy W, Devine S, Ucker D, Deans R, Moseley A, Hoffman R. Mesenchymal stem cells suppress lymphocyte proliferation in vitro and prolong skin graft survival in vivo. Exp Hematol. 2002;30:42–8.
4. Benichou G, Tocco G. The road to transplant tolerance is paved with good dendritic cells. Eur J Immunol. 2013 Mar;43(3):584–8.
5. Bernardo ME, Ball LM, Cometa AM, Roelofs H, Zecca M, Avanzini MA, Bertaina A, Vinti L, Lankester A, Maccario R, Ringden O, Le Blanc K, Egeler RM, Fibbe WE, Locatelli F. Co-infusion of ex vivo-expanded, parental MSCs prevents life-threatening acute GVHD, but does not reduce the risk of graft failure in pediatric patients undergoing allogeneic umbilical cord blood transplantation. Bone Marrow Transplant. 2011;46:200–7.
6. Bianco P. Back to the future: moving beyond "mesenchymal stem cells";. J Cell Biochem. 2011;112:1713–21.
7. Bianco P, Robey PG, Simmons PJ. 2008. Mesenchymal stem cells: revisiting history, concepts, and assays. Cell Stem Cell. 2:313–9.
8. Bianco P, Cao X, Frenette PS, Mao JJ, Robey PG, Simmons PJ, Wang CY. The meaning, the sense and the significance: translating the science of mesenchymal stem cells into medicine. Nat Med. 2013;19:35–42.
9. Brandacher G, Margreiter R, Fuchs D. Implications of IFN-gamma-mediated tryptophan catabolism on solid organ transplantation. Curr Drug Metab. 2007;8:273–82.
10. Caplan AI. Mesenchymal stem cells. J Orthop Res. 1991;9:641–50.
11. Caplan AI, Dennis JE. Mesenchymal stem cells as trophic mediators. J Cell Biochem. 2006;98:1076–84.
12. Casiraghi F, Azzollini N, Cassis P, Imberti B, Morigi M, Cugini D, Cavinato RA, Todeschini M, Solini S, Sonzogni A, Perico N, Remuzzi, G. Noris, M. Pretransplant infusion of mesenchymal stem cells prolongs the survival of a semiallogeneic heart transplant through the generation of regulatory T cells. J Immunol. 2008;181:3933–46.
13. Casiraghi F, Azzollini N, Todeschini M, Cavinato RA, Cassis P, Solini S, Rota C, Morigi M, Introna M, Maranta R, Perico N, Remuzzi, G. Noris, M. Localization of mesenchymal stromal cells dictates their immune or proinflammatory effects in kidney transplantation. Am J Transplant. 2012;12:2373–83.
14. Casiraghi F, Perico N, Remuzzi G. Mesenchymal stromal cells to promote solid organ transplantation tolerance. Curr Opin Organ Transplant. 2013;18:51–8.
15. Chabannes D, Hill M, Merieau E, Rossignol J, Brion R, Soulillou JP, Anegon I, Cuturi MC. A role for heme oxygenase-1 in the immunosuppressive effect of adult rat and human mesenchymal stem cells. Blood. 2007;110:3691–4.
16. Chang EI, Bonillas RG, El-Ftesi S, Ceradini DJ, Vial IN, Chan DA, Michaels JT, Gurtner GC. Tissue engineering using autologous microcirculatory beds as vascularized bioscaffolds. FASEB J. 2009;23:906–15.
17. Chen W, Li M, Cheng H, Yan Z, Cao J, Pan B, Sang W, Wu Q, Zeng L, Li Z, Xu K. Overexpression of the mesenchymal stem cell Cxcr4 gene in irradiated mice increases the homing capacity of these cells. Cell Biochem Biophys. 2013;28:28.
18. Deng W, Han Q, Liao L, You S, Deng H, Zhao RC. Effects of allogeneic bone marrow-derived mesenchymal stem cells on T and B lymphocytes from BXSB mice. DNA Cell Biol. 2005;24:458–63.
19. Di Nicola M, Carlo-Stella C, Magni M, Milanesi M, Longoni PD, Matteucci P, Grisanti S, Gianni AM. Human bone marrow stromal cells suppress T-lymphocyte proliferation induced by cellular or nonspecific mitogenic stimuli. Blood. 2002;99:3838–43.
20. Ding Y, Xu D, Feng G, Bushell A, Muschel RJ, Wood KJ. Mesenchymal stem cells prevent the rejection of fully allogenic islet grafts by the immunosuppressive activity of matrix metalloproteinase-2 and -9. Diabetes. 2009;58:1797-06
21. Djouad F, Charbonnier LM, Bouffi C, Louis-Plence P, Bony C, Apparailly F, Cantos C, Jorgensen C, Noel D. Mesenchymal stem cells inhibit the differentiation of dendritic cells through an interleukin-6-dependent mechanism. Stem Cells. 2007;25:2025–32.

22. Eggenhofer E, Renner P, Soeder Y, Popp FC, Hoogduijn MJ, Geissler EK, Schlitt HJ, Dahlke MH. Features of synergism between mesenchymal stem cells and immunosuppressive drugs in a murine heart transplantation model. Transpl Immunol. 2011;25:141–7.
23. El Haj AJ, Glossop JR, Sura HS, Lees MR, Hu B, Wolbank S, Van Griensven M, Redl H, Dobson J. An *in vitro* model of mesenchymal stem cell targeting using magnetic particle labelling. J Tissue Eng Regen Med. 2012. DOI:10.1002/term.1636
24. Eltzschig HK, Eckle T. Ischemia and reperfusion–from mechanism to translation. Nat Med. 2011;17:1391–401.
25. English K, Barry FP, Field-Corbett CP, Mahon BP. IFN-gamma and TNF-alpha differentially regulate immunomodulation by murine mesenchymal stem cells. Immunol Lett. 2007;110:91–100.
26. English K, Barry FP, Mahon BP. Murine mesenchymal stem cells suppress dendritic cell migration, maturation and antigen presentation. Immunol Lett. 2008;115:50–8.
27. English K, French A, Wood KJ. Mesenchymal stromal cells: facilitators of successful transplantation? Cell Stem Cell. 2010;7:431–42.
28. English K, Wood KJ. Mesenchymal stromal cells in transplantation rejection and tolerance. Cold Spring Harb Perspect Med. 2013;3:a015560.
29. Friedenstein AJ, Piatetzky S, II, Petrakova KV. Osteogenesis in transplants of bone marrow cells. J Embryol Exp Morphol. 1966;16:381–90.
30. Friedenstein AJ, Petrakova KV, Kurolesova AI, Frolova GP. Heterotopic of bone marrow. Analysis of precursor cells for osteogenic and hematopoietic tissues. Transplantation. 1968;6:230–47.
31. Friedenstein AJ, Chailakhyan RK, Latsinik NV, Panasyuk AF, Keiliss-Borok IV. Stromal cells responsible for transferring the microenvironment of the hemopoietic tissues. Cloning *in vitro* and retransplantation *in vivo*. Transplantation. 1974;17:331–40.
32. Gatti S, Bruno S, Deregibus MC, Sordi A, Cantaluppi V, Tetta C, Camussi G. Microvesicles derived from human adult mesenchymal stem cells protect against ischaemia-reperfusion-induced acute and chronic kidney injury. Nephrol Dial Transplant. 2011;26:1474–83.
33. Ge W, Jiang J, Arp J, Liu W, Garcia B, Wang H. Regulatory t-cell generation and kidney allograft tolerance induced by mesenchymal stem cells associated with indoleamine 2,3-dioxygenase expression. Transplantation. 2010;90:1312–20
34. Ge W, Jiang J, Baroja ML, Arp J, Zassoko R, Liu W, Bartholomew A, Garcia B, Wang H. Infusion of mesenchymal stem cells and rapamycin synergize to attenuate alloimmune responses and promote cardiac allograft tolerance. Am J Transplant. 2009;9:1760–72.
35. Gjorgieva D, Zaidman N, Bosnakovski D. Mesenchymal stem cells for anti-cancer drug delivery. Recent Pat Anticancer Drug Discov. 2013;8:310–8.
36. Hannoush EJ, Sifri ZC, Elhassan IO, Mohr AM, Alzate WD, Offin M, Livingston DH. Impact of enhanced mobilization of bone marrow derived cells to site of injury. J Trauma. 2011;71:283–9; discussion 289–91.
37. Hara Y, Stolk M, Ringe J, Dehne T, Ladhoff J, Kotsch K, Reutzel-Selke A, Reinke P, Volk HD, Seifert M. *In vivo* effect of bone marrow-derived mesenchymal stem cells in a rat kidney transplantation model with prolonged cold ischemia. Transplant Int. 2011;24:1112–23.
38. Hare JM, Fishman JE, Gerstenblith G, Difede Velazquez DL, Zambrano JP, Suncion VY, Tracy M, Ghersin E, Johnston PV, Brinker JA, Breton E, Davis-Sproul J, Schulman IH, Byrnes J, Mendizabal AM, Lowery MH, Rouy D, Altman P, Wong Po Foo C, Ruiz P, Amador A, Da Silva J, Mcniece IK, Heldman AW. Comparison of allogeneic vs autologous bone marrow-derived mesenchymal stem cells delivered by transendocardial injection in patients with ischemic cardiomyopathy: the POSEIDON randomized trial. JAMA. 2012;308:2369–79.
39. Haynesworth SE, Baber MA, Caplan AI. Cytokine expression by human marrow-derived mesenchymal progenitor cells *in vitro*: effects of dexamethasone and Il-1 alpha. J Cell Physiol. 1996;166:585–92.
40. Heidt S, Segundo DS, Chadha R, Wood KJ. The impact of Th17 cells on transplant rejection and the induction of tolerance. Curr Opin Organ Transplant. 2010;15: 456–61.

41. Hoogduijn MJ, Dor FJ. Mesenchymal stem cells in transplantation and tissue regeneration. Front Immunol. 2011;2:84.
42. Hoogduijn MJ, Roemeling-Van Rhijn M, Korevaar SS, Engela AU, Weimar W, Baan CC. Immunological aspects of allogeneic and autologous mesenchymal stem cell therapies. Hum Gene Ther. 2011;22:1587–91.
43. Horwitz EM, Le Blanc K, Dominici M, Mueller I, Slaper-Cortenbach I, Marini FC, Deans RJ, Krause DS, Keating A. Clarification of the nomenclature for MSC: the international society for cellular therapy position statement. Cytotherapy. 2005;7:393–5.
44. Ildstad ST, Sachs DH. Reconstitution with syngeneic plus allogeneic or xenogeneic bone marrow leads to specific acceptance of allografts or xenografts. Nature. 1984;307:168–70.
45. Inoue S, Popp FC, Koehl GE, Piso P, Schlitt HJ, Geissler EK, Dahlke MH. Immunomodulatory effects of mesenchymal stem cells in a rat organ transplant model. Transplantation. 2006;81:1589–95.
46. Jiang Y, Jahagirdar BN, Reinhardt RL, Schwartz RE, Keene CD, Ortiz-Gonzalez XR, Reyes M, Lenvik T, Lund T, Blackstad M, Du J, Aldrich S, Lisberg A, Low WC, Largaespada DA, Verfaillie CM. Pluripotency of mesenchymal stem cells derived from adult marrow. Nature. 2002;418:41–9.
47. Karp JM, Leng Teo GS. Mesenchymal stem cell homing: the devil is in the details. Cell Stem Cell. 2009;4:206–16.
48. Keating A. Mesenchymal stromal cells: new directions. Cell Stem Cell. 2012;10;709–16.
49. Kern S, Eichler H, Stoeve J, Kluter H, Bieback K. Comparative analysis of mesenchymal stem cells from bone marrow, umbilical cord blood, or adipose tissue. Stem Cells. 2006;24:1294–301.
50. Kiel MJ, Morrison SJ. Uncertainty in the niches that maintain haematopoietic stem cells. Nat Rev Immunol. 2008;8:290–301.
51. Kim IK, Bedi DS, Denecke C, Ge X, Tullius SG. Impact of innate and adaptive immunity on rejection and tolerance. Transplantation. 2008;86:889–94.
52. Kim YH, Wee YM, Choi MY, Lim DG, Kim SC, Han DJ. Interleukin (il)-10 induced by cd11b(+) cells and il-10-activated regulatory t cells play a role in immune modulation of mesenchymal stem cells in rat islet allografts. Mol Med. 2011;17:697–08
53. Kim EJ, Kim N, Cho SG. The potential use of mesenchymal stem cells in hematopoietic stem cell transplantation. Exp Mol Med. 2013;45:e2.
54. Kinnaird T, Stabile E, Burnett MS, Lee CW, Barr S, Fuchs S, Epstein SE. Marrow-derived stromal cells express genes encoding a broad spectrum of arteriogenic cytokines and promote *in vitro* and *in vivo* arteriogenesis through paracrine mechanisms. Circ Res. 2004a;94:678–85.
55. Kinnaird T, Stabile E, Burnett MS, Shou M, Lee CW, Barr S, Fuchs S, Epstein SE. Local delivery of marrow-derived stromal cells augments collateral perfusion through paracrine mechanisms. Circulation. 2004b;109:1543–9.
56. Krampera M, Cosmi L, Angeli R, Pasini A, Liotta F, Andreini A, Santarlasci V, Mazzinghi B, Pizzolo G, Vinante F, Romagnani P, Maggi E, Romagnani S, Annunziato F. Role for interferon-gamma in the immunomodulatory activity of human bone marrow mesenchymal stem cells. Stem Cells. 2006a;24:386–98.
57. Krampera M, Pasini A, Pizzolo G, Cosmi L, Romagnani S, Annunziato F. Regenerative and immunomodulatory potential of mesenchymal stem cells. Curr Opin Pharmacol. 2006b;6:435–41.
58. Kroemer A, Edtinger K, Li XC. The innate natural killer cells in transplant rejection and tolerance induction. Curr Opin Organ Transplant. 2008;13:339–43.
59. Kuo YR, Goto S, Shih HS, Wang FS, Lin CC, Wang CT, Huang EY, Chen CL, Wei FC, Zheng XX, Lee WP. Mesenchymal stem cells prolong composite tissue allotransplant survival in a swine model. Transplantation. 2009;87:1769–77.
60. Kuo YR, Chen CC, Goto S, Lee IT, Huang CW, Tsai CC, Wang CT, Chen CL. Modulation of immune response and T-cell regulation by donor adipose-derived stem cells in a rodent hind-limb allotransplant model. Plast Reconstr Surg. 2011a;128:661e–72e.

61. Kuo YR, Chen CC, Shih HS, Goto S, Huang CW, Wang CT, Chen CL, Wei FC. Prolongation of composite tissue allotransplant survival by treatment with bone marrow mesenchymal stem cells is correlated with T-cell regulation in a swine hind-limb model. Plast Reconstr Surg. 2011b;127:569–79.
62. Kuo YR, Chen CC, Shih HS, Goto S, Huang CW, Wang CT, Chen CL, Wei FC. Prolongation of composite tissue allotransplant survival by treatment with bone marrow mesenchymal stem cells is correlated with T-cell regulation in a swine hind-limb model. Plast Reconstr Surg. 2011c;127:569–79.
63. Kuo YR, Chen CC, Goto S, Huang YT, Wang CT, Tsai CC, Chen CL. Immunomodulatory effects of bone marrow-derived mesenchymal stem cells in a swine hemi-facial allotransplantation model. PLoS One. 2012;7:e35459.
64. Kuzmina LA, Petinati NA, Parovichnikova EN, Lubimova LS, Gribanova EO, Gaponova TV, Shipounova IN, Zhironkina OA, Bigildeev AE, Svinareva DA, Drize NJ, Savchenko VG. Multipotent mesenchymal stromal cells for the prophylaxis of acute graft-versus-host disease-A phase II study. Stem Cells Int. 2012;968213.
65. Land WG. The role of postischemic reperfusion injury and other nonantigen-dependent inflammatory pathways in transplantation. Transplantation. 2005;79:505–14.
66. Le Blanc K, Tammik C, Rosendahl K, Zetterberg E, Ringden O. HLA expression and immunologic properties of differentiated and undifferentiated mesenchymal stem cells. Exp Hematol. 2003a;31:890–6.
67. Le Blanc K, Tammik L, Sundberg B, Haynesworth SE, Ringden O. Mesenchymal stem cells inhibit and stimulate mixed lymphocyte cultures and mitogenic responses independently of the major histocompatibility complex. Scand J Immunol. 2003b;57:11–20.
68. Le Blanc K, Frassoni F, Ball L, Locatelli F, Roelofs H, Lewis I, Lanino E, Sundberg B, Bernardo ME, Remberger M, Dini G, Egeler RM, Bacigalupo A, Fibbe W, Ringden O. Mesenchymal stem cells for treatment of steroid-resistant, severe, acute graft-versus-host disease: a phase II study. Lancet. 2008;371:1579–86.
69. Li FR, Wang XG, Deng CY, Qi H, Ren LL, Zhou HX. Immune modulation of co-transplantation mesenchymal stem cells with islet on t and dendritic cells. Clin Exp Immunol. 2010;161:357–63
70. Liechty KW, Mackenzie TC, Shaaban AF, Radu A, Moseley AM, Deans R, Marshak DR, Flake AW. Human mesenchymal stem cells engraft and demonstrate site-specific differentiation after in utero transplantation in sheep. Nat Med. 2000;6:1282–6.
71. Liotta F, Angeli R, Cosmi L, Fili L, Manuelli C, Frosali F, Mazzinghi B, Maggi L, Pasini A, Lisi V, Santarlasci V, Consoloni L, Angelotti ML, Romagnani P, Parronchi P, Krampera M, Maggi E, Romagnani S, Annunziato F. Toll-like receptors 3 and 4 are expressed by human bone marrow-derived mesenchymal stem cells and can inhibit their T-cell modulatory activity by impairing Notch signaling. Stem Cells. 2008;26:279–89.
72. Maggini J, Mirkin G, Bognanni I, Holmberg J, Piazzon IM, Nepomnaschy I, Costa H, Canones C, Raiden S, Vermeulen M, Geffner JR. Mouse bone marrow-derived mesenchymal stromal cells turn activated macrophages into a regulatory-like profile. PloS One. 2010;5:e9252.
73. Maitra B, Szekely E, Gjini K, Laughlin MJ, Dennis J, Haynesworth SE, Koc ON. Human mesenchymal stem cells support unrelated donor hematopoietic stem cells and suppress T-cell activation. Bone Marrow Transplant. 2004;33:597–604.
74. Mantovani A, Biswas SK, Galdiero MR, Sica A, Locati M. Macrophage plasticity and polarization in tissue repair and remodelling. J Pathol. 2013;229:176–85.
75. Martin-Fontecha A, Thomsen LL, Brett S, Gerard C, Lipp M, Lanzavecchia A, Sallusto F. Induced recruitment of NK cells to lymph nodes provides IFN-gamma for T(H)1 priming. Nat Immunol. 2004;5:1260–5.
76. Mcrae M, Ham M, Soares M, Wilson S, Lotfi P, Lalezarzadeh F, Ojo C, Saadeh P, Ceradini D. *Ex vivo* mesenchymal stem cell therapy protects vascular networks from ischemia reperfusion injury. Plast Reconstr Surg. 2013;131:105. 10.1097/01.prs.0000430081.33685.43.

77. Mendez-Ferrer S, Michurina TV, Ferraro F, Mazloom AR, Macarthur BD, Lira SA, Scadden DT, Ma'ayan A, Enikolopov GN, Frenette PS. Mesenchymal and haematopoietic stem cells form a unique bone marrow niche. Nature. 2010;466:829–34.
78. Michaels JV, Levine JP, Hazen A, Ceradini DJ, Galiano RD, Soltanian H, Gurtner GC. Biologic brachytherapy: *ex vivo* transduction of microvascular beds for efficient, targeted gene therapy. Plast Reconstr Surg. 2006;118:54–65. 10.1097/01.prs.0000220466.27521.22.
79. Muguruma Y, Yahata T, Miyatake H, Sato T, Uno T, Itoh J, Kato S, Ito M, Hotta T. Ando K. Reconstitution of the functional human hematopoietic microenvironment derived from human mesenchymal stem cells in the murine bone marrow compartment. Blood. 2006;107:1878–87.
80. Najar M, Raicevic G, Boufker HI, Fayyad Kazan H, De Bruyn C, Meuleman N, Bron D, Toungouz M, Lagneaux L. Mesenchymal stromal cells use PGE2 to modulate activation and proliferation of lymphocyte subsets: combined comparison of adipose tissue, Wharton's Jelly and bone marrow sources. Cell Immunol. 2010;264:171–9.
81. Nauta AJ, Fibbe WE. Immunomodulatory properties of mesenchymal stromal cells. Blood. 2007;110:3499–506.
82. Opitz CA, Litzenburger UM, Lutz C, Lanz TV, Tritschler I, Koppel A, Tolosa E, Hoberg M, Anderl J, Aicher WK, Weller M, Wick W, Platten M. Toll-like receptor engagement enhances the immunosuppressive properties of human bone marrow-derived mesenchymal stem cells by inducing indoleamine-2,3-dioxygenase-1 via interferon-beta and protein kinase R. Stem Cells. 2009;27:909–19.
83. Perico N, Casiraghi F, Introna M, Gotti E, Todeschini M, Cavinato RA, Capelli C, Rambaldi A, Cassis P, Rizzo P, Cortinovis M, Marasa M, Golay J, Noris M, Remuzzi G. Autologous mesenchymal stromal cells and kidney transplantation: a pilot study of safety and clinical feasibility. Clin J Am Soc Nephrol. 2011;6:412–22.
84. Perico N, Casiraghi F, Gotti E, Introna M, Todeschini M, Cavinato RA, Capelli C, Rambaldi A, Cassis P, Rizzo P, Cortinovis M, Noris M, Remuzzi G. Mesenchymal stromal cells and kidney transplantation: pretransplant infusion protects from graft dysfunction while fostering immunoregulation. Transplant Int. 2013;26(9):867–78.
85. Pittenger MF. Multilineage potential of adult human mesenchymal stem cells. Science. 1999;284:143–7.
86. Pittenger MF, Martin BJ. Mesenchymal stem cells and their potential as cardiac therapeutics. Circ Res. 2004;95:9–20.
87. Ponte AL, Marais E, Gallay N, Langonne A, Delorme B, Herault O, Charbord P, Domenech J. The *in vitro* migration capacity of human bone marrow mesenchymal stem cells: comparison of chemokine and growth factor chemotactic activities. Stem Cells. 2004;25:1737–45.
88. Popp FC, Eggenhofer E, Renner P, Slowik P, Lang SA, Kaspar H, Geissler EK, Piso P, Schlitt HJ, Dahlke MH. Mesenchymal stem cells can induce long-term acceptance of solid organ allografts in synergy with low-dose mycophenolate. Transplant Immunol. 2008;20:55–60.
89. Popp FC, Fillenberg B, Eggenhofer E, Renner P, Dillmann J, Benseler V, Schnitzbauer AA, Hutchinson J, Deans R, Ladenheim D, Graveen CA, Zeman F, Koller M, Hoogduijn MJ, Geissler EK, Schlitt HJ, Dahlke MH. Safety and feasibility of third-party multipotent adult progenitor cells for immunomodulation therapy after liver transplantation–a phase I study (Misot-I). J Transl Med. 2011;9:124.
90. Potian JA, Aviv H, Ponzio NM, Harrison JS, Rameshwar P. Veto-like activity of mesenchymal stem cells: functional discrimination between cellular responses to alloantigens and recall antigens. J Immunol. 2003;171:3426–34.
91. Prockop DJ, Oh JY. Mesenchymal stem/stromal cells (MSCs): role as guardians of inflammation. Mol Ther. 2012;20:14–20.
92. Raffaghello L, Bianchi G, Bertolotto M, Montecucco F, Busca A, Dallegri F, Ottonello L, Pistoia V. Human mesenchymal stem cells inhibit neutrophil apoptosis: A model for neutrophil preservation in the bone marrow niche. Stem Cells. 2008;26:151–62
93. Ren G, Zhang L, Zhao X, Xu G, Zhang Y, Roberts AI, Zhao RC, Shi Y. Mesenchymal stem cell-mediated immunosuppression occurs via concerted action of chemokines and nitric oxide. Cell Stem Cell. 2008;2:141–50.

94. Roemeling-Van Rhijn M, Weimar W, Hoogduijn MJ. Mesenchymal stem cells. Curr Opin Organ Transplant. 2012;17:55–62.
95. Sacchetti B, Funari A, Michienzi S, Di Cesare S, Piersanti S, Saggio I, Tagliafico E, Ferrari S, Robey PG, Riminucci M, Bianco P. Self-renewing osteoprogenitors in bone marrow sinusoids can organize a hematopoietic microenvironment. Cell. 2007;131:324–36.
96. Satija NK, Gurudutta GU, Sharma S, Afrin F, Gupta P, Verma YK, Singh VK, Tripathi RP. Mesenchymal stem cells: molecular targets for tissue engineering. Stem Cells Dev. 2007;16:7–23.
97. Sato K, Ozaki K, Oh I, Meguro A, Hatanaka K, Nagai T, Muroi K, Ozawa K. Nitric oxide plays a critical role in suppression of T-cell proliferation by mesenchymal stem cells. Blood. 2007;109:228–34.
98. Sica A, Mantovani A. Macrophage plasticity and polarization: in vivo veritas. J Clin Invest. 2012;122:787–95.
99. Siemionow M, Klimczak A. Basics of immune responses in transplantation in preparation for application of composite tissue allografts in plastic and reconstructive surgery: part I. Plast Reconstr Surg. 2008;121:4e–12e.
100. Soares M, Low Y, Hwang L, Ham M, Ezeamuzie O, Saadeh P, Ceradini D. Tolerizing allografts with *ex-vivo* MSC immunotherapy. Plast Reconstr Surg. 2012;130:51. 10.1097/01.prs.0000416148.24037.3d.
101. Soares MA, Low Y, Ham M, Lalezarzadeh F, Wilson S, Ojo C, Lotfi P, Saadeh P, Ceradini D. Indoleamine-2,3-dioxygenase is critical to mesenchymal stem cell-based allograft tolerization. Plast Reconstr Surg. 2013;131:89. 10.1097/01.prs.0000430057.82105.aa.
102. Solari MG, Srinivasan S, Boumaza I, Unadkat J, Harb G, Garcia-Ocana A, Feili-Hariri M. Marginal mass islet transplantation with autologous mesenchymal stem cells promotes long-term islet allograft survival and sustained normoglycemia. J Autoimmun. 2009;32:116–24
103. Spaggiari GM, Capobianco A, Abdelrazik H, Becchetti F, Mingari MC, Moretta L. Mesenchymal stem cells inhibit natural killer-cell proliferation, cytotoxicity, and cytokine production: role of indoleamine 2,3-dioxygenase and prostaglandin E2. Blood. 2008;111:1327–33.
104. Spaggiari GM, Abdelrazik H, Becchetti F, Moretta L. MSCs inhibit monocyte-derived DC maturation and function by selectively interfering with the generation of immature DCs: central role of MSC-derived prostaglandin E2. Blood. 2009;113:6576–83.
105. Steinert AF, Rackwitz L, Gilbert F, Noth U, Tuan RS. Concise review: the clinical application of mesenchymal stem cells for musculoskeletal regeneration: current status and perspectives. Stem Cells Transl Med. 2012;1:237–47.
106. Tan J, Wu W, Xu X, Liao L, Zheng F, Messinger S, Sun X, Chen J, Yang S, Cai J, Gao X, Pileggi A, Ricordi C. Induction therapy with autologous mesenchymal stem cells in living-related kidney transplants: a randomized controlled trial. JAMA. 2012;307:1169–77.
107. Tatara R, Ozaki K, Kikuchi Y, Hatanaka K, Oh I, Meguro A, Matsu H, Sato K, Ozawa K. Mesenchymal stromal cells inhibit Th17 but not regulatory T-cell differentiation. Cytotherapy. 2011;13:686–94.
108. Toledano Furman NE, Lupu Haber Y, Bronshtein T, Kaneti L, Letko N, Weinstein E, Baruch L, Machluf M. Reconstructed stem cell nano-ghosts: a natural tumor targeting platform. Nano Lett. 2013;20:20.
109. Tse WT, Pendleton JD, Beyer WM, Egalka MC, Guinan EC. Suppression of allogeneic T-cell proliferation by human marrow stromal cells: implications in transplantation. Transplantation. 2003;75:389–97.
110. Tyndall A, Walker UA, Cope A, Dazzi F, De Bari C, Fibbe W, Guiducci S, Jones S, Jorgensen C, Le Blanc K, Luyten F, Mcgonagle D, Martin I, Bocelli-Tyndall C, Pennesi G, Pistoia V, Pitzalis C, Uccelli A, Wulffraat N, Feldmann M. Immunomodulatory properties of mesenchymal stem cells: a review based on an interdisciplinary meeting held at the Kennedy Institute of Rheumatology Division, London, UK, 31 October 2005. Arthritis Res Ther. 2007;9:301.
111. Uccelli A, Moretta L, Pistoia V. Immunoregulatory function of mesenchymal stem cells. Eur J Immunol. 2006;36:2566–73.

112. Unutmaz D, Waterman RS, Tomchuck SL, Henkle SL, Betancourt AM. A new mesenchymal stem cell (MSC) paradigm: polarization into a pro-inflammatory MSC1 or an Immunosuppressive MSC2 Phenotype. PloS One. 2010;5:e10088.
113. Van Raemdonck D, Neyrinck A, Rega F, Devos T, Pirenne J. Machine perfusion in organ transplantation: a tool for *ex-vivo* graft conditioning with mesenchymal stem cells? Curr Opin Organ Transplant. 2013;18:24–33.
114. Vanecek V, Zablotskii V, Forostyak S, Ruzicka J, Herynek V, Babic M, Jendelova P, Kubinova S, Dejneka A, Sykova E. Highly efficient magnetic targeting of mesenchymal stem cells in spinal cord injury. Int J Nanomed. 2012;7:3719–30.
115. Wagner W, Wein F, Seckinger A, Frankhauser M, Wirkner U, Krause U, Blake J, Schwager C, Eckstein V, Ansorge W, Ho AD. Comparative characteristics of mesenchymal stem cells from human bone marrow, adipose tissue, and umbilical cord blood. Exp Hematol. 2005;33:1402–16.
116. Wang Y, Zhang A, Ye Z, Xie H, Zheng S. Bone marrow-derived mesenchymal stem cells inhibit acute rejection of rat liver allografts in association with regulatory T-cell expansion. Transplant Proc. 2009;41:4352–6.
117. Wilson A, Trumpp A. Bone-marrow haematopoietic-stem-cell niches. Nat Rev Immunol. 2006;6:93–106.
118. Wood KJ, Bushell A, Hester J. Regulatory immune cells in transplantation. Nat Rev Immunol. 2012;12:417–30.
119. Wu GD, Bowdish ME, Jin YS, Zhu H, Mitsuhashi N, Barsky LW, Barr ML. Contribution of mesenchymal progenitor cells to tissue repair in rat cardiac allografts undergoing chronic rejection. J Heart Lung Transplant. 2005;24:2160–9.
120. Wyburn KR, Jose MD, Wu H, Atkins RC, Chadban SJ. The role of macrophages in allograft rejection. Transplantation. 2005;80:1641–7.
121. Xu DM, Yu XF, Zhang D, Zhang MX, Zhou JF, Tan PH, Ding YC. Mesenchymal stem cells differentially mediate regulatory t cells and conventional effector t cells to protect fully allogeneic islet grafts in mice. Diabetologia. 2012;55:1091–02
122. Yu X, Lu C, Liu H, Rao S, Cai J, Liu S, Kriegel AJ, Greene AS, Liang M, Ding X. Hypoxic preconditioning with cobalt of bone marrow mesenchymal stem cells improves cell migration and enhances therapy for treatment of ischemic acute kidney injury. PloS One. 2013;8(5):e62703.
123. Zeiser R, Nguyen VH, Beilhack A, Buess M, Schulz S, Baker J, Contag CH, Negrin RS. Inhibition of cd4+cd25+ regulatory t-cell function by calcineurin-dependent interleukin-2 production. Blood. 2006;108:390–9
124. Zhou HP, Yi DH, Yu SQ, Sun GC, Cui Q, Zhu HL, Liu JC, Zhang JZ, Wu TJ. Administration of donor-derived mesenchymal stem cells can prolong the survival of rat cardiac allograft. Transplant Proc. 2006;38:3046–51.
125. Zonta S, De Martino M, Bedino G, Piotti G, Rampino T, Gregorini M, Frassoni F, Dal Canton A, Dionigi P, Alessiani M. Which is the most suitable and effective route of administration for mesenchymal stem cell-based immunomodulation therapy in experimental kidney transplantation: endovenous or arterial? Transplant Proc. 2010;42:1336–40.

Chapter 18
Strategies for Gene Transfer to Vascularized Composite Allografts

Denver Lough and Damon S. Cooney

Introduction

Over the last half century, the increasing sophistication of molecular biology has allowed clinicians and researchers to identify, repair, and replace genes causing human disease in the laboratory and in the clinic. The ability to reprogram the genetic code in cases of derangement or disease is a profound concept that holds great promise for all aspects of medicine. Unfortunately, in practice, the results of the initial clinical trials of gene therapy have been somewhat disappointing. Perhaps too much was expected of the initial trials or perhaps they were undertaken to soon. Despite the tarnished reputation that gene therapy received from the failure of these initial high-profile cases, the field has made steady progress in its newest generation of clinical trials and seems poised to make real contributions to the treatment of several diseases. It is clear that gene therapy in one form or another, on its own or in combination with cellular therapies, will be an important tool in the clinician's armamentarium in the coming century.

The use of gene therapy in allotransplantation, and in particular in vascularized composite allotransplantation (VCA), is highly attractive for a number of reasons. There are many ways transplant surgeon or physicians would like to immunologically "reprogram" the graft or its recipient. The essential difference between the graft and the patient is a genetic one, and the basic problem of transplantation is that of the genetic difference between the donor and recipient at several important alleles. By directly modifying the expression of these genes in the graft, or the response to these genes by the recipient, gene therapy could potentially offer an attractive long-term solution to the problem of allograft rejection. Additional possibilities for the use of genetic modification include targeting ischemia–reperfusion injury, allowing longer

D. S. Cooney (✉) · D. Lough
Department of Plastic and Reconstructive Surgery, The Johns Hopkins University School of Medicine, 601 N. Caroline Street, Baltimore, MD 21202, USA
e-mail: dcooney2@jhmi.edu

G. Brandacher (ed.), *The Science of Reconstructive Transplantation*,
Stem Cell Biology and Regenerative Medicine, DOI 10.1007/978-1-4939-2071-6_18

ischemia times, and thus a wider donor pool. Another aspect of transplantation that makes gene therapy a particularly attractive technique is the ability to perfuse the allograft in isolation from the patient as part of the transplant procedure. The most poignant failures of gene therapy trials to date have been due to the systemic inflammatory toxicity and immunogenicity of the vector agents [1, 2].

The safety of these vectors is enhanced dramatically when they are used on an isolated allograft and removed before transplantation. Modern methods of *ex vivo* organ perfusion provide just such an opportunity and are the focus of many treatment proposals. By limiting the effect of the therapy and the subsequent genetic modification to the graft itself not only enhance safety but also open therapeutic opportunities based on the localized effect of the modifications [3]. Examples of this include modulating the local cytokine milieu, expressing biologic immunosuppressants at locally higher concentrations, and then systemically and potentially expressing advance warning markers of impending rejection prior to clinical symptoms. Many of the issues relating to gene therapy in VCA are similar to that of solid organ transplantation, and our discussion draws heavily on work that has been done in that field. However, while there are numerous similarities and reasons for comparison, there are several unique advantages posed by VCAs.

First, the nonlife-saving character of many VCAs such as in upper extremity transplantation shifts the ethical balance toward experimenting with novel protocols to decrease toxicities of immunosuppression in these otherwise healthy patients. We feel this makes advanced trials such as gene therapy even more attractive in these patients than in solid organ recipients. Second, the ability to directly access the transplant and take skin biopsies allows researchers in clinical trials to monitor the effects of these novel agents in a way not possible with the interior solid organ. Finally, the skin component of the VCA allows the use of novel gene therapy techniques such as transdermal nanoparticles and gene gun delivery of genetic material which is hardly possible in solid organ transplants.

In the following chapter, we explore possible targets for gene therapy in the arena of VCA. We begin by discussing in more detail specific vectors which can be used to deliver genetic therapies. Then we examine the wealth of studies relating to solid organ models and then finish with a brief discussion of the pioneering work specifically relating to VCA models.

Vectors for Gene Delivery

The continued improvement in vector design, production, and delivery has enhanced transfection efficiency and optimized gene expression over the last few decades. Having developed successful transplant models, clinicians and researchers in transplant immunology now have the opportunity to orient these applications toward preventing acute rejection, minimizing ischemia/reperfusion, toxicities associated with chronic immune suppression, and those features associated with progressive functional loss of the transplanted organ [4]. When selecting from among the wide array of vectors available, the features of each delivery system must be carefully weighed. There is no "best vector," and each system has advantages and disadvantages that must be optimized for each gene therapy application.

The basic characteristic of any vector is that it must be able to deliver genetic material (its payload) to a specific cell type (its target) with certain efficiency. The details vary greatly in the type and size of payload, the affinity for certain target cells and tissues, and efficiency and safety profiles. According to an in-depth review on the evolving field of gene manipulation, there is a relatively standard and vital set of factors that one must take into consideration when selecting both the transfer vector and cellular target within applications of gene delivery technology [4–6]. Some of these considerations are summarized in the table below (Table 18.1).

Once the critical features needed for an application have been determined, a type or class of vector can be selected. Until recently, gene therapy delivery agents were classified into two basic categories which were defined as (1) viral agent vectors and (2) nonviral vectors. Those gene transfer means which operated within viral delivery constructs employed the use of adenoviruses, oncoretroviruses, lentiviruses, and adeno-associated viruses or AAVs [4, 6]. Although evolutionarily adept at the

Table 18.1 Desired feature of delivery system and description of feature. Developed from delivery of gene and cellular therapies for heart disease [7]

Desired feature of delivery system	Description of feature
Safety	Achieving desired effect with minimal morbidity
Practicality (ease)	Readily adoptable by broad range of users, while also maintaining patient safety
Minimal invasiveness	Limited procedural trauma may be more readily translated into patients with advanced disease, e.g., heart failure
Achieves delivery at a critical concentration	Allows delivery of a biological agent at a threshold level to ensure effect
Appropriate regional distribution	Dependent on clinical need, provides either regional or global tissue/organ delivery
Homogeneity of expression and biologic effect	Ensures that all cells within an targeted area are impacted (rather than patchy distribution)
Limited systemic exposure and/or toxicity	Minimizes induction of systemic responses (e.g., immunologic) or off-target consequences of accumulation in nontarget tissue
Cost effectiveness	Determined by technical and equipment aspects
Procedural repeatability	Determined by technical aspects such as invasiveness and cost in addition to biologic responses that limit the effect of repeat exposure

delivery of genetic material into both eukaryotic and prokaryotic cells, viruses as bioengineered vectors for the transfer of therapeutic deoxyribonucleic acid (DNA) sequences are not without significant concerns, which has directed the field toward alternative vectors ranging from naked DNA sequences to liposomal delivery structures and even pretransplantation infusion of allogenic donor cells termed donor-specific transfusion (DST) which are thought to induce anergy and T cell clonal deletion.

Naked DNA

Perhaps the simplest method of gene transfer which does not require a viral vector or additional conjugate carrier is a form of so called "naked" transfection. Here, the DNA itself is both the transport element and coding sequence which is delivered to target cells often through a more stabilized circular plasmid structure (pDNA), which prevents rapid degradation of the DNA once within the cell. An array of techniques have been developed in order to optimize the transfection of naked DNA into cells, all of which have shown relatively limited efficacy when it comes to *in vivo* application.

The common theme of delivery involves the transient induction of increased cell membrane permeability through the application of either physical stress or simple chemical compounds on cells which challenge the cell membrane integrity for short periods of time. Basic physical or nonchemical methods encompass electroporation, micro-fluid platforms cell squeezing, and sonoporation. New advances in electroporation as well as gene administration techniques have increased naked DNA transfection efficiencies [4, 8].

Intravascular delivery of DNA into mice and nonhuman primates has been demonstrated to result in increased transgene delivery to kidney and muscle cells of up to 50% [4, 9]. The use of electroporation produces a tenfold increase in intramuscular DNA delivery gene expression. However, this enhanced delivery is accompanied by cell death, as evidenced by increases in serum creatinine kinase levels [6, 10].

One of the major drawbacks of naked DNA therapy is the transient nature of expression that lasts approximately no longer than 2–10 days [4, 6, 9]. Increasing injection volume with rapid delivery can prolong gene expression up to 12 weeks. However, the volume requirement of 100 ml/kg body weight delivered over 15 s makes the administration method difficult for clinical implementation [4, 6, 11]. The induction of cell damage with current delivery techniques makes naked DNA gene therapy use problematic in transplantation. The increased cell death can lead to activation of the immune response, prompting allograft rejection and causing significant organ dysfunction. Improvement in naked DNA delivery system makes this a possible option viable for use in solid organ transplantation and VCA.

Cationic Polymer Vectors

An alternative to the transfection of naked DNA, which can augment the efficacy of delivery, is through the application of carrier which is complexed with the DNA sequence that can increase the uptake by target cells.

At this time, in order to gain sufficient and reproducible delivery of pDNA into the nucleus of the desired human cells both in a safe and an efficient manner, pDNA requires a carrier. A spectrum of chemically synthesized cationic polymer vectors has been developed by different groups which have been assayed to determine if certain complexes possess an advantage over another and if any can compete with the more effective viral-directed gene transfer [12]. Cationic polymer vectors or, often termed, polyplexes (polycation/DNA complexes) are formed between cationic polymers and DNA through electrostatic interactions, which physically compact the DNA into a smaller structure. These types of compacted poly-charged structures are by far the most widely used nonviral gene delivery vector system [13].

A myriad of complex variables affect gene transfection efficiency when using cationic polymer vectors such as molecular weight, surface charge, charge density, hydrophilicity of the complex, structure of cationic polymers/polyplex, and those specific features which those cells or tissues being target for gene therapy possess [4, 6, 13]. Understanding the dynamic features of cationic polymers and their interaction with both coding DNA and target cellular elements requires optimization of *each* polyplex for *each* target system in order to optimize gene transfection efficiency for that specific assay. Currently, there are several cationic polymeric vectors most commonly used in model systems gene transfer. Among these cationic polymeric vectors are:

1. Polyethylenimine (PEI): A cationic polymer vector that maintains a repeating subunit which contains both an NH_2 amine group separated by two carbon aliphatic CH_2CH_2 spacers. Often utilized within liquid adhesives detergents, bonding agents, and cosmetics, PEI's positive charged polymer properties allows it to electrostatically interact with the net negative charge of cell membranes. This interaction permits PEI-coated materials to bind cells weakly for attachment. Additionally, this attraction allows PEI complexed with DNA to interact with the anionic cell membrane, allowing for endocytosis of the polyplex. Following endocytosis, the polyplex carrier within the endocytic vesicle utilizes NH_2 amine groups as sites of protonation. This alteration in pH of the vesicle offsets osmotic equilibrium by altering the flux of ions, and subsequently, the hypoosmotic vesicles become swollen with an influx of equilibrating water influx, leading to rupture. The DNA is then released and can migrate into the nucleus [13]. Although capable of simple delivery of a DNA polyplex into cells, the constructs through its mechanism of delivery are cytotoxic at both the cell and mitochondrial membrane levels among a spectrum of mammalian cells at given carrier concentrations [14].
2. Poly-(l-Lysine) (PLL): Similar to PEI, this polycationic vector has demonstrated a capacity to induce apoptosis in a range of human cell lines (Jurkat T cells;

epithelial cells transformed with SV40 large T antigen (THLE3-hepatocytes); human umbilical vein endothelial cell, HUVEC). However, the mechanism of apoptosis differs in PLL-treated cells when PEI to those which received PEI. PLL is thought to induce apoptosis through cytochrome C accumulation as well as subsequent activation of protein kinases within recipient cells [15, 16].

3. Chitosan: A polycationic polymeric structure isolated from specific sugars within crustaceans skeletons. Following Na(OH) processing, the linearly oriented polysaccharides is composed of glucosamine and *N*-acetyl-glucosamine units, which has allowed biomedical companies to utilize this substrate in wound care and hemostasis agents. For example, forms of chitosan have been approved as a wound dressing (Tegasorb® by 3M) and hemostatic patch (Hem-con® by HemCon) as well as within other fields of skin, nerve, cartilage, and bone regeneration [17]. Current application and research have not only shown that the use of chitosan nanoparticles greatly increases the transport of drugs through tissues but that this polysaccharide is also able to coat DNA for cellular delivery. Attractive qualities associated with chitosan-derived vectors pivot around the fact that these complexes are naturally derived, biocompatible, biodegradable, mucoadhesive, and nontoxic in polymer form [5].
4. Polyamidoamine (PAMAM): First described in 1985 as "A New Class of Polymers: Starburst-Dendritic Macromolecules," they are now the most common class of dendrimers in biotechnology applications and materials science engineering [18]. Structurally consisting of a varying alkyl-diamine core and branching side chains containing tertiary amines, these dendrimers are thought to result in higher gene transfection efficiency and lower cytotoxicity compared with other cationic polymers [19]. PAMAM, initially thought to have too complicated synthesis technology to be clinically useful, now has a more promising future with the advent of hyperbranched polyamidoamine (h-PAMAM). This form is synthesized by a simpler one-pot method yet has similar gene delivery properties with PAMAM. In addition, through a polyethylene glycol (PEG)ylation modification, the previously described cytotoxicity has been reduced [20].

At the present time, polycationic vectors are not as efficient as viral vector delivery systems within gene transfer modeling [13]. Researchers in the field have seen an increase in transfection efficiency with PEGylated polycationic polymeric forms which allows an increased total quantity of amount of pDNA uptake by target cells. A wide range of polymeric vectors have been utilized to deliver therapeutic genes *in vivo*. The modification of polymeric vectors has also shown successful improvements in achieving target-specific delivery and in promoting intracellular gene transfer efficiency. Various systemic and cellular barriers, including serum proteins in blood stream, cell membrane, endosomal compartment, and nuclear membrane, were successfully circumvented by designing polymer carriers having a smart molecular structure [15]. Cationic polymers display less toxicity associated with cytokine induction compared to their cationic lipid counterparts, which often elicit form of innate immune response through fatty acid inflammatory reactivity [4, 5].

Lipid Carriers

Another form of nonviral delivery vectors utilized in gene transfer was described in a 1963 publication of *Nature* as a negatively charged self-forming lipid membrane; this form of carrier has developed a spectrum of lipid-derived vectors such as liposomes, micelles, and nanoemulsions. Alterations in organic synthesis have allowed biochemists to manipulate these lipid elements to become net positive or cationic at the head region of the lipid, which allows for interaction with both the negatively charged DNA backbone and target cell membranes. As early as the 1980s, researchers were describing these synthetic cationic lipids such as N-[1-(2,3-dioleyloxy) propyl]-N, N,N-trimethylammonium chloride (DOTMA). At those times, studies described small unilamellar liposomes containing DOTMA which were capable of spontaneous interaction and incorporation with DNA. This interaction appeared to result in a lipid–DNA complex that was capable of 100% DNA entrapment. Furthermore, DOTMA-enriched liposome/DNA complex could facilitate the fusion of the complex with the plasma membrane of cells, resulting in both uptake and expression of the DNA [21].

And since the discovery that phospholipids could form lamellar bilayer structures in aqueous systems, liposomes have become a prominent topic of research in gene delivery. The structure of these cationic lipid vectors defines their simplicity, versatility, and biocompatibility. Many fields, particularly medicine, have yielded successful breakthroughs through liposome delivery mechanisms with over 12 liposomal-based drug therapies on the market and over 20 in clinical trials worldwide that the vector has proven adequate safety and efficacy within human drug therapy [22]. From a production standpoint, the relatively simple preparation and various structural aspects of the liposome and lipid carriers have given rise to a reproduce mode of the internalization of a wide variety of biomolecules such as drugs, DNA, RNA, and even imaging probes [23].

Drawbacks concerning liposomal-mediated gene transfer have been published citing liposomes as having the capability of eliciting an immune response within the recipient tissues systems due to the expression of cytosine poly-guanine (CpG) bacterial motifs [24]. Moreover, in some cell-based and tissue-based systems, studies have described the necessity of delivering relatively high liposome construct concentrations in order to facilitate an adequate cellular transfection, resulting in measurable transgene expression. Here, limited transduction efficiency coupled with a reactive immune response to potentially inflammatory components of lipid-conjugated molecules suggests prospective concerns for irreversible cellular toxicity and subsequent systemic damage. Researchers investigating gene transfer in pancreatic islets cells, through liposomal–DNA complex vector delivery, described a concentration phenomenon which is both clinically relevant and concerning to gene modification allotransplantation efforts. This phenomenon describes an association of where the concentration of liposomal cations vectors necessary to achieve effective gene transfer directly results in the death of 50% of islet cells receiving the therapy, in addition to impaired *in vitro* insulin release of the surviving transduced islets [4].

As the field continues to evolve further away from conventional vesicles toward yet another new generation liposomes, such as cationic liposomes, temperature-sensitive liposomes, and virosomes, researchers have begun to find ways to improve lipid carriers' sustainability within a given environment while also increasing target-directed specificity. To achieve better selective targeting by PEG-coated liposomes or other particulates, targeting ligands were attached to nanocarriers via the PEG spacer arm, so that the ligand is extended outside of the dense PEG brush, excluding steric hindrances for its binding to the target receptors [23]. With the increasing momentum behind lipid carrier applications for drug and gene therapy, in addition to limited toxicity and ease of synthesis, contemporary gene therapy may be based in a lipid vector delivery system very soon.

Viral Vectors

With regard to any gene therapy delivery vector, in order to achieve therapeutic success, the transfer vehicles must first be capable of transducing target cells. Second, the administration, transduction, integration, and replication processes should ideally have no impact on nontarget cells. Despite improvements over the last three decades and the most effective transduction methods within gene therapy delivery systems, the viral vectors continue to suffer from a tropism to therapeutic need mismatch [25].

The promise of viral vector-based gene therapy has become integrated within many medical and surgical fields. With tremendous focus and significant accomplishments in both oncology and medical genetics (particularly among those patients suffering from a specific enzyme-deficient pathway), one may ask what is the role in viral vector-based gene therapy in the management of solid organ transplantation or VCA [26]. Heart transplantation and liver transplantation have provided some advances in vector development and target improvement. This has allowed clinical progression within the field of transplantation as realistic and applicable form of therapy for transplant patients. Furthermore, review of the current literature will show how viral vector technology has proved to be better than nonviral vectors at delivering therapeutic genes to cells. With growing interest and ability within the field of viral gene delivery, transplant researchers are beginning to come to a realization of both the benefits and potential risks in viral delivery systems for gene-based therapy in transplantation [26] (Table 18.2).

Adenovirus and Adeno-Associated Virus

Adenovirus is a member of the *Adenoviridae* family and defined as a nonenveloped 100-nm-diameter virus with a double-stranded DNA (dsDNA) genome (insertional genetic capacity of approximately 7–8 kb) which in its natural state typically targets membranes within the gastrointestinal, hepatobiliary, and respiratory systems

Table 18.2 Summary of gene therapy viral vectors and actors associated with delivery and expression [6]. Modified from [93]

	Plasmid	Oncoretrovirus	Lentivirus	Foamy	Herpes	Adenovirus	AAV
Genetic material	DNA	RNA	RNA	RNA	DNA	DNA	DNA
Genetic material packaging capacity	No limitation	9 kb	10 kb	12 kb	>30 kb	30 kb	4.7 kb
Duration of expression	Transient	Long	Long	Long	Transient	Transient	Long in postmitotic tissues
Genome integration	Yes	Yes	Yes	No	No	Rarely	Rarely
Transduction of postmitotic cells	No	Low	Low	High	High	Moderate	Moderate

***kb* kilobase, *AAV* adeno-associated virus**

[27]. Of those 51 serotypes endemic throughout the world, typically serotype 2 and serotype 5 are classified as potential vectors for gene transfer. The virus typically utilizes the endocytic pathway and, once within the cell, does not integrate into the host genome but instead remains an episome within the nuclear envelope. Because of this, there is a reduced risk of insertion-associated mutagenesis, which is unfortunately seen with other genome-integrating viral vectors.

There has continued to be significant improvement in the design of the adenoviral vector which has subsequently led to momentum in technologies that increase the utilization and broaden the application of the adenoviruses as a clinically tangible gene therapy vector construct. There are a multitude of features that adenoviral vectors possess which make them forefront candidates for modification strategies in transplant-related gene therapeutics. Among the promising transplant-oriented elements, most adenovirus subtypes (specifically serotypes 2 and 5) are not found to be associated with severe human illnesses. Because of this, serotypes 2 and 5 retain only limited risk for replication-competent viruses, causing acute infectious pathology in patients receiving adenoviral vector-based therapy [28, 29].

Since adenoviral vectors can effectively transduce nonreplicating and replicating cells, unlike many retroviruses which require active replication of cells in order to propagate their delivered transgene and viral life cycle, they offer a variety of options for genetic engineering in the incongruent proliferative phases seen in cells from donor, recipient, and chimeric tissues, seen in allotransplantation. Beyond an ability to transduce cells undergoing transient proliferative variability, the adenoviral vectors can easily provide systems with high-titer production along with an ability to concentrate based on certain tropism characteristics within each serotype. It remains the combination of all of these characteristics which suggests clinical feasibility within VCA and solid organ transplant. A tangible example of adenoviral application within liver transplant is demonstrated with the capability of adenoviral vector serotype 5 to localize to liver parenchyma and concentrate within hepatocytes following intravenous administration. The absolute value of these findings beyond the setting of conventional orthotopic liver transplantation cannot begin to be quantified at this time since the liver, no matter what solid organ or composite allograft, plays vital roles in acute inflammatory processes and disposal of toxic metabolites, which if genetically altered could promote the survival of any transplant element. The insertional genetic capacity of adenoviral vectors is approximately 7–8 kb, and they have high-level transgene expression. The ability of adenoviral vector serotype 5 to localize and concentrate in the liver after intravenous administration in mice allows for the development of *in vivo* strategies that can have profound systemic effects.

A drawback of the adenoviral vector associated with nongenomic integration is that the transgene expression within a host cell is relatively short lived, despite efficient transduction and high titer levels that can be achieved with the virus making gene transfer more likely. Additionally, direct intravenous delivery of the virus results in nearly 90 % of the vector being degraded, while the immune system, namely T cells, destroy those cells infected with adenoviral particles. Furthermore, the adenoviral serotype used to infect the cells will develop an antibody-derived immune response to protein expressed on the capsid as such antibodies will be directed to the transgene expressed. These antibodies will remain present and prevent further therapy by that specific adenoviral vector.

Concern over the death of a young man in 1999 at the University of Pennsylvania while being a research participant in an ornithine transcarbamylase (OTC) deficiency transgene research study initiated an investigation and report by the National Institutes of Health (NIH) Recombinant DNA Advisory Committee on the assessment of toxicity and safety when using the adenovirus vector for gene therapy [30]. At the conclusion of the report, the NIH committee made recommendations on general safety aspects such as consent, monitoring, and postmortem examination.

Within the field of solid organ transplantation, free tissue transfer, and VCA, the adenovirus has remained a staple in viral vector delivery of genetic material. Researchers have not only shown that interleukin (IL)-10 transgene delivery by a adenoviral to donors can reduce ischemia–reperfusion injury [31], but that adenoviral vectors can also be used to pre-sensitize a patient by targeting host cells (such as the liver) with allo-major histocompatibility complex (MHC) in order to induce pre-allotransplant tolerance [32].

Adeno-Associated Virus (AAV)

The single-stranded DNA AAV was defined over 20 years ago and was administered to the first human subject in 1995 [33]. A member of the *Parvoviridae* family (genus *Dependoparvovirus*), these viral subtypes have been found to not produce human infectious disease pathology within clinical applications of gene transfer [34–36]. While similar to the adenovirus and lentivirus vectors in their spectrum of cellular targets, secondary to advantageous tropisms, the AAV harbors a smaller genome and genomic insertion capability (4–5 kb) as well as possesses a requirement for AAV helper viruses for replication and transgene expression within host cells [37].

Members of the AAV group have been shown to be a useful viral vector in clinical trials, with now 11 in the process for approval in humans. Of those, AAV for the cystic fibrosis gene therapy is perhaps the most well-known application using this specific viral vector. Small, approximately 20 nm in diameter, the virus is currently thought to be nonpathogenic. However, one of the most attractive features of this virus is the specificity of genome integration. Although random insertions can occur at low levels, the AAV virus has the consistent ability to integrate at a region on chromosome 19 termed the AAVS1 site [38]. Mediated by the Rep78 and Rep68 proteins (though involve endonuclease, helicase, and transcriptase functions), the AAV virus is able to replicate, integrate, and rescue the provirus from the genome to undergo both latent and lytic infection. Furthermore, with the assistance of a helper virus such as an adenovirus or herpesvirus, the AAV can initiate productive viral infections with long transgene expression. Other than the rep genes, the AAV 4.7 kb genome contains inverted terminal repeat (ITR) regions, open reading frames (ORFs), and a sequence region containing "cap" genes, which code for capsid and structural proteins. Between these two coding regions, the plasmid or gene of interest can be inserted and the virus can be used to deliver a sequence of interest [38].

Among the 11 AAV serotypes defined, the AAV2 or (serotype 2) has developed a natural viral tropism directed toward smooth muscle, neurons, and skeletal muscle, which remain targets of interest in VCA modeling [39]. Furthermore, one of the most difficult challenges in optimizing viral vectors for gene therapy relates to the immune response of the host toward the viral vector. Interestingly, AAV vectors are associated with low immunogenicity and toxicity, resulting in vector persistence and long-term transgene expression, again suggesting a role in VCA [40]. Additionally, although patients can potentially mount an immune response against the input virions, transient immunosuppression during the vector uncoating phase could be sufficient in blocking this response [41, 42].

Retroviruses and Lentivirus

The *Retroviridae* family or retroviruses has classically been defined as an enveloped virion particle approximately 100 nm in diameter which possess a single-stranded messenger RNA (mRNA) genome. This obligate cellular parasite was identified by its unique ability to transcribe its two copies of identical single-stranded RNA viral genome into dsDNA following delivery of the pre-integration complex after fusion of the virion to the host cell through binding and fusion proteins. The 10-kb genome of a retrovirus commonly contains four coding regions which are seen as ORFs: gag (viral capsid core proteins), pr (proteases for cleavage), pol (polymerase, synthase, and integration), and env (proteins for viral entry into a cell). Integration of the retro-transcribed dsDNA into the host genome defines the provirus form, where it can replicate and be incorporated into the genome of cellular progeny through somatic mitosis or germ line meiosis [4, 6, 43, 44].

Sequence variations of the "env" gene, known as viral tropism, is what determines the ability of envelope proteins of a given virus to bind and fuse with a defined host or target cell [43, 44]. From a research perspective, manipulation and replacement of the env gene allows one to dictate potential targets of an engineered viral vector and is known as pseudotyping. The practice of viral envelope engineering through pseudotyping, or altering the sequence of the viral "env" gene, not only alters the target cell but also can expand the potential target range to multiple cell types.

With a basic understanding of the unique retrovirus life cycle, one can begin to see the underlying set of weakness in this specific vector gene therapy delivery system. First, cells must be actively replicating and/or undergoing transcription for the virus genome (including the therapeutic gene) to propagate. Second, there is no site-specific region within the genome for viral integration of the provirus. This lack of definitive integration is why provirus insertion within an oncogene can lead to neoplasm. Finally, in addition to building an acquired immune response against virions released from host cells, retroviruses are also uniquely sensitive to the c1 protein of the complement cascade, leading to an inability to maintain prolonged transgene expression.

A genus within the *Retroviridae* family known as *Lentivirus* is a subclass that contains five documented serogroups ranging from 80 to 100 nm in diameter. Perhaps known best for its primate infectious lentivirus group, which contains HIV1, HIV2, and simian immunodeficiency virus (SIV), the others are bovine, equine, feline, and ovine. Containing an additional six protein coding regions within its genome, the lentivirus not only has the capacity to deliver more viral RNA into a host cell, but the genus also has an advantage over the standard retrovirus in that it can effectively infect nondividing cells and remain functional during long inactive incubation periods. Consequently, these features make them effective at delivering larger sequence gene therapy options in addition to a broader spectrum of cells based on their relative proliferation kinetics.

As early as the 1990s, there were several clinical trials which utilized retroviral vectors as a means of gene transfer. These studies were frequently related to the treatment of inherited monogenetic disorders and AIDS [4, 45, 46]. However, since the optimal retroviral vector targets are those cells, which have higher tendency to undergo division, which permits viral gene incorporation into the host genome, their application to allotransplant gene therapeutics is limited at this time [4, 6]. Consequent to the lower transduction efficiency of the retroviral vectors, less cells will be able to express the delivered gene. Limited relative expression of the transgene makes the pathways involving anti-apoptotic or ischemia/reperfusion less effective targets for retroviral-directed genetic modification, since both pathways require sufficient molecular opposition to prevent their initiating elements from causing irreversible damage to the cell [4–6].

Despite concerns over gene propagation within hypo-proliferative cell and tissue systems, retroviral-directed gene therapy remains a viable option for genetic transfer. Research into the ability of retroviral delivery of immunosuppressive cytokines, where focal accumulation of such cytokines could reduce immune response elements within the graft without altering the systemic profile is particularly appealing. The stable gene expression of retroviral vectors can be viewed as either advantageous or detrimental, depending on the therapeutic goals one is trying to accomplish. In allotransplantation, stable gene expression is likely required for the prevention of acute cellular rejection. On the other hand, stable linear expression of the transgene may not be necessary to prevent ischemia/reperfusion injury seen at the onset following engraftment [4–6]. Because of these fluid and sometimes transient requirements for gene expression, the lentivirus subtype of retroviruses would be required which could propagate transgene expression in "injured" grafts that may not be orienting their metabolic pathways toward cellular proliferation (nondividing cells) but rather toward cell and tissue repair and disposal of toxic products of ischemia such as free radicals, lactic acid, and inflammatory lipid components [47–49].

SI RNAs

Descriptions of relatively short noncoding RNA elements (~25 nucleotides) within cells became relevant in the 1990s, when researchers found them to function as a type of nucleotide sequence-specific defense system which could target both cellular and viral mRNAs for destruction [50]. Small interfering RNA or (siRNA) is a form of dsRNA with the capacity to induce posttranscriptional gene silencing within cells through antisense binding to active form of mRNA and subsequently inducing endonuclease activity against the mRNA transcript/siRNA complex. Since the discovery of this mechanism of RNA interference using siRNA, researchers have begun to develop a spectrum of powerful tools to downregulate mRNA levels and even silence genes involved in the pathogenesis of various diseases associated with a known genetic background [51].

Although there are relatively limited studies on VCA and siRNA applications, the field of solid organ transplantation has developed basic therapeutic constructs and models. Researchers in renal transplantation have sought siRNA technologies therapy directed toward not only rejection but also ischemic–reperfusion injury. In order to prevent ischemia–reperfusion injury and damage to allograft tissues, siRNA technology is being actively studied [52]. However at the current time, most of these attempts at either naked or vector-based siRNA therapy within the arena of solid organ transplantation have been performed *in vitro*, with very limited studies progressing toward more advanced *in vivo* settings or animal models. Of the delivery mechanisms employed for siRNA therapeutic application, hydrodynamic intravenous injection of naked or carrier-bound forms of siRNA are most commonly the route for delivery of these RNA interference constructs [4–6, 52]. Of the gene targets thought to be prime targets for siRNA interference therapy, roughly 50 have been tested in transplantation-related models. Most of these mRNA transcripts belong to genes that are either related to apoptosis or involved in immunomodulatory networks. Curiously, secondary to the small size of the interfering RNA, researchers have seen opportunity to combine multiple forms of siRNA in order to downregulate multiple targets. These separate siRNA elements can be transported within the same delivery vector or injected at the same time and by targeting more than one pathway, or by hitting the same pathways within two different key points, will augment the effects of each other [52].

Cells as Gene Therapy Vectors

DST, although discovered over 30 years ago as a potential tool to induce donor-specific immunological tolerance in renal transplant, the protocol has widely been abandoned in contemporary clinical solid organ transplantation. Despite researchers having shown some encouraging results with regard to immunosuppressive drug minimization in human patient subsets and potential induction of immunological tolerance in some animal model systems, the mechanism underlying tolerance induction to recipient tissue remained vague and associative at best [53]. Not until recently did transplant immunologist begin to understand the immunomodulatory mechanisms of DST or donor bone marrow infusion for inducing tolerance, particularly within skin containing VCA. Building on previously published concepts, which indicated that increased levels of regulatory T cells (Tregs) in recipients of an allograft prevented both acute and chronic rejection, investigators then identified the subset of cells dictating the immunomodulatory processes involved. Results of these studies show tolerance is mediated by an interaction among $CD11b^{+}$ expressing cells (DCs, dendritic cells, and macrophages) augmented indoleamine 2,3-dioxygenase (IDO) and IL-10 expression and subsequent $Foxp3^{+}$ $CD4^{+}$ $CD25^{+}$expressing Treg induction [53, 54].

Transdermal Gene Delivery (Gene Gun)

Described in a letter to *Nature* in 1987, researchers presented a novel phenomenon, where nucleic acids could be delivered into plant cells using high-velocity microprojectiles through what was described as a gene gun. This and prior research were conducted in the hope of circumventing some of the inherent limitations of existing methods for delivering DNA into plant cells, namely *Agrobacterium tumefaciens,* which was limited to a specific target host set of plants [55].

The mechanism of delivery utilized small tungsten particles, termed microprojectiles, which where accelerated through a pressurized air gun and used to pierce cell walls and membranes. The microparticles, although penetrating the exterior of the cell, did not kill the organism. Following this success, the microprojectiles were used to carry RNA or DNA into epidermal tissue of an onion for subsequent gene expression analysis.

Since the inception of the gene gun delivery of genes complex with microparticles, the method of delivery has been an option within the search for a more efficient pDNA gene transfer system. Realizing that gene therapy offers a novel approach for the prevention and treatment of a variety of diseases and perhaps a mechanism to promote tolerance in VCA and solid organ transplantation, there remains a paucity of accepted methods because of reoccurring issues concerning either efficacy of delivery or toxicity of vector constructs. Viral vectors, although shown to have higher efficiency of gene transfer, are limited by certain elements of toxicity, host-developed immunity, and length of time related to incubation and transgene expression. While other options of nonviral vector-based delivery systems of pDNA are relatively safe, the efficiency of delivery is significantly lower when compared to viral vector gene delivery [56].

Today, the gene gun, and other similar intradermal delivery devices, takes advantage of the knowledge that skin is known to be a highly immunogenic site for vaccination and other intradermal gene delivery application devices have been developed. Many of these technologies have been shown to administer materials within the skin by noninvasive or minimally invasive techniques. Those platforms utilizing noninvasive methods include high-velocity powder and liquid jet injection, as well as diffusion-based patches in combination with skin abrasion, thermal ablation, ultrasound, electroporation, and chemical enhancers. Additional "minimally invasive" approaches are largely based on microneedle injections [57].

In summary, the choice of a vector to deliver the genetic material of interest is almost as important as what is being delivered. Critical factors include timing, target tissue type, expression level, and safety. The VCA scenario is more challenging due to the variety of tissue types that must be successfully modified, but offers unique opportunities such as the ability to monitor and deliver genes directly through the skin. In general, most groups studying gene therapy in VCA have utilized vectors successfully used for solid organ and systemic gene therapy approaches as we will see below (Table 18.3).

The Utility of Gene Therapy in VCA and Solid Organ Transplantation

Gene therapy within solid organ transplant or VCA, although not as well understood as the role of gene therapy in cancer, there still remains promise and tangible cell and pathway targets, which within the realm of transplant immunology is well studied. Because of this, many in the field believe that if a proper vector could be developed for delivery, immunosuppression and cellular rejection would be the first targets for genetic alteration.

Historically, following the advent of contemporary immunosuppressive therapies, fulminant rejection has been less of a concern while the side effects of lifelong systemic immunosuppression has taken over as the more common posttransplant

Table 18.3 Advantages and disadvantages of gene delivery strategies. Table from [4]

Vectors	Advantages	Disadvantages
Retroviruses	Broad cell tropism	Infects only dividing cells
–	Stable gene expression[a]	Risk of insertional mutagenesis
–	High-titer production (10^7 cfu/ml)	Susceptible to complement degradation
–	Large insertion capacity 10 kb	Risk of recombination with human endogenous retroviruses
–	Low levels of gene expression[a]	Risk of competent virus formation
–	–	Low levels of gene expression[a]
–	–	Stable gene expression[a]
Lentivirus	Broad cell tropism with retroviral and VSV G pseudotyping	Risk of insertional mutagenesis
–	Can infect nondividing cells	Pseudotyped vectors are susceptible to complement degradation
–	Stable gene expression[a]	Risk of recombination with human endogenous retroviruses
–	High-titer production (10^7 cfu/ml)	Risk of competent virus formation
–	Large insertion capacity 10 kb	Serum conversion with HIV-based vectors
–	–	Stable gene expression[a]
Adenovirus	High levels of gene expression[a]	High levels of gene expression[a]
–	Infect nondividing cells	Transient gene expression[a]
–	Very high titers (10^{12} pfu/ml)	Immunogenic
–	Large insertion capacity 8 kb	*In vivo* delivery minimized by host immune response
–	Transient gene expression[a]	–
AAV	Infects nondividing cells	Stable gene expression[a]
–	Stable gene expression[a]	Requires a helper virus for replication
–	High titers (10^{10} cfu/ml)	Small insertional capacity (5 kb)
–	Nonpathogenic	Risk of insertional mutagenesis
–	–	Low transduction levels
Foamy virus	Infects dividing cells	Suboptimally infects nondividing cells[a]
–	Suboptimally infects nondividing cells[a]	Stable gene expression[a]
–	Stable gene expression[a]	Risk of insertional mutagenesis
–	Broad cell tropism with retroviral and VSV G pseudotyping	Risk of recombination with human endogenous retroviruses
–	Large insertion capacity	Risk of competent virus formation
–	Resistant to complement-mediated lysis	Serum conversion to foamy virus
Cationic liposomes	Noninfectious	Toxicity dose dependent and most evident in nonreplicating cells
–	No limit to size of DNA insert	Transient gene expression[a]
–	Transient gene expression[a]	Low transduction levels
–	–	Difficult to implement clinically

Table 18.3 (continued)

Vectors	Advantages	Disadvantages
Naked DNA	Noninfectious	Transient gene expression[a]
–	No limit to size of DNA insert	Low transduction levels
–	Transient gene expression[a]	Stimulation of host immune responses due to bacterial CpG

AAV adeno-associated virus, *CpG* cytosine poly-guanine, *VSV G* vesicular stomatitis virus G
[a] Can be viewed as advantageous or detrimental depending on therapeutic goal

difficulty seen in patients. Transplant immunologists, surgeons, and scientists believe that gene therapy may possess the potential to eliminate these side effects associated with chronic immunosuppression. This concept entails target gene therapy within donor grafts, which could permit the expression of specific immunomodulators within a focal region rather than throughout the systemic circuit. Alternatively, gene therapy approaches could eliminate the requirement for general immunosuppression by allowing the induction of donor-specific tolerance as well as prevent graft damage owing to nonimmune-mediated graft loss or injury.

Research groups in gene transfer and vector development technologies within heart transplantation have remained at the forefront not only in pragmatic application of gene therapy but also in defining those optimal features for its application in human trials.

Among transplant-related gene transfer studies, viral vector-based therapies show promise in that a virus continues to deliver genetic material most efficiently and predictably. With regard to viral vector constructs, the retrovirus family members have been utilized to modify various cell types and address challenges in transplantation successfully. Certain retroviruses such as HIV, feline immunodeficiency virus (FIV), and murine leukemia virus (MLV) have been shown to effectively transduce islet cells at efficiencies of 1.9, 13.7, and 0.9%, respectively [4, 58]. These studies suggest that genetic engineering of islets before transplantation with protective genes may potentially enhance their posttransplantation survival, leading to improved functional outcomes at both the cellular and patient level. Furthermore, the results of islet cell-directed gene therapy via viral vectors showed that transduction was nontoxic and remained an efficient method to genetically modify this cell type for transplantation [58, 59].

Cellular entities within the liver have also been effectively transduced. In 2002, using a pLXSN-CTLA4Ig plasmid construct within a pLXSN retrovirus vector, researchers were able to transduce the fusion gene cytotoxic T lymphocyte antigen-4-immunoglobulin (CTLA4-Ig) after *in vivo* perfusion of a murine model with retroviral vectors [27]. The results of this study describe enhanced liver regeneration, CTLA4-Ig transcript production, and a sustained fluorescent CTLA4 protein, which was genetically labeled with a fluorescent protein. Moreover, gene transduction of CTLA4-Ig conferred no adverse effect on the regeneration of the liver graft or the general health status of the animals [4, 27].

Specific Targets for Gene Therapy in Transplantation

The potential to alter cell interactions and immune responses, within the realm of solid organ transplantation and VCA, through gene therapy is unique in that there remains a wide spectrum of targets and pathways that not only alter the graft function and survival but also those that effect the recipients' morbidity and mortality. Within this dual or chimeric system, the modification of gene expression, through gene transfer technologies, has the potential to improve outcomes following cell, tissue, and/or solid organ allotransplantation. Additionally, since the allograft is transferred from one person to another, there is a unique opportunity in transplantation, where therapies could be applied to either donor, recipient, or chimeric systems. In fact, gene therapies need not be restricted to only *in vivo* use; rather, there is a tremendous prospect for *ex vivo* gene modification of cells and/or organs during recovery [4–6].

Since the discovery and clinical application of potent immunosuppressive agents, such as calcineurin inhibitors, nucleic acid synthesis blockers, and monoclonal antibodies, directed at T cells or IL-2 receptor α (IL-2Rα), the transplant community has seen a significant reduction in the incidence of acute rejection with a subsequent improvement in 1-year allograft survival rates from about 60 %, in the early 1980s, to the current levels of 80–95 % [4, 6]. However, patient noncompliance with transplant medication poses a serious challenge, as it accounts for a significant number of acute rejection episodes as well as for graft loss. Furthermore, allografts do not always exhibit optimal function. Nonimmune mechanisms, such as ischemia/reperfusion injury, drug toxicity, and immune mechanisms, such as chronic allograft rejection, all contribute to the loss of allografts [4, 6, 60].

Rejection and Alloantigens

Alloantigen recognition occurs primarily through a cell-to-cell interaction that takes place between circulating T cells and tissue-based antigen-presenting cells (APC), which will acquire and present these nonself-antigens via extracellular MHC structures. These antigens can elicit a range of immune response; however for simplicity, these responses have been categorized at the most basic levels as either major or minor with regard to response intensity and length [4, 6]. Alloantigen presentation can occur through both MHC class I and MHC class II structures. MHC class I antigens are present on all nucleated cells while MHC class II antigens are typically limited to APCs (monocytes, macrophages, Langerhans cells, DCs), lymphocytes (B lymphocytes, activated T cells), as well as a variety of endothelial and epithelial cell lineages [4, 6, 11]. $CD4^{+}$ T cell/MHCII and $CD8^{+}$ T cell/MHCI interactions leading to alloantigen recognition can occur through essentially two basic pathways:

1. The direct pathway: Recognition of donor alloantigen involves recognition of the donor's MHC class I or MHC class II antigenic peptides, by recipient's T cells, when expressed on the donor APC.

2. The indirect pathways: Recognition of donor alloantigen by the recipient immune system occurs through recipient APC capture, processing, and presentation of donor-derived MHC class I or MHC class II peptide fragments to recipient T cells.

It is through the recognition of donor MHC class I or MHC class II peptides by recipient immune cells, the recipient immune system will initiate a "major" immune response in an effort to prevent donor immune cells from commencing an immune response against native recipient tissues, commonly seen in graft-versus-host disease (GVHD) [4, 6, 61, 62].

Understanding how immune response element kinetics command and propagate a variety of reactions to the presence of alloantigen within the system and how certain successive on/off rejection pathway switches coordinate either rejection or tolerance provides those advocating for gene therapy with initial targets of interest. For example, the induction of mixed lymphohematopoietic chimerism through allogenic bone marrow transplantation is one of the most stable and predictable methods in establishing transplant tolerance [4, 42–44]. Appreciative of the limits seen with bone marrow transplantation (toxicity of conditioning regimens, failure to engraft, or risk of GVHD), gene therapy may provide an attractive complementary strategy of circumventing some of these shortfalls of establishing chimerism in patients.

Murine transduction studies involving retroviral vectors delivering forms of MHC class I antigens to autologous bone marrow cells were able to produce a permanent state of MHC molecular chimerism within recipients, resulting in allografted skin to undergo long-term acceptance [45]. The success of models utilizing MHC-based bone marrow genetic modification platforms to induce tolerance between a transgene MHC and T cell interface appears to correlate with not only the level of gene expression but also transfection efficiency, which validates the need to optimize specific vectors for gene delivery [46–48]. Genetic engineering that results in reduced MHC expression leads to the development of T cell hyporesponsiveness without tolerance induction, demonstrating the importance of the dose effect [63–70].

Gene transfer of MHC complex may permit a form of allograft acceptance without having to induce mixed chimerism through a donor bone marrow transplant. The suggested mechanism of MHC class I gene transfer promotes tolerance induction through negative thymic selection of those $CD8^+$ T cells which recognize the MHC encoded by the transgene [69, 70]. Relative expression levels of the MHC molecules also appear to play a role in the success of the allograft. In fact, when hyper-expression of MHC class I transgene occurs, subsequent recipient immune suppression results in an elevated and maintained level of donor-specific MHC [71, 72].

The MHC molecule found on nucleated cells can also be present within a system as a separate soluble form. Such forms of the MHC complex has been used to protect liver, skin, heart, and renal allotransplant tissues from scenarios of hyperacute rejection seen in patients who have previously been sensitized and therefore have preformed antibodies to alloantigenic components of MHC molecules [71–74].

Rejection and Antigen-Presenting Cells

Naïve recipient T cells are poised to determine initial immune response elements following of nonself minor or major alloantigens through interactions with APCs and nonimmune nucleated cells. From this initiating immune response element, the native immune system will act to tolerate or reject tissues based off of the peptides a target expression. The initial interaction in a solid organ transplant or VCA occurs between immature DCs, which are primed for antigen capture, processing, and presentation (following maturation) through augmented genes affecting MHC affinity and quantity of complexes produced [75–77]. During maturation of the DC, costimulatory molecules CD80, CD86, and CD40 are upregulated [4–6]. Upon the formation of the cell-to-cell interface, between the matured DC and naïve T cell, two signals are required for T cell activation against the alloantigen being presented. The initial signal event employs the T cell receptor (TCR) or CD3 engaging the antigen containing MHC molecule, which defines these cells as "capable" of binding that specific peptide [4, 6]. The second signal event involves the complexed cells to make a "decision" on whether to see the peptide as self, tolerate the peptides presence, or determine it foreign and therefore promote elimination of the threat. This second signal is dependent on costimulatory signaling through CD28 and CD80/CD86 interaction [4, 6, 78, 79].

The interface that occurs between the peptide containing APC and T cell is perhaps the most pinnacle cell-to-cell interaction which occurs through induced adaptive immunity. Positive signaling events between MHC/TCR and CD80 and CD86/CD28 result in an exponential increase in downstream cellular proliferative and synthesis effects. Such alterations augment and activate (1) the transcription factor NFκB, (2) the T cell stimulatory IL-2, (3) the inhibition of T cell apoptosis, and (4) the initiation of T cell clonal expansion.

When binding between the APC and TCR occurs without the additional costimulatory signal, the T cell will undergo anergy induction and either develop a general hyporesponsiveness or proceed with apoptosis. Additionally, whether there is a void or direct blockade of the costimulation signal, the APC will evolve an enhanced ability to perform ligand-induced T cell apoptosis, resulting in termination of the antigen-binding capable T cell. Researchers, through advanced studies of this APC/T cell initiating interface, have been able to apply their knowledge and use cytotoxic T lymphocyte antigen 4 CTLA-4 (CD 152) to manipulate the abrogation of MHC-/antigen-directed T cell activation. CTLA-4, a molecule classically found on T cells, can bind costimulatory complex members CD80, and CD86 (displayed on the involved APC) can abrogate T cell activation [4, 6, 78, 79].

Methods for developing forms of blockade against inciting antigen recognition events and downstream effector pathways which can provide immune tolerance and T cell blockade is critical to the prevention of allograft rejection with any transplant system. Researchers in transplant immunology have put forth tremendous effort in search for technologies to assist in the genetic modification of either the allograft, the APC (primarily DCs), and/or T cell lymphocytes in order to prevent the interface and engagement of the costimulatory complex from forming or activating. By altering the orientation, molecular interactions, active sites, or, potentially, production of costimulatory complex molecules, the need for pharmaceutical immunosuppressive therapy could theoretically be eliminated. Through the prevention of T cell activation, gene therapy targeting the costimulatory protein complex interface could result in the promotion of donor-specific hyporesponsiveness [4, 6, 78, 79].

Applied modeling technologies which focus on the targeting of the costimulation protein complex and the interface occurring between the APC and T cell lymphocyte has been an area of promising research in both solid organ transplant and VCA. Some methods of antibody-mediated costimulatory blockade against CD28, CD80, CD154, and CD40 have been used in both animal and human studies to prevent graft rejection. Depending on the target and method and timing of application, antibody-mediated blockade has led to varying degrees of donor-specific hyporesponsiveness [80–83].

Utilization of monoclonal anti-CD154 antibodies in combination with lymphocyte or splenocyte infusions has led to prolonged murine islet and cardiac allograft survival, suggesting proficiency in both the tolerance of cells and tissue allograft constructs [84, 85]. Studies employing antibody-directed costimulatory blockade against multiple cell surface proteins augments the efficacy of therapy and subsequently allograft survival when compared to single antibody therapy [85]. Published studies have shown that simultaneous CD40/CD154 and CD80/CD28 antibody blockade can provide recipients with graft acceptance in cardiac and cutaneous allotransplant murine models. Moreover, the addition of CTLA4-Ig to multi-target costimulatory blockade promotes synergism among therapies. Gene therapy could be applied to targeted costimulatory blockade in a similar manner as direct antibody administration while circumventing the need to treat patients with multiple doses of antibody therapy, which over time degrades, while genes would sustain antibody production [4, 6, 67].

One way to implement gene production of antibodies against costimulatory complex targets is to have the gene transduced within donor allograft, rather than within the recipient system. Studies which involved anti-CD40 transgene expression showed that murine liver allografts could be maintained indefinitely. In addition, long-term survivors were able to receive donor-specific skin grafts, while rejecting third-party grafts [66]. Bone marrow-derived DCs modified to express CTLA4-Ig are found to have impaired allogenic T cell proliferation and induce alloantigen-specific hyporesponsiveness [67]. *Ex vivo* treatment of donor corneas with adCTLA4-Ig as well as *in vivo* pretreatment of the recipient has been reported to prolong corneal graft survival. In the murine liver transplantation model, *ex vivo* transduction of liver allografts with adCTLA4-Ig has also been shown to lead to indefinite graft survival [68]. *Ex vivo* treatment of rat islets with adCTLA4-Ig protected the islets not only from autoimmune destruction but also from alloimmune destruction in the spontaneously diabetic biobreeding (BB) rats [69].

Another potential strategy to inhibit T cell activation is to prevent DC maturation and maturity-associated upregulation of CD80 and CD86. Immature DCs are poor stimulators of naïve T cells and induce alloantigen-specific hyporesponsiveness

[70–72]. Pretreatment of mice with immature donor-derived DCs has been found to prolong cardiac allograft survival [73]. Among known inhibitors of DC maturation are IL-10, vitamin D, propagation in low-dose granulocyte–macrophage colony-stimulating factor (GM-CSF), and transforming growth factor (TGF)-β1 [71, 74]. DCs, cultured in low-dose GM-CSF and TGF-β1, have decreased expression of costimulatory molecules CD80, CD86, and CD40 and low T cell allostimulatory activity [75].

Genetic engineering that leads to increased TGF-β1 expression can promote maintenance of DCs in an immature state. Gene expression can be targeted for localized delivery via either genetic modification of DCs or the transplant organ or utilized for systemic delivery by treating the recipient. Independent of its effect on DC maturation, TGF-β1 is an immunosuppressive cytokine for B cells, T cells, and natural killer (NK) cells [76]. The potent immunosuppressive effects of TGF-β1 are best demonstrated by the data that mice deficient in TGF-β1 die within few weeks of birth from multifocal inflammatory disease. Transduction of rhesus monkey monocyte-derived DCs with active rhesus TGF-β1 led to inhibition *in vitro* of $CD4^+$ and $CD8^+$ cellular proliferation [77]. Modification of donor-derived DCs with adenovirus TGF-β1 (adTGF-β1) has been reported to prolong cardiac allograft survival [78]. The successful use of TGF-β1 in mice and nonhuman primate models shows promise as a potential strategy to prevent allograft rejection in the clinic.

Cytokine Activity

$CD4^+$ T cells can be classified into subsets based on their cytokine signature. Naïve T helper cells (Th0) produce primarily IL-2. Th1 T cells express high levels of interferon-γ (IFN-γ), IL-2, and tumor necrosis factor (TNF)-α, and Th1 cytokine release results in the activation of cytotoxic $CD8^+$ T cells and antibody production by B cells [79]. Th2 cells secrete primarily IL-4, IL-5, IL-9, IL-10, and IL-13, and they are involved in eosinophil activation and humoral activation that may lead to graft destruction. Th3 or Tr1 T cells are regulatory cells believed to be involved in suppression of the immune response through IL-10 and TGF-β1 production [79, 80].

Cytokines play a central role in T cell differentiation. Production of IL-12 and IFN-γ by DCs leads to Th1 T cell differentiation and expansion. IL-12 propagates the differentiation of Th1 cells by stimulating IFN-γ and inhibiting the production of IL-4 [81]. IFN-γ has direct antiproliferative effects on Th2 T cells, and IL-4 serves as a negative inhibitor by blocking further IFN-γ production.

Genetic engineering to downregulate expression or signaling of inflammatory cytokines can lead to inhibition of the alloimmune response. This strategy would require targeting of multiple cell types, including APCs, NK cells, and T cells. Gene therapy-mediated inhibition of the cytokine inflammatory response can also be achieved by the use of soluble receptors that prevent cytokine signal transduction and cellular activation. Candidate genes for receptor-mediated cytokine blockade include TNF-α, IL-4, and IL-2. TNF-α is a potent mediator of innate immunity primarily by neutrophil activation and mononuclear cell cytokine production [80]. Mice deficient in TNF-α receptor-1 (TNFR-1) have prolonged cardiac allograft survival compared to wild-type mice [81]. T cells lacking TNFR-1 have also been found to exhibit diminished responses *in vitro* to alloantigens, as characterized by reduced proliferation and decreased type I cytokine production and cytolytic function [82]. Drugs that inhibit TNF-α prolong murine cardiac allograft survival and in combination with rapamycin lead to indefinite allograft survival. Gene therapy modification leading to localized delivery and inhibition of TNF-α to prevent allograft dysfunction need to be further explored, as it would eliminate the need for systemic immunosuppression and its associated side effects.

Another gene therapy approach is to increase circulating levels of immunosuppressive cytokines, such as TGF-β1 and IL-10, leading to T cell polarization towards a Th2 T cell response. TGF-β1 is an antiproliferative agent that inhibits activation of both B and T cells [80]. This immunosuppressive cytokine is also involved in wound healing, collagen formation, and fibrosis. Intramuscular administration of adenoviral TGF-β1 in rat lung recipients led to increased serum protein levels with significant improvement of lung oxygenation [82, 83]. Perfusion of murine donor hearts with liposomal DNA complexes containing active human TGF-β1 was found to result in prolonged allograft survival, an effect amplified by $CD8^+$ T cell depletion [84]. Interestingly, perfusion of donor hearts with adenoviral TGF-β1 did not show similar graft prolongation, illustrating the importance of mode of administration and vector use in maximizing transplant outcome. Genetic engineering targeted at producing an anti-inflammatory cytokine environment is well suited for *ex vivo* modification strategies such as transduction of DCs or donor organs with localized and limited delivery of the transgene of interest.

Chemokine Activity

Chemokines are members of a superfamily of small proteins involved in the recruitment and trafficking of hematopoietic cells to sites of inflammation. Besides their role of initiating chemotaxis in a variety of cells, these proteins also function as signaling mediators for tissue homeostasis. Chemokines are divided into four subfamilies (C, CC, CXC, and CX3C) based on the presence and positioning of their conserved cysteine residues. Cell surface expression of chemokine receptors varies from cell to cell, and any given cell may express multiple chemokine receptors. Many of these receptors have the ability to bind more than one ligand, illustrating the redundancy that exists within the system. In transplantation, the initiation of chemokine production is mostly derived from inflammation of the transplanted organ. In the early posttransplant period, chemokine production is triggered by tissue injury sustained during the surgical procedure such as trauma and ischemia–reperfusion. This initial activation cascade leads to recruitment of primarily macrophages and neutrophils, which is shortly followed by increased levels of chemokines directed at increasing trafficking of alloantigen-activated T cells into the graft. The cellular source of chemokine production plays an important role in initiating the alloimmune response. Transplantation of donor hearts derived from IFN-inducing protein-10 (IP-10) knockout mice results in prolongation of allograft survival [85, 86]. In contrast, IP-10-deficient recipient mice have rejection rates equal to that of wild-type mice. Targeted deletions of chemokine receptors have also been shown to be beneficial in preventing allograft rejection. Recipient deletion of the chemokine receptor CCR1 which has as its ligand macrophage inflammatory protein (MIP)-1α, MIP-1β, and RANTES (regulated upon activation, normal T cell expressed, and presumably secreted) has been shown to double allograft survival, and CCR1 deletion in combination with low-dose cyclosporine results in indefinite survival [86].

The redundancy that exists within the chemokine signaling system makes the development of an inhibitory anti-chemokine rejection strategy challenging. Ideally, molecules that are capable of binding and antagonizing multiple chemokine receptors would be the most effective candidates for successful antirejection gene therapy. In transplantation, the use of gene therapy to alter the chemokine-signaling response is quite limited. Recently, two viral proteins that have antagonistic properties against different CC and CXC chemokine receptors, designated vMIP-II and MC148, were used to prevent acute allograft rejection. The protein vMIP-II is a human herpesvirus product that blocks the calcium signaling associated with chemokine receptor activation [87]. Similarly, MC148 is derived from *Molluscum contagiosum* and functions as a chemotaxis inhibitor [88]. Cardiac allografts injected with pDNA encoding for these viral proteins demonstrated prolonged graft survival compared to untreated grafts [89]. In this model, coadministration of the vector with viral IL-10 led to further prolongation of graft survival demonstrating that localized chemokine blockade is of benefit. The transient nature of the transfection method may account for the lack of indefinite graft survival or tolerance observed in the study. Gene transfer of chemokine antagonists and soluble chemokine receptors to prevent transplant rejection should be further explored.

Antibodies

The presence of preformed antibodies against donor MHC class I or II antigens can lead to rapid and acute rejection of the transplanted allograft. Humoral rejection accounts for 20–30 % of acute rejection cases [90]. Formation of alloantibodies most frequently occurs in patients sensitized by prior transplants, transfusions, or pregnancies but can also develop in the nonsensitized transplant recipient by mechanisms not yet well understood.

In patients with prior sensitization, the initial exposure to alloantigen leads to the development of alloreactive B cells that present donor-specific allopeptides on their MHC II complexes. Upon subsequent antigen reexposure, activated T cells provide the additional signals necessary to induce B cell replication and antibody formation [91]. Antibody coating of the transplanted graft initiates destruction of the cells through complement, NK, and cell infiltrate-mediated pathways. In severe cases, hyperacute rejection with thrombosis of the large vessels leads to necrosis of the allograft.

Alloantibodies initiate complement-mediated destruction of the graft. Donor-specific antibodies trigger polymerization of C5–C9 complement molecules with formation of the cell membrane attack complex (MAC). The complex creates a pore in the membrane that ultimately results in cell death and allograft dysfunction. The importance of complement activation in allograft rejection has been demonstrated in C6-deficient mice [92]. Cardiac allograft survival has been shown to be significantly prolonged in mice lacking C6 complement compared to their counterparts with normal complement systems. In addition, pre-sensitized kidney transplant recipients with severe allograft rejection have been found to have C4d deposition in their biopsies. C4d is an inactive fragment of complement, and its presence in allograft biopsies correlates with increased graft loss and refractoriness to conventional T-cell-targeted immunosuppressive therapy [93].

Antibody-mediated cell toxicity can also occur by Fc receptor signaling of NK cells. Cross-linking of NK cell receptors leads to granzyme B and perforin release, followed by target cell apoptosis. NK cells possess receptors for the complement component C3b, and in the presence of complement activation, there is further activation of NK cells [94].

Preformed antibodies and complement activation also play a role in the rejection of xenotransplants. There is significant interest in inhibiting xenogeneic immune responses, as xenotransplantation can address the severe shortage in human organs that currently exists. Complement activation is the primary component of hyperacute rejection of xenogeneic organ transplants.

In xenogeneic transplants, activation of the complement system occurs via both classical and alternative pathways, providing a potent defense against the acceptance of xenografts. Activation of the alternative complement pathway via factor H is a significant barrier to xenotransplantation. In autologous cells, complement regulatory proteins are responsible for inhibition of complement activation and MAC formation. These membrane-bound proteins are involved in self–nonself recognition. The absence of cell surface expression of complement regulatory proteins CD55, CD46, and CD59 might account for the increased susceptibility of xenografts to complement-mediated lysis [95]. Xenografts from swine transgenic for complement regulatory proteins were found to be less susceptible to complement-mediated injury in a swine-to-primate cardiac xenograft model [96]. In addition to membrane-bound complement inhibitors, there are also soluble proteins, such as C1 inhibitor and clusterin, that are responsible for inhibiting complement activation in body fluids.

The presence of natural xenoreactive antibodies is responsible for initiating the classical pathway of complement activation. These xenogeneic antibodies are designated as "natural," as all immunocompetent mammals have them in circulation without any prior exposure to the foreign antigen. There are at least two types of natural xenogeneic antibodies with monoreactive anti-carbohydrate antibodies playing a significant role in xenogeneic graft rejection. The anti-Gal α1,3 Gal monoreactive anti-carbohydrate antibody is thought to play a major role in human xenograft rejection. These carbohydrates are present in the cells of lower mammals, and binding of anti-Gal α1,3 Gal antibodies leads complement activation.

The development of gene therapy strategies that inhibit complement activation and B cell antibody production could reduce acute allorejection in sensitized patients as well as block xenograft rejection. Soluble inhibitory complement receptors, such as sCR1 (a potent regulator of C3 and C5 activation), provide a strategy for reducing complement-mediated rejection. The use of a soluble CR1 has been shown to be beneficial in pig lung allotransplantation [97]. Administration of sCR1 via retroviral and naked DNA gene therapy prevents the progression of collagen-induced arthritis [98].

Engineering of xenografts to express human membrane-associated proteins can also prevent complement-mediated rejection. Transfection of porcine endothelial cells with human CD59 was able to protect the cells from complement-mediated destruction [99]. Baboons transplanted with hearts, derived from CD55 transgenic pigs, were protected from hyperacute rejection [100]. Gene therapy strategies that modify cell surface expression of the transplanted graft to reduce complement-mediated rejection need to be further explored.

Ischemia–Reperfusion Injury, Graft Dysfunction, and Graft Failure

Ischemia–reperfusion injury is a common source of morbidity, graft loss, and mortality in patients undergoing allotransplantation. Gene therapy directed at inhibitors of receptors and/or chemokines which allow acute reactive leukocytes (neutrophils, macrophages, and NKs) as well the augmentation of free radial scavengers and certain toll-like receptors (TLR-1 and TLR-4) could reduce the stress of ischemic–reperfusion injury on allotransplant organs [100, 101].

Ischemia/reperfusion injury adversely affects graft function and survival. Therapies aimed at reducing cellular ischemia/reperfusion injury would expand the available donor pool to include non-heart-beating donors, thereby increasing the number of organs available for transplantation. *Ex vivo* genetic modification approaches as well as transient gene expression are ideal for maximizing clinical outcome.

In both cellular and organ transplantation, ischemia/reperfusion is an important cause of early nonimmune graft dysfunction and failure. In organ transplantation, the ischemic period begins with the in situ normothermic perfusion at the time of organ harvest, followed by hypothermic cold preservation, and ending with the process of vascular reanastomosis and reperfusion in the transplant recipient. The process of reperfusion through ischemic tissues triggers an inflammatory process characterized by complement deposition, cytokine release, and infiltration of inflammatory cells [101, 102]. The lack of available adenosine triphosphate (ATP) energy sources within the graft generates ionic and membrane potential differences that cause cell swelling and damage. The degree of tissue injury is dependent on the duration of the ischemic event, and endothelial cells play a major role in the inflammatory process [102].

Ischemic endothelial cell production of oxygen-free radicals initiates intracellular signaling that can cause apoptotic cell death. This increased oxygen-free radical production enhances the recruitment of leukocytes into the graft by reducing the availability of the anti-adhesive and vasodilatory protein nitric oxide [103]. During reperfusion, reduced levels of nitric oxide as well as the increased tissue swelling lead to impaired blood flow with enhanced leukocyte adhesion and activation [102]. This process of enhanced leukocyte adhesion occurs within minutes of reperfusion and persists for several hours [104]. This process is further perpetuated by increased endothelial cell cytokine expression of TNF-α, ILs, as well as platelet-activating factor (PAF) and leukotrienes.

The observation that well-matched cadaver donors have a poorer outcome than nonmatched living donors has raised interest in the role ischemia/reperfusion plays in stimulating the innate immune response [94, 101]. Complement activation has been shown to play an important role in the inflammatory response to ischemic injury. Inhibition of complement activation during ischemia/reperfusion via genetic deletion of C3 or C4 complement components causes a significant reduction in skeletal muscle injury [101]. Immunoglobulin knockout mice had their susceptibility to ischemia/reperfusion restored after administration of normal mice serum, suggesting the involvement of natural antibodies in ischemia/reperfusion injury [101]. Complement activation via increased C-reactive protein as well as mannose-binding lectin has been implicated as potential mediators of ischemia/reperfusion injury [94].

Overexpression of anti-apoptotic genes as well as free radical scavengers can protect the graft from ischemia/reperfusion injury. Pretreatment of liver grafts with anti-apoptotic Bcl-2 adenoviral vectors has been shown to significantly decrease organ injury and apoptosis [105]. Endothelial cells genetically modified to express caspase-resistant Bcl-2 demonstrate increased resistance to apoptosis and cytotoxic T-cell-mediated lysis [106]. In addition, expression of the anti-apoptotic proteins Bcl-xL and A20 prevents antibody-induced transplant atherosclerosis in a cardiac allograft model [107].

Free radical production blockade via genetic transfer of superoxide dismutase and heme oxygenase-1 has also been beneficial in several ischemia/reperfusion experimental models. Overexpression of heme oxygenase-1 protects rat livers from ischemia/reperfusion-induced necrosis and apoptosis [108]. Adenoviral-mediated expression of copper–zinc superoxide dismutase has also been shown to protect liver from ischemia/reperfusion injury [109]. In a similar fashion, increased manganese superoxide dismutase production via *ex vivo* adenoviral transfection has been shown to protect cardiac allografts from ischemia/reperfusion injury [110].

The implementation of anti-leukocyte adhesion genetic engineering strategies has also been shown to be beneficial in reducing ischemia/reperfusion injury. Blockade of intercellular adhesion molecule (ICAM)-1, with the use of antisense oligonucleotides, leads to a reduction in rat cardiac allograft reperfusion injury [111, 112]. Increased gene expression of the anti-inflammatory cytokines has also been shown to ameliorate the effects of ischemic–reperfusion injury. Intratracheal adenoviral administration of IL-10 reduces lung ischemia–reperfusion injury [113].

In a cold ischemia model, IL-13 adenoviral gene expression resulted in an increase in liver graft survival rates from 50 to 100 % [114]. As mentioned earlier, intramuscular delivery of TGF-β1 via adenoviral vector results in enhanced oxygenation and improved functioning of lung allografts [115].

Gene Therapy in VCA

The field of VCA owes much to findings from the larger field of transplantation, and all of the studies outlined above from solid organ transplantation have findings directly applicable to VCA transplants as well. The data and experience learned from these studies will directly inform the next generation of trials in composite allotransplantation. In addition to the much larger volume of work in the solid organ transplant field, there have been a small number of important studies in VCA models directly examining the use of gene therapy in these transplants to date.

The first set of important experiments in this field was work performed not on "true" allografts but on free flaps instead. The importance to the field of VCA is apparent when one considers that a free flap is a vascularized composite autotransplant which shares many of the physiological properties of a VCA without the allotype mismatch. Using *ex vivo* perfusion of the vascular bed of the free flap with adenovirus viral vector system, Dr. Gurtner and his group have published a series of studies in which they express first a marker gene (LacZ) and then several physiologically active substances, including the angiogenesis inhibitor endostatin, the human cathelicidin antimicrobial peptide-LL37, and IL-12 [116–118]. This group coined the term "biologic brachytherapy" to describe the localized delivery of a biologically active molecule using gene transduction of a graft. They were able to demonstrate the ability of the antimicrobial peptide to decrease bacterial loads, and even more exciting from the transplant perspective, they were able to decrease breast cancer growth rates by expressing IL-12 and increasing the inflammatory environment [61]. Presumably, if gene therapy can be used to express pro-inflammatory cytokines, it should be possible to do the reverse by expressing inhibitory cytokines.

This technique was then extended into a true VCA model by a series of experiments by the senior author of this chapter. By utilizing the rat hind limb model, a series of experiments was performed comparing various methods of gene delivery into the hind limb allograft during *ex vivo* isolated graft perfusion. The methods tested included using naked DNA, cationic polymer/DNA complex transfection, and adenoviral vector transduction to express a marker gene (luciferase).

Although all methods could produce detectible expression, the adenoviral transduction was far superior in expression level and different tissue types expressing the marker gene [44]. The graft was transduced with *ex vivo* perfusion and the vector was washed out of the graft before transplantation. This resulted in expression located only in the grafted limb itself (Fig. 18.1). Controls infusing intact animals or transplants after completion of the surgery showed high levels of expression in other tissues, including lung, liver, and spleen. Importantly, there was no effect on

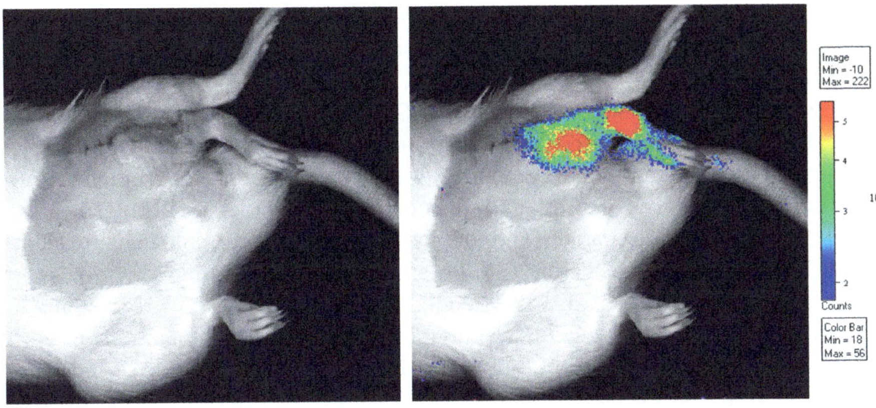

Fig. 18.1 Orthotopic rat hind limb transplant showing luciferase transgene expression (*false color scale*) isolated to transplanted graft only

the rejection of the graft from the presence of the virus. Although expression levels were relatively high and the expression was detectible quite rapidly, the expression was stable for only 14 days. This was to be expected due to the episomal (extrachromosomal) location of DNA introduced by the adenoviral system. Clearly, this limits the effectiveness of this method of gene delivery as discussed above. Further work as used retroviral systems such as lentivirus to extend the duration of expression. Leto-Barone et al. [2] used transdermal injection of luc-eGFP containing lentiviral vector and were able to support marker gene expression >150 days in a face and hand transplant rat model. Interestingly, this group found no expression when this vector was infused intra-arterially in contrast to the findings of the adenoviral studies. This work showed that by using retrovirus, sustained expression in skin and composite tissue is possible using retroviral vectors capable of genetic integration. The collective work of these groups has been shown that gene therapy can be effective in composite grafts both through intradermal and *ex vivo* perfusion of viral vectors. This can be safely done in the transplant setting and limits the expression of the transgene to the area of the graft itself. Thus, the stage is set to attack the potential molecular targets that we have discussed in the section above, but now in the setting of a VCA transplant.

Utilizing these tools, a small number of groups have attempted to improve graft survival or decrease immunosuppression using gene therapy. One approach is to regulate the expression of pro-tolerogenic cytokines. This has been used with some success in solid organ models, and the combination of IL-10 and TGF-β can cause the formation of Treg cells in the presence of allogenic stimulation *in vitro* [62]. The rat hind limb transplant model was used to test the ability of the adenoviral expression system driving the overexpression of these two cytokines to prolong graft survival. Unfortunately, this combination did not result in more than a few days of prolongation before the grafts were rejected [44]. More work is necessary to determine why this was not sufficient to prolong survival. It may be that adenoviral expression

(14 days) was too transitory to establish a robust Tregs population. Alternatively, the activity of these cytokines *in vitro* and in solid organ/cell transplant differs from the situation in the hind limb VCA model. An alternative approach was used in work by a group under the direction of Dr. Shuzhong Guo. They have attempted to use gene therapy for the local administration of biological immunosuppressive molecules. Utilizing a nonbone containing VCA model in rats, this group has used adenoviral vectors to overexpress both the costimulatory blocking agents CTLA4-Ig and OX40-Ig [82] in separate experiments. Both agents were able to cause small but statistically significant increases in graft survival when combined with a short course of rapamycin (survival was extended 5 days for OX40-IG expression and 35 days for CTLA4-Ig expression). Although modest, these prolongations in graft survival are proof of the concept that upregulating anti-inflammatory or immunosuppressive molecules using gene therapy in VCA can be effective. An alternative approach taken by Dr. Ceradini has been to attempt to downregulate the expression of the allogenic MHC molecules that are the target of the T cell direct antigen recognition pathway. To accomplish this goal, their group has utilized the transfection of small interfering mRNA to the MHC-1 molecule (siMHC1). By transfecting the vascular endothelium with siMHC1 during *ex vivo* perfusion, they were able to decrease MHC expression by 87% [119]. Further studies are necessary to see the effect this will have on the ability of the immune system to recognize these grafts.

In summary, gene therapy has been shown to offer great potential in the field of VCA. Multiple approaches and vectors have been used to express genes and limit the expression of these genes to the graft. This should make gene therapy relatively safe in these transplants and open the way for therapies that attack of many therapeutic targets. Cytokines and immunosuppressive molecules have been upregulated in VCAs but have shown only modest effects. This may be due to the increased hurdle of achieving tolerance in skin containing graphs. However, it is very early in these studies and perhaps we have simply not targeted the ideal pathway yet. Alternative approaches using small interfering mRNAs are becoming increasingly popular in solid organ transplant and have been shown to be a possible target VCA as well. Although there are no current clinical trials in VCA using gene therapy, it may not be long before we are able to "reprogram" the graft and recipient to live together without the need for immunosuppression.

References

1. Sarhane KA, Tuffaha SH, Broyles JM, Ibrahim AE, Khalifian S, Baltodano P, Santiago GF, Alrakan M, Ibrahim Z. A critical analysis of rejection in vascularized composite allotransplantation: clinical, cellular and molecular aspects, current challenges, and novel concepts. Front Immunol. 2013;4:406. doi:10.3389/fimmu.2013.00406. PubMed PMID: 24324470; PubMed Central PMCID: PMC3839257.
2. Leto Barone AA, Zhou ZY, Hughes MW, Park R, Schulman RM, Lee S, Vidar EN, Shiba TL, Weber EL, Cetrulo CL Jr. Lentiviral transduction of face and limb flaps: implica-

tions for immunomodulation of vascularized composite allografts. Plast Reconstr Surg. 2012;129(2):391–400. doi:10.1097/PRS.0b013e31823aeaeb. PubMed PMID: 22286422.
3. Siemionow M, Klimczak A. Tolerance and future directions for composite tissue allograft transplants: part II. Plast Reconstr Surg. 2009;123(1):7e–17e. doi:10.1097/PRS.0b013e318193467d. PubMed PMID: 19116508.
4. Thomas D, Suthanthiran M. Optimal modes and targets of gene therapy in transplantation. Immunol Rev. 2003;196:161–75. PubMed PMID: 14617204
5. Al-Dosari MS, Gao X. Nonviral gene delivery: principle, limitations, and recent progress. AAPS J. 2009;11(4):671–81. doi:10.1208/s12248-009-9143-y. PubMed PMID: 19834816; PubMed Central PMCID: PMC2782077.
6. Transplantation of Composite Tissue Allografts, Charles W. Hewitt, Chad R. Gordon, W.P. Andrew Lee Springer Science & Business Media, Dec 20, 2007
7. Mariani JA, Kaye DM. Delivery of gene and cellular therapies for heart disease. J Cardiovasc Transl Res. 2010;3(4):417–26. doi:10.1007/s12265-010-9190-x. PubMed PMID: 20559776.
8. Herweijer H, Wolff JA. Progress and prospects: naked DNA gene transfer and therapy. Gene Ther. 2003;10(6):453–8. PubMed PMID: 12621449.
9. Maruyama H, Higuchi N, Nishikawa Y, Hirahara H, Iino N, Kameda S, Kawachi H, Yaoita E, Gejyo F, Miyazaki J. Kidney-targeted naked DNA transfer by retrograde renal vein injection in rats. Hum Gene Ther. 2002;13(3):455–68. PubMed PMID: 11860712.
10. Hartikka J, Sukhu L, Buchner C, Hazard D, Bozoukova V, Margalith M, Nishioka WK, Wheeler CJ, Manthorp M, Sawdey M. Electroporation-facilitated delivery of plasmid DNA in skeletal muscle: plasmid dependence of muscle damage and effect of poloxamer 188. Mol Ther. 2001;4(5):407–15. PubMed PMID: 11708877.
11. Klein J, Sato A. The HLA system. Second of two parts. N Engl J Med. 2000;343(11):782–6. (Erratum in: N Engl J Med. 2000 Nov 16;343(20):1504. PubMed PMID: 10984567).
12. De Smedt SC, Demeester J, Hennink WE. Cationic polymer based gene delivery systems. Pharm Res. 2000;17(2):113–26. PubMed PMID: 10751024.
13. Sun X, Zhang N. Cationic polymer optimization for efficient gene delivery. Mini Rev Med Chem. 2010;10(2):108–25. PubMed PMID: 20408796.
14. Boussif O, Lezoualc'h F, Zanta MA, Mergny MD, Scherman D, Demeneix B, Behr JP. A versatile vector for gene and oligonucleotide transfer into cells in culture and *in vivo*: polyethylenimine. Proc Natl Acad Sci U S A. 1995;92(16):7297–301. PubMed PMID: 7638184; PubMed Central PMCID: PMC41326.
15. Park TG, Jeong JH, Kim SW. Current status of polymeric gene delivery systems. Adv Drug Deliv Rev. 2006;58(4):467–86. PubMed PMID: 16781003.
16. Hunter AC. Molecular hurdles in polyfectin design and mechanistic background to polycation induced cytotoxicity. Adv Drug Deliv Rev. 2006;58(14):1523–31. PubMed PMID: 17079050.
17. Buschmann MD, Merzouki A, Lavertu M, Thibault M, Jean M, Darras V. Chitosans for delivery of nucleic acids. Adv Drug Deliv Rev. 2013;65(9):1234–70. doi:10.1016/j.addr.2013.07.005. PubMed PMID: 23872012.
18. Tomalia DA. Dendrimer research. Science. 1991;252(5010):1231. PubMed PMID: 17842936.
19. Daneshvar N, Abdullah R, Shamsabadi FT, How CW, Mh MA, Mehrbod P. PAMAM dendrimer roles in gene delivery methods and stem cell research. Cell Biol Int. 2013;37(5):415–9. doi:10.1002/cbin.10051. PubMed PMID: 23504853.
20. Fant K, Esbjörner EK, Jenkins A, Grossel MC, Lincoln P, Nordén B. Effects of PEGylation and acetylation of PAMAM dendrimers on DNA binding, cytotoxicity and *in vitro* transfection efficiency. Mol Pharm. 2010;7(5):1734–46. doi:10.1021/mp1001312. PubMed PMID: 20695423.
21. Felgner PL, Gadek TR, Holm M, Roman R, Chan HW, Wenz M, Northrop JP, Ringold GM, Danielsen M. Lipofection: a highly efficient, lipid-mediated DNA-transfection procedure. Proc Natl Acad Sci U S A. 1987;84(21):7413–7. PubMed PMID: 2823261; PubMed Central PMCID: PMC299306.

22. Chang HI, Yeh MK. Clinical development of liposome-based drugs: formulation, characterization, and therapeutic efficacy. Int J Nanomedicine. 2012;7:49–60. doi:10.2147/IJN.S26766. PubMed PMID: 22275822; PubMed Central PMCID: PMC3260950.
23. Mallick S, Choi JS. Liposomes: versatile and biocompatible nanovesicles for efficient biomolecules delivery. J Nanosci Nanotechnol. 2014;14(1):755–65. PubMed PMID: 24730295.
24. Krieg AM, Yi AK, Matson S, Waldschmidt TJ, Bishop GA, Teasdale R, Koretzky GA, Klinman DM. CpG motifs in bacterial DNA trigger direct B-cell activation. Nature. 1995;374(6522):546–9. PubMed PMID: 7700380.
25. Waehler R, Russell SJ, Curiel DT. Engineering targeted viral vectors for gene therapy. Nat Rev Genet. 2007;8(8):573–87. PubMed PMID: 17607305.
26. Tan PH, Tan PL, George AJ, Chan CL. Gene therapy for transplantation with viral vectors—how much of the promise has been realised? Expert Opin Biol Ther. 2006;6(8):759–72. PubMed PMID: 16856798.
27. Cheung ST, Tsui TY, Wang WL, Yang ZF, Wong SY, Ip YC, Luk J, Fan ST. Liver as an ideal target for gene therapy: expression of CTLA4Ig by retroviral gene transfer. J Gastroenterol Hepatol. 2002;17(9):1008–14. PubMed PMID: 12167123.
28. Stratford-Perricaudet LD, Levrero M, Chasse JF, Perricaudet M, Briand P. Evaluation of the transfer and expression in mice of an enzyme-encoding gene using a human adenovirus vector. Hum Gene Ther. 1990;1(3):241–56. PubMed PMID: 2081192.
29. Jaffe HA, et al. Adenovirus-mediated *in vivo* gene transfer and expression in normal rat liver. Nat Genet 1992;1:372–8.
30. NIH Report. Assessment of adenoviral vector safety and toxicity: report of the National Institutes of Health Recombinant DNA Advisory Committee. Hum Gene Ther. 2002;13(1):3–13. PubMed PMID: 11779411.
31. Yeung JC, Wagnetz D, Cypel M, Rubacha M, Koike T, Chun YM, Hu J, Waddell TK, Hwang DM, Liu M, Keshavjee S. *Ex vivo* adenoviral vector gene delivery results in decreased vector-associated inflammation pre- and post-lung transplantation in the pig. Mol Ther. 2012;20(6):1204–11. doi:10.1038/mt.2012.57. PubMed PMID: 22453765; PubMed Central PMCID: PMC3369301.
32. Cunningham EC, Tay SS, Wang C, Rtshiladze M, Wang ZZ, McGuffog C, Cubitt J, McCaughan GW, Alexander IE, Bertolino P, Sharland AF, Bowen DG, Bishop GA. Gene therapy for tolerance: high-level expression of donor major histocompatibility complex in the liver overcomes naive and memory alloresponses to skin grafts. Transplantation. 2013;95(1):70–7. doi:10.1097/TP.0b013e318278d39a. PubMed PMID: 23263501.
33. Carter BJ. Adeno-associated virus vectors in clinical trials. Hum Gene Ther. 2005;16(5):541–50. PubMed PMID: 15916479.
34. Blacklow NR, Hoggan MD, Kapikian AZ. Epidemiology of adenovirus-associated virus infection in a nursery population. Am J Epidemiol. 1968;88:368–78.
35. Lebkowski JS, McNally MM, Okarma TB, Lerch LB. Adeno-associated virus: a vector system for efficient introduction and integration of DNA into a variety of mammalian cell types. Mol Cell Biol. 1988;8:3988–96.
36. Cheung AK, Hoggan MD, Hauswirth WW, Berns KI. Integration of the adeno-associated virus genome into cellular DNA in latently infected human Detroit 6 cells. J Virol 1980;33:739–48.
37. Dong JY, Fun PD, Frizzell RA. Quantitive analysis of the packaging capacity of adeno-associated virus. Hum Gene Ther. 1996;7:2101–12.
38. Surosky RT, Urabe M, Godwin SG, McQuiston SA, Kurtzman GJ, Ozawa K, Natsoulis G. Adeno-associated virus Rep proteins target DNA sequences to a unique locus in the human genome. J Virol. 1997;71(10):7951–9. PubMed PMID: 9311886; PubMed Central PMCID: PMC192153.
39. Manno CS, Chew AJ, Hutchison S, Larson PJ, Herzog RW, Arruda VR, Tai SJ, Ragni MV, Thompson A, Ozelo M, Couto LB, Leonard DG, Johnson FA, McClelland A, Scallan C, Skarsgard E, Flake AW, Kay MA, High KA, Glader B. AAV-mediated factor IX gene transfer

to skeletal muscle in patients with severe hemophilia B. Blood. 2003;101(8):2963–72. PubMed PMID: 12515715.
40. Zaiss AK, Muruve DA. Immune responses to adeno-associated virus vectors. Curr Gene Ther. 2005;5(3):323–31. PubMed PMID: 15975009.
41. Bozulic LD, Breidenbach WC, Ildstad ST. Past, present, and future prospects for inducing donor-specific transplantation tolerance for composite tissue allotransplantation. Semin Plast Surg. 2007;21(4):213–25. doi:10.1055/s-2007-991191. PubMed PMID: 20567674; PubMed Central PMCID: PMC2884849.
42. Lowenstein PR. The case for immunosuppression in clinical gene transfer. Mol Ther. 2005;12(2):185–6. PubMed PMID: 16002339.
43. Morizono K, Chen IS. Receptors and tropisms of envelope viruses. Curr Opin Virol. 2011;1(1):13–8. doi:10.1016/j.coviro.2011.05.001. PubMed PMID: 21804908; PubMed Central PMCID: PMC3144558.
44. Cooney DS, Chambers CB, Suchy H, Hornick MC, Reichensperger J, Neumeister MW. Transgene expression in a model of composite tissue allotransplantation. Plast Reconstr Surg. 2010;125(3):837–45. doi:10.1097/PRS.0b013e3181cb6389. PubMed PMID: 20009792.
45. Romano G, Pacilio C, Giordano A. Gene transfer technology in therapy: current applications and future goals. Stem Cells. 1999;17(4):191–202. PubMed PMID: 10437982.
46. Romano G, Pacilio C, Giordano A. Gene transfer technology in therapy: current applications and future goals. Oncologist. 1998;3(4):225–36. PubMed PMID: 10388109.
47. Zufferey R, et al. Self-inactivating lentivirus vector for safe and efficient *in vivo* gene delivery. J Virol. 1998;72:9873–80.
48. Johnston JC, et al. Minimum requirements for efficient transduction of dividing and nondividing cells by feline immunodeficiency virus vectors. J Virol. 1999;73:4991–5000.
49. Johnston J, Power C. Productive infection of human peripheral blood mononuclear cells by feline immunodeficiency virus: implications for vector development. J Virol. 1999;73:2491–8.
50. Baulcombe DC. RNA as a target and an initiator of post-transcriptional gene silencing in transgenic plants. Plant Mol Biol. 1996;32(1–2):79–88. PubMed PMID: 8980475.
51. Koutsilieri E, Rethwilm A, Scheller C. The therapeutic potential of siRNA in gene therapy of neurodegenerative disorders. J Neural Transm Suppl. 2007;(72):43–9. PubMed PMID: 17982877.
52. Glebova K, Reznik ON, Reznik AO, Mehta R, Galkin A, Baranova A, Skoblov M. siRNA Technology in Kidney Transplantation: Current Status and Future Potential. BioDrugs. 2014;28(4):345–61. doi:10.1007/s40259-014-0087-0. PubMed PMID: 24573958.
53. Joffre O, Santolaria T, Calise D, Al Saati T, Hudrisier D, Romagnoli P, van Meerwijk JP. Prevention of acute and chronic allograft rejection with CD4+ CD25+ Foxp3+ regulatory T lymphocytes. Nat Med. 2008;14(1):88–92. PubMed PMID: 18066074; PubMed Central PMCID: PMC2443705.
54. Ikemoto T, Takita M, Levy MF, Shimada M, Naziruddin B. CD11b(+) cells in donor-specific transfusion prolonged allogenic skin graft survival through indoleamine 2,3-dioxygenase. Cell Immunol. 2013;283(1–2):81–90. doi:10.1016/j.cellimm.2013.06.004. PubMed PMID: 23933136.
55. Klein RM, Wolf ED, Wu R, Sanford JC. High-velocity microprojectiles for delivering nucleic acids into living cells, 1987. Biotechnology. 1992;24:384–6. PubMed PMID: 1422046.
56. Taniyama Y, Azuma J, Kunugiza Y, Iekushi K, Rakugi H, Morishita R. Therapeutic option of plasmid-DNA based gene transfer. Curr Top Med Chem. 2012;12(15):1630–7. PubMed PMID: 23061881.
57. Kim YC, Prausnitz MR. Enabling skin vaccination using new delivery technologies. Drug Deliv Transl Res. 2011;1(1):7–12. PubMed PMID: 21799951; PubMed Central PMCID: PMC3143039.
58. Curran MA, Ochoa MS, Molano RD, Pileggi A, Inverardi L, Kenyon NS, Nolan GP, Ricordi C, Fenjves ES. Efficient transduction of pancreatic islets by feline immunodeficiency virus vectors1. Transplantation. 2002;74(3):299–306. PubMed PMID: 12177606.

59. He Z, Wang F, Kumagai-Braesch M, Permert J, Holgersson J. Long-term gene expression and metabolic control exerted by lentivirus-transduced pancreatic islets. Xenotransplantation. 2006;13(3):195–203. PubMed PMID: 16756562.
60. Pascual M, Theruvath T, Kawai T, Tolkoff-Rubin N, Cosimi AB. Strategies to improve long-term outcomes after renal transplantation. N Engl J Med. 2002;346(8):580–90. PubMed PMID: 11856798.
61. Dempsey MP, Hamou C, Michaels J 5th, Ghali S, Jazayeri L, Grogan RH, Gurtner GC. Using genetically modified microvascular free flaps to deliver local cancer immunotherapy with minimal systemic toxicity. Plast Reconstr Surg. 2008;121(5):1541–53. doi:10.1097/PRS.0b013e31816ff6aa. PubMed PMID: 18453976.
62. Deng S, Ketchum RJ, Kucher T, Weber M, Shaked A, Naji A, Brayman KL. IL-10 and TGF-beta gene transfer for xenogeneic islet transplantation: comparison of effect in concordant vs discordant combination. Transplant Proc. 1997;29(4):2204–5. PubMed PMID: 9193591.
63. Sykes M, Sachs DH. Bone marrow transplantation as a means of inducing tolerance. Semin Immunol. 1990;2(6):401–17. PubMed PMID: 2104278.
64. Sykes M, Hoyles KA, Romick ML, Sachs DH. *In vitro* and *in vivo* analysis of bone marrow-derived CD3+, CD4−, CD8−, NK1.1+ cell lines. Cell Immunol. 1990;129(2):478–93. PubMed PMID: 2143439.
65. Bagley J, Tian C, Sachs DH, Iacomini J. Induction of T-cell tolerance to an MHC class I alloantigen by gene therapy. Blood. 2002;99(12):4394–9. PubMed PMID: 12036867.
66. Fraser CC, Sykes M, Lee RS, Sachs DH, LeGuern C. Specific unresponsiveness to a retrovirally-transferred class I antigen is controlled through the helper pathway. J Immunol. 1995;154(4):1587–95. PubMed PMID: 7836744.
67. Mayfield RS, Hayashi H, Sawada T, Bergen K, LeGuern C, Sykes M, Sachs DH, Iacomini J. The mechanism of specific prolongation of class I-mismatched skin grafts induced by retroviral gene therapy. Eur J Immunol. 1997;27(5):1177–81. PubMed PMID: 9174608.
68. Rivière I, Brose K, Mulligan RC. Effects of retroviral vector design on expression of human adenosine deaminase in murine bone marrow transplant recipients engrafted with genetically modified cells. Proc Natl Acad Sci U S A. 1995;92(15):6733–7. PubMed PMID: 7624312; PubMed Central PMCID: PMC41403.
69. Tian C, Bagley J, Iacomini J. Expression of antigen on mature lymphocytes is required to induce T cell tolerance by gene therapy. J Immunol. 2002;169(7):3771–6. PubMed PMID: 12244171.
70. Kang ES, Iacomini J. Induction of central deletional T cell tolerance by gene therapy. J Immunol. 2002;169(4):1930–5. PubMed PMID: 12165518.
71. Graeb C, Justl M, Scherer MN, Andrassy J, Frank E, Zuelke C, Jauch KW, Geissler EK. Use of an adenoviral vector to express soluble donor-major histocompatibility complex molecules capable of suppressing the immune response in rat transplant recipients. Hum Immunol. 2002;63(10):844–52. PubMed PMID: 12368036.
72. Graeb C, Scherer MN, Kroemer A, Jauch KW, Geissler EK. Hepatocyte expression of soluble donor MHC class I antigen via gene transfer inhibits multiple aspects of the antidonor immune response in fully sensitized rat transplant recipients. Hum Immunol. 2002;63(10):948–54. PubMed PMID: 12368047.
73. Sumimoto R, Shinomiya T. Examination of serum class I antigen in liver-transplanted rats. Clin Exp Immunol 1991;85:114–20.
74. Kamada N, Yamaguchi A, Yoshimura S, Sumimoto R, Teramoto K, Baquerizo A. Mechanisms of immunosuppression by serum from liver grafted rats: demonstration of suppressor cell activity *in vivo* and *in vitro*. Transplant Proc. 1990;22:1973–4.
75. Lough D, Abdo J, Guerra-Castro JF, Matsumoto C, Kaufman S, Shetty K, Kwon YK, Zasloff M, Fishbein TM. Abnormal CX3CR1^{+} lamina propria myeloid cells fromintestinal transplant recipients with NOD2 mutations. Am J Transpl. 2012;12(4):992–1003. doi:10.1111/j.1600-6143.2011.03897.x. PubMed PMID: 22233287.
76. Lechler R, Ng WF, Steinman RM. Dendritic cells in transplantation—friend or foe? Immunity. 2001;14: 357–68.

77. Steinman RM, Inaba K, Turley S, Pierre P, Mellman I. Antigen capture, processing, and presentation by dendritic cells: recent cell biological studies. Hum Immunol. 1999;60:562–7.
78. Watts TH, DeBenedette MA. T cell co-stimulatory molecules other than CD28. Curr Opin Immunol. 1999;11:286–93.
79. Lenschow DJ, Walunas TL, Bluestone JA. CD28/B7 system of T cell costimulation. Annu Rev Immunol. 1996;14:233–258.
80. Ravindra K, Haeberle M, Levin LS, Ildstad ST. Immunology of vascularized composite allotransplantation: a primer for hand surgeons. J Hand Surg Am. 2012;37(4):842–50. doi: 10.1016/j.jhsa.2012.01.042. PubMed PMID: 22464235.
81. Whitelegg A, Barber LD. The structural basis of T-cell allorecognition. Tissue Antigens. 2004;63(2):101–8. PubMed PMID: 14705981.
82. Fu S, Yang Y, Xiao B, Li Y, Yi CG, Xia W, Guo SZ. Use of genetically modified allograft to deliver local immunomodulatory molecule with minimal systemic toxicity in a rat model of allogeneic skin flap transplantation. Transplant Proc. 2010;42(9):3815–9. doi:10.1016/j.transproceed.2010.08.040. PubMed PMID: 21094862.
83. Eun SC. Composite tissue allotransplantation immunology. Arch Plast Surg. 2013;40(2):141–53. doi:10.5999/aps.2013.40.2.141. PubMed PMID: 23529264; PubMed Central PMCID: PMC3605559.
84. Parker DC, et al. Survival of mouse pancreatic islet allografts in recipients treated with allogeneic small lymphocytes and antibody to CD40 ligand. Proc Natl Acad Sci U S A. 1995;92:9560–4.
85. Sebille F, et al. Tolerance induction in rats, using a combination of anti-CD154 and donor splenocytes, given once on the day of transplantation. Transplantation 2003;75:169–72.
86. Gao W, Topham PS, King JA, Smiley ST, Csizmadia V, Lu B, Gerard CJ, Hancock WW. Targeting of the chemokine receptor CCR1 suppresses development of acute and chronic cardiac allograft rejection. J Clin Invest. 2000;105(1):35–44. PubMed PMID: 10619859; PubMed Central PMCID: PMC382589.
87. Kledal TN, Rosenkilde MM, Coulin F, Simmons G, Johnsen AH, Alouani S, Power CA, Lüttichau HR, Gerstoft J, Clapham PR, Clark-Lewis I, Wells TN, Schwartz TW. A broad-spectrum chemokine antagonist encoded by Kaposi's sarcoma-associated herpesvirus. Science. 1997;277(5332):1656–9. PubMed PMID: 9287217.
88. Damon I, Murphy PM, Moss B. Broad spectrum chemokine antagonistic activity of a human poxvirus chemokine homolog. Proc Natl Acad Sci U S A. 1998;95(11):6403–7. PubMed PMID: 9600978; PubMed Central PMCID: PMC27746.
89. DeBruyne LA, Li K, Bishop DK, Bromberg JS. Gene transfer of virally encoded chemokine antagonists vMIP-II and MC148 prolongs cardiac allograft survival and inhibits donor-specific immunity. Gene Ther. 2000;7(7):575–82. PubMed PMID: 10819572.
90. Mauiyyedi S, Colvin RB. Humoral rejection in kidney transplantation: new concepts in diagnosis and treatment. Curr Opin Nephrol Hypertens. 2002;11(6):609–18. PubMed PMID: 12394606.
91. Le Moine A, Goldman M, Abramowicz D. Multiple pathways to allograft rejection. Transplantation. 2002;73(9):1373–81. PubMed PMID: 12023610.
92. Merten S, Chen JC, Ha H, Plain K, Boyd RA, Penny MJ, Leenaerts P, Hall BM. The cellular basis of cardiac allograft rejection: VIII. Mechanisms underlying delayed allograft rejection in PVG C6-deficient rats. Transplantation. 1998;65(9):1152–8. PubMed PMID: 9603160.
93. Feucht HE, Schneeberger H, Hillebrand G, Burkhardt K, Weiss M, Riethmüller G, Land W, Albert E. Capillary deposition of C4d complement fragment and early renal graft loss. Kidney Int. 1993;43(6):1333–8. PubMed PMID: 8315947.
94. Baldwin WM 3rd, Larsen CP, Fairchild RL. Innate immune responses to transplants: a significant variable with cadaver donors. Immunity. 2001;14(4):369–76. PubMed PMID: 11336682.
95. Platt JL, Vercellotti GM, Dalmasso AP, Matas AJ, Bolman RM, Najarian JS, Bach FH. Transplantation of discordant xenografts: a review of progress. Immunol Today. 1990;11(12):450–6; discussion 456–7. PubMed PMID: 2073317.

96. McCurry KR, Kooyman DL, Alvarado CG, Cotterell AH, Martin MJ, Logan JS, Platt JL. Human complement regulatory proteins protect swine-to-primate cardiac xenografts from humoral injury. Nat Med. 1995;1(5):423–7. PubMed PMID: 7585088.
97. Pierre AF, Xavier AM, Liu M, Cassivi SD, Lindsay TF, Marsh HC, Slutsky AS, Keshavjee SH. Effect of complement inhibition with soluble complement receptor 1 on pig allotransplant lung function. Transplantation. 1998;66(6):723–32. PubMed PMID: 9771835.
98. Dreja H, Annenkov A, Chernajovsky Y. Soluble complement receptor 1 (CD35) delivered by retrovirally infected syngeneic cells or by naked DNA injection prevents the progression of collagen-induced arthritis. Arthritis Rheum. 2000;43(8):1698–709. PubMed PMID: 10943859.
99. Heckl-Ostreicher B, Wosnik A, Kirschfink M. Protection of porcine endothelial cells from complement-mediated cytotoxicity by the human complement regulators CD59, C1 inhibitor, and soluble complement receptor type 1. Analysis in a pig-to-human *in vitro* model relevant to hyperacute xenograft rejection. Transplantation. 1996;62(11):1693–6. PubMed PMID: 8970634.
100. Schlegel M, Granja T, Kaiser S, Körner A, Henes J, König K, Straub A, Rosenberger P, Mirakaj V. Inhibition of neogenin dampens hepatic ischemia-reperfusion injury. Crit Care Med. 2014;42(9):e610–9. doi:10.1097/CCM.0000000000000485. PubMed PMID: 25029243.
101. Meimei H, Dejin M, Erzhen C, Minmin S, Songyao J, Jianfang L, Hao C. Inhibiting the toll-like receptor 4 Toll/interleukin-1 receptor domain protects against hepatic warm ischemia and reperfusion injury in mice. Crit Care Med. 2014;42(2):e123–31. doi:10.1097/CCM.0b013e3182a647c1. PubMed PMID: 24231763.
102. Massberg S, Messmer K. The nature of ischemia/reperfusion injury. Transpl Proc. 1998;30:4217–23.
103. Anaya-Prado R, Toledo-Pereyra LH, Lentsch AB, Ward PA. Ischemia/reperfusion injury. J Surg Res. 2002;105:248–58.
104. Gute DC, Ishida T, Yarimizu K, Korthuis RJ. Inflammatory responses to ischemia and reperfusion in skeletal muscle. Mol Cell Biochem. 1998;179:169–87.
105. Bilbao G, et al. Reduction of ischemia-reperfusion injury of the liver by *in vivo* adenovirus-mediated gene transfer of the antiapoptotic Bcl-2 gene. Ann Surg. 1999;230:185–93.
106. Zheng L, et al. Cytoprotection of human umbilical vein endothelial cells against apoptosis and CTL-mediated lysis provided by caspase-resistant Bcl-2 without alterations in growth or activation responses. J Immunol. 2000;164:4665–71.
107. Hancock WW, Buelow R, Sayegh MH, Turka LA. Antibody-induced transplant arteriosclerosis is prevented by graft expression of anti-oxidant and anti-apoptotic genes. Nat Med. 1998;4:1392–6.
108. Amersi F, et al. Upregulation of heme oxygenase-1 protects genetically fat Zucker rat livers from ischemia/reperfusion injury. J Clin Invest. 1999;104:1631–9.
109. Lehmann TG, et al. Gene delivery of Cu/Zn-superoxide dismutase improves graft function after transplantation of fatty livers in the rat. Hepatology. 2000;32:1255–64.
110. Abunasra HJ, Smolenski RT, Yap J, Sheppard M, O'Brien T, Yacoub MH. Multigene adenoviral therapy for the attenuation of ischemia-reperfusion injury after preservation for cardiac transplantation. J Thorac Cardiovasc Surg. 2003;125:998–1006.
111. Feeley BT, et al. Optimization of *ex vivo* pressure mediated delivery of antisense oligodeoxynucleotides to ICAM-1 reduces reperfusion injury in rat cardiac allografts. Transplantation. 2000;69:1067–74.
112. Gourishankar S, Halloran PF. Late deterioration of organ transplants: a problem in injury and homeostasis. Curr Opin Immunol. 2002;14:576–83.
113. Fischer S, et al. *In vivo* transtracheal adenovirus-mediated transfer of human interleukin-10 gene to donor lungs ameliorates ischemia-reperfusion injury and improves early post-transplant graft function in the rat. Hum Gene Ther. 2001;12:1513–26.
114. Ke B, et al. Interleukin-13 gene transfer protects rat livers from antigen-independent injury induced by ischemia and reperfusion. Transplantation. 2003;75:1118–23.

115. Suda T, D'Ovidio F, Daddi N, Ritter JH, Mohanakumar T, Patterson GA. Recipient intramuscular gene transfer of active transforming growth factor-beta1 attenuates acute lung rejection. Ann Thorac Surg. 2001;71:1651–6.
116. Ghali S, Bhatt KA, Dempsey MP, Jones DM, Singh S, Aarabi S, Butler PE, Gallo RL, Gurtner GC. Treating chronic wound infections with genetically modified free flaps. Plast Reconstr Surg. 2009;123(4):1157–68. doi:10.1097/PRS.0b013e31819f25a4. Erratum in: Plast Reconstr Surg. 2009 Aug124(2):672. Arabi, Shahram [corrected to Aarabi, Shahram]. Plast Reconstr Surg. 2009;123(5):1640. PubMed PMID: 19337084.
117. Michaels J 5th, Levine JP, Hazen A, Ceradini DJ, Galiano RD, Soltanian H, Gurtner GC. Biologic brachytherapy: *ex vivo* transduction of microvascular beds for efficient, targeted gene therapy. Plast Reconstr Surg. 2006;118(1):54–65; discussion 66–8. PubMed PMID: 16816674.
118. Michaels J 5th, Dobryansky M, Galiano RD, Ceradini DJ, Bonillas R, Jones D, Seiser N, Levine JP, Gurtner GC. *Ex vivo* transduction of microvascular free flaps for localized peptide delivery. Ann Plast Surg. 2004;52(6):581–4. PubMed PMID: 15166989.
119. Soares, MA; Low, YC; Henderson, RC; Chang, JB; Hwang, LH; Ezeamuzie, OC; Ham, MJ; Saadeh, PB; Ceradini, DJ Now you see it, Now you don't: MHC1 knockdown promotes allograft tolerance by evading immune surveillance. Plast Reconstr Surg. 2012;130:81. doi:10.1097/01.prs.0000416195.06112.4c.

Chapter 19
Experimental Models and Clinical Tools to Assess Nerve Regeneration and Functional Outcomes

Sami H. Tuffaha, Justin M. Broyles and Jaimie T. Shores

Introduction

Vascularized composite tissue allotransplantation (VCA) is different from visceral transplantation in that graft function is dependent on adequate peripheral nerve regeneration. Following transplantation, the recipient's peripheral nerve axons must regenerate into the graft so as to innervate the transplanted muscle and skin. This process allows the recipient to establish motor control and receive sensory input from the graft. Without adequate innervation, a graft remains inanimate and insensate and provides little if any benefit to the recipient. Over time, lack of innervation will result in progressive, permanent atrophy within the graft. The importance of adequate graft innervation applies to all types of VCA; in upper extremity transplantation, meaningful hand function is dependent on the recipient's axons reaching the intrinsic muscles and skin of the transplanted hand; in facial transplantation, graft innervation is necessary for everything from facial expression to preventing drooling.

Because so much of the benefit derived from VCA is derived from peripheral nerve regeneration, systematically monitoring this process clinically is of utmost importance, particularly with regards to determining prognosis. Furthermore, greater attention must be placed on establishing new experimental treatment modalities that can be used to enhance the process of nerve regeneration and graft innervation so as to maximize the benefit and viability of VCA.

J. T. Shores (✉)
Johns Hopkins Outpatient Center, 601 N. Caroline Street,
Suite 8161, Baltimore, MD 21287, USA
e-mail: jshores3@jhmi.edu

S. H. Tuffaha · J. M. Broyles
Department of Plastic and Reconstructive Surgery, The Johns Hopkins
University School of Medicine, Baltimore, MD, USA

G. Brandacher (ed.), *The Science of Reconstructive Transplantation*,
Stem Cell Biology and Regenerative Medicine, DOI 10.1007/978-1-4939-2071-6_19

Experimental Models to Assess Nerve Regeneration in VCA

Available strategies to improve peripheral nerve regeneration in the setting of VCA are limited. Although the commonly used immunosuppressive drug tacrolimus is known to have secondary positive effects on axonal regeneration [1–4], therapeutic strategies aimed specifically at enhancing peripheral nerve regeneration and functional outcomes are lacking. Functional outcomes following hand transplantation have been encouraging [5, 6] and superior to those achievable with currently available prostheses [7], but there is still much room for improvement. As such, more attention must be placed on developing new strategies to enhance peripheral nerve regeneration and graft innervation. To achieve this goal, a number of experimental models and assessment tools are available to develop and test new therapeutic options prior to clinical translation

The most widely used animal model to assess peripheral nerve regeneration and functional outcomes in the setting of VCA is the rat orthotopic hind-limb transplant model [8, 9]. The model involves vascularized transplantation of the hind limb at the level of the mid-femur. After fixating the femur with an intramedullary rod, the femoral artery and vein of the graft are microsurgically anastomosed to the recipient's femoral vessels to perfuse the graft. The utility of the model as a tool for assessing peripheral nerve regeneration and graft innervation comes from the ability to neurotize the transplanted limb. This is normally achieved by approximating the recipient and donor sciatic nerve stumps primarily with an epineurial suture. The femoral nerve can be utilized as well, but is typically left in discontinuity. The location of sciatic nerve approximation is typically 1–2 cm proximal to the trifurcation of the sciatic nerve into the tibial, peroneal, and sural nerves (Fig. 19.1). This repair site allows for reinnervation of all muscles distal to the knee, including those responsible for ankle plantar flexion and dorsiflexion, as well as the intrinsic muscles of the foot.

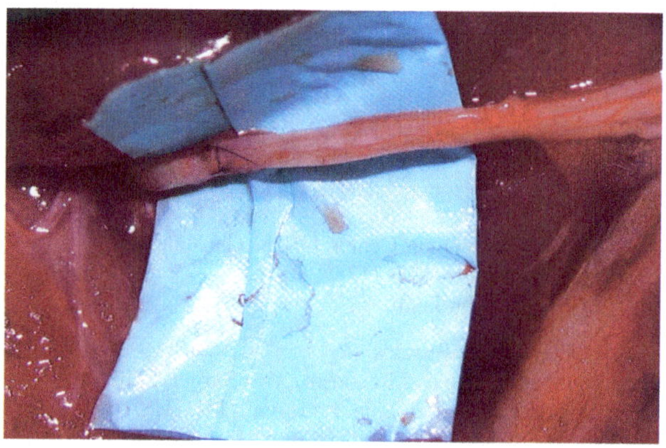

Fig. 19.1 Rat orthotopic hind-limb transplant model for peripheral nerve regeneration. The sciatic nerve is repaired with 10–0 epineurial suture 1 cm proximal to the trifurcation into the tibial, peroneal, and sural nerves

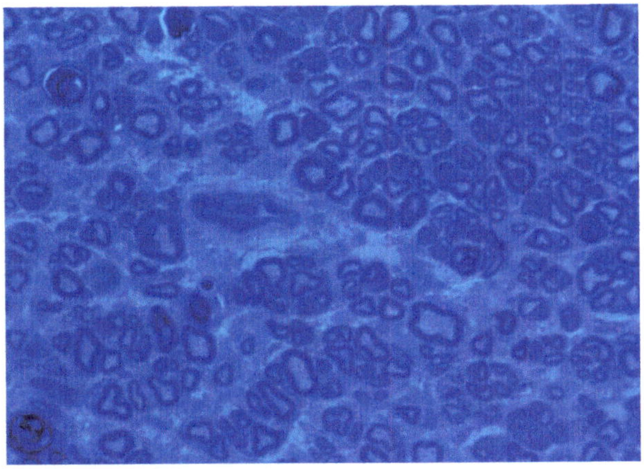

Fig. 19.2 Nerve cross-sectional histomorphometry demonstrating concentration of myelinated axons allowing for objective quantification

The rat orthotopic hind-limb transplant model with sciatic nerve approximation provides the opportunity to utilize a number of assessment tools for the measurement of peripheral nerve regeneration, muscle innervation, and functional outcomes. Nerve histomorphometry allows for direct measurement of axonal regeneration beyond the repair site [10]. At the time of sacrifice, the sciatic nerve harvested and fixed, typically in glutaraldehyde. Ultrafine cross sections taken distal to the approximation site are then stained for myelin, allowing for visualization and quantification of myelinated axons. Measurement parameters include total axon number, width, and density, among others. Results from nerve histomorphometry taken in isolation can be misleading, as it is impossible to differentiate meaningful axonal regeneration resulting in end-organ innervation from axonal sprouting which is of little benefit [11]. For this reason, nerve histomorphometric results should be considered in the context of other data related to end-organ reinnervation and functional outcomes (Fig. 19.2).

Retrograde labeling techniques allow for measurement of the number of cell bodies within the recipient's spinal cord that are contributing to axonal regeneration into the graft [12, 13]. This technique involves severing the sciatic nerve distal to the approximation site and applying fluorescent tracer to the proximal nerve end. Over the course of the next 2–3 days, any viable axons that have regenerated into the graft will transport the tracer in a retrograde fashion into their cell bodies within the spinal cord ventral horn or dorsal root, for motor or sensory axons, respectively. At the time of sacrifice, the fixed spinal cord can then be sectioned and imaged for quantification of the number of fluorescing cell bodies (Fig. 19.3).

Direct visualization and quantification of neuromuscular junctions can also be performed as a measure of muscle reinnervation [14]. To do this, motor end plates are stained with α-bungarotoxin and axonal neurofilaments stained with synaptophysin. Alternatively, transgenic rats that express axonal green fluorescent protein (GFP) can be used for this purpose. Following staining, the number of neuromuscular junctions, indicated by overlapping staining of neurofilament and motor end plate, can be counted (Fig. 19.4).

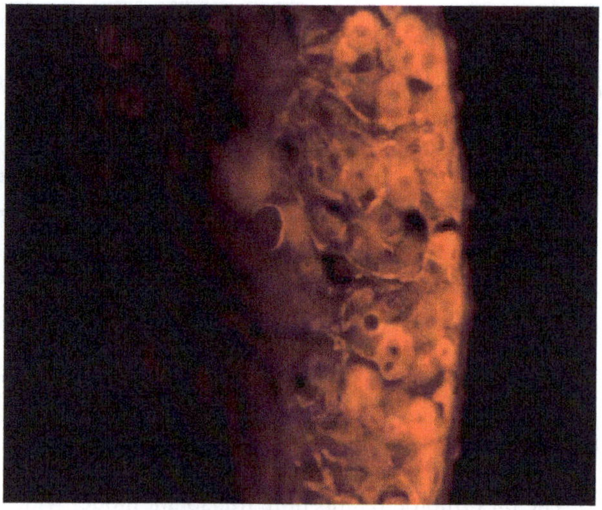

Fig. 19.3 Example of retrograde labeling technique demonstrating cell bodies in the ventral root of the spinal cord which exhibit fluorescently radiolabeled substrate administered at the site of repair

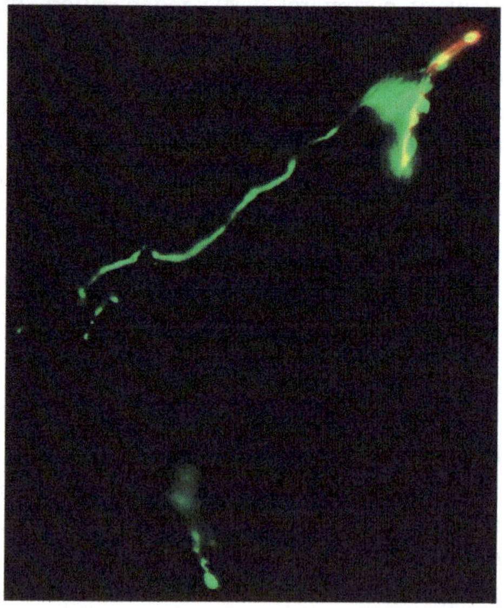

Fig. 19.4 Neuromuscular junction staining demonstrating reinnervation at the motor end plate (stained in red with alpha-bungaratoxin) by terminal axons (stained in green with synaptophysin)

Perhaps the most important outcome measures related to peripheral nerve regeneration are those that measure the ultimate functional outcomes resulting from the end-organ reinnervation. A number of functional assessments are behavioral in nature, in that they require rats to perform a specific activity. The sciatic functional index (SFI) is a measure of distal muscle innervation [15, 16]. It is based on the premise that the progressive reinnervation of ankle plantar flexors and the intrinsic muscles of the foot will gradually affect the footprint left during walking in a predictable manner. Specifically, the footprint length should progressively shorten with improved plantar flexion, and toe spread should increase with reinnervation of the intrinsic muscles of the foot. These measurements are used to calculate the SFI, which is reported as a ratio of the operated to the unoperated side. This technique is limited by a number of potential complicating factors. Relying or a rat to walk in a consistent manner without dragging the injured leg can be difficult. Furthermore, foot contractures occur frequently as the extrinsic muscles of the foot become reinnervated, and these contractures preclude the use of the SFI because the toes remain fully flexed, obscuring the toe spread that would otherwise be observed with distal reinnervation [17]. Lastly, this technique typically involves applying paint, or another type of dye, to the feet of the rat and can therefore be complicated by smudging and smearing of the paint.

More recently, a number of dynamic gait-analysis techniques have been developed that allow for measurement of a number of gait-related parameters. Some of these methods involve video recording the rat's gait with sensors in place on each hind-limb joint. The location of each sensor can then be measured in relation to the other sensors over time. From these measurements, a number of derivative calculations can be made regarding angular velocity and displacement at each joint at different phases of the gait cycle [18]. Calculation of the time the foot is in contact with floor throughout the gait cycle has also been shown to correlate well with reinnervation [19, 20]. The application of this technology can be difficult due to the behavioral variables involved. Variability in gait can occur for a number of reasons unrelated to nerve regeneration and achieving consistent and reliable results from gait analysis is challenging, often requiring behavioral training. However, some have been able to achieve reliable and consistent data using these techniques [21].

Electrophysiologic neuromuscular testing provides another useful option for assessing functional outcomes in animal models [22, 23]. As opposed to other modalities used to assess a function, electrophysiologic studies do not rely on behavioral input from the animal and benefit from the lack of behavioral variability. Nerve conduction studies (NCS) can be used to measure compound nerve action potentials (CNAPs) [24]. CNAP latency is a measure of the conduction velocity generated by the fasted conducting fibers, and CNAP amplitude is a measure of the number of conducting axons. Electromyographic measurement of compound muscle action potentials (CMAPs) can also be obtained to assess the relative number of functional neuromuscular junctions that have been formed. The results, however, must be interpreted with caution as motor unit size can increase following injury resulting in larger CMAP readings. This phenomenon only masks differences in the number of reinnervated motor units up to ~25% [25, 26]. Muscle contractile force can also be performed to assess the number of functional neuromuscular units present. This can be done by dis-inserting the gastrocnemius muscle and attaching it to a force transducer. The tibial nerve is then stimulated and the generated twitch and tetanic force are recorded [27].

Assessment of sensory recovery resulting from reinnervation of cutaneous sensory organs is performed by measuring the behavioral response to mechanical and thermal cutaneous stimuli [28, 29]. Following sensory reinnervation, rats demonstrate an involuntary withdrawal response to noxious stimuli. The withdrawal response can be recorded in a binary fashion to set stimuli, or alternatively, the threshold of stimulus required to elicit withdrawal can be measured. In the setting of VCA, complete sensory integration of the graft relies not only on peripheral innervation of target sensory organs but also on cortical plasticity. The process of cortical sensory reintegration can be measured utilizing a rat hemifacial transplant model [30]. This involves stimulating individual whiskers following transplantation and assessing for neural signals in the corresponding areas of the somatosensory cortex.

The effects of the acute rejection on peripheral nerve regeneration are still poorly understood in the setting of VCA. Utilizing a full major histocompatibility complex (MHC) mismatch between donor and recipient rats (i.e., Brown Norway to Lewis) provides an opportunity to assess the alloimmune response seen clinically in VCA. With withdrawal and reintroduction of immunosuppression, rejection episodes can be introduced in a controlled manner and the resulting effects on peripheral nerve regeneration and functional outcomes can be assessed [31].

Clinical Evaluation of Nerve Regeneration in VCA

Peripheral nerve regeneration continues to be one of the most difficult problems faced in the regenerative medicine today [32, 33]. Traditional axioms have set the bar low for functional expectations after severe nerve injuries and/or reconstructions, though more recent clinical experience and newer techniques may begin to challenge this notion.

Clinical evaluation of nerve regeneration is not as straightforward as one may think due to the multiple problems that nonfunctioning nerves may cause. A nonfunctional nerve does not only cause problems from its *absence* of function but may impart its own impairment from pain, sympathetic, and/or sudomotor dysfunction. The clinical evaluation of nerve regeneration should combine functional testing, patient outcome reported testing, and more objective neurophysiologic and/or imaging-based testing (Table 19.1). Additional information that is helpful in understanding and predicting clinical outcomes in nerve regeneration are the underlying principles of the speed of nerve regeneration, expected time course for possible motor reinnervation, and expected time course for possible sensory reinnervation.

On a fundamental level, we estimate the speed of nerve regeneration to occur at approximately 1 mm/day in healthy young adults [34, 35]. This rate has not been objectively or prospectively studied and can vary greatly. Nature of the injury, location, wound bed, age of the patient, and technique used for repair or reconstruction have all been demonstrated to affect velocity or nerve regeneration [36, 37]. This process does not begin until several weeks after nerve injury, when inflammatory and axonal regenerative processes occur to result in endoneural tube formation that allows neurite growth cone progression. This process may take 3–6 weeks to occur.

Clinical evaluation of nerve regeneration begins with thorough history, physical examination, and evaluation of all studies already performed to understand or diagnose the nerve injury in question and interpret the findings within the context of and

Table 19.1 Clinical evaluation of nerve regeneration

Category	Test
Functional testing	*Sensory*: Semmes–Weinstein monofilaments, static and moving two-point discrimination, vibration, temperature, British MRC score of sensory recovery, sharp/dull discrimination, texture and shape discrimination, etc *Motor*: British MRC score of motor recovery, specific strength measurements per body part affected (e.g.,. pinch and grip strength for forearm/hand) *Sudomotor:* quantitative sudomotor axon reflex test (QSART), thermoregulatory sweat test
Pain	Visual analogue scale (VAS), numerical rating for pain, McGill pain Questionnaire, cold intolerance questionnaires
Patient reported/outcomes testing	Self-reported disability and pain measures (e.g., disabilities of the hand, arm, and shoulder (DASH) for upper extremity nerve injuries, Michigan Hand Outcomes Questionnaire, etc)
Objective testing	*Neurophysiologic:* Nerve conduction studies (NCS), electromyography (EMG) *Imaging:* Myelography, magnetic resonance neurography (MRN) with and without diffusion tension imaging (DTI), high-resolution ultrasound (HRUS)

MRC Medical Research Council

the time course from injury. In general, motor nerve fibers must reinnervate motor end plates within 12–18 months before an irreversible loss of the motor end plate occurs preventing further muscle fiber reinnervation [26, 38]. Sensory reinnervation was long though to not involve a time-dependent mechanism, but has now been recognized to have diminishing ability for sensory mechanoreceptor reinnervation after approximately 5 years. Other factors that help establish the "context" of the injury include the suitability of the wound bed, history of radiation, infection, the suitability of distal reinnervation targets, patient age, comorbidities, and the mechanism of injury to the nerve.

Analysis of the ongoing regeneration process typically consists of observation or return of function with application of the measures listed in Table 19.1. In addition, utilization of the advancing "Tinel's" sign has long been considered a standard measure of nerve regeneration progression [39–41]. The "Tinel" sign is simply tapping over the regenerating nerve and the point at which maximal paresthesias can be elicited into the distribution of the nerve being tested is considered that length to which axons have regenerated (Fig. 19.5). However, focal pain is frequently misinterpreted for paresthesias, and many patients exhibit paresthesias over a long distance, not just a focal location which can lead to an erroneous misinterpretation of regeneration.

Sensory recovery measurement can be superficially tested by simply ascertaining the presence of sensation to a light touch or sharp, painful stimulation. However, more standardized or objective measurements serve to give more meaningful and reproducible data. Objective sensory testing can evaluate sensory *threshold* or sensory *density*. Threshold testing typically utilizes measures such as the Semmes–

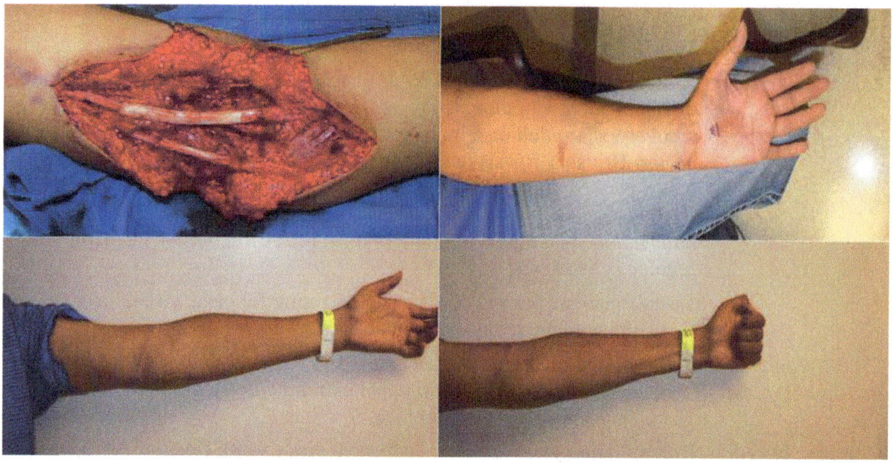

Fig. 19.5 Advancing Tinel's sign of regenerating axons 12 months after a nerve graft repair of a proximal ulnar and median nerve injury. (*Above, Left*) Repair of proximal median and ulnar nerve with nerve allografts after resection of neuroma-in-continuity approximately 12 months after a traumatic laceration. (*Above, Right*) Advancing Tinel's sign marked in proximal palm demonstrating reinnervation of median and ulnar nerves. (*Below*) Functional return of median and ulnar nerves demonstrating ability to form composite fist

Weinstein monofilaments which measure the threshold of perceptible pressure exerted against the skin of an innervated area. Standard limits for what is considered normal perception of light touch, diminished normal perception but protective sensation, diminished protective sensation, loss of protective sensation, and presence/absence of deep-pressure perception have already been established [42].

Sensory density discrimination can be performed with static and moving two-point discrimination. Two-point discrimination is the measurement of the minimal distance at which two points can be discriminated as separate sites simultaneously. This can be done gently pushing the measurement device (such as calipers) onto the skin until the skin blanches (static two-point discrimination, S2PD) or by moving the calipers of the skin surface at a set distance (moving two-point discrimination M2PD). The British Medical Research Council (MRC) has further defined overall "functional" grades of sensory recovery that have been since modified by Mackinnon and Dellon [43–45] (Table 19.2).

Further, less frequently used sensory testing modalities that can be employed to test vibration perception, thickness, texture, shape recognition; sharp and dull discrimination; and temperature perception (warm and cold objects), which in the authors' clinical experience, may be the first noticeable sign sensory recovery. Patients frequently report noticing the cold temperature of an object they are holding or touching as their first conscious return of sensory perception.

Recovery of motor function has also been defined by the British MRC into a graded and reproducibly testable classification. This system uses both the examiners observation of muscle contraction as well as defined movements in and out of

Table 19.2 British medical research council grades of functional sensory recovery

Grade	Sensation present
S0	Absence of sensibility in the isolated nerve region of sensation
S1	Recovery of deep cutaneous pain and tactile sensibility
S1+	Recovery of superficial pain sensibility
S2	Recovery of some degree of superficial cutaneous pain and tactile sensibility
S2+	Recovery of some degree of superficial cutaneous pain and tactile sensibility, but with over response
S3	Return of pain and tactile sensibility with disappearance of over response; S2PD >15 mm, M2PD >7 mm
S3+	Return of sensibility as in S3 with some recovery of two-point discrimination; S2PD: 7–15 mm, M2PD: 4–7 mm
S4	Complete recovery; S2PD: 2–6 mm, M2PD: 2–3 mm

Table 19.3 Modified medical research council muscle strength scale

Grade	British MRC measurement	Modified British MRC measurement	–
–	–	Range of motion	Resistance
M0	No contraction perceptible by examiner	None	No palpable contraction
M1	Flicker or trace perceptible contraction	None	Perceptible contraction by examiner
M2	Full active range of motion *out*-of-the plane of gravity (gravity eliminated)	Reduced	None
M3	Active motion against gravity present	Normal	None
M4	Active movement against additional resistance, but not full strength	Normal	Reduced
M5	Normal strength	Normal	Normal

MRC Medical Research Council

the plane of gravity and against resistance of stratify motor recovery. It has been further modified to incorporate both the visible range of motion of the muscle group as well as the strength or resistance demonstrated (i.e., motion without resistance such as out of plan of gravity, against reduced resistance such as gravity or light external resistance, or against full examiner external resistance such as seen with a normal muscle group; 46; Table 19.3).

More objective testing may be performed using electrophysiologic testing utilizing NCS and electromyography. During these tests, the examiner uses analysis of electrical signals applied between two points of a known distance on a peripheral nerve. The conduction velocity, amplitude, and analysis of its waveform can demonstrate important findings along both sensory and motor pathways. Electromyography further helps clarify the electrophysiologic analysis of the nerve by testing its motor end organ for effect. A denervated muscle will demonstrate characteristic fibrillations and/or spontaneous activity, depending upon the length of time since denervation [47]. In addition, the ability for motor unit potential recruitment either actively by the patient or passively via electrical stimulation, as well as the magnitude and organization of the response, are also very telling. As objective as these findings may seem, there can be great interobserver variability and even factors such as skin surface temperature may alter some results [48]. It is therefore important to conduct testing in a uniform manner every time. Additionally, the largest and most well-myelinated axons will be preferentially tested during NCS testing as the electrical current will follow the path of least resistance. These many reasons may be why carpal tunnel syndrome, which is one of the most common uses of NCS testing, still has a false negative rate approximating 12 % in some studies [49].

More recent objective testing using newer imaging modalities incorporate high-frequency ultrasound and magnetic resonance (MR) neurography. High-frequency ultrasound is a noninvasive modality that can provide *dynamic* imaging. That is, assessment of a nerve in motion, which may be a better indicator of nerve pathology than a static nerve which is not in motion [50–52]. MR neurography on high-field-strength scanners (3 T or greater) has the advantage of demonstrating peripheral nerve tracts easily but also evaluates the surrounding structures which can influence the diagnosis and treatment of some nerve disorders [53].

The nerve components of a transplanted VCA are similar to nerve allografts as they provide an intact nerve architecture. Therefore, clinical testing of regenerative capacity should follow similar guidelines to those of any other nerve repair. However, there are several unique features of nerve regeneration in VCA when compared with primary neural and/or allograft repair. In a VCA, there is no distal nerve stump and no source of host Schwann cells initially in the allograft. Therefore, if the allograft is to regenerate and maintain structure, it will rely on either host cell Schwann cell migration or complete donor Schwann cell survival. The ability for donor Schwann cell survival is dependent on myriad of immunologic factors and immunosuppression must be adequate as to not produce rejection episodes which would significantly impair the Schwann cell population [54]. Therefore, the loss of donor Schwann cells due to multiple, subclinical episodes of allograft rejection poses a significant barrier to long-term functional recovery and every effort must be made to prevent acute rejection episodes. Future strategies to enhance nerve regeneration in VCA should account for the delicate relationships between donor and host Schwann cells and promote strategies to maintain the donor cell population until host cell lines can arrive.

There is a tremendous amount of research being performed on how to minimize immunosuppression in a VCA to mitigate the deleterious side effects of a chronic, long-term immunosuppressive drug regimen. As progress is ongoing with the development of therapeutic tolerance, it should be noted that different drug regimens may differentially affect Tacrolimus's ability to enhance nerve regeneration. Animal studies using therapeutic levels of Tacrolimus with anti-CD40 ligand costimulatory blockade demonstrate that Tacrolimus's ability to stimulate nerve regeneration was eliminated by this combination of drugs. However, the combination of anti-CD40 ligand costimulatory blockade and subtheraputic doses of Tacrolimus maintained the drug's enhancing effects on nerve regeneration [55]. Therefore, the specific drugs and dosages in each immunosuppressive regimen should be taken into account when evaluating the transplanted extremity as they may have differential impacts on nerve regeneration.

Thus, the clinical assessment of nerve regeneration in VCA involves multiple modalities, of which the most important is an in-depth history and physical examination. While objective findings may aid in diagnosis, the physician must frequently rely upon the complaints and physical findings of patients to provide an accurate diagnosis. Furthermore, accurate documentation and communication of these findings are of utmost importance as nerve regeneration is a dynamic, time-dependent process. If detailed, reproducible documentation is not made, one has no way of determining whether or not regeneration is actually occurring.

Conclusions

As we endeavor to treat peripheral nerve injuries and enter an era where VCA is becoming increasingly more common, we are reminded that the ultimate functionality of any limb is completely dependent upon its motor and sensory innervation for its many functions. While clinical assessments are based in traditional history and physical examination with supplementation by more advanced digital testing, we are constantly reminded of our clinical short-comings in regard to both diagnosis and treatment. Greater attention must be placed on establishing new experimental treatment modalities that can be used to enhance the process of nerve regeneration and graft innervation so as to maximize the benefit and viability of VCA.

References

1. Gold BG, Storm-Dickerson T, Austin DR. The immunosuppressant FK506 increases functional recovery and nerve regeneration following peripheral nerve injury. Restor Neurol Neurosci. 1994;6:287–96.
2. Udina E, Rodriguez FJ, Verdu E, et al. FK506 enhances regeneration of axons across long peripheral nerve gaps repaired with collagen guides seeded with allogeneic Schwann cells. Glia. 2004;47:120–9.
3. Doolabh VB, Mackinnon SE. FK506 accelerates functional recovery following nerve grafting in a rat model. Plast Reconstr Surg. 1999;103:1928–36.
4. Moore AM, Borschel GH, Santosa KA, et al. A transgenic rat expressing green fluorescent protein (GFP) in peripheral nerves provides a new hindlimb model for the study of nerve injury and regeneration. J Neurosci Methods 2012;204:19–27.
5. Lanzetta M, Petruzzo P, Margreiter R, et al. The international registry on hand and composite tissue transplantation. Transplantation. 2005;79:1210–4.
6. Lanzetta M, Petruzzo P, Dubernard JM, et al. Second report (1998–2006) of the international registry of hand and composite tissue transplantation. Transpl Immunol. 2007;18:1–6.

7. Ravindra KV, Buell JF, Kaufman CL, et al. Hand transplantation in the United States: experience with 3 patients. Surgery. 2008;144:638–43; discussion 643–34.
8. Sucher R, Oberhuber R, Margreiter C, et al. Orthotopic hind-limb transplantation in rats. J Vis Exp. 2010 Jul 12;(41).
9. Tuffaha S, Quigley M, Ng T, et al. The effect of chondroitinase on nerve regeneration following composite tissue allotransplantation. J Hand Surg. 2011;36:1447–52.
10. Hunter DA, Moradzadeh A, Whitlock EL, et al. Binary imaging analysis for comprehensive quantitative histomorphometry of peripheral nerve. J Neurosci Methods. 2007;166:116–24.
11. Mackinnon SE, Dellon AL, O'Brien JP. Changes in nerve fiber numbers distal to a nerve repair in the rat sciatic nerve model. Muscle Nerve. 1991;14:1116–22.
12. Brushart TM. Motor axons preferentially reinnervate motor pathways. J Neurosci. 1993;13:2730–8.
13. Brushart TM, Mesulam MM. Alteration in connections between muscle and anterior horn motoneurons after peripheral nerve repair. Science. 1980;208:603–5.
14. Magill CK, Tong A, Kawamura D, et al. Reinnervation of the tibialis anterior following sciatic nerve crush injury: a confocal microscopic study in transgenic mice. Exp Neurol. 2007;207:64–74.
15. de Medinaceli L, Freed WJ, Wyatt RJ. An index of the functional condition of rat sciatic nerve based on measurements made from walking tracks. Exp Neurol. 1982;77:634–43.
16. Bain JR, Mackinnon SE, Hunter DA. Functional evaluation of complete sciatic, peroneal, and posterior tibial nerve lesions in the rat. Plast Reconstr Surg. 1989;83:129–38.
17. Dellon AL, Mackinnon SE. Sciatic nerve regeneration in the rat. validity of walking track assessment in the presence of chronic contractures. Microsurgery. 1989;10:220–5.
18. Santos PM, Williams SL, Thomas SS. Neuromuscular evaluation using rat gait analysis. J Neurosci Methods. 1995;61:79–84.
19. Walker JL, Evans JM, Meade P, et al. Gait-stance duration as a measure of injury and recovery in the rat sciatic nerve model. J Neurosci Methods. 1994;52:47–52.
20. Lin FM, Pan YC, Hom C, et al. Ankle stance angle: a functional index for the evaluation of sciatic nerve recovery after complete transection. J Reconstr Microsurg 1996;12:173–7.
21. de Ruiter GC, Spinner RJ, Alaid AO, et al. Two-dimensional digital video ankle motion analysis for assessment of function in the rat sciatic nerve model. J Peripher Nerv Syst 2007;12:216–22.
22. Egeland BM, Urbanchek MG, Peramo A, et al. *In vivo* electrical conductivity across critical nerve gaps using poly(3,4-ethylenedioxythiophene)-coated neural interfaces. Plast Reconstr Surg 2010;126:1865–73.
23. Werdin F, Grussinger H, Jaminet P, et al. An improved electrophysiological method to study peripheral nerve regeneration in rats. J Neurosci Methods. 2009;182:71–77.
24. Navarro X, Udina E. Chapter 6: methods and protocols in peripheral nerve regeneration experimental research: part III-electrophysiological evaluation. Intl Rev Neurobiol. 2009;87:105–26.
25. Fu SY, Gordon T. Contributing factors to poor functional recovery after delayed nerve repair: prolonged axotomy. J Neurosci 1995;15:3876–85.
26. Fu SY, Gordon T. Contributing factors to poor functional recovery after delayed nerve repair: prolonged denervation. J Neurosci 1995;15:3886–95.
27. Li QT, Zhang PX, Yin XF, et al. Functional recovery of denervated skeletal muscle with sensory or mixed nerve protection: a pilot study. PLoS One. 2013;8:e79746.
28. Kemp SW, Webb AA, Dhaliwal S, et al. Dose and duration of nerve growth factor (NGF) administration determine the extent of behavioral recovery following peripheral nerve injury in the rat. Exp Neurol. 2011;229:460–70.
29. Koka R, Hadlock TA. Quantification of functional recovery following rat sciatic nerve transection. Exp Neurol. 2001;168:192–5.
30. Washington KM, Solari MG, Sacks JM, et al. A model for functional recovery and cortical reintegration after hemifacial composite tissue allotransplantation. Plast Reconstr Surg. 2009;123:26S–33S.

31. Unadkat JV, Bourbeau D, Afrooz PN, et al. Functional outcomes following multiple acute rejections in experimental vascularized composite allotransplantation. Plast Reconstr Surg 2013;131:720e–30e.
32. Lundborg G. A 25-year perspective of peripheral nerve surgery: evolving neuroscientific concepts and clinical significance. J Hand Surg Am. 2000; 25:391–414.
33. Sumarto A. Peripheral nerve injuries: an international survey of current treatments and future perspectives. J Reconstr Microsurg. 2009;25:339–44.
34. Sunderland S. Nerve and nerve injuries. 2nd edition. Edinburgh: Churchill Livingstone; 1978.
35. Wujek JR, Lasek RJ. Correlation of axonal regeneration and slow component B in two branches of a single axon. J Neurosci. 1983;2:243–51.
36. Sunderland S. Factors influencing the course of regeneration and quality of recovery after nerve suture. Brain. 1952;75:19–54.
37. Brushart TM. Nerve repair. New York, NY: Oxford University Press; 2011.
38. Gordon T, Tyreman N, Raji MA. The basis for diminished functional recovery after delayed peripheral nerve repair. J Neurosci. 2011;31:5325–34.
39. Szabo RM, Slater RR, Farver TB, et al. The value of diagnostic testing in carpal tunnel syndrome. J Hand Surg Am. 1999 Jul;24(4):704–14.
40. Shores JT, Lee WP. An evidence-based approach to carpal tunnel syndrome. Plast Reconstr Surg. 2010 Dec;126(6):2196–204.
41. Keith MW, Masear V, Chung K, et al. Diagnosis of carpal tunnel syndrome. J Am Acad Orthop Surg. 2009;17:389–96.
42. Jerosch-Herold C. Assessment of sensibility after nerve injury and repair: a systematic review of evidence for validity, reliability and responsiveness of tests. J Hand Surg Br. 2005 Jun;30(3):252–64.
43. Rosen B, Lundborg G. A model instrument for the documentation of outcome after nerve repair. J Hand Surg Am. 2000;25:535–43.
44. Novak CB, Kelly L, Mackinnon SE. Sensory recovery after median nerve grafting. J Hand Surg Am. 1992;17:59–68.
45. Leechavengvongs S, Ngamlamiat K, Malungpaishrope K, et al. End-to-side radial sensory to median nerve transfer to restore sen- sation and relieve pain in C5 and C6 nerve root avulsion. J Hand Surg Am. 2011;36:209–15.
46. Brandsma JW, Schreuders TA, Birke JA, et al. Manual muscle strength testing: intraobserver and interobserver reliabilities for the intrinsic muscles of the hand. J Hand Ther. 1995;8:185–90.
47. Concannon MJ, Petroski GF, Puckett CL. The predictive value of electrodiagnostic studies in carpal tunnel syndrome. Plast Reconstr Surg. 1997;100:1452–8.
48. Chaudhry V, Cornblath DR, Mellitis ED, et al. Inter- and intra-examiner reliability of nerve conduction measurements in normal subjects. Ann Neurol. 1991;30:841–3.
49. Strickland JW, Gozani SN. Accuracy of in-office nerve conduction studies for median neuropathy: a meta-analysis. J Hand Surg Am. 2011 Jan;36(1):52–60.
50. Kara M, Ozcakar L, De Muynck M, et al. Musculoskeletal ultrasound for peripheral nerve lesions. Eur J Phys Rehabil Med. 2012;48:665–74; quiz 708.
51. Boon AJ, Smith J, Harper CM. Ultrasound applications in electrodiagnosis. Phys Med Rehab 2012;4: 37–49.
52. Koenig RW, Pedro MT, Heinen CP, et al. High-resolution ultrasonography in evaluating peripheral nerve entrapment and trauma. Neurosurg Focus. 2009;26:E13.
53. Chalian M, Behzadi AH, Williams EH, et al. High-resolution magnetic resonance neurography in upper extremity neuropathy. Neuroimaging Clin N Am. 2014;24:109–25.
54. Moore AM, Ray WZ, Chenard KE, et al. Nerve allotransplantation as it pertains to composite tissue transplantation. Hand (NY). 2009;4:239–44.
55. Brenner MJ, Mackinnon SE, Rickman SR, et al. FK506 and anti-CD40 ligand in peripheral nerve allotransplantation. Restor Neurol Neurosci. 2005;23:237–49.

Chapter 20
Mesenchymal and Adipose Stem Cell Strategies for Peripheral Nerve Regeneration

Riccardo Schweizer, Sudheer K. Ravuri, Jan A. Plock, Kacey G. Marra and Vijay S. Gorantla

Introduction

After peripheral nerve injury (PNI) motor, sensory as well as autonomic functionality of the respective nerve are impaired or lost. The incidence of PNI is given with 3–5 % in the population [1, 2] and is mainly a consequence of motor vehicle accidents, sharp lacerations, as well as other traumas and gunshot wounds. The consequences of PNI are important, being related to high treatment costs, loss of function, associated morbidities and having a serious impact on both quality of life and employment status [1].

After injury, the distal axon degenerates and an inflammation cascade is activated, known as Wallerian degeneration [3, 4]. Schwann cells (SC) located distal from the injury site and thus denervated are activated to secrete neurotropic and neurotrophic growth factors, creating a microenvironment that eventually promotes peripheral

Schweizer and Sudheer contributed equally in this chapter.

V. S. Gorantla (✉) · K. G. Marra
Department of Plastic Surgery, University of Pittsburgh Medical Center, Suite 6B, Scaife Hall, 3550 Terrace Street, 15261 Pittsburgh, PA, USA
e-mail: gorantlavs@upmc.edu

R. Schweizer
Vascularized Composite Allotransplantation (VCA) Laboratory, Department of Plastic Surgery, University of Pittsburgh, 200 Lothrop St, 15213 Pittsburgh, PA, USA

S. K. Ravuri
Adipose Stem Cell Center, Department of Plastic Surgery, University of Pittsburgh, 200 Lothrop St, 15213 Pittsburgh, PA, USA

J. A. Plock
Department of Plastic and Hand Surgery, University Hospital Zurich, Raemistrasse 100CH-8091, Zurich, Switzerland

G. Brandacher (ed.), *The Science of Reconstructive Transplantation*,
Stem Cell Biology and Regenerative Medicine, DOI 10.1007/978-1-4939-2071-6_20

nerve regeneration (PNR). Together with inflammatory cells like macrophages, SC phagocytize debris, e.g. remaining myelin and degenerating axon, which are not useful anymore and may obstruct new axonal growth [4]. In this phase, SC clear the path, allowing the damaged axon to regrow from the proximal end to the distal one. If working properly, the Wallerian cascade is a prerequisite which finally may lead to nerve recovery. However, if this process is disturbed, the nerve will probably fail to regenerate. Moreover, unfortunately, SC often are not able to sustain the positive neurotrophic milieu long enough if the nerve gap exceeds a certain critical length, resulting in decreased or absent functional recovery as frequently observed in humans [4].

One of the most important factors predicting the functional recovery after PNI is the time for the axon to regrow from the injury site to the target organ (e.g. muscle), and in humans, that is one of the main reasons for the poor functional outcomes even though the peripheral nerve system has the ability to regenerate [5]. When denervated and not functional anymore, the target muscles rapidly begin to atrophy; thus, even after successful nerve regeneration, there is still an important functional impairment if the nerve is not recovering timely.

Novel therapeutic options are needed to improve nerve regeneration and accelerate the functional recovery. This is of great interest because it would enable the patients to regain quality of life and restore their employable status [1]. Successful treatment for such repair should combine strategies of neuro-protection (SC creating a protective and pro-regenerative microenvironment), be able to yield axonal regeneration (eliminating inhibitory factors and reducing scar formation) and eventually increase the final neuro-rehabilitation and functional recovery [6, 7].

Although the peripheral nervous system has the ability to regenerate, often after PNI a nerve gap remains, which hinders the axon proximal to the injury site from regrowing towards the distal nerve stump and find the previous motoric and sensory innervation site. The state-of-the-art therapy after PNI with a remaining defect is still the implantation of autologous nerve grafts for bridging the gap, which involves harvesting of a graft on some other location on the patient's body and correlates to morbidity and loss of function at the specific harvest site and has contrasting outcomes. To avoid these drawbacks in the last decades, many studies have been searching for alternatives using different kinds of artificial or biological grafts and conduits coupled to growth factors and/or cell-based therapeutic strategies, including embedding of SC and multipotent mesenchymal stromal cells (MSC) of various types [8–20] (Fig. 20.1).

However, the use of autologous cultured SC for the treatment of acute injuries may be impractical, owing to the technical difficulties, and time required, in harvesting and expanding such cells. In this scenario, the ideal 'transplantable cells' should be easily accessible, capable of rapid expansion in culture, immunologically inert, capable of long-term survival and integration in the host tissue and amenable to stable transfection and expression of exogenous genes [24]. In a search for such cells, attention has been drawn to the possible use of stem cells. Stem cells are

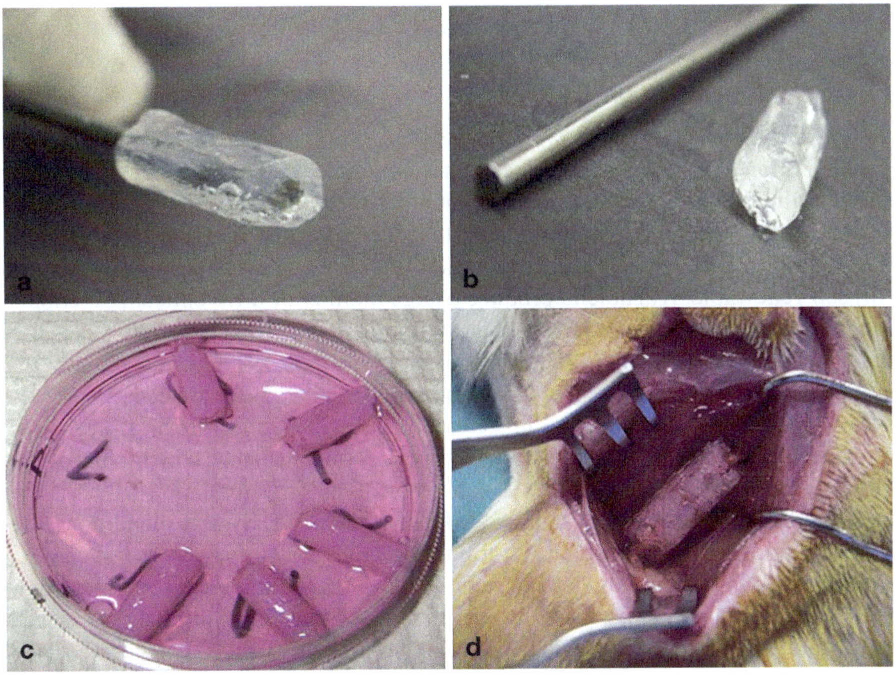

Fig. 20.1 Preparation of fibrin conduits. **a** The fibrin conduit (Tisseel) with stainless steel wire inserted in the lumen as a spacer while the conduit is store before implantation. **b** The fibrin conduit shows an intact lumen after removal of the stainless steel wire. **c** The fibrin conduits are stored in DMEM medium before being seeded with regenerative cells. **d** A fibrin conduit implanted *in vivo* to repair a rat sciatic nerve injury. *DMEM* Dulbecco modified Eagle's medium

distinguished from other cell types by two important characteristics. First, they are unspecialized cells capable of renewing themselves through cell division, sometimes after long periods of inactivity. Second, under certain physiologic or experimental conditions, they can be induced to become tissue- or organ-specific cells with special functions.

In this context, MSC seem to be a promising option for cellular therapy, as they are relatively easy to harvest, higher aliquots available and grow faster if culture expanded. MSC have originally been harvested from bone marrow mesenchymal stem cells (BM-MSC), and so for many years, yet other MSC sources have been found such as placenta, skin, thymus, umbilical cord, amniotic fluid and Wharton's jelly, including the abundant and easy accessible subcutaneous fat tissue, the so-called adipose tissue-derived stem cells (ADSC) [25]. ADSC are of great interest because of recent studies showing they might have similar phenotypic, genetic and differentiation characteristics to BM-MSC but are obviously easier to harvest and more abundantly available [26]. Therefore, herein we focus on the application of both BM-MSC and ADSC to promote nerve regeneration after PNI.

Cellular Therapy for PNR

Basic Biology of Stem Cells

The definition of a stem cell describes it as a clonogenic cell capable of self-renewal and multilineage differentiation [27]. A number of criteria have been proposed to help identify a stem cell. These dictate that the cell must be (i) undifferentiated (i.e. lacking a tissue-specific differentiation marker), (ii) capable of proliferation, (iii) self-renewable, (iv) able to produce a large number of differentiated functional progeny and (v) able to regenerate tissue after injury [28]. The proportion and activity of stem cells within adult tissue depends on tissue type, i.e. depending on whether the tissue is renewing, e.g. intestinal epithelium, or static, e.g. the central nervous system [29]. With such diversity in stem cell activity within different tissues, it is obviously important to understand the regulatory mechanisms which control the two key properties of stem cells, i.e. self-renewal and differentiation which is the most fundamental property of stem cells. Briefly, under steady-state conditions, the stem cell pool remains constant, but following injury or during disease, it can expand rapidly. The haematopoietic system displays regenerative capacity more dramatically than any other tissue, as huge numbers of cells are continuously produced throughout life, a property which would be impossible if there was a fixed, and hence depletable, number of progenitors of any particular haematopoietic subset. Cells mobilized from bone marrow, peripheral blood and cord blood are capable of re-establishing the entire haematopoietic compartment following complete ablation, as is seen following cytotoxic therapy for haematological malignancies; indeed, it is this property which has been exploited to successfully treat such conditions [30]. Embryonic stem cells can be propagated in culture [31]; however, obtaining such cells presents practical, ethical and legislative difficulties [32, 33]. The identification, isolation and expansion of truly multipotent adult-derived stem cells remain problematic. Toma et al. [34] fairly recently described the isolation and expansion of multipotent stem cells from skin; however, it was not possible to determine the origin of the cells with absolute certainty. The problem of isolating clonal stem cells from adult tissues is probably the greatest obstacle faced in this field [35, 36].

There are a number of paradigms to describe regulatory mechanisms in stem cell division and self-renewal. The mechanistic of stem cell expansion and division have been described as either ‘asymmetric’ or ‘symmetric’. In asymmetric division, each stem cell divides into a daughter stem cell and a differentiated cell, and thus, there is no change in the total number of stem cells in the stem cell pool. A number of genes have been implicated in the control of asymmetric cell kinetics, such as the p53 tumour-suppressor gene [37] and the p21 cyclin-dependent kinase inhibitor gene [38]. In symmetric division, half the time stem cells divide into either two daughter stem cells or two daughter differentiated progeny, leaving the steady-state number of stem cells constant, similar to the situation in asymmetric division [39].

A third model involves the role of the microenvironment. Depending on environmental cues, such as injury, a stem cell may produce two daughter cells, which will either remain as stem cells or differentiate, depending on the environment.

The control of self-renewal and differentiation is most probably influenced by extrinsic factors, i.e. the environment or niche, and intrinsic cellular factors. The environment influences the biochemical and morphological properties of stem cells. Stem cells adjust their properties according to their surroundings and select specific lineages according to the cues they receive from their niche [40]. Extrinsic factors can be divided into cell–cell interactions, locally secreted factors and the extracellular matrix. Cell–cell interactions mediated by integral membrane proteins have been shown to influence the maintenance of self-renewing potential and differentiation. For example, Notch-related receptors activated by ligands only found on neighbouring cells maintain stem cells in an undifferentiated and self-renewing state, and upon removal of the ligand, cells begin to differentiate [41, 42]. Locally secreted factors or growth factors play an important role in regulating stem cell proliferation within the niche. Growth factors, such as epidermal growth factor and basic fibroblast growth factor, stimulate stem cell proliferation [43, 44]. They may act selectively on specific progenitors after stem cell division or instructively by stimulating a specific group of stem cells to differentiate into a specific progeny [45, 46]. Within the extracellular matrix, cell adhesion is mediated by cellular integrin expression. The extracellular matrix can modulate the local concentration of growth factors and hence indirectly influence stem cell proliferation and differentiation within the stem cell niche [47]. Extracellular matrix proteins influence the expression of integrins, alterations of which can lead to departure of a cell from the stem cell niche towards differentiation [48].

Intrinsic factors determining self-renewal and division exist. Under similar conditions, it has been observed that sub-populations of haematopoietic stem cells have different capacities for self-renewal depending on telomerase activity, with increased telomerase expression correlating with an increased capacity for self-renewal [49]. Maintenance of the undifferentiated state may be determined by the asymmetric inheritance of certain proteins which influence gene expression; for example, the nuclear protein pharynx and intestine in excess-1 (PIE-1) inhibits embryonic genes which are responsible for somatic lineage differentiation [50]. The role *in vivo* of asymmetric kinetic genes may be of fundamental importance as intrinsic factors controlling self-renewal; however, it is believed the predominant form of self-renewal in mammals detectable so far is symmetric division [51]. Manipulation of asymmetric genes *in vitro* may enable great advancement in stem cell expansion [52]. The role of systemic cytokines and growth factors in stem cell regulation is less clear. Although a large number of growth factors have been identified that are mitogenic for haematopoietic stem cells such as granulocyte colony-stimulating factor [53], interleukin-1 (IL-1), IL-3 and IL-6 [54], leukaemia inhibitory factor [55] and basic fibroblast growth factor [56], long-range regulatory mechanisms for tissue-specific stem cells are unclear, as these cells are thought to be primarily responsive to local changes in the stem cell niche as their regenerative capacity is usually called upon in circumstances of local tissue damage. However, the control of differentiation is a complex event requiring exit of the cell from the undifferentiated state and entry into a determined developmental pathway. At present, cell determination is seen as a stochastic event initiated by intrinsic factors with an outcome biased by extrinsic factors. The challenge at present is to identify further intrinsic and extrinsic regulatory mechanisms which determine stem cell fate and to define the relationship between these systems. Developmentally, the components of the peripheral nervous system originate from the neural crest. In looking for alternatives to SC transplantation in bioengineered conduits, it is useful to focus on utilizing neural progenitor cells in PNR.

Sources of Stem Cells for PNR

Emphasis has been placed on exploring stem or progenitor cells that are easily accessible, rapidly expandable in culture, capable of survival and integration within the host tissue and amenable to stable transfection and expression of exogenous genes. Table 20.1 summarizes some of the studies conducted on nerve repair mechanisms to date. Embryonic neural stem cells have been used to repair nerve injuries with demonstration of regenerative success [57, 58, 59] but suffer the drawback of being somewhat difficult to obtain. On the other hand, adult stem cells (ASC) have the advantage of being available from relatively non-invasive, autologous harvest methods, and are likely the most promising choice for the majority of clinical nerve injuries. Bone marrow stromal cells have attracted the attention of several groups interested in cellular strategies to supplement nerve repair [60–63]. The mesenchymal stem cells are harvested from the long bones, and when placed in culture medium containing the appropriate cytokine cocktail, transdifferentiate into an adherent SC-like phenotype expressing S100 protein, glial fibrillary acid protein (GFAP) and p75. Several studies have been used with artificial conduits and cellular grafts, where they have contributed to improved electrophysiological, morphometric, and/or behavioural recovery outcomes versus vehicle controls. Although their potential to produce functional myelin *in vivo* has been questioned,[64] others have shown that the BM-MSC-derived SC are at very least capable of myelinating cultured PC12 cells *in vitro* further highlighting their therapeutic potential.

Neural Progenitor Cells

The transplantation of primary SC has been shown to improve nerve regeneration through bioengineered conduits. However, this approach has a number of pragmatic difficulties if it is to be applied to the clinical setting, the greatest of which

Table 20.1 Selected studies of transplanted stem cells for peripheral nerve repair

Cell source/type	Authors and year	Donor/host animal	No. of cells injected	Delivery method	Cell survival time	% Survival	Phenotype	Regenerative advantage conferred over vehicle
Bone marrow aspirate/mesenchymal stem cell	Hu et al. 2007	Rhesus monkey in	2×10^7	Proximal/distal side of acellular allograft	ND	ND	ND	↑ No. of NF + axons, improved CMAP amp/latency
–	Keilhoff et al. 2006[25]	Wistar rat	2×10^6/ml	Devitalized muscle conduits	6 wks	ND	MBP+, bipolar morphology in predifferentiated cells only	↑ No. of myelinated fibers; faster return of thermosensitivity
–	Chen et al. 2007	Sprague-Dawley rat	10^6 cells	In gelatin w/in lumen of silicone tube; 15-mm gap	Unable to detect due to label loss	ND	Express neurotrophins; not P0, PMP22	Improved SFI, improved CMAP amp/latency
–	Dezawa et al. 2001	Wistar rat	$1–2 \times 10^7$ cells/ml	In Matrigel, w/in hollow fibers; 15-mm gap	3 wks	MD	MAG+; produced myelin	↑ Axonal outgrowth achieved w/ predifferentiated cells
–	Zhang et al. 2004	Sprague-Dawley rat	10^7	Microinjected into crush-injured sciatic nerve	Up to 3 wks	ND	Limited expression of GFAP, S100, p75	ND
–	Shimizu et al. 2007	Human/Wistar rat	$1–2 \times 10^6$ cells/ml	In Matrigel, w/in transpermeable tube; 10-mm gap	3 wks	$< 12.6 \pm 2.98\,\%$ of all MAG+ SCs	MAG+; enveloped re-generating axons; many phagocytozed	Slight ↑ SFI conferred by transdifferentiated cells over

Table 20.1 (continued)

Cell source/type	Authors and year	Donor/host animal	No. of cells injected	Delivery method	Cell survival time	% Survival	Phenotype	Regenerative advantage conferred over vehicle
–	Tohill and Terenghi 2004	Sprague-Dawley rat	8×10^6 cells/ml	w/in PHB conduits; 10-mm gap	Up to 15 days	ND	Some differentiated to S100+ SCs	Naive ↑ outgrowth
C17.2 neonatal cerebellar granule cells ± overexpression of GDNF	Heine et al. 2004	Mouse cell line/Sprague-Dawley rat	5×10^5 cells	Subepineural injection into chronically denervated nerve	Up to 4 mos	0.5–1 %	Most remained distal to repair site, very few GFAP or NF+; mesenchymal tumour	↑ No. of axons, improved CMAP amp/latency
Hippocampal E17 neuronal progenitor cells	Murakami et al. 2003	Fischer rat	10^5 cells	In collagen gel w/in silicone tube; 15-mm gap	Up to 10 wks	ND	some cells positive for S100/p75	Superior electrophysiological recovery
E11 DRG/boundary cap neural crest stem cells	Aquino et al. 2006	Rosa 26 mouse (lac-Z+)/Sprague-Dawley rat	4×10^3 cells	Intact nerve; cultured in 12-mm silicone tube & implanted in nerve gap	Up to 90 days; only predifferentiated cells survived	ND	69.7–94.6 % GFAP+ after 13 & 60 days, respectively; MBP+ transplanted cells ensheathed axons in tube	ND
Neonatal skin/neural crestlike precursors	Marchesi et al. 2007	Wistar rat	10^6 cells	In PBS in lumen of collagen guide; 16-mm gap	Up to 2 mos	25–38 %	4.5 % S100+, 6.1 % GFAP+	Improved CMAP, SFI, no. of myelinated fibres

Table 20.1 (continued)

Cell source/type	Authors and year	Donor/host animal	No. of cells injected	Delivery method	Cell survival time	% Survival	Phenotype	Regenerative advantage conferred over vehicle
–	McKenzie et al., 2006	Rodent or human/shiverer mouse	$1–4 \times 10^5$ cells	Microinjected distal to crush injury	³6 wks	~6.5 %	70.4 % of transplanted cells MBP+, associated w/ NFM+ axons	MBP positive myelin on shiverer axons
Vibrissal follicles	Amoh et al. 2005	C57/B6-GFP/C57/B6 mouse	ND	Transplanted btwn severed sciatic/tibial nerve stumps	Detected after 2 mos	ND	GFAP+; envelop β III tubulin+ axons	Improved SFI, contraction of gastrocnemius
Amniotic fluid/mesenchymal stem cells	Murakami et al. 2003	Sprague-Dawley rat	1.5×10^4 cells	In fibrin glue around crushed sciatic nerve	Up to 10 days, none at 4 wks	ND	NT-3 and CNTF+; no expression of GFAP/S100β	Motor function recovery, improved CMAP

CMAP compound muscle action potential, *CNTF* ciliary neurotrophic factor, *DRG* dorsal root ganglion, *GDNF* glial cell line-derived neurotrophic factor, *MAG* myelin-associated glycoprotein, *ND* not described, *NF* neurofilament, *NFM* neurofilament (medium chain), *NT-3* neurotrophin-3, *PHB* poly-3-hydroxybutyrate, *P0* myelin protein 0, *PMP22* peripheral myelin protein 22, *SC* Schwann cell, *SFI* sciatic functional index, + positive, ↑ increase in is accessing a source of primary SC. Such cells would have to be autologous to prevent graft-versus-host reactions, as the use of immunosuppressants is not considered ethically acceptable in the case of nerve injury, on account of their systemic side effects. Hence, transplantable cells would have to be donor derived, being harvested and expanded from the individual with nerve injury prior to transplantation of the conduit. Primary human SC are technically difficult to purify, culture and expand to the numbers' required optimal regeneration inferred from experimental studies in a rat model. The time frame of culture to appropriate numbers can be up to 10 weeks. Such a delay in performing a primary nerve repair following injury would be deleterious to clinical outcome. Indeed, it has been shown that the longer the time lag between injury and repair, the greater the neuronal cell death in the dorsal root ganglia and therefore a reduced potential for recovery [65]. To circumvent these practical problems of SC transplantation, attention has been turned towards stem and progenitor cell transplantation. Possible candidates for peripheral nerve transplantation are neural progenitor cells, olfactory ensheathing cells (OECs) and MSC-derived cells. There are a limited number of reports describing the use of stem and progenitor cells in peripheral nerve transplantation. SC are derived from the neural crest of the neuroectoderm and differentiate and migrate into the peripheral nervous system with development [66]. The use of neurally derived progenitors from the fetal rat hippocampus to seed a conduit spanning a 15-mm gap in the rat sciatic nerve has been described. Significant morphological and functional evidence of regeneration, with integration and differentiation of transplanted neural progenitors into SC-like cells, was reported [67]. Neural stem cells have been identified in the adult central nervous system [68]. Such adult-derived stem cells and those isolated from fetal tissues have been successfully propagated and differentiated *in vitro* into all major cell types of the nervous system [69]. The use of fetal or adult-derived neural progenitors fails to avoid the problems encountered with SC, i.e. the source of donor tissue and the time taken to expand such cells.

Bone Marrow Stromal Cells

Until recently, dogma dictated that organ-specific stem cells were restricted to differentiate only into cell types from the tissue from which they originate. Unlike embryonic cells, organ-specific stem cells were believed to have lost their capacity to generate other somatic lineages. However, recent reports have suggested that stem cells from one tissue can cross lineage boundaries to differentiate into cells of other lineages either *in vitro* or *in vivo* after transplantation. This plasticity, or ability for cells to transdifferentiate, has aroused great interest for its therapeutic potential in tissue engineering. A promising candidate to display such plasticity is the bone marrow stromal cell.

Bone marrow contains two distinct populations of progenitor cells: haematopoietic progenitors and bone marrow stromal progenitors. Bone marrow stroma has been identified as the site of origin for mesenchymal progenitors for bone, cartilage, tendon, adipose tissue and muscle [70]. In addition, the stroma has been identified as a supportive tissue for the haematopoietic system, enhancing cytokine-induced proliferation of haematopoietic precursors [71]. A semantic issue confuses the description of the progenitor cells of the stroma, as they have been defined historically as colony-forming unit fibroblasts, marrow stromal fibroblasts, marrow stromal cells (MSCs), mesenchymal stem cells and mesenchymal progenitor cells. Unless such cells can be shown to be clonogenic and undifferentiated prior to their use in experimental studies, the putative use of the term 'stem cell' should be avoided. MSCs are easily accessible through the aspiration of the bone marrow cavity. They readily adhere in tissue culture in comparison with their non-adherent haematopoietic counterparts [72]. A number of mitogenic factors have been identified which stimulate colony formation and proliferation such as platelet-derived growth factor, epidermal growth factor, basic fibroblast growth factor, transforming growth factor b and insulin-like growth factor-I (IGF1) [73]. Bone marrow stromal cells have been shown to be inherently heterogeneous in terms of growth kinetics, morphology and phenotype. This may in part be due to the fact that the actual number of true stem cells is small, being estimated at 2–5 per 106 mononuclear cells, with the majority of cell growth arising from expanding differentiated colonies under mitogenic influence. With the development of specific antibodies such as STRO-1 [74] and characterization (CD45-ve, CD34-ve, CD105+ve, CD73+ve) [75] for human mesenchymal stem cells, their expansion and properties can be studied and defined with more certainty.

Interestingly, MSCs differentiate according to a hierarchical paradigm into osteoblastic, chondroblastic and adipocytic progenitors [76]. Orthodox plasticity between these lineages has been reported from adipocytic and chondrocytic lineages to osteogenic differentiation [77, 78]. Recent reports have described unorthodox plasticity of haematopoietic progenitors [79, 80] and bone marrow stromal cells in that they have been shown to cross oligolineage boundaries which were previously thought to be uncrossable. The potential of MSCs to transdifferentiate from mesenchymal lineages has aroused great interest. An early report by Ferrari et al. [81] described the migration of labelled MSCs to areas of damaged skeletal muscle. Transdifferentiation was detected, owing to the stromal cell expression of myogenic markers and their participation in the regeneration of damaged muscle fibres. Gojo et al. [82] demonstrated mesenchymal differentiation into cardiomyocytes, endothelial cells and smooth muscle cells following direct injection into adult heart; it has also been shown that these cells will migrate to zones of myocardial injury following systemic delivery [83], although neither group of authors described any functional improvement. Jiang et al. [84] identified marrow-derived cells in all somatic lineages in chimeric mice following injection of labelled cells into 3–5-day-old blastocysts and engraftment into many adult tissues following systemic infusion. There is evidence that MSCs are capable of neuronal antigen expression *in vitro* [85–87] and *in vivo* [88, 89]. They have been shown to differentiate into astrocytes following direct transplantation into the rodent brain [88]. Studies have also reported the ability of such cells to transdifferentiate into cells displaying a characterization similar to that of neuroectodermal cells [88, 89]. Akiyama et al. [90] described re-myelination of spinal cord lesions following intravenous delivery of MSCs, and Hofstetter et al. [91] showed that local delivery of MSCs at the site of spinal cord injury was associated with the formation of neurofilament bundles at the interface between scar tissue and graft.

Similarities and Differences of BM-MSC and ADSC

MSC from different sources seem to have overlapping characteristics but also differences in their phenotype (e.g. cytokine secretion properties and receptor profiles) as well as differentiation and expansion potential [92–95]. BM-MSC have been studied for many decades now and are the best-known and characterized type of MSC [96–98]. In animal studies as well as clinical trials, BM-MSC have shown beneficial outcomes in a variety of diseases ranging from ischemic and inflammatory disorders [99–101], through wound healing [102] and lung diseases, to tissue engineering and regenerative medicine, including nerve regeneration [103]. BM-MSC are easily aspirated from bone marrow, but associated with discomfort (pain) and potential morbidities after bone marrow punction.

ADSC bear some appealing characteristics in respect to BM-MSC and are thus rapidly gaining the interest of researchers. The main advantages are the ease of harvesting, higher availability of subcutaneous tissue, higher cell aliquots [104, 105], less harvest-site morbidity and the apparently higher expansion potential [106, 107], suggesting them as an ideal candidate for clinical application in the typical scenario of PNI.

There is plenty of subcutaneous fat tissue in humans which can deliver already high cell numbers after harvesting, shortening the culture time span required to achieve the desired cell aliquots [108]. In fact, since during culture MSC lose cytokines and receptors with increasing passage number, influencing their differentiation and homing capabilities (due to increasing cell size and change in receptor and chemokine profile), shorter culture times seem to be beneficial for clinical application [109, 110]. However, the beneficial effect of ADSC, compared to BM-MSC, has to be further investigated.

BM-MSC have been shown to decrease their expansion capabilities after passage 3–4 [111], showing signs of senescence, whereas ADSC do not, which together with the fact that bone marrow contains fewer MSC highlights the superiority of ADSC for achieving high cell yields shortly.

Some authors suggested that MSC from different sources may have different potential of differentiation towards the mesenchymal lineages [112], which could also be true for non-mesenchymal lineages as for neural tissues, whereas other groups found no significant differences between their differentiation potential

[113]. Furthermore, it has been shown that BM-MSC and ADSC might differ in their homing capability [114].

Mesenchymal Stem Cells in the Peripheral Nervous System

Two recent reports have described the use of MSC transplantation in models of PNI. Dezawa et al. [115] described the *in vitro* expression of the glial cell markers p75 and S100 by rat MSC following exposure to a cocktail of growth factors, and integration of such cells in the regenerating growth cone upon transplantation into a blind-ending tube grafted to the proximal stump of the rat sciatic nerve. Cuevas et al. [116] described the migration and differentiation of MSCs following the injection of cultured undifferentiated cells into the site of a sciatic nerve axotomy repair. In our laboratory, we have studied rat marrow stromal cell differentiation *in vitro* and *in vivo* following transplantation into a model of PNI. Following *in vitro* exposure of MSCs to glial growth factor, a powerful SC mitogen [117] which has also been shown to stimulate neural crest stem cells to differentiate into SC, MSCs were found to exhibit glial cell immunoreactivity to markers, including GFAP and S100. In addition, a small percentage of cells were also found to exhibit some morphological characteristics of SC. We have studied the effect of transplanting such glial-differentiated MSCs into a rat model of PNI and have found that they can confer a beneficial effect on SC growth within regenerating nerves, indicating either a stimulatory or a supportive role.

Immunotolerance to Allogeneic Stem Cells

An important problem which faces all studies of transplantation is that of host rejection of transplanted tissue. In the clinical setting of organ transplantation, this problem is overcome with the use of immunosuppressants, with the benefits of a functioning heart, lung, liver or kidney outweighing the potentially serious systemic side effects of immunosuppression, such as susceptibility to opportunistic infection and skin cancer. A number of experimental reports have shown that there is an unexpected and fortuitous level of immune tolerance to transplanted stem and progenitor cells. MSCs have been shown to block the proliferation of allogeneic T cells *in vitro* and prolong skin graft survival *in vivo* [118]. Saito et al. [119] demonstrated immune tolerance to xenogeneic cells following transplantation in addition to retaining their ability to engraft and differentiate. The ability of MSCs to induce tolerance may be due to cytokine (such as transforming growth factor b) secretion or due to their phenotypical immaturity, as they lack major histocompatibility complex (MHC) class II and other T cell stimulatory surface proteins [120, 121]. With respect to studies in the nervous system, Hori et al. [122] have shown that central nervous system progenitor cells may not express MHC class I and II *in vitro* and that no short-term evidence of rejection is seen after allogeneic transplantation within the central nervous system. These observations are seemingly a local effect of stem cell transplantation and are distinct from the recognized phenomenon that infusion of bone marrow donor cells can improve allograft survival and reduce the need for short-term immunosuppression. Such macrochimeric models require significant host conditioning, such as central and peripheral T cell depletion [123, 124], prior to transplantation.

One problem which may be faced when considering host tolerance to progenitor cell allografts is the consequence of the up-regulation of immunoreactive surface markers following *in vitro* or *in vivo* differentiation due to the effects of growth factors or cytokines. Such effects may offset the benefits of immune tolerance to immature cells. MSCs might be used as primary engrafting cells for differentiation into cells of a de novo tissue, and they might also be used as cellular platforms for transgene expression for the production growth factors or cytokines which could ameliorate the immune response to more differentiated and functionally mature allogeneic cells [125, 126].

In Vitro Studies

Isolation of Stem Cells

Bone marrow-derived mesenchymal stem cells can be harvested from femoral bones [34] and cultured in mesenchymal cell growth medium (modified Eagle's medium, MEM) supplemented with 10 % fetal bovine serum (FBS; v/v and 1 % penicillin streptomycin). Briefly, a 21-guage needle can be used to run the medium in and out of bone to get the cell suspension. The resulting cell suspension is triturated, filtered through a 70-mm Falcon filter and centrifuged for 10 min at 600 g. The supernatant is aspirated, the cell bolus re-suspended in mesenchymal cell growth medium and the cells plated in 75-cm^2 tissue culture flasks and incubated in 5 % CO_2 at 37 °C. Haematopoietic cells are eliminated by washing daily with Dulbecco modified Eagle's medium (DMEM) until all the non-adherent cells were removed. The cells were then allowed to grow to confluence.

ADSC can be harvested from fat tissue by enzymatically digested for 2 h at 37 °C using 0.15 % (w/v) collagenase type I. Briefly, the digested solution is passed through a 70-mm filter to remove undissociated tissue, neutralised by adding modified Eagle's medium (a-MEM) containing 10 % (v/v) FBS and centrifuged at 800 × g for 10 min. The stromal cell pellet is re-suspended in MEM containing 10 % (v/v) FBS and 1 % (v/v) penicillin/streptomycin solution. The cultures are maintained at sub-confluent levels in a 37 °C incubator with 5 % CO_2 and passaged with trypsin/ethylene diamine tetra acetic acid (EDTA)-mediated subculturing methods.

Microsphere Technology: Implication for Nerve Repair

Several studies have focused on alternate methods to administer neurotrophic and growth factors for their sustained release in peripheral nerve repair applications in the past. Currently, implantation of a slow release pump device or biodegradable system is of great interest to deliver or induce neurotrophic factors for longer time points in order to achieve success nerve repair. In that context, biodegradable microsphere technology, derived from utilization of poly-dl-lactide-coglycolide (PLGA), the frequent material that has been tested in studies and showed sustained release of fabricated factors for long-term. Kokai et al. showed that implantation of nerve guides generated based on microsphere technology across a 1.5-cm defect in a rat sciatic nerve gap resulted in an increase in tissue integration in both the proximal and distal segments of the lumen of the nerve guide after 6 weeks [127]. Marra et al. have pioneered in this technology and showed improved tissue integration within neurotrophic and/or growth factor releasing nerve guides when compared to negative controls and also showed the presence of nerve fibres across the entire length of nerve guides [128]. For instance, the glial-derived neurotrophic factor (GDNF) sustained release systems have also been tested in models of glaucoma. Multiple injections of GDNF slow release PLGA microspheres showed up to 3.5 times greater retinal ganglion cell (RGC) density than untreated mice at 15 months' survival [129]. Moreover, intravitreal injection of GDNF/Vit E PLGA microspheres exhibited significant improvement of RGC axon and soma survival 8 weeks after intraocular pressure (IOP) elevation [130]. It is clearly evident that microsphere technology is superior in terms of its reproducibility, easy fabrication, quality control and cost effective. However, there is a minimum challenge in terms of repeated administration of fabricated material into nerve guide. Hence the recent studies are focused on achieving sustained release of proteins/drugs for several months to bypass re-administration.

Viral Vector-mediated Gene Delivery System

Adenoviral (AdV) vector system was first demonstrated as a vehicle to deliver genes that express neurotrophic factors in nerve repair [131]. In general, AdV vector encode genes of interest from a strong viral promoter (cytomegalovirus); Studies showed that the over expression of recombinant protein could trigger cellular toxicity and in turn lead to immune reaction, and early AdV system was reported to show strong cytotoxicity. Hence, studies have focused on alternative vector systems like adeno-associated virus (AAV) to deliver genes expressing neurotrophic factors like GDNF. AAV mediated gene delivery to retinal cells demonstrated statistically fewer apoptotic cells in the RGC layer [132]. Currently, AAV is the approved viral vector system for clinical gene therapy, which was confirmed harmless to human body and seems promising for peripheral nerve repair (PNR) applications in clinical set up. Another widely used viral vector is lentivirus (LV) system, for basic science research but not for translational research due to Food and Drug Administration (FDA) regulations on safety concerns. However, it is a very powerful tool to engineer difficult cells like MSCs for the expression and evaluation of the effects of neurotrophic and/or growth factors for PNR applications.

Lentiviral Vectors as a Gene Delivery System

LV (a slow virus) belongs to the *Retroviridae* family of viruses, characterized by their long incubation Period (latency). Unlike other retroviruses, LV can be engineered into gene therapy vehicle by deleting pathogenic genes from its genome and to make it avirulent. The resulted recombinant LV could be used as the most efficient gene delivery system which contains the necessary cellular and molecular components required for viral packaging and stable expression of transgene in transduced cells. Human Immunodeficiency Virus (HIV) is the best example of LV that is generally engineered to produce recombinant LV to transduce difficult cell types like nerve cells, stem cells, macrophages and different cell lines. Around a decade and half ago, research studies on evaluation of LV vector system in PNR had begun [133] and studies showed that LV is characteristic by less immune response with larger cloning capacity when compared to AdV and AAV [134] and more stable expression [135] in neuronal cells. However, despite major advances in excellent cellular transduction, biosafety of LV vectors still remains a major concern and limitation in clinical translation due to potential genetic manipulation, recombination, genome integration, oncogenic potential etc., More studies are certainly needed to overcome the barrier for clinical application in PNR. Fig. 20.2 depicts the example of transgene cloning and expression in engineered cells. (Original figure on Lentiviral vector production was published by Frank Park in Physiological genomics, 2007)

SC-Like Differentiation of MSC

Both BM-MSC and ADSC have been shown to be able to differentiate to SC-like cells *in vitro* if appropriately induced during culture [136]. Caddick et al. induced BM-MSC in culture to differentiate along a SC-like phenotype. After differentiation, the cultured cells were expressing high levels of S100, P-75 and GFAP [137]. In similar experiments, Kingham et al. differentiated ADSC to SC-like cells also expressing typical SC markers such as S-100 and P-75. Others have successfully differentiated BM-MSC to glial cell-like cells secreting neurotropic factors [138]. Wei et al. could elegantly show how SC can induce ADSC to get a SC-like phenotype with typical marker expression through indirect co-culture techniques [139]. Liu et al. successfully differentiated ADSC (positive S100 and P-75 expression) *in vitro* using culture media preconditioned for 2 days by a rat sciatic nerve.

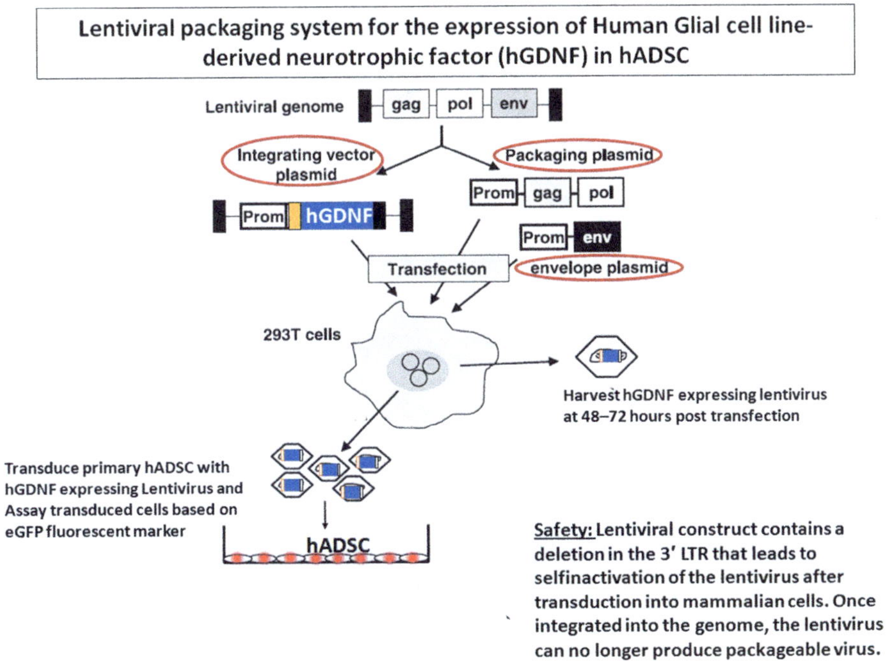

Fig. 20.2 Schematic representation of lentiviral vector construction and virus production. In general, 2nd generation system contains three plasmids (transgene, packaging and envelope plasmids) to generate recombinant lentiviruses to transduce primary stem cells to utilize the dual activities and functionalities of stem cells and transgene. The system usually contains a downstream GFP reporter for flow cytometry selection of transgene positive cells and for *in vivo* tracking. (GFP green fluorescent protein). *(Original picture by Frank Park, Physiological genomics, 2007)*

Neurotropic Effects of MSC

The positive effects of MSC on nerve regeneration have already been investigated *in vitro* by many groups, and there is common agreement that this is achieved in a paracrine fashion, often after differentiation to SC-like cells [140–142].

MSC were found to improve neurite growth *in vitro* and express SC markers as well as showing their phenotype characteristics. Kingham et al. found SC-like differentiated ADSC-promoting neurite growth *in vitro*. A study of Kalbermatten et al. showed that human ADSC secrete many neurotrophic factors as nerve growth factor, brain-derived neurotrophic factor and GDNF (real-time polymerase chain reaction, rt-PCR, and enzyme-linked immunosorbent assay, ELISA) [143]. Moreover, ADSC were able to significantly promote neurite outgrowth *in vitro*. Interestingly, cells harvested from superficial abdominal adipose layers performed better in terms of cell proliferation as well as nerve growth enhancement. Kaewkhaw et al. found that the source of ADSC seem to be determinant in the SC marker expression, differentiation potential and final neurite growth improvement *in vitro* using three different fat sources on rats (subcutaneous, perinephric and epididymal).

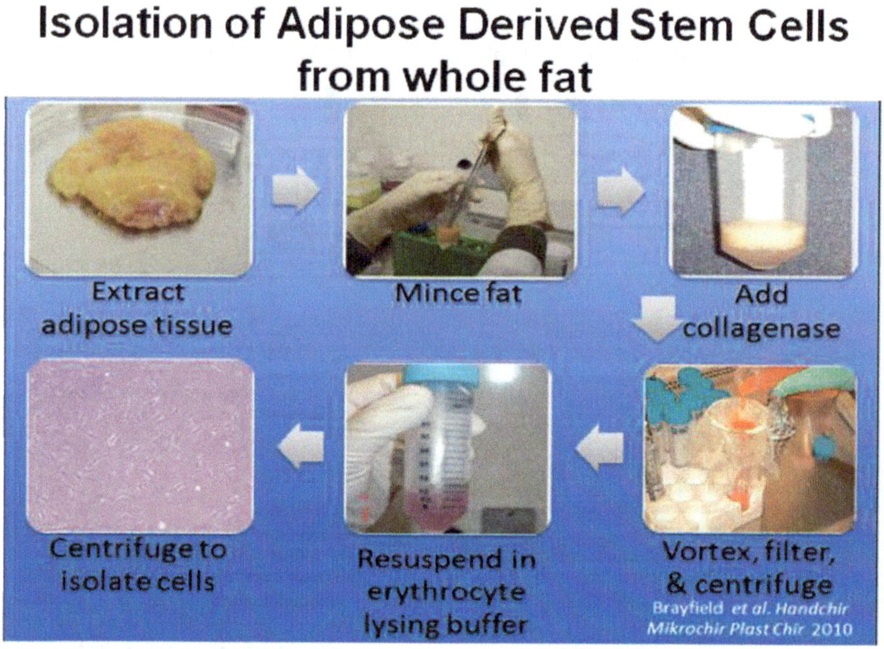

Fig. 20.3 Schematic representation of ASCs isolation from whole fat: Stem cells are isolated as per laboratory developed and optimized standard protocol by collagenase digestion and centrifugation methods. The final yield of cells is called as stromal vascular fraction (SVF) that is heterogeneous and contains sub-populations. The primary sub-population found in SVF is preadipocytes that are CD34+, CD31− and highly proliferative. These are considered to be stem cells and could be differentiated to different cell types. *ASC* adult stem cells

Subcutaneous and perinephric ADSC showed superior neurotrophic potential [144]. However, other groups found no different neurotrophic potential of ADSC from different donor region as well as different ages [145].

In Vivo Studies

Route and Time of Administration

Almost all studies reviewed have used MSC delivered locally using grafts (e.g. vein, artery, muscle or acellular nerve grafts) or different kinds of conduits (e.g. chitosan, collagen or silicon), as well as direct cell depots. There is only one study which used intravenously injected ADSC (Fig. 20.3), 1 week after sciatic nerve crush injury [143]. Several groups have shown that MSC are able to home to sites of tissue injury and/or inflammation [144–147], suggesting the cells are attracted by different chemo- and cytokines and thus will find the location where they are needed when applied systemically. However, using this route of administration has been shown to bear some potential drawbacks, though, as the injected cells have to trespass the capillaries of organs like lungs, spleen and liver and may get entrapped reducing considerably the final amount of cells reaching the tissue of interest. Eggenhofer et al. recently showed that most of the intravenously infused MSC were retained in the lung and liver capillaries [148] and had shortened survival. During culture, MSC have been shown to increase their size, which can explain differences in homing abilities [149]. Nevertheless, there is a group which found MSC to be able to exert beneficial effects also 'remotely' on the humoral way [150]. This may be true for inflammatory and ischemic diseases, but could not be enough for nerve regeneration after PNI as the sprouting axons need a local, precise patterning of cytokines and growth factors to guide them to the distal target tissues [151].

Furthermore, even if MSC home to the regenerating tissue and are able to be locally active, using a sole systemical MSC application in case of a nerve gap could be still problematic because even if promoted, no 'guide' will lead the regrowing nerve to the target muscle insertion if the distal nerve degenerates. To the best of our knowledge, there is no work assessing the effect of systemic MSC combined to the aid of nerve grafts or conduits.

Methods for Tracking Fate of Transplanted Stem Cells

Given the evidence presented above, it is apparent that studies exploring stem cell transplantation for PNR should give careful thought on strategies to track the fate of transplanted cells over time. There is often little importance placed on pre-labelling cells prior to delivery into the injured nerve, and as such authors cannot comment on the mechanism of any advantage conferred by cell therapy. Others have used labelling techniques that are not sufficiently robust or long-lasting to be detected at the study end points. Chemical markers such as bisbenzimide and PKH26 have been used to label SC delivered to peripheral nerve injuries, but their usefulness is limited to the short term and may in fact affect the viability and phenotype of transplanted cells. Genetic labelling with either lacZ or fluorescent proteins such as green fluorescent protein (GFP) is increasingly popular and appears to be a relatively long-lasting method that is not deleterious to transduced cells. The use of lipophilic carbocyanine derivative CellTracker CM-DiI (Molecular Probes) is significantly reliable to label stem cells within a variety of nerve injury models with no dilution or loss of signal for $\geq$ 10 weeks following transplantation. These dyes have the advantage of being technically simple to use, rapid and resistant to leakage to nearby cells. Emerging technologies such as quantum dots offer an exciting alternative to traditional cell labelling methods. These nanoparticles are available in a wide range of photostable colours and are resistant to chemical and metabolic degradation, making them ideal for use in long-term fate tracking of transplanted stem cells. The use of PKH26 (red fluorescence) and LV-mediated GFP (green) labelling is also widely used for successful *in vivo* tracking of transplanted cells.

Dosage

The cell yields transplanted in rodent small-animal models of PNI generally ranged from 1×10^5 to 2×10^7 cells per conduit or graft with some exceptions, where higher [151] and lower amounts were used [152]. One has to differentiate if undifferentiated MSC are used instead of SC-like differentiated MSC, since it is likely that only a part of applied undifferentiated cells will differentiate *in vivo*.

Some authors proposed that when applied locally, too high MSC counts would kill parts of the cells because there would be not enough nutrients for them to survive and too few cells would not have the desired effect. Using the above-mentioned range of MSC for local delivery, many groups found the cells survive for many months [153–157] and had improved PNR, although others could not find any transplanted MSC alive after 14 days.

Safety of MSC Application

There are several groups following up the animals for several months, up to 6 or even 12 months [158–164], especially using large animals, and so far there is no report on adverse effects of local MSC delivery or any incidence of neoplasm, at least locally, during histological examination. In a recent study, Hu et al. went deeper and evaluated the safety of local MSC therapy for PNR during 12 months, investigating blood samples (with heart, liver, metabolic, lipid, blood count and tumour marker panels) every 3 months and tissue specimens of heart, liver, spleen, lymph nodes, kidney and lung [165]. Furthermore, since MSC have also immunomodulatory properties, the authors observed the systemical effects on the blood immunoglobulin E profile and did not find any abnormal change.

When applied systemically, MSC potentially bear some safety problems as the risk of embolization (lung, liver, spleen) [166] or systemical adverse effects (e.g. allergic/anaphylactic reactions). This depends on the animal species used as a PNI model as well as characteristics, especially the size, of the injected cells. Generally, there are many animal and clinical studies using systemical MSC injection without any acute adverse effect [167–170].

Outcomes

The improvements on nerve growth and regeneration on histological level are mainly investigated through qualitative and quantitative analysis of nerve fibres and their myelination using imaging tools like electron [168, 170], confocal [171], light and fluorescence microscopy (FM). Myelin sheath thickness, number of myelinated fibres, number of new fibres and growth distance of proximal nerve stump are a few examples of values that can be assessed and give a picture of the nerve regeneration process [172–177].

Using immunohistochemistry (IHC) and FM, survival of previously labelled MSC can be monitored, as well as their transdifferentiation to other cell types and their localization in the tissue [178]. Common target factors for recognition of SC-like transdifferentiation through IHC and FM are GFAP, S-100 and nerve growth factor receptor (NGFR) P-75 [179]. Some groups investigated growth factor expression by MSC in nerve tissue specimens using Western blots [180].

For indirect assessment of nerve innervation, ratios have been used for the assessment of target muscle atrophy during PNR [178, 179, 181–183]. Other groups measured muscle tension after harvesting [184].

Besides, there are functional test to investigate the overall nerve recovery using tools like electrophysiology and behavioural tests. Electrophysiological exams include measurement of latency of the electric signal and their amplitude using compound muscle action potentials (CMAP) and nerve conduction velocity (NCV), all giving a clue about the myelination of nerve fibres and their number [185, 186].

Behavioural examination for example comprise ladder-climbing and walking track analysis, especially for rodent PNR model, measuring different parameters (e.g. toe spread and foot length) out of which indexes (e.g. the sciatic function index, SFI) can be calculated [186, 187]. Furthermore, pinch test, thermo-sensitiveness, whisking, as well as grasping and finger opposition/motion abilities in non-human primates have been assessed [187, 188].

Small-Animal Models

The sciatic nerve gap model has been mostly used and is a standard for the study of PNR in the last decade. All studies on small animals have been performed on rodents, merely on rats, with a few exceptions using murine models [174–176, 189].

In one of the first studies on the topic, Dezeawa et al. found BM-MSC to transdifferentiate to SC-like cells and increase nerve fibre growth as well as myelination in a rodent sciatic gap model. Mimura et al. and many other groups had similar findings later [190]. McGrath et al. used human BM-MSC in rat sciatic gap model and after 3 weeks found only enhanced nerve regeneration if combined with cyclosporin A. Their assumption was that through immunosuppression the transplanted human cells are protected and otherwise the host would kill them.

Some experimental protocols have compared undifferentiated MSC to SC-like differentiated MSC [184, 185] and/or cultivated SC [190]. Keilhoff et al. 6 weeks after BM-MSC transplantation in a sciatic gap injury model found a significant enhanced functional outcome (toe spreading and thermo-sensitiveness) and an increased myelination of nerve fibres in animals receiving SC-like differentiated BM-MSC and SC, compared to undifferentiated BM-MSC [8]. Other groups had similar findings [184], whereas Costa et al. found BM-MSC therapy to be inferior to SC application in a rat facial gap model [185]. One group found no outcome differences using differentiated BM-MSC rather than differentiated ADSC [184], but still significant improvements were found compared to acellular controls, which is in line with findings of Di Summa et al. [10].

Due to increased interest in ADSC-based therapy, in the last years many groups further addressed the question if these cells are promising candidates for PNR. Generally, similar to BM-MSC therapy, ADSC are able to promote PNR in rodent models at a histological level as well as on functional recovery [10, 11, 181, 182, 187, 191]. Orbay et al. found undifferentiated ADSC to improve functional outcome measured with SFI and electrophysiological test as good as differentiated ADSC.

Using intravenous MSC infusion on PNR, Marconi et al. found the cells home to site of PNI and accelerate the nerve fibre as well as functional recovery. Additionally, they found MSC having an immunomodulatory effect with decreased infiltration of inflammatory cells into the injured nerve and surviving until end point (5 weeks), similar to the findings of Pereira et al. in a murine model [171]. The proposed mechanism is both immunomodulation and paracrine mechanism (neurotropic, stimulation of SC to secrete GDNF) in addition to some extend of transdifferentiation. Also Carriel et al. propose an increased stimulation of resident SC through ADSC; moreover, they found also an improved angiogenesis in site of PNR [154].

As putative mechanisms of the beneficial effects of MSC, there is a general agreement on the paracrine function of MSC in PNR enabling a pro-regenerative and neurotrophic microenvironment and modulating other cells like SC (lit) to improve nerve recovery. However, some authors found MSC not only to act as a local mediator, but also differentiate *in vivo* to SC-like cells as suggested by S100 and Hoechst dyes on nerve tissue samples [14, 154, 156, 192], whereas others failed to observe such transdifferentiation [154, 156].

Large-animal Models

There are not many large-animal studies on PNR so far; the most experiments have been using peripheral nerve gap models on non-human primates and dogs and had follow-ups of up to 12 months [161, 162–165, 188, 192, 193]. In non-human primates, for the assessment of functional recovery after nerve regeneration, models using forearm nerves have been preferred because behaviours like thumb-finger opposition and grasping are ideal tools to that purpose and the recovery of the specific nerves can be assessed more precisely [188]. Most of the large-animal studies after MSC application found an increased nerve regeneration (IHC and histomorphometry) compared to control [192, 194] with no significant difference compared to autograft transplantation or SC use, suggesting BM-MSC and ADSC as potential alternative to difficult-to-cultivate SC in PNR.

Almost all reviewed large-animal studies used undifferentiated BM-MSC injected in biological or artificial conduits [193–194] or embedded in acellular nerve grafts [192]. Wang et al. compared undifferentiated BM-MSC to SC-like differentiated BM-MSC in a rabbit facial nerve gap model and found the latter to increase functional outcome as well as enhancing histomorphometric findings indicating a significant enhanced nerve regeneration [195]. Only one group applied undifferentiated ADSC in a large-animal model injected in a conduit bridging a facial nerve gap on dogs [193]. Interestingly enough, after 3 months, the authors found no significant promoted nerve regeneration looking at histomorphometry and histological samples; only the electrophysiological findings were improved. This is surprising since in small-animal models, ADSC were generally as beneficial as BM-MSC.

To our knowledge, a large-animal study evaluating the effects of systemical MSC application on PNR is missing so far.

Conclusions

With growing evidence to support the presence of stem cells within adult tissues and the apparent plasticity of tissue-specific stem cells, it seems likely that the MSCs/ADSC can become important components of bioengineered systems in the future. MSCs, in particular, have received much attention because of their accessibility, abundance and ease of culture *in vitro*. There is no doubt that such cells can express neuronal cell immunoreactivity under specific mitogenic influences both *in vitro* and *in vivo*. To explain such phenomena, we may have to look at the molecular characterization of such transdifferentiated MSCs prior to and following differentiation. Microarray analysis has revealed over 2000 genes expressed by MSCs, many of these genes being unrelated to mesenchymal lineages. It is important to examine the effects of cellular microenvironments on cell differentiation, as *in vitro* experiments have identified a mitogenic effect of a large number of unrelated cytokines and growth factors on MSCs. It is also necessary to identify the factors influencing differentiation, such as intrinsic and extrinsic mechanisms, and the interaction between the two. Notwithstanding these challenges, stem cell therapies are becoming an exciting development in the field of peripheral nerve tissue engineering.

Animal studies have demonstrated that transplantation of stem and precursor cells has the potential to serve as an adjunct therapy to common practices of surgical nerve repair. Although the application of cell-based strategies in a clinical setting is promising, optimization of cell delivery and careful investigation of the fate of transplanted cells are required to guarantee the safety and maximum efficacy of these therapies. As discussed in this chapter, it will be important to determine the ideal number and method of cell delivery, and elucidate the extent of transplant cell survival and differentiation that is required to elicit a therapeutic effect. Future studies should place emphasis on using reliable labelling methods to track the long-term fate of transplanted cell. Finally, while many cell types have been investigated for their potential use in cell replacement therapy, few studies have directly compared the utility of different stem cells in augmenting PNR. It is believed that cells that are easily isolated from autologous sources such as fat (adipose tissue) and that can survive and differentiate to a glial phenotype within the milieu of the injured nerve provide the most promise.

To date, BM-MSC and ADSC both share similar beneficial effects *in vitro* as well as in animal studies of PNR. There is a trend showing better results after implantation of previously *in vitro* SC-like differentiated MSC, with findings similar to SC application or even autograft transplantation. Being ADSC easier to harvest and culture to high cell numbers, they seem promising candidates to future clinical applications. Possible mechanisms include *in vivo* transdifferentiation of undifferentiated MSC and paracrine mechanisms in terms of SC modulation, local immunomodulation and direct neurotropic and neurotrophic effects on growing nerve fibres, as well as local angiogenesis. Clinical studies on PNR are needed to further investigate the real potential of MSC for treatment of peripheral nerve injuries.

References

1. Thorsen F, et al. Digital nerve injuries: epidemiology, results, costs, and impact on daily life. J Plast Surg H and Surg. 2012;46(3–4):184–90.
2. Taylor CA, et al. The incidence of peripheral nerve injury in extremity trauma. Am J Phys Med Rehabil. 2008;87(5):381–5.
3. Gaudet AD, Popovich PG, Ramer MS. Wallerian degeneration: gaining perspective on inflammatory events after peripheral nerve injury. J Neuroinflammation. 2011;8:110.
4. Rotshenker S. Wallerian degeneration: the innate-immune response to traumatic nerve injury. J Neuroinflammation. 2011;8:109.
5. Hoke A. Mechanisms of disease: what factors limit the success of peripheral nerve regeneration in humans? Nat Clin Pract Neurol. 2006;2(8):448–54.
6. Allodi I, Udina E, Navarro X. Navarro, Specificity of peripheral nerve regeneration: interactions at the axon level. Prog Neurobiol. 2012;98(1):16–37.
7. Bosse F. Extrinsic cellular and molecular mediators of peripheral axonal regeneration. Cell Tissue Res. 2012;349(1):5–14.
8. Keilhoff G, et al. Peripheral nerve tissue engineering: autologous Schwann cells vs. transdifferentiated mesenchymal stem cells. Tissue Eng. 2006;12(6):1451–65.
9. Kingham PJ, et al. Adipose-derived stem cells differentiate into a Schwann cell phenotype and promote neurite outgrowth *in vitro*. Exp Neurol. 2007;207(2):267–74.
10. di Summa PG, et al. Adipose-derived stem cells enhance peripheral nerve regeneration. J Plast Reconstr Aesthet Surg. 2010;63(9):1544–52.
11. Erba P, et al. Regeneration potential and survival of transplanted undifferentiated adipose tissue-derived stem cells in peripheral nerve conduits. J Plast Reconstr Aesthet Surg. 2010;63(12):e811–7.
12. Zheng L, Cui HF. Cui, Use of chitosan conduit combined with bone marrow mesenchymal stem cells for promoting peripheral nerve regeneration. J Mater Sci Mater Med. 2010;21(5):1713–20.
13. Sun F, et al. Repair of facial nerve defects with decellularized artery allografts containing autologous adipose-derived stem cells in a rat model. Neurosci Lett. 2011;499(2):104–8.
14. Yang Y, et al. Repair of rat sciatic nerve gap by a silk fibroin-based scaffold added with bone marrow mesenchymal stem cells. Tissue Eng Part A. 2011;17(17–18):2231–44.
15. Bain JR, et al. The peripheral nerve allograft: an assessment of regeneration across nerve allografts in rats immunosuppressed with cyclosporin A. Plast Reconstr Surg. 1988;82(6): 1052–66.
16. Lukaszuk M, et al. Repair of the peripheral nerve gap with epineural sheath conduit to prevent muscle denervation atrophy in the diabetic rat model. Pol Przegl Chir. 2013;85(7): 387–94.
17. Lin YC, Oh SJ, Marra KG. Synergistic lithium chloride and glial cell line-derived neurotrophic factor delivery for peripheral nerve repair in a rodent sciatic nerve injury model. Plast Reconstr Surg. 2013;132(2):251e–62e.
18. Wang S, et al. Acceleration effect of basic fibroblast growth factor on the regeneration of peripheral nerve through a 15-mm gap. J Biomed Mater Res A. 2003;66(3):522–31.

19. Ohta M, et al. Novel heparin/alginate gel combined with basic fibroblast growth factor promotes nerve regeneration in rat sciatic nerve. J Biomed Mater Res A. 2004;71(4):661–8.
20. Pierucci A, de Duek EA, de Oliveira AL. Peripheral nerve regeneration through biodegradable conduits prepared using solvent evaporation. Tissue Eng Part A. 2008;14(5):595–606.
21. Guenard V, Kleitman N, Morrissey TK, Bunge RP, Aebischer PJ. Syngeneic Schwann cells derived from adult nerves seeded in semipermeable guidance channels enhance peripheral nerve regeneration. Neurosci. 1992;12:3310–20
22. Mosahebi A, Woodward B, Wiberg M, Martin R, Terenghi G. Retroviral labeling of Schwann cells: *in vitro* characterization and *in vivo* transplantation to improve peripheral nerve regeneration. Glia. 2001;34:8–17.
23. Rodriguez FJ, Verdu E, Ceballos D, Navarro X. Nerve guides seeded with autologous schwann cells improve nerve regeneration. Exp Neurol. 2000;161:571–84.
24. Azizi SA, Stokes D, Augelli BJ, DiGirolamo C, Prockop DJ. Engraftment and migration of human bone marrow stromal cells implanted in the brains of albino rats–similarities to astrocyte grafts. Proc Natl Acad Sci U S A 1998;95:3908–13.
25. da Silva Meirelles L, Chagastelles PC, Nardi NB. Mesenchymal stem cells reside in virtually all post-natal organs and tissues. J Cell Sci. 2006;119(Pt 11):2204–13.
26. Strem BM, et al. Multipotential differentiation of adipose tissue-derived stem cells. Keio J Med. 2005;54(3):132–41.
27. Till JE, McCulloch EA. A direct measurement of the radiation sensitivity of normal mouse bone marrow cells. Rad Res. 1961;14:213–22.
28. Loeffler M, Potten CS. Stem cells and cellular pedigrees: a conceptual introduction. In: Potten CS, editor. Stem Cells. London: Academic; 1997. pp. 1–27.
29. Marshak DR, Gottlieb D, Gardner RL. Introduction: stem cell biology. In: Marshak DR, Gardner RL, Gottlieb D, Editors. Stem Cell Biology. Cold Spring Harbor: Cold Spring Harbor Laboratory; 2001. pp. 1–16.
30. Thomas ED. Frontiers in bone marrow transplantation. Blood Cells. 1991;17:259–67.
31. Thompson JA, Itskovitz-Eldor J, Shapiro SS, Waknitz MA, Swiergiel JJ, Marshal VS, Jones JM. Embryonic stem cell lines derived from human blastocysts. Science. 1998;282:1145–47.
32. McLaren A. Ethical and social considerations of stem cell research. Nature. 2000;414:129–31.
33. Shapiro HT. Ethical dilemmas and stem cell research. Science. 1999;285:2065.
34. Toma JG, Akhaven M, Fernandes KJ. Isolation of multipotent adult stem cells from the dermis of mammalian skin. Nat Cell Biol. 2001;3:778–84.
35. Donovan PJ, Gearhart J. The end of the beginning for pluripotent stem cells. Nature. 2001;414:92–7.
36. Vogel G. Stem cell policy. Can adult stem cells suffice? Science. 2001;292:1820–22.
37. Sherley JL, Stadler PB, Johnson DR. Expression of the wild-type p53 antioncogene induces guanine nucleotide-dependent stem cell division kinetics. Proc Natl Acad Sci U S A. 1995:92(1):136–40.
38. Cheng T, Rodrigues N, Shen H. Hematopoietic stem cell quiescence maintained by p21cip1/waf1. Science. 2000;287(5459):1804–8.
39. Potten CS, Loeffler M. Stem cells: attributes, cycles, spirals, pitfalls and uncertainties. Lessons for and from the crypt. Development. 1990;110:1001–20.
40. Fuchs E, Segre JA. Stem cells: a new lease on life. Cell. 2000;100:143–55.
41. Henderson ST, Gao D, Lambie EJ, Kimble J. lag-2 may encode a signaling ligand for the GLP-1 and LIN-12 receptors of C. elegans. Development. 1994;120(10):2913–24.
42. Kopan R, Nye JS, Weintraub H. The intracellular domain of mouse Notch: a constitutively activated repressor of myogenesis directed at the basic helix-loop-helix region of MyoD. Development. 1994;120(9):2385–96.
43. Reynolds BA, Weiss S. Clonal and population analyses demonstrate that an EGF-responsive mammalian embryonic CNS precursor is a stem cell. Dev Biol. 1996;175:1–13.
44. Gritti A, Parati EA, Cova L, Frolichsthal P, Galli R, Wanke E, Faravelli L, Morassutti DJ, Roisen F, Nickel DD, Vescovi AL. Multipotential stem cells from the adult mouse

brain proliferate and self-renew in response to basic fibroblast growth factor. J Neurosci. 1996;16(3):1091–100.
45. Shah NM, Marchionni MA, Stroobant P, Anderson DJ. Glial growth factor restricts mammalian neural crest stem cells to a glial fate. Cell. 1994;77(3):349–60.
46. Shah NM, Groves A, Anderson DJ. Alternative neural crest cell fates are instructively promoted by TGFbeta superfamily members. Cell. 1996;85(3):331–43.
47. Quesenberry PJ, Becker PS. Stem cell homing: rolling, crawling, and nesting. Proc Natl Acad Sci U S A 1998;95(26):15155–157.
48. Jensen UB, Lowell S, Watt FM. The spatial relationship between stem cells and their progeny in the basal layer of human epidermis: a new view based on whole-mount labelling and lineage analysis. Development. 1999;126(11):2409–18.
49. Morrison SJ, Prowse KR, Ho P, Weissman IL. Telomerase activity in hematopoietic cells is associated with self-renewal potential. Immunity. 1996;5(3):207–16.
50. Seydoux G, Mello CC, Pettitt J, Wood WB, Priess JR, Fire A. Repression of gene expression in the embryonic germ lineage of C. Nature. 1996;382(6593):713–6.
51. Morrison SJ, Shah NM, Anderson DJ. Regulatory mechanisms in stem cell biology. Cell. 1997;88(3):287–98.
52. Sherley JL. Asymmetric cell kinetics genes: the key to expansion of adult stem cells in culture. Stem Cells. 2002;20:561–72.
53. Ikebuchi K, Clark SC, Ihle JN, Souza LM, Ogawa M. Granulocyte colony-stimulating factor enhances interleukin 3-dependent proliferation of multipotential hemopoietic progenitors. Proc Natl Acad Sci U S A 1988;85(10):3445–51.
54. Leary AG, Ikebuchi K, Hirai Y, Wong GG, Yang YC, Clark SC, Ogawa M. Synergism between interleukin-6 and interleukin-3 in supporting proliferation of human hematopoietic stem cells: comparison with interleukin-1 alpha. Blood. 1988;71(6):1759–65.
55. Verfaillie C, McGlave P. Leukemia inhibitory factor/human interleukin for DA cells: a growth factor that stimulates the *in vitro* development of multipotential human hematopoietic progenitors. Blood. 1991;77(2):263–70.
56. Szynkaruk M, Kemp SWP, Wood MMD, Gordon T, Borschel GH. Experimental and clinical evidence for use of decellularized nerve allografts in peripheral nerve gap reconstruction. Tissue Eng Part B: Rev. 2013;19(1):83–96.
57. Jia H, Wang Y, Tong X, Liu G, Li Q, Zhang L, Sun X. Sciatic nerve repair by acellular nerve xenografts implanted with BMSCs in rats xenograft combined with BMSCs. Synapse. 2012;66(3):256–69
58. Murakami T, Fujimoto Y, Yasunaga Y, et al. Transplanted neuronal progenitor cells in a peripheral nerve gap promote nerve repair. Brain Res. 2003;974(1–2):17–24.
59. Chen CJ, Ou YC, Liao SL, et al. Transplantation of bone marrow stromal cells for peripheral nerve repair. Exp Neurol. 2007;204(1):443–53.
60. Dezawa M, Takahashi I, Esaki M, et al. Sciatic nerve regeneration in rats induced by transplantation of *in vitro* differentiated bone-marrow stromal cells. Eur J Neurosci. 2001;14(11):1771–76.
61. Hu J, Zhu QT, Liu XL, et al. Repair of extended peripheral nerve lesions in rhesus monkeys using acellular allogenic nerve grafts implanted with autologous mesenchymal stem cells. Exp Neurol. 2007;204(2):658–66.
62. Keilhoff G, Stang F, Goihl A, et al. Transdifferentiated mesenchymal stem cells as alternative therapy in supporting nerve regeneration and myelination. Cell Mol Neurobiol. 2006;26(7–8):1235–52.
63. Tohill M, Mantovani C, Wiberg M, et al. Rat bone marrow mesenchymal stem cells express glial markers and stimulate nerve regeneration. Neurosci Lett. 2004;362(3):200–3.
64. Vitry S, Bertrand JY, Cumano A, et al. Primordial hematopoietic stem cells generate microglia but not myelin-forming cells in a neural environment. J Neurosci. 2003;23(33):10724–31.
65. Hart AM, Brannstorm T, Wiberg M, Terenghi G. Primary sensory neurons and satellite cells after peripheral axotomy in the adult rat: timecourse of cell death and elimination. Exp Brain Res. 2002;142(3):308–18.

66. Le Douarin N. Glial cell lineages in the neural crest. Glia. 1991;4(2):175–84.
67. Murakami T, Fujimoto Y, Yasunaga Y, Ishida O, Tanaka N, Ikuta Y, Ochi M. Transplanted neuronal progenitor cells in a peripheral nerve gap promote nerve repair. Brain Res. 2003;974(1–2):17–24.
68. Reynolds BA, Weiss S. Generation of neurons and astrocytes from isolated cells of the adult mammalian central nervous system. Science. 1992;255(5052):1707–10.
69. Morrison SJ, White PM, Zock C, Anderson DJ. Cell. Prospective identification, isolation by flow cytometry, and *in vivo* self-renewal of multipotent mammalian neural crest stem cells. 1999;96(5):737–49.
70. Caplan AI. The mesengenic process. Clin Plastic Surg. 1994;21(3):429–35.
71. Verfaillie CM. Soluble factor(s) produced by human bone marrow stroma increase cytokine-induced proliferation and maturation of primitive hematopoietic progenitors while preventing their terminal differentiation. Blood. 1993;82(7):2045–53.
72. Bianco P, Riminucci M, Gronthos S, Gehron Robey P. Bone marrow stromal stem cells: nature, biology, and potential applications. Stem Cells. 2001;19(3):180–92.
73. Gronthos S, Simmons PJ. The growth factor requirements of STRO-1-positive human bone marrow stromal precursors under serum-deprived conditions *in vitro*. Blood. 1995;85(4):929–40.
74. Simmons PJ, Torok-Storb B. Identification of stromal cell precursors in human bone marrow by a novel monoclonal antibody, STRO-1. Blood. 1991;78(1):55–62.
75. Pittenger MF, Mackay AM, Beck SC, Jaiswal RK, Douglas R, Mosca JD, Moorman MA, Simonetti DW, Craig S, Marshak DR. Multilineage potential of adult human mesenchymal stem cells. Science 1999;284:143–7.
76. Muraglia A, Cancedda R, Quarto R. Clonal mesenchymal progenitors from human bone marrow differentiate *in vitro* according to a hierarchical model. J Cell Sci. 2000;113:1161–6.
77. Beresford JN, Bennett JH, Devlin C, Leboy PS, Owen ME. Evidence for an inverse relationship between the differentiation of adipocytic and osteogenic cells in rat marrow stromal cell cultures. J Cell Sci. 1992;102:341–51.
78. Gentili C, Bianco P, Neri M, Malpeli M, Campanile G, Castagnola P, Cancedda R, Cancedda FD. Cell proliferation, extracellular matrix mineralization, and ovotransferrin transient expression during *in vitro* differentiation of chick hypertrophic chondrocytes into osteoblast-like cells. J Cell Biol. 1993;122(3):703–12.
79. Orlic D, Kajstura J, Chimenti S, Jakoniuk I, Anderson SM, Li B, Pickel J, McKay R, Nadal-Ginard B, Bodine BM, Leri A, Anversa P. Bone marrow cells regenerate infarcted myocardium. Nature. 2001;410(6829):701–5.
80. Lagasse E, Connors H, Al-Dhalimy M, Reitsma M, Dohse M, Osbourne L, Wang Z, Finegold M, Weissman IL, Grompe M. Purified hematopoietic stem cells can differentiate into hepatocytes *in vivo*. Nature. 2000;6(11):1229–34.
81. Ferrari G, Cusella-De Angelis G, Coletta M, Paolucci E, Storniuolo A, Cossu G, Mavilio F. Muscle regeneration by bone marrow-derived myogenic progenitors. Science. 1998;279(5356):1528–30.
82. Gojo S, Gojo N, Mori T, Abe H, Kyo S, Hata J, Umezawa A. *In vivo* cardiovasculogenesis by direct injection of isolated adult mesenchymal stem cells. Exp Cell Res. 2003;288(1):51–9.
83. Barbash IM, Chouraqui P, Baron J, Feinberg MS, Etzion S, Tessone A, Miller L, Guetta E, Zipori D, Kedes LH, Kloner RA, Leor J. Systemic delivery of bone marrow-derived mesenchymal stem cells to the infarcted myocardium: feasibility, cell migration, and body distribution. Circulation. 2003;108(7):863–8.
84. Jiang Y, Jahagirdar BN, Reinhardt RL, Schwartz RE, Keene CD, Ortiz-Gonzalez XR, Reyes M, Lenvik T, Lund T, Blakstad M, et al. Pluripotency of mesenchymal stem cells derived from adult marrow. Nature. 2002;418(6893):41–9.
85. Kim BJ, Seo JH, Bubien JK, Young SO. Differentiation of adult bone marrow stem cells into neuroprogenitor cells *in vitro*. Neuroreport. 2002;13(9):1185–8.
86. Wislet-Gendebien S, Leprince P, Moonen G, Rogister B. Regulation of neural markers nestin and GFAP expression by cultivated bone marrow stromal cells. J Cell Sci. 2003;116(Pt 16):3295–3302.

87. Dezawa M, Takahashi I, Esaki M, Takano M, Sawada H. Sciatic nerve regeneration in rats induced by transplantation of *in vitro* differentiated bone-marrow stromal cells. Eur J Neurosci. 2001;14(11):1771–6.
88. Kopen GC, Prockop DJ, Phinney DG.Marrow stromal cells migrate throughout forebrain and cerebellum, and they differentiate into astrocytes after injection into neonatal mouse brains. Proc Natl Acad Sci U S A 1999;96(19):10711–6.
89. Mezey E, Chandross KJ, Harta G, Maki RA, McKercher SR. Turning blood into brain: cells bearing neuronal antigens generated *in vivo* from bone marrow. Science. 2000;290(5497):1779–82.
90. Akiyama Y, Radtke C, Honmou O, Kocsis JD. Remyelination of the spinal cord following intravenous delivery of bone marrow cells. Glia. 2002;39(3):229–36.
91. Hofstetter CP, Schwarz EJ, Hess D, Widenfalk J, El Manira A, Prockop DJ, Olson L. Marrow stromal cells form guiding strands in the injured spinal cord and promote recovery. Proc Natl Acad Sci U S A 2002;99(4):2199–2204.
92. Parekkadan B, Milwid JM. Mesenchymal stem cells as therapeutics. Annu Rev Biomed Eng. 2010;12:87–117.
93. Ugarte DA, et al. Differential expression of stem cell mobilization-associated molecules on multi-lineage cells from adipose tissue and bone marrow. Immunol Lett. 2003;89(2–3):267–70.
94. Ugarte DA, et al. Comparison of multi-lineage cells from human adipose tissue and bone marrow. Cells Tissues Organs. 2003;174(3):101–9.
95. Mantovani C, et al. Morphological, molecular and functional differences of adult bone marrow- and adipose-derived stem cells isolated from rats of different ages. Exp Cell Res. 2012;318(16):2034–48.
96. Im GI, Shin YW, Lee KB. Do adipose tissue-derived mesenchymal stem cells have the same osteogenic and chondrogenic potential as bone marrow-derived cells? Osteoarthritis Cartilage. 2005;13(10):845–53.
97. Boxall SA, Jones E. Markers for characterization of bone marrow multipotential stromal cells Stem Cells Int. 2012;2012:975871.
98. Caplan AI. Review: mesenchymal stem cells: cell-based reconstructive therapy in orthopedics. Review: Tissue Eng. 2005;11(7–8):1198–211.
99. Caplan AI, Correa D. The MSC: an injury drugstore. Cell stem cell. 2011;9(1):11–5.
100. Griffin MD, et al. Adult mesenchymal stromal cell therapy for inflammatory diseases: how well are we joining the dots? Stem Cells. 2013.
101. Min JY, et al. Significant improvement of heart function by cotransplantation of human mesenchymal stem cells and fetal cardiomyocytes in postinfarcted pigs. Ann Thorac Surg. 2002;74(5):1568–75.
102. Tateishi-Yuyama E, et al. Therapeutic angiogenesis for patients with limb ischaemia by autologous transplantation of bone-marrow cells: a pilot study and a randomised controlled trial. Lancet. 2002;360(9331):427–35.
103. Dabiri G, Heiner D, Falanga V. The emerging use of bone marrow-derived mesenchymal stem cells in the treatment of human chronic wounds. Expert Opin Emerg Drugs. 2013;18(4):405–19.
104. Caplan AI. Adult mesenchymal stem cells for tissue engineering versus regenerative medicine. J Cellular Physiol. 2007;213(2):341–7.
105. Kern S, et al. Comparative analysis of mesenchymal stem cells from bone marrow, umbilical cord blood, or adipose tissue. Stem Cells. 2006;24(5):1294–301.
106. Lee RH, et al. Characterization and expression analysis of mesenchymal stem cells from human bone marrow and adipose tissue. Cell Physiol Biochem. 2004;14(4–6):311–24.
107. James AW, et al. An abundant perivascular source of stem cells for bone tissue engineering. Stem Cells Transl Med. 2012;1(9):673–84.
108. Rombouts WJ, Ploemacher RE. Primary murine MSC show highly efficient homing to the bone marrow but lose homing ability following culture. Leukemia. 2003;17(1):160–70.
109. Dmitrieva RI, et al. Bone marrow- and subcutaneous adipose tissue-derived mesenchymal stem cells: differences and similarities. Cell Cycle. 2012;11(2):377–83.

110. Al-Nbaheen M, et al. Human stromal (mesenchymal) stem cells from bone marrow, adipose tissue and skin exhibit differences in molecular phenotype and differentiation potential. Stem Cell Rev. 2013;9(1):32–43.
111. Bayati V, et al. Expression of surface markers and myogenic potential of rat bone marrow- and adipose-derived stem cells: a comparative study. Anat Cell Biol. 2013;46(2):113–21.
112. De Ugarte DA, et al. Differential expression of stem cell mobilization-associated molecules on multi-lineage cells from adipose tissue and bone marrow. Immunol Lett. 2003;89(2–3):267–70.
113. Tohill M, et al. Rat bone marrow mesenchymal stem cells express glial markers and stimulate nerve regeneration. Neurosci Lett. 2004;362(3):200–3.
114. Dezawa M, et al. Sciatic nerve regeneration in rats induced by transplantation of *in vitro* differentiated bone-marrow stromal cells. Eur J Neurosci. 2001;14(11):1771–6.
115. Dezawa M, Takahashi I, Esaki M, Takano M, Sawada H. Sciatic nerve regeneration in rats induced by transplantation of *in vitro* differentiated bone-marrow stromal cells. Eur J Neurosci. 2001;14(11):1771–6.
116. Cuevas P, Carceller F, Dujovny M, Garcia-Gomez I, Cuevas B, Gonzalez-Corrochan R, Diaz-Gonzalez D, Reimers D. Peripheral nerve regeneration by bone marrow stromal cells. Neurolog Res. 2002;24(7):634–8.
117. Mahanthappa NK, Anton ES, Matthew WD. Glial growth factor 2, a soluble neuregulin, directly increases Schwann cell motility and indirectly promotes neurite outgrowth. J Neurosci. 1996;16(15):4673–83.
118. Bartholomew A. Mesenchymal stem cells suppress lymphocyte proliferation *in vitro* and prolong skin graft survival *in vivo*. Exp Haematol. 2002;30(1):42–8.
119. Saito T, Kuang JQ, Bittra B, Al-Khaldi A, Chin RC. Xenotransplant cardiac chimera: immune tolerance of adult stem cells. Ann Thorac Surg. 2002;74(1):19–24.
120. Deans RJ, Moseley AB. Mesenchymal stem cells: biology and potential clinical uses. Exp Haematol. 2000;28(8):875–84.
121. Tse WT, Pendleton D, Beyer W, D'Andrea A, Guinan EC. Suppression of allogeneic T-cell proliferation by human marrow stromal cells: implications in transplantation. Transplantation. 2003;75(3):389–97.
122. Hori J, Ng TF, Shatos M, Klassen H, Streilein JW, Young MJ. Neural progenitor cells lack immunogenicity and resist destruction as allografts. Stem Cells. 2003;21(4):405–16.
123. Rifle G, Mousson C. Donor-derived hematopoietic cells in organ transplantation: a major step toward allograft tolerance? Transplantation. 2003;75(9 Suppl):3S–7S.
124. Wekerle T, Blaha P, Koporc Z, Bigenzahn S, Pusch M, Muehlbacher F. Mechanisms of tolerance induction through the transplantation of donor hematopoietic stem cells: central versus peripheral tolerance. Transplantation. 2003;75(9 Suppl):21S–5S.
125. Bartholomew A, Patil S, Mackay A, Nelson M, Buyaner D, Hardy W, Mosca J, Sturgeon C, Siatska M, Mahmud N, et al. Baboon mesenchymal stem cells can be genetically modified to secrete human erythropoietin *in vivo*. Hum Gene Ther. 2001;12(12):1527–41.
126. Studeny M, Marini FC, Champlin RE, Zompetta C, Fidler IJ, Andreeff M. Bone marrow-derived mesenchymal stem cells as vehicles for interferon-beta delivery into tumors. Cancer Res. 2002;62(13):3603–8.
127. Kokai LE, Ghaznavi AM, Marra KG. Incorporation of double-walled microspheres into polymer nerve guides for the sustained delivery of glial cell line-derived neurotrophic factor. Biomaterials. 2010;31(8):2313–22.
128. Kokai LE, Bourbeau D, Weber D, McAtee J, Marra KG. Sustained growth factor delivery promotes axonal regeneration in long gap peripheral nerve repair. Tissue Eng Part A. 2011;17(9–10):1263–75.
129. Ward MS, Khoobehi A, Lavik EB, Langer R, Young MJ. Neuroprotection of retinal ganglion cells in DBA/2J mice with GDNF-loaded biodegradable microspheres. J Pharm Sci. 2007;96(3):558–68.
130. Checa-Casalengua P, Jiang C, Bravo-Osuna I, Tucker BA, Molina-Martinez IT, Young MJ, Herrero-Vanrell R. Retinal ganglion cells survival in a glaucoma model by GDNF/Vit E

PLGA microspheres prepared according to a novel microencapsulation procedure. J Control Release. 2011;156(1):92–100.
131. Schmeer C, Straten G, Kugler S, Gravel C, Bahr M, Isenmann S. Dose-dependent rescue of axotomized rat retinal ganglion cells by adenovirus-mediated expression of glial cell-line derived neurotrophic factor *in vivo*. Eur J Neurosci. 2002;15(4):637–43.
132. Wu WC, Lai CC, Chen SL, Sun MH, Xiao X, Chen TL, Tsai RJ, Kuo SW, Tsao YP. GDNF gene therapy attenuates retinal ischemic injuries in rats. Mol Vis. 2004;10:93–102.
133. Miyoshi H TM, Gage FH, Verma IM. Stable and efficient gene transfer into the retina using an HIV-based lentiviral vector. Proc Natl Acad Sci U S A. 1997;94(19):10319–23.
134. Hermens WT, Verhaagen J. Viral vectors, tools for gene transfer in the nervous system. Prog Neurobiol. 1998;55(4):399–432.
135. Bainbridge JW, Stephens C, Parsley K, Demaison C, Halfyard A, Thrasher AJ, Ali RR. *In vivo* gene transfer to the mouse eye using an HIV-based lentiviral vector; efficient long-term transduction of corneal endothelium and retinal pigment epithelium. Gene Ther. 2001;8(21):1665–8.
136. Liu Y, et al. A new method for Schwann-like cell differentiation of adipose derived stem cells. Neurosci Lett. 2013;551:79–83.
137. Caddick J, et al. Phenotypic and functional characteristics of mesenchymal stem cells differentiated along a Schwann cell lineage. Glia. 2006;54(8):840–9.
138. Dadon-Nachum M, et al. Differentiated mesenchymal stem cells for sciatic nerve injury. Stem Cell Rev. 2011;7(3):664–71.
139. Wei Y, et al. Schwann-like cell differentiation of rat adipose-derived stem cells by indirect co-culture with Schwann cells *in vitro*. Cell Prolif. 2010;43(6):606–16.
140. Kalbermatten DF, et al. Neurotrophic activity of human adipose stem cells isolated from deep and superficial layers of abdominal fat. Cell Tissue Res. 2011;344(2):251–60.
141. Kaewkhaw R, Scutt AM, Haycock JW. Haycock, Anatomical site influences the differentiation of adipose-derived stem cells for Schwann-cell phenotype and function. Glia. 2011;59(5):734–49.
142. Sowa Y, et al. Adipose-derived stem cells produce factors enhancing peripheral nerve regeneration: influence of age and anatomic site of origin. Stem Cells Dev. 2012;21(11): 1852–62.
143. Marconi S, et al. Human adipose-derived mesenchymal stem cells systemically injected promote peripheral nerve regeneration in the mouse model of sciatic crush. Tissue Eng Part A. 2012;18(11–12):1264–72.
144. Schlosser S, et al. Paracrine effects of mesenchymal stem cells enhance vascular regeneration in ischemic murine skin. Microvascular. 2012
145. Omori Y, et al. Optimization of a therapeutic protocol for intravenous injection of human mesenchymal stem cells after cerebral ischemia in adult rats. Brain Res. 2008;1236:30–8.
146. Assis AC, et al. Time-dependent migration of systemically delivered bone marrow mesenchymal stem cells to the infarcted heart. Cell Transplant. 2010;19(2):219–30.
147. Gao J, et al. The dynamic *in vivo* distribution of bone marrow-derived mesenchymal stem cells after infusion. Cells Tissues Organs. 2001;169(1):12–20.
148. Eggenhofer E, et al. Mesenchymal stem cells are short-lived and do not migrate beyond the lungs after intravenous infusion. Front Immunol. 2012;3:297.
149. Karp JM, Teo GS. Teo, Mesenchymal stem cell homing: the devil is in the details. Stem Cell. 2009;4(3):206–16.
150. Lee RH, et al. Intravenous hMSCs improve myocardial infarction in mice because cells embolized in lung are activated to secrete the anti-inflammatory protein TSG-6. Stem Cell. 2009;5(1):54–63.
151. McGrath AM, et al. Fibrin conduit supplemented with human mesenchymal stem cells and immunosuppressive treatment enhances regeneration after peripheral nerve injury. Neurosci Lett. 2012;516(2):171–6.
152. Carriel V, et al. Combination of fibrin-agarose hydrogels and adipose-derived mesenchymal stem cells for peripheral nerve regeneration. J Neural Eng. 2013;10(2):026022.

153. Cuevas P, et al. Peripheral nerve regeneration by bone marrow stromal cells. Neurol Res. 2002;24(7):634–8.
154. Santiago LY, et al. Delivery of adipose-derived precursor cells for peripheral nerve repair. Cell Transplant. 2009;18(2):145–58.
155. Liu BS, Yang YC, Shen CC. Regenerative effect of adipose tissue-derived stem cells transplantation using nerve conduit therapy on sciatic nerve injury in rats. J Tissue Eng Regen Med. 2012;8(5):337–50.
156. Chen X, et al. Study of *in vivo* differentiation of rat bone marrow stromal cells into schwann cell-like cells. Microsurgery. 2006;26(2):111–5.
157. Keilhoff G, et al. Transdifferentiated mesenchymal stem cells as alternative therapy in supporting nerve regeneration and myelination. Cell Mol Neurobiol. 2006;26(7–8):1235–52.
158. Mimura T, et al. Peripheral nerve regeneration by transplantation of bone marrow stromal cell-derived Schwann cells in adult rats. J Neurosurg. 2004;101(5):806–12.
159. Hu N, et al. Long-term outcome of the repair of 50 mm long median nerve defects in rhesus monkeys with marrow mesenchymal stem cells-containing, chitosan-based tissue engineered nerve grafts. Biomaterials. 2013;34(1):100–11.
160. Orbay H, et al. Differentiated and undifferentiated adipose-derived stem cells improve function in rats with peripheral nerve gaps. J Plast Reconstr Aesthet Surg. 2012;65(5): 657–64.
161. Hu J, et al. Repair of extended peripheral nerve lesions in rhesus monkeys using acellular allogenic nerve grafts implanted with autologous mesenchymal stem cells. Exp Neurol. 2007;204(2):658–66.
162. Ding F, et al. Use of tissue-engineered nerve grafts consisting of a chitosan/poly(lactic-co-glycolic acid)-based scaffold included with bone marrow mesenchymal cells for bridging 50-mm dog sciatic nerve gaps. Tissue Eng Part A. 2010;16(12):3779–90.
163. Wakao S, et al. Long-term observation of auto-cell transplantation in non-human primate reveals safety and efficiency of bone marrow stromal cell-derived Schwann cells in peripheral nerve regeneration. Exp Neurol. 2010;223(2):537–47.
164. Xue C, et al. Joint use of a chitosan/PLGA scaffold and MSCs to bridge an extra large gap in dog sciatic nerve. Neurorehabil Neural Repair. 2012;26(1):96–106.
165. Fischer UM, et al. Pulmonary passage is a major obstacle for intravenous stem cell delivery: the pulmonary first-pass effect. Stem Cells Dev. 2009;18(5):683–92.
166. Ball LM, et al. Multiple infusions of mesenchymal stromal cells induce sustained remission in children with steroid-refractory, grade III-IV acute graft-versus-host disease. Br J Haematol. 2013;163:501–9.
167. Forbes GM, et al. A phase 2 study of allogeneic mesenchymal stromal cells for luminal Crohn's disease refractory to biologic therapy. Clin Gastroenterol Hepatol. 2014;12(1): 64–71.
168. Hao H, et al. Multiple intravenous infusions of bone marrow mesenchymal stem cells reverse hyperglycemia in experimental type 2 diabetes rats. Biochem Biophys Res Commun. 2013;436(3):418–23.
169. Wang H, et al. Hematopoietic stem cell transplantation with umbilical cord multipotent stromal cell infusion for the treatment of aplastic anemia-a single-center experience. Cytotherapy. 2013;15(9):1118–25.
170. Pereira Lopes FR, et al. Bone marrow stromal cells and resorbable collagen guidance tubes enhance sciatic nerve regeneration in mice. Exp Neurol. 2006;198(2):457–68.
171. Zhang P, et al. Bridging small-gap peripheral nerve defects using biodegradable chitin conduits with cultured schwann and bone marrow stromal cells in rats. J Reconstr Microsurg. 2005;21(8):565–71.
172. Zheng L, Cui HF. Enhancement of nerve regeneration along a chitosan conduit combined with bone marrow mesenchymal stem cells. J Mater Sci Mater Med. 2012;23(9):2291–302.
173. Ao Q, et al. The regeneration of transected sciatic nerves of adult rats using chitosan nerve conduits seeded with bone marrow stromal cell-derived Schwann cells. Biomaterials. 2011;32(3):787–96.

174. Zhao Z, et al. Repair of nerve defect with acellular nerve graft supplemented by bone marrow stromal cells in mice. Microsurgery. 2011;31(5):388–94.
175. Oliveira JT, et al. Mesenchymal stem cells in a polycaprolactone conduit enhance median-nerve regeneration, prevent decrease of creatine phosphokinase levels in muscle, and improve functional recovery in mice. Neuroscience. 2010;170(4):1295–303.
176. Pereira Lopes FR, et al. Transplantation of bone-marrow-derived cells into a nerve guide resulted in transdifferentiation into Schwann cells and effective regeneration of transected mouse sciatic nerve. Micron. 2010;41(7):783–90.
177. Shimizu S, et al. Peripheral nerve regeneration by the *in vitro* differentiated-human bone marrow stromal cells with Schwann cell property. Biochem Biophys Res Commun. 2007;359(4):915–20.
178. Frattini F, et al. Mesenchymal stem cells in a polycaprolactone conduit promote sciatic nerve regeneration and sensory neuron survival after nerve injury. Tissue Eng Part A. 2012;18(19–20):2030–9.
179. Ladak A, et al. Differentiation of mesenchymal stem cells to support peripheral nerve regeneration in a rat model. Exp Neurol. 2011;228(2):242–52.
180. Wang J, et al. Bone marrow mesenchymal stem cells promote cell proliferation and neurotrophic function of Schwann cells *in vitro* and *in vivo*. Brain Res. 2009;1262:7–15.
181. Liu G, et al. Transplantation of adipose-derived stem cells for peripheral nerve repair. Int J Mol Med. 2011;28(4):565–72.
182. Liu GB, et al. Adipose-derived stem cells promote peripheral nerve repair. Arch Med Sci. 2011;7(4):592–6.
183. Chen CJ, et al. Transplantation of bone marrow stromal cells for peripheral nerve repair. Exp Neurol. 2007;204(1):443–53.
184. Wang Y, et al. Recellularized nerve allografts with differentiated mesenchymal stem cells promote peripheral nerve regeneration. Neurosci Lett. 2012;514(1):96–101.
185. Costa HJ, et al. Mesenchymal bone marrow stem cells within polyglycolic acid tube observed *in vivo* after six weeks enhance facial nerve regeneration. Brain Res. 2013;1510:10–21.
186. Zarbakhsh S, et al. The effects of schwann and bone marrow stromal stem cells on sciatic nerve injury in rat: a comparison of functional recovery. Cell J. 2012;14(1):39–46.
187. Shen CC, Yang YC, Liu BS. Peripheral nerve repair of transplanted undifferentiated adipose tissue-derived stem cells in a biodegradable reinforced nerve conduit. J Biomed Mater Res A. 2012;100(1):48–63.
188. Wang D, et al. Repairing large radial nerve defects by acellular nerve allografts seeded with autologous bone marrow stromal cells in a monkey model. J Neurotrauma. 2010;27(10):1935–43.
189. Lopatina T, et al. Adipose-derived stem cells stimulate regeneration of peripheral nerves: BDNF secreted by these cells promotes nerve healing and axon growth de novo. PLoS One. 2011;6(3):e17899.
190. Hou SY, et al. Tissue-engineered peripheral nerve grafting by differentiated bone marrow stromal cells. Neuroscience. 2006;140(1):101–10.
191. Reid AJ, et al. Nerve repair with adipose-derived stem cells protects dorsal root ganglia neurons from apoptosis. Neuroscience. 2011;199:515–22.
192. Wang D, et al. Bridging small-gap peripheral nerve defects using acellular nerve allograft implanted with autologous bone marrow stromal cells in primates. Brain Res. 2008;1188:44–53.
193. Ghoreishian M, et al. Facial nerve repair with Gore-Tex tube and adipose-derived stem cells: an animal study in dogs. J Oral Maxillofac Surg. 2013;71(3):577–87.
194. Choi BH, et al. Transplantation of cultured bone marrow stromal cells to improve peripheral nerve regeneration. Int J Oral Maxillofac Surg. 2005;34(5):537–42.
195. Wang X, et al. Schwann-like mesenchymal stem cells within vein graft facilitate facial nerve regeneration and remyelination. Brain Res. 2011;1383:71–80.

Chapter 21
Why Brain Science is Essential to the Success of Hand Allotransplantation

Scott H. Frey

Introduction

As a neuroscientist, my first reaction when learning that chronic amputees were receiving hand transplants was concern: Having undergone extensive reorganizational changes following years or even decades of hand absence, was there really any chance that a fully mature brain could readapt to support dexterous hand control and weave coherent perceptual experiences from sensory signals arriving over long-disused sensory pathways? The subsequent years working with several of these remarkable individuals have profoundly altered my view on the potential to reverse amputation, and have forced me to reconsider some accepted tenets regarding the effects of experience (or lack thereof) on the mature brain. The emerging picture is one in which consequences of limb amputation on the human brain may be far more reversible than could ever have been imagined; a possibility that elevates hope of recovery not only for amputees but also a wide range of patients who have experienced injuries to the body, spinal cord, or brain.

Why the Brain?

To understand the role of the brain in recovery from peripheral nerve injury more generally, consider first the repair of a transected median nerve in the forearm. While it is often possible to unite the severed fascicles surgically, function depends on regenerating axons on the proximal side forming neuromuscular junctions and reaching sensory organelles on the distal end. To exploit a tree metaphor, in expert hands the repair can ensure that the proper axons are directed along the correct limbs. However, growth along the smaller branches and twigs is not under the surgeon's control, and

S. H. Frey (✉)
Brain Imaging Center & Rehabilitation Neuroscience Laboratory, University of Missouri, 205a Melvin H. Marx Building, 1416 Carrie Francke Drive, 65211 Columbia, MO, USA
e-mail: freys@missouri.edu

G. Brandacher (ed.), *The Science of Reconstructive Transplantation*,
Stem Cell Biology and Regenerative Medicine, DOI 10.1007/978-1-4939-2071-6_21

therefore some reinnervation errors are going to occur, i.e., axons from the regenerating side are unlikely to find their original targets. As detailed in what follows, the brain is capable of reorganizing in response to changes in stimulation, and it is this capacity for *experience-dependent plasticity* that provides the potential to *compensate* for these reinnervation errors by acquiring new central representations for motor and sensory functions. The need for central adaptations in functional recovery would seem to be even greater in cases where long-standing amputations are reversed through allogeneic hand transplants. Determining how best to exploit this remarkable capacity is absolutely essential to optimizing recovery of function and is therefore vital to the success of composite tissue allotransplant efforts. In what follows, I review some of what is presently understood about the effect of upper extremity amputation on brain organization and the central response to its reversal through allogeneic transplantation.

Amputation and the Brain

Changes in the sensory and especially motor nerves of the residual limb of amputees are widely recognized. Perhaps less known is that injuries that disrupt the flow of afferent and/or efferent signals between brain and body (including limb amputation) precipitate significant reorganization of function at all levels of the central nervous system, including the spinal cord, brain stem, thalamus, and the primary sensory (S1) and motor (M1) cortices. While the majority of research has focused on the cortex, it is important to recognize that changes here could reflect an amplification of reorganization occurring at any, or all, of these levels [1–5]. Far from subtle, these experience-dependent changes in the functional organization of the cortical representations have been dubbed "massive" [6, 7], and their discovery in the early 1980s [4, 8–10] was pivotal in reversing the prevailing view of the mature brain as lacking the capacity for plasticity [11].

Experience-Dependent Cortical Organization

Primary sensory and motor cortices contain somatotopically organized maps of the body with the representation of the hand flanked laterally by the face and medially by the upper arm [12]. These representations play a critical role in sensation and motor control, and *their organization is maintained through competitive interactions that depend on afferent and efferent activity* [3, 13]. Studies in nonhuman primates using cortical sensory unit recording or motor microstimulation techniques reveal that injuries (e.g., limb amputation, brachial plexus lesion, or spinal cord injury) that disrupt the flow of signals between brain and body therefore cause pronounced cortical reorganization [14]. Specifically, there is an *expansion of adjacent maps* into the cortical territory formerly devoted to the affected body part [15, 16]. In S1, these changes begin almost immediately following hand amputation, continue for months or longer, and culminate in substantial changes in neurons' receptive fields [6, 17]. More precisely, some S1 neurons—located within the hemisphere contralateral to the amputation—that were formerly responsive to stimulation of the amputated hand, can be activated by stimulation of the somatotopically adjacent face and/or residual forearm [2, 18].

Similar changes occur in M1 contralateral to the amputated hand. Years after amputation, microstimulation of neurons in areas that formerly targeted amputated hand muscles can evoke neuromuscular activity in the residual limb or shoulder [19]. Multiple mechanisms appear to contribute to both S1 and M1 changes over different time scales including dynamic functional rebalancing of inhibitory and excitatory synapses [14], as well as longer term growth of intracortical (but not thalamocortical) connections [20].

The advent of noninvasive functional brain mapping techniques makes it possible to entertain whether similar *intrahemispheric changes* occur in the human brain following limb amputation. Though lacking the resolution of the invasive techniques discussed above, studies of human upper extremity amputees provide consistent evidence for reorganizational changes in the gross somatotopic organization (on the scale of multiple millimeters) of both S1 and M1contralateral to the amputated hand. These include an expanded representation of muscles of the residual limb [21, 22], and in some cases a shift in the representation of muscles of the face into the former hand territory [23]. Likewise, the sensory representation of axial body surfaces [24] and/or the face [25–27] may intrude into the former S1 hand territory. Consistent with animal models [3], transcranial magnetic stimulation (TMS) studies suggest that reduced levels of intrahemispheric inhibition post-injury may play a key role in M1 reorganization [28, 29]. To date, there is no evidence for macroscopic changes in cortical gray matter following amputation [30, 31], although thalamus volume may be decreased [30].

In addition to these changes occurring within the hemisphere contralateral to the amputation, studies involving animal models [32, 33] and human amputees [34] provide evidence for *interhemispheric changes* following unilateral deafferentation. Specifically, movements of the intact hand by amputees are associated with increased activity within the sensorimotor cortex ipsilateral to the amputation [35] and more precisely within the former M1/S1 hand territory [34]. These changes appear to be attributable, at least in part, to a reduction in the normal levels of interhemispheric inhibition that exists between cortical hand representations [21], and this is likely the mechanism responsible for the fact that rodents with unilateral deafferentation reveal increased functional connectivity between forepaw representations in the left and right cerebral hemispheres when at rest [33]. Recent work in our group also suggests that increased activity within the former hand territory during use of the intact hand may play a functional role in precision control [36].

Adaptive, Maladaptive, or Functionally Irrelevant?

From a clinical standpoint, the key question is whether intra and/or interhemispheric changes in functional organization following limb amputation impact quality of life. The common assumption is that these reorganizational changes must be *functionally relevant.* This issue is, however, far from settled. Some investigations contend that *decrease-related reorganizational changes* have *maladaptive relevance*; i.e., they contribute to the neuropathic pain experienced following limb amputation [26], spinal cord injury [37], or by those who develop complex regional pain syndrome [38, 39]. This hypothesis has motivated a variety of behavioral, neurostimulatory, and pharmacological interventions targeting the reversal of these reorganizational changes [40]. However, not all evidence from amputees supports this hypothesis [29, 41]. A recent report presents data that pain is actually associated with preservation of function and structure within the former sensorimotor hand territory following amputation [31]; a result that questions the idea that cortical reorganization is maladaptive (cf. [42]). A possible limitation of this work is that it did not differentiate between pain in the residual limb versus neuropathic pain in the phantom, which may have different underlying mechanisms. It is also worth noting that nearly all neuroscientific investigations of human amputees suffer from small sample sizes and that may contribute to type I and II statistical errors [43]

Another candidate for maladaptive plasticity is the persistent reorganization in S1 observed in mature monkeys following experimental nerve section and repair [44, 45]. Although direct tests of the relationship between cortical map reorganization and sensory functions are lacking, similar changes may underlie poor functional outcomes in human adults following surgical repairs of peripheral nerves in the hand or forearm. I will return to this issue shortly.

Conversely, an argument can be made that expansions of adjacent cortical maps into the former hand territory should have *adaptive relevance*. The reason is that neurons formerly devoted to the now absent body part are functionally "reassigned" to intact regions of the body thereby increasing its cortical representation. Indeed some older studies do report behavioral improvements in sensory functions of amputees' residual limbs [46–48]. However, more recent studies do not support this view [49, 50]. In a recent unpublished investigation of unilateral upper extremity amputees, we found significant expansion of the face representations into the former hand territory contralateral to the amputated limb using functional magnetic resonance imaging (fMRI). However, we too were unsuccessful in obtaining any evidence for changes in sensory thresholds or discrimination when comparing the ipsilateral and contralateral sides of the face (Fig. 21.1). Furthermore, despite reorganization of primary sensorimotor regions, we find that chronic unilateral amputees retain the ability to plan actions of the missing hand with a level of accuracy comparable to the unaffected side and to controls [51]. This suggests retention of an internal model of the affected limb's biomechanical properties despite long-term absence of sensory feedback.

On the basis of these data, one might be tempted to conclude that reorganizational changes in cortical maps due to decreases in activity are *functionally irrelevant*. Consistent with this perspective, neurophysiological recordings and microstimulation in residual limb peripheral nerves suggest that even chronic amputees retain functional sensory [49, 52] and motor [53–56] representations of their missing hands.

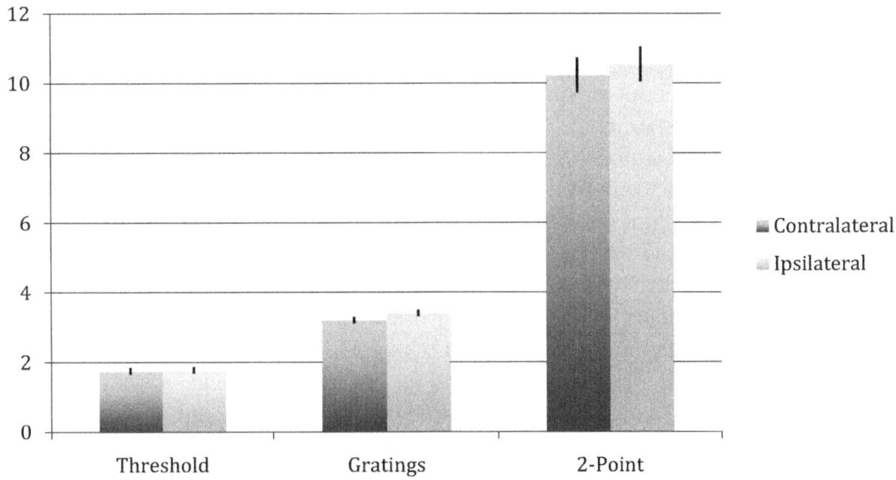

Fig. 21.1 Performance of 16 unilateral upper extremity amputees across three tests of sensory function. In each test, stimuli were applied to the perioral regions located contra- or ipsilateral to the amputation. We failed to detect any significant differences between sides in the SemmesWein-stein test of sensory thresholds, grating orientation test, or static 2-point discrimination

Another perspective on the functional relevance of amputation-related cortical reorganization comes from attempts to determine whether it can be reversed, and to identify associated behavioral consequences.

Can Amputation-Related Cortical Reorganization Be Reversed?

This question will remain unanswerable in humans. Rightfully, we never will have the opportunity to map the brain of healthy individuals prior to their traumatic amputations. Further, there is considerable intra and interindividual variation in brain structural and functional organization. As illustrated below, this places limitations on the precision of comparing cortical representations of the affected versus unaffected hand, and between patients and healthy controls. We can, however, hope to answer a number of other important questions: Is the representation of the transplanted hand located where we would expect it to be based on normative data from carefully matched controls? Do stimulation and/or use of the hand recruit other brain regions more heavily? Is there any persistent evidence for intrusion of representations of other body parts into the normatively defined former hand territory? And, perhaps of greatest clinical importance, do any of the changes that we observe at the central level correlate with current or future recovery of function?

Nerve Repair and Cortical Reorganization

What little we know about the reversibility of reorganizational changes in the cortex comes from classic experiments of sensory nerve injury and repair in monkeys. Following axotomy and surgical repair of the fascicles, there is an initial silencing of S1 cortical neurons whose receptive fields lie below the lesion. As the nerve regenerates at approximately 2–3 mm per day [57], the S1 neurons whose receptive fields were formerly devoted to the deafferented region gradually redevelop a map. However, in adult monkeys, map topology remains disorganized and includes islands of responsive regions amongst areas of unresponsive neurons [45]. Fetal monkeys experiencing the nerve injury and repair recover what appears to be a typically organized S1 map [58], which is likely attributable to the increased capacity for experience-dependent plasticity during the period of rapid cortical development. Importantly, if the nerve is crushed rather than severed, the S1 map in adult monkeys also recovers its normal topography [10]. Similarly, adult whisker amputation alters the receptive field properties of barrel cells of the rodent sensory cortex, but appears to leave the architecture of the cortical map intact and thus recoverable once the whisker regenerates, reinstating afferent signals [59, 60]. This indicates that it is not the period of deafferentation that is problematic for sensory map recovery, but rather the reinnervation errors that occur through nerve regeneration [3]. As noted above, these errors and the resulting changes in cortical maps may account for the persistent sensory difficulties experienced by patients following adulthood nerve repairs [44, 45].

Recipients of hand replants provide an opportunity to investigate the reversibility of reorganizational changes in the human cortex, a complete loss of afferent and efferent traffic between the brain and regions distal to the injury. Although hand replantation is done within 24 h of amputation, the brains of these patients also undergo an extended period of time during which activity between the hand and brain is first absent and then diminished due to the pace of nerve regeneration. Compared with hand transplants, there may be greater potential for reinnervation errors due to the fact that the surgery involves a traumatically amputated limb. Nevertheless, the few fMRI investigations of hand replant recipients conducted indicate increased activity within M1–S1 during hand movement [61–63]. Interactions between these primary areas and higher-level motor regions may evolve along with hand function. Functional connectivity analyses of these fMRI data reveal increased contributions of premotor cortex and increased inhibition from ipsilateral M1 during movements of the affected hand, which may be compensatory and related to the increased difficulty in sensorimotor control [63].

Chronic Deafferentation and Cortical Reorganization

In comparison to replant patients, transplant recipients have typically been without their hand(s) for years, or even decades, before transplanted. Despite substantial variation among cases and the small number of published functional neuroimaging studies, it appears that volitional movement can elicit increased activity in the contralateral primary sensorimotor cortex of allogeneic transplant recipients [64–66]. It is important, however, to recognize that in most (if not all) cases, these tasks involved use of forearm flexors and extensors, which were not transplanted. This fact constrains what should be concluded from these data with regard to the reversal of amputation-related cortical reorganization. It does seem reasonable to expect that forearm muscles would be considerably less active following distal hand loss, and this could certainly contribute to changes in activity-dependent representations. However, it would be a mistake to treat these changes as reflecting the emergence of cortical representations of transplanted muscles.

To address the cortical representation of transplanted muscles one would need a task that isolates movements of intrinsic hand muscles (e.g., finger abduction–adduction). In practice, however, this is challenging because few patients appear to develop volitional control of muscles intrinsic to the transplanted hand. When this does develop, it appears to occur late in the recovery process [67]. An alternative is to determine whether TMS of M1 can elicit motor-evoked potentials (MEPs) from transplanted muscles. This approach was applied successfully to two double-hand transplant recipients and demonstrated substantial interindividual variation, as well as significant differences within individuals between the time to re-establish M1 muscle representations on the left and right sides [68]. Because TMS is a means of artificially stimulating M1, the question remains as to when (or if) these representations will become relevant to hand control.

Whether or not cortical responses arise from transplanted tissue is a persistent issue that plagues motor (really "sensorimotor" due to the afferent feedback that accompanies movement) studies of transplants. This challenge would seem to be overcome in studies employing localized sensory stimuli. In the first such published investigation, brushing of the transplanted insensate palm with a coarse sponge increased activity within the contralateral postcentral (putative S1) and precentral (putative M1) gyri a mere 4 days postsurgery [69]. As appreciated by the authors, there is no known physiological mechanism that can account for such rapid recovery. What could produce these data? My suspicion is that this seemingly impossible result most likely arises from the inadvertent induction of neuromuscular activity in the flexors and extensors of the residual forelimb during application of the tactile stimulus. A similar criticism can be levied against our earlier work on sensory recovery 4 months posttransplant in a right hand recipient. In this case, however, the patient had regained sensitivity in the thenar eminence and wrist [70]. It is clear that more well-controlled systems for the delivery of sensory stimuli during fMRI are required. Along these lines, we have developed a computer-controlled, pneumatic system to deliver cutaneous sensory stimuli to any location of the body including the individual fingers and the face [71]. Some preliminary findings from this technique are presented below.

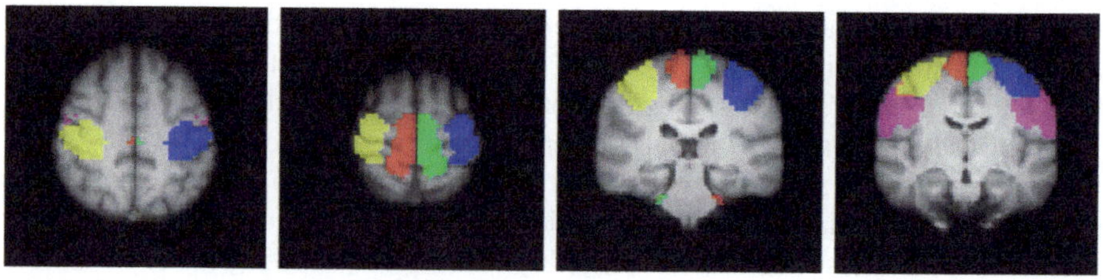

Fig. 21.2 Functionally defined normative sensorimotor representations based on fMRI data from 17 healthy, right-handed adults. *Blue*=right hand, *yellow*=left hand, *green*=right foot, *red*=left foot, *violet*=lips

Recruitment of the Normatively Defined Former Hand Territory

While it may be impossible to ever determine whether amputation-related reorganizational changes can be fully reversed, progress is being made on answering a number of other important, clinically relevant questions. Is the representation of the transplanted hand located in the location we would expect based on normative data from carefully matched controls? Do stimulation and/or use of the hand recruit other brain regions more heavily? Is there any persisting evidence for intrusion of representations of other body parts into the normatively defined former hand territory? And, perhaps of greatest clinical importance, do any of the changes that we observe at the central level correlate with current or future recovery of function?

Along with our collaborators at the Christine M. Kleinert Institute, we have developed a strategy to address what happens to the former hand territory after amputation and after transplantation. A distinguishing feature of our approach is that fMRI data from a control group matched to patients on the basis of age, sex, and hand-dominance is used to functionally define the regions of cortex responsive to sensory and/or motor stimulation of the hands (i.e, the normative hand representations) in a standardized space (Fig. 21.2 [72]). Then, using nonlinear techniques, we can warp the coregistered structural and functional brain images of our patients into this template space and undertake quantitative comparisons of activity within the probabilistically defined former hand territory during various manipulations. By way of illustration, consider DS, who was a unilateral right hand amputee for 35 years before receiving his transplant 4 months prior to our testing. His level of hand functional recovery at this early stage was limited (Carroll score=57, disabilities of the arm, shoulder, and hand (DASH) score=12.5). Sensory threshold testing revealed a localized patch of low normal sensitivity on the thenar eminence, while the rest of the hand remained insensitive at this early stage. Because we did not test this individual prior to his transplant, the amputated state was estimated based on data from a cohort of unilateral, below elbow, pretransplant amputees (details can be found in [70]).

Together with the research introduced earlier, these findings are consistent with two working hypotheses. First, areas of the sensory and motor cortex devoted to

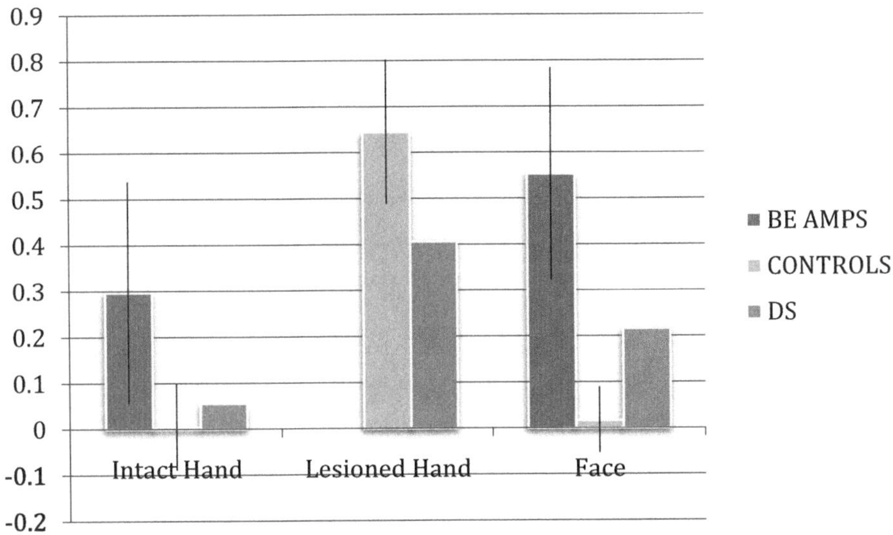

Fig. 21.3 Percent change in the fMRI signal relative to rest within the normatively defined former hand sensorimotor hand territory. Finger flexion–extension of DS's transplanted right hand 4 months postsurgery are associated with increased activation within the former hand territory, albeit below what is detected on an average in the group of matched controls. Like the below-elbow (BE) amputee group mean, DS shows a tendency for increased activity within the former hand territory during movements of the face, which is not detected in healthy controls. Error bars represent 95 % confidence intervals around group means [70]

representing the hand prior to amputation come to represent the transplanted hand. Volitional finger extension–flexion movements of the transplanted hand increased activity within the contralateral primary sensorimotor areas and specifically within the normatively defined former sensorimotor hand territory (Fig. 21.3).

Second, changes in cortical organization associated with unilateral hand amputation may not be fully reversed. Like the pretransplant amputees, DS shows some evidence for persistent intrahemispheric reorganization. Movements of the face are associated with increased activity within the former sensorimotor hand territory (Fig. 21.3). This is likely due to an expansion of the somatotopically adjacent face representation into the former hand territory following limb loss. The critical question is whether these effects relate to the recovery of hand function, and if so how. One possibility is that complete dissipation of responses within the hand territory to facial movements will require experience-dependent changes driven by extended increased use of the hand; an idea to which I return in the final section.

As alluded to earlier, we have also been developing techniques that use high-resolution fMRI and a novel, computer-controlled, sensory stimulation system to assess details of the S1 hand representation in healthy adults, as well as hand replant and transplant recipients (Fig. 21.4; see details in [71]). This system makes use of a bank of computer-controlled switches to deliver trains of pulsed, compressed air to specific targets on the body. I have illustrated one system for delivery of cutaneous

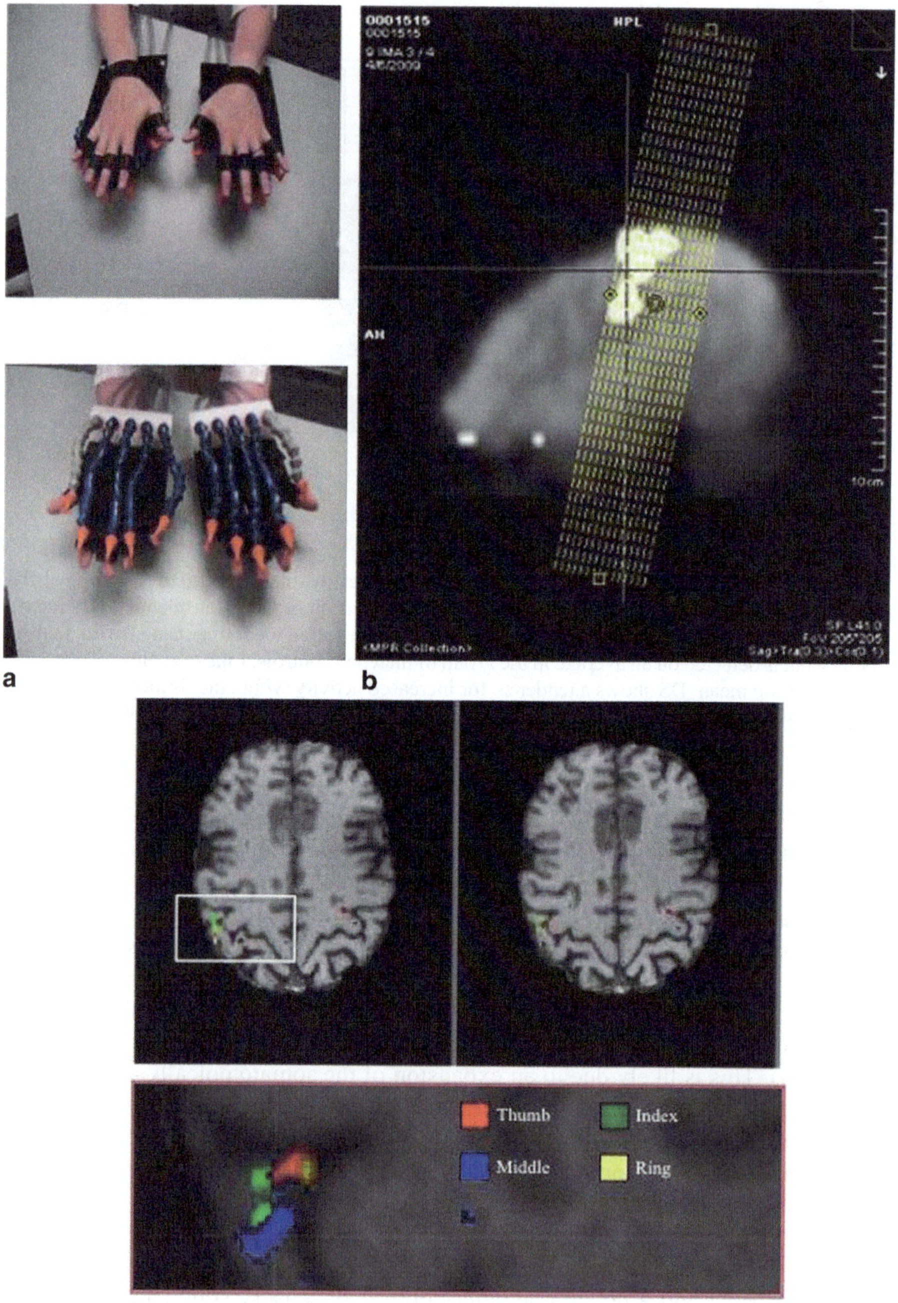

a b c

Fig. 21.4 High-resolution mapping of digit representations in S1 area 3b. **a** MRI-compatible pneumatic stimulators that deliver air pulses to the individual digit tips. **b** Positioning of MRI slices based on use of a sensorimotor localizer. Fifteen slices are positioned to allow coverage of primary and secondary somatosensory cortices defined functionally for each individual [71]. *MRI* magnetic resonance imaging stimulation to the digit tips (Panel A). A short fMRI localizer is first acquired to identify regions that show increased activity during simple movements of the hands and face. This data is then used to strategically position 15 (1.5 mm in-plane resolution) slices for optimal coverage of the primary and secondary somatosensory cortices (Panel B). This allows us to identify peak responses to individual digit stimulation located within the putative area 3b along the rostral bank of the central sulcus, defined based on probabilistic atlas derived from postmortem analyses of human brain micro-architecture [73]. At this level of resolution, there is considerable individual variation in the structural and functional organizational details of the cortex. For this reason, while it is feasible to determine whether responses fall within the likely former hand territory of area 3b, defining the precise locations of normative digit representations through current noninvasive imaging techniques is impossible. Details of findings from patients will appear in an upcoming publication.

To date, these neuroimaging investigations of hand transplant recipients have focused exclusively on simple repetitive movements or passive sensory stimulation. If the goal is to understand clinical relevance, however, then we need to move toward evaluations that address mechanisms underlying functional use of the hand. Doing so during fMRI is challenging for many reasons including the small space, limtations of magnetic resonance (MR)-compatible hardware for stimulus delivery, and the considerable potential to induce artifacts through movements of the head. Nevertheless, it is possible to study mechanisms involved in goal-directed actions such as manual reaching and grasping that involve not only primary sensorimotor regions but also parietal, premotor, and a number of subcortical structures [74–76]. This provides an opportunity to understand how these structures respond to changes in afferent and efferent activity, and determine any compensatory roles that may play in recovery of function. Our team is presently engaged in a project designed to address these very issues in both hand transplant and replant recipients.

Reconciling Reorganization, and Recovery: A Working Model

On the one hand, I have reviewed some of the large body of evidence that loss of a limb perpetuates experience-dependent changes in the organization of the central nervous system, and that these effects are especially pronounced in the activity-dependent maps of the cerebral cortex. These changes appear to involve multiple mechanisms operating over time scales ranging from hours to years. On the other hand, I drew on the available, albeit limited, data that showing that a hand transplanted years after amputation comes to be represented within the former S1 and M1 hand territories. Do these findings suggest that the changes seen postamputation are functionally irrelevant and epiphenomenal? I believe that the answer is no, and that these results instead suggest the existence of two stages of recovery with distinct mechanisms and time courses.

A key mechanism underlying the changes in cortical maps that occur following amputation is the dynamic functional rebalancing of inhibitory and excitatory synapses after deafferentation [14]. It has been suggested that these functional changes occur while leaving the original pattern of cortical connectivity structurally intact [52, 77]. If this is correct, then one might expect that re-establishing afferent and efferent activity between the brain and hand may reverse these changes, returning the cortex to a state of functional organization similar to what was in place prior to the injury [70]. A parallel might be drawn here with the evidence mentioned earlier showing that the S1 hand map returns to its original state of organization as monkeys recover from a crushed nerve [10]. Put differently, the cortex may maintain a coarse-level, latent somatotopy even years after limb loss that may account for gross recovery of the cortical somatotopy (e.g., responses within the former hand territory during sensory stimulation or active use of the transplanted hand). These changes should emerge in synchrony with peripheral nerve regeneration and be fully completed within the same time frame.

This initial stage may explain the recovery of gross sensory and motor functions in a transplanted hand. However, recovery of more refined sensory and motor functions requires experience-dependent, cortical-level adaptations to compensate for the inevitable peripheral nerve reinnervation errors, as well as reorganization at the spinal and brain stem levels. Recovery of intrinsic S1 hand territory (critical for sensory localization), and the development of representations within the M1 (necessary for nonsynergistic control of intrinsic hand musculature), depend on experience-driven changes, i.e., learning. This is a more long-term process that accounts for the fact that these functions appear to continue to improve many years after hand transplantation [67]. It is in this latter stage of recovery that rehabilitation guided by advances in modern neuroscience has the greatest potential for impact.

Though beyond the scope of this chapter, progress is being made on a number of potentially relevant fronts ranging from identification of specific single nucleotide polymorphisms that predispose the brains of certain carriers to have greater reorganizational potential to brain stimulation and pharmacological manipulations that hold promise for placing the brain in a more pliable state, better able to respond adaptively to therapeutic interventions. Ultimately, the success of the allogeneic transplant program will be judged on patient outcomes, and what brain science has to offer with respect to this goal is growing.

Acknowledgments In alphabetical order, the author wishes to acknowledge the essential contributions of Drs. Sergei Bogdanov, Warren Breidenbach III, Christina Kaufman, Joseph Kutz, and Jolinda Smith to the research described. Preparation of this chapter was supported by the USAMRMC (W81XWH-10-1-1020), and the NIH (NS083377) to S.H.F.

References

1. Florence SL, Kaas JH. Large-scale reorganization at multiple levels of the somatosensory pathway follows therapeutic amputation of the hand in monkeys. J Neurosci. 1995;15(12):8083–95.
2. Jones EG. Cortical and subcortical contributions to activity-dependent plasticity in primate somatosensory cortex. Annu Rev Neurosci. 2000;23:1–37.

3. Kaas JH. Plasticity of sensory and motor maps in adult mammals. Annu Rev Neurosci. 1991;14:137–67.
4. Kaas JH, Merzenich MM, Killackey HP. The reorganization of somatosensory cortex fol lowing peripheral nerve damage in adult and developing mammals. Annu Rev Neurosci. 1983;6:325–56.
5. Wu CW, Kaas JH. The effects of long-standing limb loss on anatomical reorganization of the somatosensory afferents in the brainstem and spinal cord. Somatosens Mot Res. 2002;19(2):153–63.
6. Pons TP, et al. Massive cortical reorganization after sensory deafferentation in adult macaques. Science. 1991;252(5014):1857–60.
7. Ramachandran VS, Rogers-Ramachandran D, Stewart M. Perceptual correlates of massive cortical reorganization. Science. 1992;258(5085):1159–60.
8. Merzenich MM, et al. Topographic reorganization of somatosensory cortical areas 3b and 1 in adult monkeys following restricted deafferentation. Neuroscience. 1983;8(1):33–55.
9. Merzenich MM, et al. Progression of change following median nerve section in the cortical representation of the hand in areas 3b and 1 in adult owl and squirrel monkeys. Neuroscience. 1983;10(3):639–65.
10. Wall JT, Felleman DJ, Kaas JH. Recovery of normal topography in the somatosensory cortex of monkeys after nerve crush and regeneration. Science. 1983;221(4612):771–3.
11. Sperry RW. The problem of central nervous reorganization after nerve regeneration and muscle transposition. Q Rev Biol. 1945;20:311–69.
12. Penfield W, Boldrey E. Somatotopic motor and sesnory representation in the cerebral cortex of man as studied by electrical stimulation. Brain. 1937;60:389–443.
13. Sanes JN, Donoghue JP. Plasticity and primary motor cortex. Annu Rev Neurosci. 2000;23:393–415.
14. Kaas JH. The reorganization of somatosensory and motor cortex after peripheral nerve or spinal cord injury in primates. Prog Brain Res. 2000;128:173–9.
15. Garraghty PE, Kaas JH. Large-scale functional reorganization in adult monkey cortex after peripheral nerve injury. Proc Natl Acad Sci U S A 1991;88(16):6976–80.
16. Merzenich MM, et al. Somatosensory cortical map changes following digit amputation in adult monkeys. J Comp Neurol. 1984;224(4):591–605.
17. Merzenich MM, Jenkins WM. Reorganization of cortical representations of the hand following alterations of skin inputs induced by nerve injury, skin island transfers, and experience. J Hand Ther. 1993;6(2):89–104.
18. Buonomano DV, Merzenich MM. Cortical plasticity: from synapses to maps. Annu Rev Neurosci. 1998;21:149–86.
19. Qi HX, Stepniewska I, Kaas JH. Reorganization of primary motor cortex in adult macaque monkeys with long-standing amputations. J Neurophysiol. 2000;84(4):2133–47.
20. Florence SL, Taub HB, Kaas JH. Large-scale sprouting of cortical connections after peripheral injury in adult macaque monkeys. Science. 1998;282(5391):1117–21.
21. Cohen LG, et al. Topographic maps of human motor cortex in normal and pathological conditions: mirror movements, amputations and spinal cord injuries. Electroencephalogr Clin Neurophysiol Suppl. 1991;43:36–50.
22. Lotze M, Flor H, Grodd W, Larbig W, Birbaumer N. Phantom movements and pain. An fMRI study in upper limb amputees. Brain. 2001;124(Pt 11):2268–77.
23. Karl A, Birbaumer N, Lutzenberger W, Cohen LG, Flor H. Reorganization of motor and somatosensory cortex in upper extremity amputees with phantom limb pain. J Neurosci. 2001;21(10):3609–18.
24. Kew JJ, et al. Abnormal access of axial vibrotactile input to deafferented somatosensory cortex in human upper limb amputees. J Neurophysiol. 1997;77(5):2753–64.
25. Elbert T, et al. Extensive reorganization of the somatosensory cortex in adult humans after nervous system injury. Neuroreport. 1994;5(18):2593–7.
26. Flor H, et al. Phantom-limb pain as a perceptual correlate of cortical reorganization following arm amputation. Nature. 1995;375(6531):482–4.
27. Yang TT, et al. Noninvasive detection of cerebral plasticity in adult human somatosensory cortex. Neuroreport. 1994;5(6):701–4.

28. Schwenkreis P, et al. Changes of cortical excitability in patients with upper limb amputation. Neurosci Lett. 2000;293(2):143–6.
29. Schwenkreis P, et al. Assessment of reorganization in the sensorimotor cortex after upper limb amputation. Clin Neurophysiol. 2001;112(4):627–35.
30. Draganski B, et al. Decrease of thalamic gray matter following limb amputation. Neuroimage. 2006;31(3):951–7.
31. Makin TR, et al. Phantom pain is associated with preserved structure and function in the former hand area. Nat Commun. 2013;4:1570.
32. Calford MB, Tweedale R. Interhemispheric transfer of plasticity in the cerebral cortex. Science. 1990;249(4970):805–7.
33. Pawela CP, et al. Interhemispheric neuroplasticity following limb deafferentation detected by resting-state functional connectivity magnetic resonance imaging (fcMRI) and functional magnetic resonance imaging (fMRI). Neuroimage. 2010;49(3):2467–78.
34. Bogdanov S, Smith J, Frey SH. Former hand territory activity increases after amputation during intact hand movements, but is unaffected by illusory visual feedback. Neurorehabil Neural Repair. 2012;26(6):604–15.
35. Kew JJ, et al. Reorganization of cortical blood flow and transcranial magnetic stimulation maps in human subjects after upper limb amputation. J Neurophysiol. 1994;72(5):2517–24.
36. Philip B, Bogdanov S, Frey, SH. Execution of a precision manual drawing task with the non-dominant hand in amputees and controls. In Annual Meeting of the Society for Neuroscience (Washington, DC); 2011.
37. Wrigley PJ, et al. Neuropathic pain and primary somatosensory cortex reorganization following spinal cord injury. Pain. 2009;141(1/2):52–9.
38. Maihofner C, et al. The motor system shows adaptive changes in complex regional pain syndrome. Brain. 2007;130(Pt 10):2671–87.
39. Maihofner C, Handwerker HO, Neundorfer B, Birklein F. Patterns of cortical reorganization in complex regional pain syndrome. Neurology. 2003;61(12):1707–15.
40. Flor H. Maladaptive plasticity, memory for pain and phantom limb pain: review and suggestions for new therapies. Expert Rev Neurother. 2008;8(5):809–18.
41. Simoes EL, et al. Functional expansion of sensorimotor representation and structural reorganization of callosal connections in lower limb amputees. J Neurosci. 2012;32(9):3211–20.
42. Flor H, Diers M, Andoh J. The neural basis of phantom limb pain. Trends Cogn Sci. 2013;17(7):307–8.
43. Button KS, et al. Power failure: why small sample size undermines the reliability of neuroscience. Nat Rev Neurosci. 2013;14(5):365–76.
44. Almquist E, Eeg-Olofsson O. Sensory-nerve-conduction velocity and two-point discrimination in sutured nerves. J Bone Joint Surg Am. 1970;52(4):791–96.
45. Wall JT, Kaas JH. Long-term cortical consequences of reinnervation errors after nerve regeneration in monkeys. Brain Res. 1986;372(2):400–4.
46. Haber WB. Reactions to loss of limb: physiological and psychological aspects. Ann N Y Acad Sci. 1958;74(1):14–24.
47. Katz D. Psychologische versuch mit amputierten. Z Psychol Physiol. 1920;85:83–117.
48. Teuber H-L, Krieger HP, Bender MB. Reorganization of sensory function in amputation stumps: two points discrimination. Fed Proc. 1949;8:156.
49. Moore CE, Schady W. Investigation of the functional correlates of reorganization within the human somatosensory cortex. Brain. 2000;123(Pt 9):1883–95.
50. Vega-Bermudez F, Johnson KO. Spatial acuity after digit amputation. Brain. 2002;125(Pt 6):1256–64.
51. Philip BA, Frey SH. Preserved grip selection planning in chronic unilateral upper extremity amputees. Exp Brain Res. 2011;214(3):437–52.
52. Dhillon GS, Lawrence SM, Hutchinson DT, Horch KW. Residual function in peripheral nerve stumps of amputees: implications for neural control of artificial limbs. J Hand Surg Am. 2004;29(4):605–15; discussion 616–608.
53. Dhillon GS, Kruger TB, Sandhu JS, Horch KW. Effects of short-term training on sensory and motor function in severed nerves of long-term human amputees. J Neurophysiol 93(5):2625–33.

54. Jia X, et al. Residual motor signal in long-term human severed peripheral nerves and feasibility of neural signal-controlled artificial limb. J Hand Surg Am. 2007;32(5):657–66.
55. Reilly KT, Mercier C, Schieber MH, Sirigu A. Persistent hand motor commands in the amputees' brain. Brain 2006;129(Pt 8):2211–23.
56. Roricht S, Meyer BU. Residual function in motor cortex contralateral to amputated hand. Neurology. 2000;54(4):984–7.
57. Stoll G, Muller HW. Nerve injury, axonal degeneration and neural regeneration: basic insights. Brain Pathol. 1999;9(2):313–25.
58. Florence SL, et al. Central reorganization of sensory pathways following peripheral nerve regeneration in fetal monkeys. Nature. 1996;381(6577):69–71.
59. Fox K, Glazewski S, Schulze S. Plasticity and stability of somatosensory maps in thalamus and cortex. Curr Opin Neurobiol. 2000;10(4):494–7.
60. Petersen CC. The functional organization of the barrel cortex. Neuron. 2007;56(2):339–55.
61. Bjorkman A, Waites A, Rosen B, Larsson EM, Lundborg G. Cortical reintegration of a replanted hand and an osseointegrated thumb prosthesis. Acta Neurochir Suppl. 2007;100:109–12.
62. Bjorkman A, Waites A, Rosen B, Lundborg G, Larsson EM. Cortical sensory and motor response in a patient whose hand has been replanted: one-year follow up with functional magnetic resonance imaging. Scand J Plast Reconstr Surg Hand Surg. 2007;41(2):70–6.
63. Eickhoff SB, et al. Central adaptation following heterotopic hand replantation probed by fMRI and effective connectivity analysis. Exp Neurol. 2008;212(1):132–44.
64. Giraux P, Sirigu A, Schneider F, Dubernard JM. Cortical reorganization in motor cortex after graft of both hands. Nat Neurosci 2001;4(7):691–2.
65. Lanzetta M, et al. Early use of artificial sensibility in hand transplantation. Scand J Plast Reconstr Surg Hand Surg. 2004;38(2):106–11.
66. Brenneis C, et al. Cortical motor activation patterns following hand transplantation and replantation. J Hand Surg. 2005;30(5):530–3.
67. Breidenbach WC, et al. Outcomes of the first 2 American hand transplants at 8 and 6 years posttransplant. J Hand Surg. 2008;33(7):1039–47.
68. Vargas CD, et al. Re-emergence of hand-muscle representations in human motor cortex after hand allograft. Proc Natl Acad Sci U S A. 2009;106(17):7197–202.
69. Neugroschl C, et al. Functional MRI activation of somatosensory and motor cortices in a hand-grafted patient with early clinical sensorimotor recovery. Eur Radiol. 2005;15(9):1806–14.
70. Frey SH, Bogdanov S, Smith JC, Watrous S, Breidenbach WC. Chronically deafferented sensory cortex recovers a grossly typical organization after allogenic hand transplantation. Curr Biol. 2008;18(19):1530–4.
71. Smith J, Bogdanov, S., Watrous, S., Frey, SH. Rapid digit mapping in the human brain at 3T in 17th Annual Meeting of the International Society for Magnet Resonance (ISMRM) in Medicine. (Honolulu); 2009.
72. Smith J, Frey, SH. Use of independent component analysis to define regions of interest for fMRI studies in 18th Annual Meeting of the International Society for Magnet Resonance (ISMRM) in Medicine. (Montreal); 2011.
73. Eickhoff SB, Grefkes C, Fink GR, Zilles K. Functional lateralization of face, hand, and trunk representation in anatomically defined human somatosensory areas. Cereb Cortex. 2008;18(12):2820–30.
74. Binkofski F, et al. Human anterior intraparietal area subserves prehension: a combined lesion and functional MRI activation study. Neurology. 1998;50(5):1253–9.
75. Culham JC, et al. Visually guided grasping produces fMRI activation in dorsal but not ventral stream brain areas. Exp Brain Res. 2003;153(2):180–9.
76. Frey SH, Vinton D, Norlund R, Grafton ST. Cortical topography of human anterior intraparietal cortex active during visually guided grasping. Brain Res Cogn Brain Res. 2005;23(2/3):397–405.
77. Schady W, Braune S, Watson S, Torebjork HE, Schmidt R. Responsiveness of the somatosensory system after nerve injury and amputation in the human hand. Ann Neurol. 1994;36(1):68–75.

Chapter 22
Ischemia–Reperfusion Injury in Reconstructive Transplantation: An Undefined Conundrum

Jerzy W. Kupiec-Weglinski and Kodi Azari

Introduction

Although vascularized composite allotransplantation (VCA) provides a means to functionally restore unreconstructable wounds in selective groups of patients, the field is in its infancy. With more than 150 VCA procedures reported during the past 15 years, including trachea, larynx, abdominal wall, face, and upper or lower extremities, this type of transplantation still remains an experimental procedure [1]. While the feasibility of the procedure has been documented with promising functional outcomes and good intermediate to long-term allograft survival, there are several obstacles that prevent VCA from enjoying widespread clinical use. For instance, there are major concerns over the damaging effects of ischemia–reperfusion injury (IRI) resulting from prolonged periods of *ex vivo* tissue cold storage, an unavoidable component of organ "procurement" insult from the cadaver sources [2]. Oxidative stress, the hallmark of IRI in any transplanted organ or tissue, triggers the release of pro-inflammatory cytokine programs, which create a deleterious local milieu promoting cell death and subsequent differentiation of rejection-mediating T effector cells, while hindering generation of tolerogenic regulatory T cells [3–5]. There is a consensus that ischemia reperfusion (IR)-induced robust local inflammation response is an essential barrier to long-term survival and the acquisition of tolerance in solid-organ transplantation [4–6]. Indeed, minimizing IRI decreases the incidence of acute allograft rejection, mitigates the severity of late chronic rejection, and improves clinical outcomes. Thus, it is plausible that better protection against IR-oxidative stress should also diminish pro-inflammatory responses

J. W. Kupiec-Weglinski (✉)
The Dumont-UCLA Transplant Center, Division of Liver and Pancreas Transplantation, Department of Surgery, David Geffen School of Medicine at University of California- Los Angeles, 77–120 CHS, 10833 Le Conte Ave, Los Angeles, CA, USA
e-mail: jkupiec@mednet.ucla.edu

K. Azari
Department of Orthopaedic Surgery, David Geffen School of Medicine at University of California- Los Angeles, Los Angeles, CA, USA

G. Brandacher (ed.), *The Science of Reconstructive Transplantation*,
Stem Cell Biology and Regenerative Medicine, DOI 10.1007/978-1-4939-2071-6_22

in VCA's divergent tissues and ameliorate host adaptive immune cascade that act in concert to facilitate VCA failure [2]. Moreover, prevention of IR-mediated VCA damage could extend the donor transfer time, allowing development of an human leukocyte antigen (HLA)–VCA national matching system. Such a system could potentially help to reduce the incidence of acute and chronic rejection and minimize immunosuppression load in prospective VCA recipients. In addition, successful prolongation of VCA preservation time should allow the expansion of the current VCA donor pool beyond local region, and provide more time to perform these complex surgical procedures. Surprisingly, however, there are major gaps in our understanding of the very basic immune mechanisms that account for IR-mediated VCA damage [2, 7], and obviously there is no therapeutic modality available to prevent and/or treat the ischemic tissue injury *in vivo*. Better appreciation of complex cellular immune events that trigger and sustain local inflammatory responses in histologically heterogeneous tissue types (e.g., skin, bone, muscle, nerves, and lymph nodes) is fundamental for developing much needed innovative therapeutic strategies for IR-stressed VCAs. Hence, both basic and translational studies dissecting cellular cross talk and molecular signaling pathways in the pathophysiology of IRI in VCAs are urgently needed. This effort should be guided by mechanistic insights garnered throughout the years from studies on tissue damage inflicted by IR in solid-organ transplantation.

As biological effects by which IR insult may affect VCAs remain largely unknown, and little if any is known about the relevant cellular events and molecular networks, in this chapter, we summarize our understanding of immune mechanisms that trigger and sustain inflammatory cascades in IR-stressed solid organs, primarily the liver. The goal is to provide a road map for future comprehensive studies exploring molecular immune IRI mechanisms in the emerging field of VCA. Our better appreciation of immune events that initiate IR-driven tissue inflammation, ultimately responsible for organ injury, is fundamental to developing innovative strategies for treating patients who have received a VCA and developed IR inflammation and transplant dysfunction.

Types and Stages of IRI

The IR insult, irrespective of the transplanted organ or tissue, is a multifaceted and dynamic process that combines elements of "warm" and "cold" injury [4, 5, 8, 9]. The "warm" IRI, initiated by parenchyma cell damage, develops in situ in low-flow states during surgery, organ retrieval, as well as in various forms of shock or trauma. The "cold" IRI, initiated by the damage to tissue endothelial cells and disruption of the microcirculation, occurs during *ex vivo* preservation, and is usually coupled with warm IRI during the transplant surgery. Although warm and cold ischemia target different non-parenchymal and parenchymal cell functions, they do share a common mechanism in the disease etiology: local inflammatory innate immune activation.

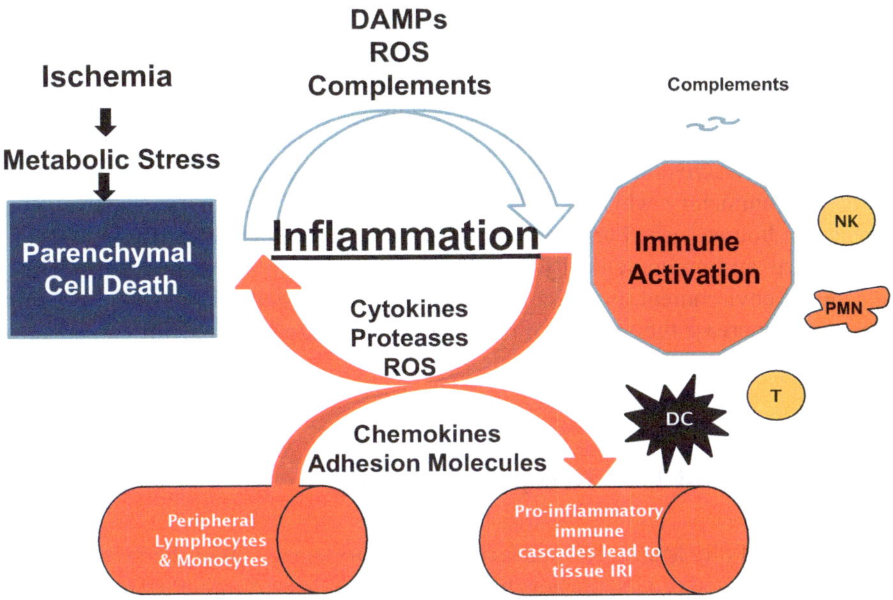

Fig. 22.1 The distinct stages of tissue IRI. The ischemic injury, a local process of metabolic disturbances, results from glycogen consumption, lack of oxygen supply, and ATP depletion. The cell death released DAMPs (alarmins), activation of complement, and oxygenation-triggered mitochondrial ROS production, all contribute to liver-immune activation after reperfusion. The process involves multiple types of nonparenchymal cells, including macrophages, dendritic cells, T cells, NK cells and neutrophils. This pro-inflammatory immune response in IR-stressed organ sustains itself by recruiting circulating peripheral immune cells from the circulation and is responsible for the ultimate reperfusion injury. *DAMPS* danger-aassociated molecular patterns, *DC* dendritic cells, *NK* natural killer cells, *PMN* polymorphonuclear cells, *ROS* reactive oxygen species, *ATP* adenosine triphosphate, *IR* ischemia reperfusion, *IRI* ischemia–reperfusion injury

The activation of tissue macrophages, neutrophils, cytokine/chemokine production, generation of reactive oxygen species (ROS), increased expression of adhesion molecules, and infiltration by circulating lymphocytes/monocytes constitute interlocked immunological cascades in both types of tissue IRI [4, 5, 9, 10]. Distinctive from alloreactive major histocompatibility complex (MHC)-disparate immune responses against organ grafts, IR-triggered tissue inflammation occurs immediately after reperfusion not only in situ or *ex vivo* but also in syngeneic grafts. It constitutes predominantly an innate immune-dominated response, mediated by a sentinel pattern recognition receptor (PRR) system. Endogenous ligands generated from cellular damage, danger-associated molecular patterns (DAMPs), rather than exogenous pathogen-associated molecular patterns (PAMPs) play the key role in IR-stressed tissue inflammation response.

Two distinctive stages of organ IRI, with unique mechanisms of tissue damage, have been identified (Fig. 22.1). The ischemic injury, a localized process of cellular metabolic disturbances, results from glycogen consumption, lack of oxygen supply, and adenosine triphosphate (ATP) depletion, leading to the parenchymal cell death.

The reperfusion injury, which immediately follows, results from both metabolic disturbances and a brisk inflammatory immune cascade that involves direct and indirect cytotoxic mechanisms. Indeed, this early, antigen nonspecific local inflammation is critical in IRI pathophysiology as prevention of immune activation uniformly ameliorates IR-mediated tissue damage. Hence, a comprehensive understanding of innate immune activation is key for identifying novel therapeutic targets to alleviate pro-inflammatory, while sparing or augmenting anti-inflammatory mechanisms needed for homeostasis. Furthermore, IR-triggered innate–adaptive cross talk readily converts an immunologically quiescent tissue to an inflammatory organ, even in a sterile environment. In direct relevance to VCA, prolonged ischemia time was reported to increase the severity of rejection in a skin flap [11] and musculocutaneous [12] rat allotransplantation models.

IR Triggers Toll-Like Receptor Signaling

Based on Dr. Polly Matzinger's concept that the principal goal of the immune system is to detect and protect the host from "danger" signals resulting from cell/tissue damage [13], Professor Walter Land introduced the "injury hypothesis" in the field of transplantation [14]. Accordingly, post-IRI activates an array of pro-inflammatory immune responses in the transplant itself, which trigger and exacerbate host adaptive immunity, ultimately progressing to graft dysfunction and ultimately rejection. All vertebrates utilize the same sentinel innate immune receptor systems, PRRs, in response to tissue damage even in the absence of infections [15–19]. Four different classes of PRRs have been recognized: Toll-like receptors (TLRs) and C-type lectin receptors (CLRs) are transmembrane proteins; Retinoic acid-inducible gene, (RIG)-I-like receptors (RLRs), and nucleotide-binding domain (NOD)-like receptors (NLRs) are cytoplasmic proteins. These PRRs, expressed primarily by activated macrophages and dendritic cells (DC), function by upregulating pro-inflammatory gene transcription programs [20].

TLRs were discovered in 1998, in mice displaying endotoxin resistance in parallel with a high susceptibility to gram-negative bacterial infections [21]. TLRs are an evolutionarily conserved group of transmembrane proteins of which, to date, 11 have been identified in humans and 13 in mice (Table 22.1; Ref. [22]). These innate receptors are central in promoting immunity against pathogens by virtue of their ability to transduce signals in response to ligation of distinctive molecular motifs, termed PAMPs. They are a major group of PRRs and are ubiquitous, being expressed on a host of both immune and nonimmune cells [23]. TLR–PAMP interactions lead to downstream cytokine and chemokine release and augmentation of co-stimulatory T cell molecule expression [24]. As TLRs are expressed on parenchyma cells, at least some of their functions are unrelated to immune-mediated pathogen destruction. Indeed, it is now apparent that endogenous, cell-derived ligands (DAMPs) from both intracellular and extracellular sources during inflammation and tissue damage do bind and facilitate TLR signaling [25]. During homeostasis, DAMPs are

Table 22.1 TLRs—their microbial, endogenous ligands and cellular distribution

Receptor	Microbial ligand(s)	Endogenous ligands	Cellular expression
TLR1	Triacyl lipopeptides	–	B cells, monocytes, macrophages, and certain dendritic cells
TLR2	Peptodoglyca, zymosan, lipoteichoic acid, and glycolipids	HSP60, HSP70, hyaluronan, HMGB1	Monocytes and macrophages, mast cells, and myeloid dendritic cells
TLR3	Double-stranded RNA, poly I:C	Messenger RNA (mRNA)	B cells, dendritic cells, and fibroblasts
TLR4	Lipopolysaccharides (LPS)	Fibrinogen, HSPs, surfactant proten A, b-defensin 2, hyaluronan, fibronectin extra domain A, heparin sulfate, HMGB-1	Monocytes and macrophages, mast cells, certain dendritic cells, B cells; intestinal epithelium and hepatocytes (low)
TLR5	Flagellin	–	Monocytes and macrophages, subset of dendritic cells; intestine
TLR6	Multiple diacyl lipopeptides on mycoplasma	–	B cells, mast cells, and macrophages
TLR7	Single-stranded RNA imidazoquinolines	RNA and protein complexes	Plasmacytoid dendritic cells, monocytesa and macrophages; B cells
TLR8	Single-stranded RNA imidazoquinolines and small synthetic compounds	–	Monocytes and macrophages; subset of dendritic cells; mast cells
TLR9	CpG oligodeoxynucleotide DNA	–	Monocytes, macrophages, and plasmacytoid dendritic cells
TLR10	Undefined	–	B cells, monocytes, and regulatory T cells
TLR11	Profilin	–	Kidney and urinary bladder epithelium
TLR12	Profilin	–	Macrophages, neurons, and dendritic cells
TLR13	Conserved bacterial 23S ribosomal RNA (rRNA) sequence	–	Monoctyes, macrophages, and dendritic cells

not expressed and remain invisible to the host immune system. However, DAMPs become released from cells are displayed on their surfaces following cellular injury, such as hypoxia. A variety of endogenous DAMPs have been described that readily engage TLRs (Table 22.1), such as heat-shock proteins [26], purines, heparan sulfate, and fibronectin degradation product, the extra domain A (EDA) domain [27].

TLR4 was the first innate immune receptor studied in organ IRI. Indeed, using murine partial hepatic warm ischemia models, data from three separate laboratories demonstrated that TLR4-deficient mice were protected from hepatic damage in liver-warm ischemia model, evidenced by markedly depressed in situ IR inflammation in the absence of TLR4 signaling [28–30]. The functional role of TLR4-specific activation in triggering IRI pathology was also confirmed in a clinically relevant orthotopic liver transplantation model, which comprises both warm and cold IR tissue damage [31] and in a steatotic liver IRI model [32]. Interestingly, donor TLR4 deficiency alone was both necessary and sufficient to confer hepatoprotection in the transplant model, and TLR4 signaling on nonparenchyma rather than parenchyma cells seems more relevant for IRI [30], although a recent study implies a unique role of TLR4 on liver parenchyma cells at the late stage of the disease process [33]. Of note, although TLR2 signaling was dispensable in the development of liver IRI [28], it was found essential in both kidneys [34] and heart [35] IRI models. In the context of solid-organ transplantation, both donor and recipient cells have the capacity to express TLR2. Notably, selective chemical ablation of the recipient TLR2 conferred protection against transplantation-associated ischemic damage in a murine renal isograft model [36], suggesting that leukocyte TLR2 not only functions in the disease pathogenesis but may also constitute a viable therapeutic target against renal IRI in transplant recipients.

All TLRs mediate signal transduction via the adapter molecule myeloid differentiation factor 88 (MyD88), apart from TLR3, which is dependent on the adapter molecule Toll/IL-1R domain-containing adapter-inducing IFN-β (TRIF) and TLR4 through which signaling is dependent on both TRIF and MyD88 [22]. Indeed, MyD88-deficient animals not only developed hepatocellular IR-damage comparable with wild-type (WT) controls, but their "signature" pro-inflammatory cytokines (IL-1, IL-6, type-I IFN) and chemokine (CXCL-10) were largely unaffected [28]. As the MyD88-independent, TRIF-dependent signaling pathway of TLR4 triggers a delayed nuclear factor (NF)-kβ activation, it seems that the MyD88-mediated early-phase NF-kβ activation is not necessary for pro-inflammatory immune response in liver IR. Indeed, this is very different from renal and heart IRI models in which either MyD88 and TRIF or only MyD88 were found operational [37–40]. The fact that severity of liver IRI peaks at 6 h of reperfusion and that of kidney and heart injury last for days may explain this discrepancy. Moreover, the liver is unique in TLR4 activation in such a way that gut-derived endotoxin may have already tolerized the hepatic innate immune system, which has been shown to target more towards the MyD88-dependent pathway [41, 42]. Both pro- and anti-inflammatory gene programs become readily induced by TLR4 activation in macrophages *in vitro* and *in vivo*. Recently, Gsk3b, a serine/threonine kinase, was shown to differentially regulate these two programs [43], and a chemical Gsk3 inhibitor selectively in-

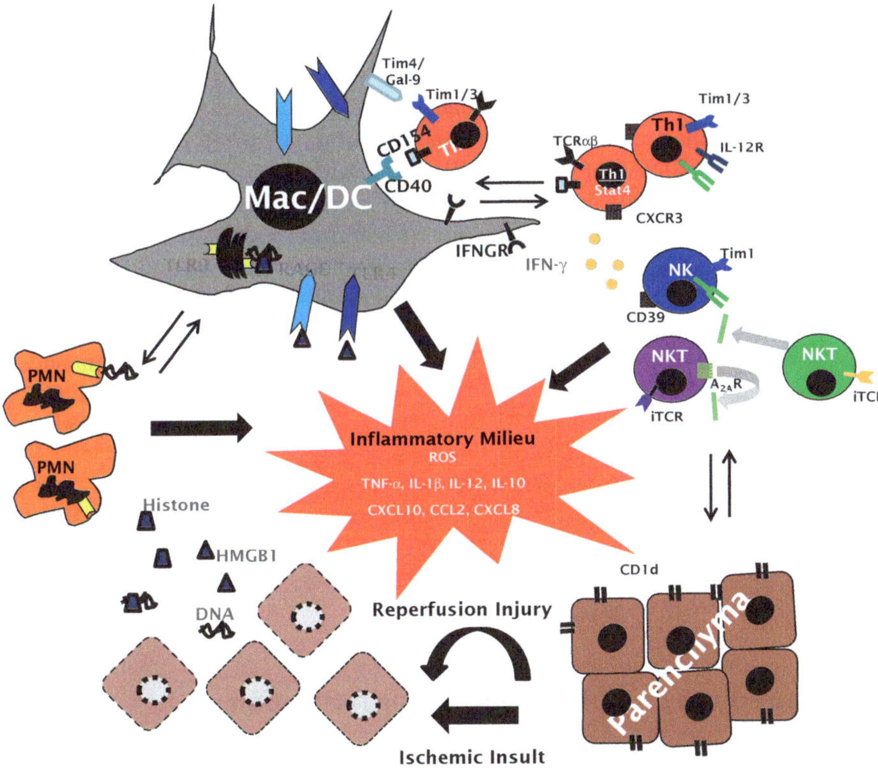

Fig. 22.2 A mechanistic scheme of immune activation in IR-stressed tissue. The ischemia insult induces necrotic cell death, which provide "danger" molecules, such as HMGB1 and DNA fragments to activate innate TLR4, RAGE, and TLR9 signaling pathways on macrophages/DC and neutrophils. CD4+Th1 effectors may also facilitate local innate immune activation via CD154–CD40 pathway, whereas IFN-γ produced by T cells, NKT, and NK cells enhances innate immune activation. In parallel, CD1d and CD39 activate NKT and NK cells, respectively. This immune activation progresses via both positive and negative regulatory loops. Pro-inflammatory TNF-α, IL-1β, IL-6, IL-12, CXCL10, CCL2, CXCL8, and ROS milieu, further activate local and recruits migrating immune cells to promote cytotoxicity against parenchymal cells. Such a sustained pro-inflammatory activation may be counter-regulated by IL-10, whereas NKT cell activation may be inhibited by adenosine receptor 2A. By inhibiting pro-inflammatory type I NKT cells, type II NKT cells may also downregulate IFN-γ production. *IR* Ischemia reperfusion, *HMGB1* high-mobility group box 1, *DC* dendritic cells, *NK* natural killer cells, *NKT* natural killer T cells, *ROS* reactive oxygen species, *TLR* toll-like receptor, *IL* interleukin, *TNF* tumor necrosis factor, *Th* T helper, *RAGE* receptor for advance glycation end products, *IFN* interferon

hibited pro-inflammatory, while simultaneously sparing immune-regulatory IL-10 gene program in IR-stressed organs [44].

The high-mobility group box 1 protein (HMGB1) represents the key endogenous TLR4 ligand responsible for IR-mediated immune activation [45]. HMGB1, released from damaged cells, may stimulate non-parenchyma cells, including macrophages and DC, through TLR4 signaling (Fig. 22.2). Hypoxic cells release HMGB1 through an active process facilitated by TLR4-dependent ROS production. In turn,

ROS induces HMGB1 release through a Calcium/calmodulin-dependent protein kinase (CaMK)-dependent mechanism, and such a positive HMGB1–TLR4 signaling promotes a sustained inflammation in IR-stressed tissue [45]. In addition to HMGB1, other DAMPs released from damaged or necrotic cells may also activate innate immune cells via an array of receptors, including S100 proteins via TLR4, RNA via TLR3, or DNA via TLR9. TLR9 was found to function in bone marrow-derived cells, particularly neutrophils in IR-stressed tissues to boost production of pro-inflammatory cytokines/chemokines. Furthermore, the inhibition of TLR9 exerted additive protective effects with concomitant HMGB1 neutralization [46]. Nuclear histone proteins were recently identified as important endogenous TLR9 ligands [47]. Thus, liver IR insult resulted in increased levels of circulating histones, whereas their neutralization was cytoprotective. Extracellular histones enhanced DNA-mediated TLR9 activation, while their infusion exacerbated IRI via TLR9 signaling. Recently, TLR3, which recognizes necrotic cell-derived RNA products, has also been shown to sustain local IR inflammation [48].

Thus, different TLR signaling pathways may function at distinct stages and in different cell types in IR-stressed solid organs. This is of importance for designing future experiments on innate activation in histologically and immune-divergent VCA tissues.

Inflammasomes in IR Innate Immune Activation

The role of other PRRs in the mechanism of tissue IRI has only recently started to be elucidated. In addition to TLRs, the necrotic cells can be sensed by the inflammasome, a caspase-1 activation platform, which regulates the secretion of L-1β, IL-18 pro-inflammatory mediators. One member of NLR family, NLRP3 (NLR family, pyrin domain containing 3) was found essential in the mechanism of polymorphonuclear neutrophil (PMN) recruitment to sites of focal hepatic necrosis in a model of sterile *in vivo* inflammation [49]. Indeed, gene silencing of NACHT, LRR and PYD domains-containing protein 3 (NALP3) attenuated tissue damage in association with reduced IL-1β, IL-18, Tumor necrosis factor α (TNF-α), and interleukin (IL)-6 levels, diminished HMGB1, and decreased local inflammatory cell infiltration [50].

Apoptosis-associated speck-like protein containing a caspase recruitment domain (ASC) plays a critical role in the activation of inflammasomes as an adaptor protein that bridges procaspase-1 and inflammasome receptors, such as NLRP3 and absent in melanoma 2 (AIM2) [51–53]. Indeed, ASC contributes to immune response through the assembly of inflammasome complexes that activate downstream effector cysteine protease caspase-1, resulting in the generation of active IL-1β and IL-18 from inactive pro-IL-1β and pro-IL-18 precursors (Fig. 22.3). Although under normal conditions ASC-associated inflammasomes are autorepressed, they become activated by a wide range of pathogen stimuli, oxidative stress, ischemia, and damage signals. The molecular mechanisms of ASC/Caspase-1/IL-1β signaling to

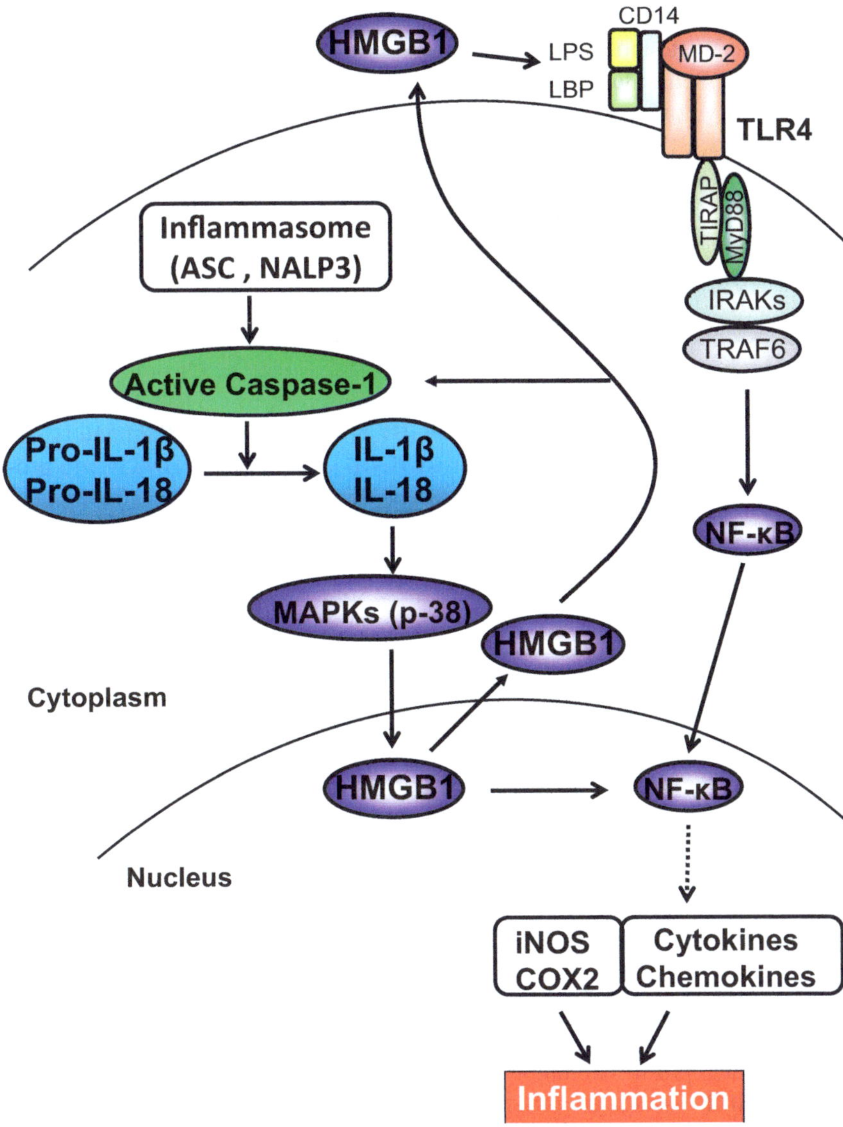

Fig. 22.3 Molecular mechanisms by which ASC/Caspase-1/IL-1β–HMGB1 axis may regulate IR-triggered inflammation. ASC activates inflammasomes, which in turn activates caspase-1 and catalyses pro-IL-1β/pro-IL-18 to mature IL-1β/IL-18. IL-18 is closely related to and shares a similar dimensional structure with IL-1β. ASC/caspase-1/IL-1 promotes HMGB1 induction through activation of p38 MAPK, which triggers TLR4 and NF-κB to program pro-inflammatory mediators. In addition, HMGB1 may provide a positive feedback mechanism to regulate caspase-1 activation. ASC/caspase-1-mediated elaboration of IL-1β and its downstream COX2 are required for the inflammatory development in the course of IRI. *ASC* Apoptosis-associated speck-like protein containing a caspase recruitment domain, *HMGB1* high-mobility group box 1, *IL* interleukin, *MAPK* mitogen-activated protein kinases, *TLR* toll-like receptor, *COX2* cyclooxygenase-2, *IRI* ischemia–reperfusion injury

program pro-inflammatory phenotype might involve activation of multiple intercellular pathways. We found disruption of ASC-inhibited HMGB1/TLR4 expression, leading to depressed induction of inflammatory mediators, suggesting ASC/Caspase-1/IL-1β plays an important role in triggering local inflammation in IR-stressed organ [54]. In fact, the adaptor ASC was initially believed to exert its effects by bridging the interaction between NLRs and caspase-1 in inflammasome complexes [55]. Activation of ASC within inflammasomes leads to the maturation of caspase-1 and processing of its IL-1β and IL-18 substrates, whereas knockout (KO) of ASC decreased caspase-1 activity and IL-1β/IL-18 production, implying the role of ASC in caspase-1/IL-1β-mediated inflammation. Although the ASC/caspase-1/IL-1β axis seems essential for the initiation of IR-inflammatory response, the molecular pathways involved in cross talk with HMGB1 remain unclear. Of note, treatment of ASC KO mice with recombinant HMGB1 increased IR tissue damage, whereas disruption of ASC without exogenous HMGB1 prevented local inflammatory development. Hence, ASC-mediated caspase-1/IL-1β axis promotes HMGB1 to produce TLR4-dependent inflammatory phenotype, leading to IR tissue inflammation and subsequent injury.

Although an array of innate PRR-targeting studies have shown promise in different animal models, the caveat is most of these studies focus on "correlation" between genetic deletion and cytoprotection rather than establishing the actual cause of the reduced tissue damage. With limited mechanistic understanding of a successful anti-IRI therapy in VCA settings, exploring multiple PRR pathways with small molecules acting preferably in a synergistic manner and/or selectively targeting positive while simultaneously promoting negative signaling may be required, while keeping in mind their different cellular sources, location specificities, and individual transcriptional kinetics.

IL-10 in IR Innate Immune Activation

Innate immune activation in IR-stressed organ is a self-limiting reaction with active regulatory mechanism by which IL-4, IL-10, and IL-13 may effectively counteract and alleviate local pro-inflammatory phenotype [56–58]. These cytokines, readily expressed in all IR tissues, are often spared or their expression even heightened in IR-resistant animals. Although generally inhibitory to IR-induced pro-inflammatory TNF-α and IL-1β "signature" when administered exogenously, the endogenous role of IL-4, IL-10, and IL-13 may not necessarily be immune regulatory. Indeed, although IL-13-deficient mice suffer from exacerbated liver injury compared with IL-13-proficient (WT) counterparts, IR-induced TNF-α and CXCL8 (MIP-2) production in IL-13 KO and WT mice was comparable in the early post-reperfusion phase [56]. Although IL-13 deficiency alters PMN distribution in IR-stressed tissues, the most profound effect of IL-13 seems to be the direct cytoprotection from ROS-induced cell death. Unlike IL-4 and IL-13, the beneficial role of IL-10 as the key immune regulatory cytokine in tissue IRI has been well documented. Hence,

IL-10 neutralization was shown both necessary and sufficient to re-create the pro-inflammatory phenotype in IR-resistant tissues of otherwise immune-suppressed or deficient recipients [59, 60]. Of note, multiple innate immune cell types, including DC, macrophages, and PMNs may all produce IL-10 and exert important autoimmune regulatory functions [61, 62]. The question of which non-parenchyma cells become IL-10 producers in response to IR insult remains to be elucidated. Recently, conventional DC have been shown to exert immune-regulatory functions by producing IL-10 via a TLR9-mediated mechanism [63]. Thus, the very same non-parenchyma cells responsible for initiation of the pro-inflammatory response against IR [64] may also terminate their own early-action function. Such a hypothesis is consistent with *in vitro* studies in which macrophages (or DC) produced both pro- and anti-inflammatory mediators in response to the very same TLR ligand supplied to the culture.

As IR activates pro- and anti-inflammatory gene programs, the question remains as to the mechanisms that determine the nature of immune responses and dictate the outcome of tissue injury. Is it merely the difference in the kinetics of innate immune gene induction or tissue/cell responsiveness to the gene products, in such a way that the pro-inflammatory cytodestructive program precedes the anti-inflammatory cytoprotective pattern, resulting in self-limited tissue damage response? Alternatively, endogenous ligands generated at different IR stages may trigger pro- and anti-inflammatory response sequentially, possibly via distinct TLR pathways and/or in different cell types. One may also envision cell–cell interactions, such as macrophage/DC–T cell, which may dictate the nature of local immune response by engaging additional activation signaling pathways. Addressing these questions in Langerhans cell-rich skin tissue should further our understanding of IRI mechanism in VCAs, and help to identify therapeutic targets to suppress pro-inflammatory without interfering with immune regulatory functions.

T Cells in IR Innate Immune Responses

Although IRI develops in syngeneic grafts, in *ex vivo* settings, or under sterile conditions, T cells, particularly of CD4 phenotype, are indispensable for the activation and regulation of pro-inflammatory immune sequelae (Fig. 22.2). The original observation that systemic immunosuppression CsA (Cyclosporin A); FK506 (Tacrolimus) attenuated peri-transplant tissue damage provided indirect evidence for T cell involvement in IRI development [65]. Experiments in T cell- (nude) and CD4-deficient mice have documented the pivotal function of CD4+T cell in the mechanism of tissue damage in several IR models [66–69]. However, the question as to how T cells may function in innate immune-driven response and in the absence of exogenous antigenic stimulation remains unanswered.

The role of T cell costimulation in promoting IRI pathology in the absence of antigen stimulation was originally shown in a study in which CD28 blockade with CTLA-4-Ig-protected rat kidneys from IRI by reducing T cell and macrophage infiltration [70]. Both CD28 and CD154 pathways were in fact essential for the development of local pro-inflammatory milieu critical to IR-mediated organ damage. Indeed, livers in CD154 KO or CD28 KO mice or in WT mice treated with anti-CD154 or CTLA-4-Ig were all IRI resistant [68]. Moreover, T helper type 1 (Th1)-type cells were shown to play a key role in IRI pathogenesis, as Stat4 KO (deficient in Th1 development), but not Stat6 KO, mice were protected from the injury, whereas reconstitution of "nude" mice with T cells from Stat6KO, but not Stat4KO, mice readily restored cardinal features of IRI [71]. Th17 cells have been also implicated in autoimmune inflammatory diseases, and their putative role in IRI has started to be unraveled. Although cellular sources of IL-17 remain to be defined, IL-17A KO mice suffer less severe IRI in parallel with reduced neutrophil infiltration. The impact of IL-17A deficiency was associated with relatively late stages of the disease and with acute IR-damage being unaffected [72]. Indeed, we consistently detect massive CD4+T cell sequestration into post-ischemic hepatic tissue well before any appreciable local neutrophil sequestration. This may occur via released IL-17, which then acts upon macrophages to produce MIP-2, a known neutrophil chemoattractant.

IL-22, an inducible cytokine of the IL-10 superfamily, is produced by selected T cell subsets (Th17, Th22, γ/δ, NKT) [73]. IL-22 is quite unique because its biological activity, unlike other cytokines, does not serve the communication between immune cells, but rather signals directly to the tissue. Its tissue action is through a heterodimer IL-10R2/IL–22R1 complex. In contrast to IL-10R2, which is ubiquitously expressed and largely dispensable, the expression of IL-22R1 is restricted to epithelial cells including hepatocytes and skin, and has not been detected in cells of the hematopoietic lineage.

By increasing tissue immunity in barrier organs such as skin, lungs, and the gastrointestinal tract, IL-22 has been associated with a number of human diseases and shown to contribute to the pathogenesis of psoriasis, rheumatoid arthritis, and Crohn's disease [73–76]. However, parallel studies in murine models of mucosal defense against pulmonary bacterial infection, inflammatory bowel disease, or acute/chronic liver failure indicate that IL-22 may exert immunoregulatory pathologic or protective functions, depending on the context in which it is expressed [77–82]. Thus, advancing our appreciation of the IL-22–IL-22R biology may yield novel therapeutic targets in multiple human diseases. Having documented that administration of rIL-22 exerted cytoprotection via STAT3 activation [83], we favor the concept that IL-22 is well positioned to orchestrate innate–adaptive immune networks by activating cell survival genes, preventing apoptosis, and promoting cell regeneration in IR-stressed organs, a novel idea directly relevant to studies on skin IRI in the emerging VCA field.

Recently discovered T cell Immunoglobulin Mucin (TIM) cell surface proteins have attracted much attention as novel regulators of host immunity [84]. T cell stimulation amplifies TIM-1, a phosphatidylserine (PS) receptor, primarily on CD4+Th2/Th1 cells, whereas TIM-4, one of the major TIM-1 ligands, is expressed mostly by macrophages and other APCs. Hence, TIM-1–TIM-4 interactions constitute a novel molecular mechanism of T cell—macrophage regulation at the innate–

adaptive interface, and may be a therapeutic target. Indeed, treatment with anti-TIM-1 mAb ameliorated liver [85] and renal [86] IR damage and was accompanied by decreased PMN infiltration/activation, inhibition of T lymphocyte/macrophage sequestration and diminished homing of TIM-1 ligand expressing TIM-4+cells in ischemic tissues. The mechanism by which TIM-1 mediates IR-triggered innate-driven inflammation is shown in Fig. 22.4a. In the "direct" pathway, TIM-1 expressed on activated Th2 cells cross-links TIM-4 to directly activate macrophages. In the "indirect" pathway, TIM-1 on activated Th1 cells triggers interferon (IFN)-γ that results in macrophage activation. Regardless of the pathway, activated macrophages do elaborate cytokine/chemokine programs that facilitate ultimate IR organ damage.

The TIM-3–Gal-9, on the other hand, constitutes a "negative" costimulation signaling between Th1 and macrophages, and has been shown to promote tolerance in transplant recipients [84]. Interestingly, TIM-3 blockade worsened the IR damage, along with increased IFN-γ but depressed IL-10 expression in IR-stressed organs [87]. One potential mechanism by which TIM-3–Gal-9 pathway controls IRI immune cascade is depicted in Fig. 22.4b. TIM-3 blockade on activated Th1 cells increases their production of IFN-γ, which in turn enhances or prolongs the activation of macrophages, DC, neutrophils, and upregulates TLR4 expression. Activated macrophages elaborate cytokine/chemokine programs through TLR4 pathway, critical to promote organ damage that can be negatively modulated via TIM-3 signaling. We favor the notion that the TIM-3 pathway may exert "protective" function by depressing IFN-γ production, and hence spare the IR-stressed organ in TLR4-dependent manner. However, although the blockade of "positive" TIM-1/TIM-4 or enhancement of "negative" TIM-3/Gal-9 costimulation might be essential, further studies are needed to accurately assess their therapeutic potential, given the opposing effects of TIM-1 and TIM-3 signaling. As PD-1– PD-L1 pathway has also been shown to promote cytoprotection [88, 89], harnessing physiological mechanisms of PD-1 negative T cell costimulation should prove instrumental for organ homeostasis by minimizing local damage and promoting IL-10-dependent cytoprotection.

In addition to "traditional" T cells, NK natural killer cells (NK) and natural killer T cells (NKT) cells may also play distinctive roles in the mechanism of IRI. Although depletion of NK1.1 cells (NK/NKT) fails to affect the severity of tissue IRI at early stages [90], it reduces the cellular damage in the later phase [91]. IR-triggered activation of NKT cells is mediated by CD1d and glycolipid Ags, released possibly by necrotic cells, to NKT cell invariant TCRs. Furthermore, NKT cell subsets may play distinctive roles *in vivo*, with type II NKT cells shown to prevent IRI when activated by specific glycolipid ligand sulfatide [92]. IR-triggered NK cell activation is dependent on CD39 to hydrolyze adenosine diphosphate (ADP) to adenosine monophosphate (AMP). Indeed, CD39-deficient organs are consistently IR-resistant with concomitantly diminished NK-derived IFN-γ production, possibly due to P2 receptor activation [93]. Thus, T cells, NKT cells, and NK cells are all involved, possibly at different stages of IR-innate activation, by providing costimulatory signaling via cell–cell interactions or cytokine stimulation to macrophages

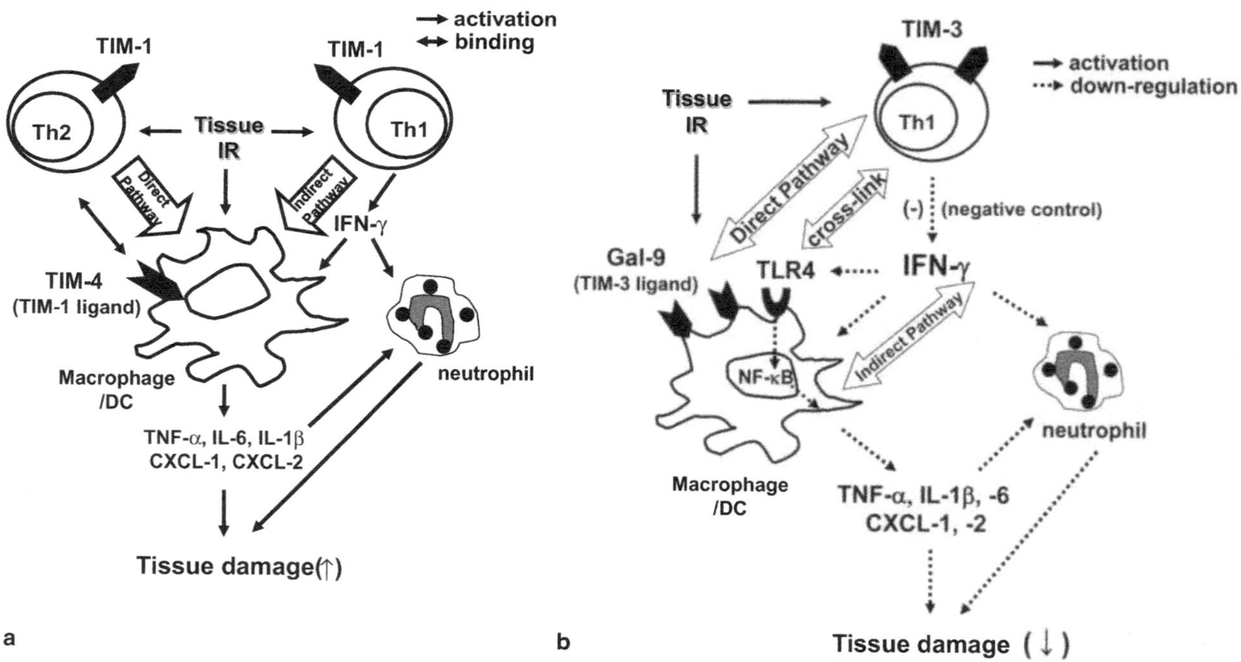

Fig. 22.4 **a** TIM-1–TIM-4 "positive" T cell costimulation in tissue IRI. Th1 and Th2 cells express TIM-1, whereas macrophages and DC express TIM-4, the TIM-1 ligand. During IR insult, TIM-1 on activated Th2 cells cross-links TIM-4 to directly activate macrophages ("direct pathway"), whereas TIM-1 on activated Th1 cells triggers IFN-γ that may also function to further stimulate macrophage activation ("indirect pathway"), evidenced by elaboration of cytokine/chemokine programs that facilitate ultimate tissue damage. *TIM* T-cell immunoglobulin and mucin, *IRI* ischemia–reperfusion injury, *Th* T helper, *IR* ischemia reperfusion, *IFN* interferon. **b** TIM-3– Gal-9 "negative" costimulation in tissue IRI. IR triggers activation of TIM-3 expression by activated macrophages and Th1 cells. TIM-3 signaling negatively regulates Th1 cells by suppressing TLR4-NF-κB pathway via IFN-γ, which in turn stimulates Gal-9 and mitigates macrophage activation. Diminished pro-inflammatory cytokine/chemokine programs ameliorate tissue damage and promote homeostasis. *TIM* T-cell immunoglobulin and mucin, *Th* T helper, *IFN* interferon, *TLR* toll-like receptor, *NF* necrosis factor, *Gal-9* galectin-9

and/or DC. This, in turn, promotes pro-inflammatory innate immune activation in IR-stressed tissue.

IRI in VCA: A New Research Frontier

There is general consensus that compared with solid organs, skin allografts are much more resistant to currently used immunosuppressive agents and tolerogenic *in vivo* strategies. Indeed, skin has been recognized as the major immunogenic component in VCA as well as the primary trigger and target of rejection response in hand or face transplants. Hence, a better understanding of skin "immunology" per se should improve our appreciation of complex immune mechanisms leading to VCA rejection or survival.

The skin of an adult human contains 10–20 billion resident memory T cells ready to respond to a variety of environmental or internal challenges. Under steady-state conditions, skin epidermis Langerhans cells (LCs) may specifically induce activation/proliferation of resident regulatory T cells (CD4+CD25+FoxP3+CD127-) able to maintain the "tolerant" state to self-antigens [94]. Upon the infectious challenge, however, the very same LCs readily trigger activation and proliferation of IFN-γ/IL-17 producing effector memory T cells. It is plausible that comparable immune patterns may operate in IR-stressed VCAs. There is evidence for the existence of two types of LCs that populate murine skin through distinct pathways [95]. Thus, under inflammatory conditions, short-term LCs, which develop from monocytic LC precursors, become recruited from the blood to the skin. In contrast, during ontogeny or in the steady state, bone marrow-derived long-term LCs may readily repopulate skin epidermis. Other mechanisms may also control the development and function of skin LCs, and hence affect their function during IR stress. A keratinocyte-derived IL-34, a ligand for colony-stimulating factor (CSF-1), has been identified as a nonredundant cytokine for LC development/homeostasis in the adult mouse and human steady-state skin [96]. Interestingly, although during local skin inflammation (such as IR insult), repopulating LCs appear to be CSF-1 dependent, once the inflammation is resolved, LC survival becomes strictly IL-34 dependent. Hence, while IL-34 is not required for monocyte recruitment and differentiation into LCs in the acute skin inflammation phase, this stroma-derived cytokine becomes crucial for LCs maintenance in the tissue-"healing" process during the homeostatic phase.

As distinct DC subsets may trigger either tolerogenic (cytoprotective) or immunogenic (cytodestructive) responses depending on the activating signal, the question arises as to whether and how skin LCs support immunogenic functions in the absence of antigen presentation by other DC subsets. Indeed, LCs exposed to diverse stimulants were committed to tolerogenic functions, and maintained a tolerogenic NFkB signature despite concomitant upregulation of costimulatory molecules CD80, CD86, and IL-12 [97]. This may explain why epithelium-containing endogenous TLR ligands are largely tolerated, whereas those that breach the epithelial basement membrane to activate dermal DCs become immune stimulators in the inflamed skin. What are putative mechanisms by which epidermal LCs may protect skin from local inflammation? In a murine cutaneous immune tolerance model, epidermal DCs were shown to confer protection by a mechanism involving anergy and deletion of allergen-specific CD8 + T cells, with simultaneous activation of ICOS + CD4 + FoxP3 + Treg cells [98]. Based on these data, one may speculate that in addition to obvious T cell phenotypic aberrations, LC deficiency or their deregulated migration patterns may contribute to skin-specific inflammatory responses, such as those in VCAs.

As molecular mechanisms and dynamics of skin damage due to either innate-mediated IR or adaptive immunity-driven rejection seem comparable with inflammatory skin conditions, these studies are of major interest to transplant researchers, especially those in the emerging field of VCA. It remains to be determined whether molecular aspects of LC function, as discussed here, may explain why skin grafts are somewhat “less antigenic” when a part of experimental VCA than skin tissue transplanted alone. Hence, dissecting innate–adaptive immune cross-regulation in clinically relevant, yet technically challenging, murine models of tissue IRI in VCAs is warranted. These studies are needed to better understand the intricate network of highly complex functional interactions among molecular targets, which can be amplified, are highly regulated, and in many cases, become flexible to be either cell or tissue-type specific. This bench research experience should translate to the bedside in continuing the effort to improve VCA function, save lives, benefit patient outcomes, and enhance the overall success of clinical transplantation.

References

1. Diaz-Siso JR, Bueno EM, Sisk GC, Marty FM, Pomahac B, Tullius SG. Vascularized composite tissue allotransplantation–state of the art. Clin Transplant. 2013;27:330–7.
2. Caterson EJ, Lopez J, Medina M, Pomahac B, Tullius SG. Ischemia-reperfusion injury in vascularized composite allotransplantation. J Craniofac Surg. 2013;24:51–6.
3. Eltzschig HK, Eckle T. Ischemia and reperfusion–from mechanism to translation. Nat Med. 2011;17:1391–401.
4. Zhai Y, Busuttil RW, Kupiec-Weglinski JW. New insights into mechanisms of innate–adaptive immune-mediated tissue inflammation. Am J Transplant. 2011;11:1563–9.
5. Zhai Y, Petrowsky H, Hong, JC, Busuttil RW, Kupiec-Weglinski JW. Ischaemia-reperfusion injury in liver transplantation–from bench to bedside. Nat Rev Gastroenterol Hepatol. 2013;10:79–89.
6. Murphy SP, Porrett PM, Turka LA. Innate immunity in transplant tolerance and rejection. Immunol Rev. 2011;241(1):39–48.
7. Wang WZ, Baynosa RC, Zamboni WA. Update on ischemia-reperfusion injury for the plastic surgeon: 2011. Plast Reconstr Surg. 2011;128(6):685e–92e.
8. Ikeda T, Yanaga K, Kishikawa K, Kakizoe S, Shimada M, Sugimachi K. Ischemic injury in liver transplantation: difference in injury sites between warm and cold ischemia in rats. Hepatology. 1992;16:454–61.
9. Fondevila C, Busuttil RW, Kupiec-Weglinski JW. Hepatic ischemia/reperfusion injury–a fresh look. Exp Mol Pathol. 2003;74:86–93.

10. Lentsch AB, Kato A, Yoshidome H, McMasters KM, Edwards MJ. Inflammatory mechanisms and therapeutic strategies for warm hepatic ischemia/reperfusion injury. Hepatology. 2000;32:169–73.
11. Shimizu F, Okamoto O, Katagiri K, Fujiwara S, Wei FC. Prolonged ischemia increases severity of rejection in skin flap allotransplantation in rats. Microsurgery. 2010;30:132–7.
12. Pradka SP, Ong YS, Zhang Y, Davis SJ, Baccarani A, Messmer C, et al. Increased signs of acute rejection with ischemic time in a rat musculocutaneous allotransplant model. Transplant Proc. 2009;41:531–6.
13. Matzinger P. Tolerance, danger, and the extended family. Annu Rev Immunol. 1994;12:991–1045.
14. Land WG. The role of postischemic reperfusion injury and other nonantigen-dependent inflammatory pathways in transplantation. Transplantation. 2005;79:505–14.
15. Beg AA. Endogenous ligands of Toll-like receptors: implications for regulating inflammatory and immune responses. Trends Immunol. 2002;23:509–12.
16. Fox-Marsh A, Harrison LC. Emerging evidence that molecules expressed by mammalian tissue grafts are recognized by the innate immune system. J Leukoc Biol. 2002;71:401–9.
17. Rifkin IR, Leadbetter EA, Busconi L, Viglianti G, Marshak-Rothstein A. Toll-like receptors, endogenous ligands, and systemic autoimmune disease. Immunol Rev. 2005;204:27–42.
18. Srikrishna G, Freeze HH. Endogenous damage-associated molecular pattern molecules at the crossroads of inflammation and cancer. Neoplasia. 2009;11:615–28.
19. Kawai T, Akira S. Toll-like receptor and RIG-I-like receptor signaling. Ann N Y Acad Sci. 2008;1143:1–20.
20. Takeuchi O, Akira, S. Pattern recognition receptors and inflammation. Cell. 2010;140:805–20.
21. Poltorak A, He X, Smirnova I, Liu MY, Van HC, Du X, et al. Defective LPS signaling in C3H/HeJ and C57BL/10ScCr mice: Mutations in Tlr4 gene. Science. 1998;282:2085–8.
22. Robson MG. Toll-like receptors and renal disease. Nephron Exp Nephrol. 2009;113:e1–7.
23. Arumugam TV, Okun E, Tang SC, Thundyil J, Taylor SM, Woodruff TM. Toll-like receptors in ischemia-reperfusion injury. Shock. 2009;32:4–16.
24. Sobek V, Birkner N, Falk I, Wurch A, Kirschning CJ, Wagner H, et al. Direct toll-like receptor 2 mediated co-stimulation of T cells in the mouse system as a basis for chronic inflammatory joint disease. Arthritis Res Ther. 2004;6:R433–46.
25. Yu L, Wang L, Chen S. Endogenous toll-like receptor ligands and their biological significance. J Cell Mol Med. 2010;14:2592–603.
26. Ohashi K, Burkart V, Flohe S, Kolb H. Cutting edge: heat shock protein 60 is a putative endogenous ligand of the toll-like receptor-4 complex. J Immunol. 2010;164:558–61.
27. Okamura Y, Watari M, Jerud ES, Young DW, Ishizaka ST, Rose J, et al. The extra domain A of fibronectin activates Toll-like receptor 4. J Biol Chem. 2001;276:10229–33.
28. Zhai Y, Shen XD, O'Connell R, Gao F, Lassman C, Busuttil RW, et al. Cutting edge: TLR4 activation mediates liver ischemia/reperfusion inflammatory response via IFN regulatory factor 3-dependent MyD88-independent pathway. J Immunol. 2004;173:7115–9.
29. Wu HS, Zhang JX, Wang L, Tian Y, Wang H, Rotstein O. Toll-like receptor 4 involvement in hepatic ischemia/reperfusion injury in mice. Hepatobiliary Pancreat Dis Int. 2004;3:250–3.
30. Tsung A, Hoffman RA, Izuishi K, Critchlow ND, Nakao A, Chan MH, et al. Hepatic ischemia/reperfusion injury involves functional TLR4 signaling in nonparenchymal cells. J Immunol. 2005;175:7661–8.
31. Shen XD, Ke B, Zhai Y, Gao F, Tsuchihashi S, Lassman CR, et al. Absence of toll-like receptor 4 (TLR4) signaling in the donor organ reduces ischemia and reperfusion injury in a murine liver transplantation model. Liver Transpl. 2007;13:1435–43.
32. Ellett JD, Evans ZP, Atkinson C, Schmidt MG, Schnellmann RG, Chavin KD. Toll-like receptor 4 is a key mediator of murine steatotic liver warm ischemia/reperfusion injury. Liver Transpl. 2009;15:1101–9.

33. Hui W, Jinxiang Z, Heshui W, Zhuoya L, Qichang Z. Bone marrow and non-bone marrow TLR4 regulates hepatic ischemia/reperfusion injury. Biochem Biophys Res Commun. 2009;389:328–32.
34. Leemans JC, Stokman G, Claessen N, Rouschop KM, Teske GJ, Kirschning CJ, et al. Renal-associated TLR2 mediates ischemia/reperfusion injury in the kidney. J Clin Invest. 2005;115:2894–903.
35. Arslan F, Smeets MB, O'Neill LA, Keogh B, McGuirk P, Timmers L, et al. Myocardial ischemia/reperfusion injury is mediated by leukocytic toll-like receptor-2 and reduced by systemic administration of a novel anti-toll-like receptor-2 antibody. Circulation. 2010; 121:80–90.
36. Farrar CA, Keogh B, McCormackW, O'Shaughnessy A, Parker, A, Reilly M, et al. Inhibition of TLR2 promotes graft function in a murine model of renal transplant ischemia-reperfusion injury. FASEB J. 2012;26:799–807.
37. Wu H, Chen G, Wyburn KR, Yin J, Bertolino P, Eris JM, et al. TLR4 activation mediates kidney ischemia/reperfusion injury. J Clin Invest. 2007;117:2847–59.
38. Pulskens WP, Teske GJ, Butter LM, Roelofs JJ, van der Poll T, Florquin S, et al. Toll-like receptor-4 coordinates the innate immune response of the kidney to renal ischemia/reperfusion injury. PLoS One. 2008;3(10):e3596. doi: 10.1371.
39. Kaczorowski DJ, Nakao A, Vallabhaneni R, Mollen KP, Sugimoto R, Kohmoto J, et al. Mechanisms of Toll-like receptor 4 (TLR4)-mediated inflammation after cold ischemia/reperfusion in the heart. Transplantation. 2009;87:1455–63.
40. Shigeoka AA, Holscher TD, King AJ, Hall FW, Kiosses WB, Tobias PS, et al. TLR2 is constitutively expressed within the kidney and participates in ischemic renal injury through both MyD88-dependent and -independent pathways. J Immunol. 2007;178:6252–8.
41. Broad A, Kirby JA, Jones DE. Toll-like receptor interactions: tolerance of MyD88-dependent cytokines but enhancement of MyD88-independent interferon-beta production. Immunology. 2007;120:103–11.
42. Biswas SK, Lopez-Collazo E. Endotoxin tolerance: new mechanisms, molecules and clinical significance. Trends Immunol. 2009;30:475–87.
43. Martin M, Rehani K, Jope RS, Michalek SM. Toll-like receptor-mediated cytokine production is differentially regulated by glycogen synthase kinase 3. Nat Immunol. 2005;6:777–84.
44. Ren F, Duan Z, Cheng Q, Shen X, Gao F, Bai L, et al. Inhibition of glycogen synthase kinase 3 beta ameliorates liver ischemia reperfusion injury by way of an interleukin-10-mediated immune regulatory mechanism. Hepatology. 2011;54:687–96.
45. Tsung A, Klune JR, Zhang X, Jeyabalan G, Cao Z, Peng X, et al. HMGB1 release induced by liver ischemia involves Toll-like receptor 4 dependent reactive oxygen species production and calcium-mediated signaling. J Exp Med. 2007;204:2913–23.
46. Bamboat ZM, Balachandran VP, Ocuin LM, Obaid H, Plitas G, DeMatteo RP. Toll-like receptor 9 inhibition confers protection from liver ischemia-reperfusion injury. Hepatology. 2010;51:621–32.
47. Huang H, Evankovich J, Yan W, Nace G, Zhang L, Ross M, et al. Endogenous histones function as alarmins in sterile inflammatory liver injury through Toll-like receptor 9 in mice. Hepatology. 2011; 54:999–1008.
48. Cavassani KA, Ishii M, Wen H, Schaller MA, Lincoln PM, Lukacs NW, et al. TLR3 is an endogenous sensor of tissue necrosis during acute inflammatory events. J Exp Med. 2008;205:2609–21.
49. McDonald B, Pittman K, Menezes GB, Hirota SA, Slaba I, Waterhouse CC, et al. Intravascular danger signals guide neutrophils to sites of sterile inflammation. Science. 2010;330:362–6.
50. Zhu P, Duan L, Chen J, Xiong A, Xu Q, Zhang H, et al. Gene silencing of NALP3 protects against liver ischemia-reperfusion injury in mice. Hum Gene Ther. 2011;22:853–64
51. Mariathasan S, Monack DM. Inflammasome adaptors and sensors: intracellular regulators of infection and inflammation. Nat Rev Immunol. 2007;7:31–40.

52. Hornung V, Ablasser A, Charrel-Dennis M, Bauernfeind F, Horvath G, Caffrey DR, et al. AIM2 recognizes cytosolic dsDNA and forms a caspase-1-activating inflammasome with ASC. Nature. 2009;458:514–8.
53. Fang R, Tsuchiya K, Kawamura I, Shen Y, Hara H, Sakai S, et al. Critical roles of ASC inflammasomes in caspase-1 activation and host innate resistance to Streptococcus pneumoniae infection. J Immunol. 2011;187:4890–9.
54. Kamo N, Ke B, Ghaffari AA, Busuttil RW, Cheng G, Kupiec-Weglinski JW. ASC/Caspase-1/IL-1b signaling triggers inflammatory responses by promoting HMGB1 induction in liver ischemia/reperfusion injury. Hepatology. 2013;58:351–62.
55. Kanneganti TD, Lamkanfi M, Núñez G. Intracellular NOD-like receptors in host defense and disease. Immunity. 2007;27:549–59.
56. Kato A, Okaya T, Lentsch AB. Endogenous IL-13 protects hepatocytes and vascular endothelial cells during ischemia/reperfusion injury. Hepatology. 2003;37:304–12.
57. Kato A, Yoshidome H, Edwards MJ, Lentsch AB. Reduced hepatic ischemia/reperfusion injury by IL-4: potential anti-inflammatory role of STAT6. Inflamm Res. 2000;49:275–9.
58. Yoshidome H, Kato A, Miyazaki M, Edwards MJ, Lentsch AB. IL-13 activates STAT6 and inhibits liver injury induced by ischemia/reperfusion. Am J Pathol. 1999;155:1059–64.
59. Ji H, Shen X, Gao F, Ke B, Freitas MC, Uchida Y, et al. Programmed death-1/B7-H1 negative costimulation protects mouse liver against ischemia and reperfusion injury. Hepatology. 2010;52:1380–9.
60. Zhai Y, Shen XD, Gao F, Zhao A, Freitas MC, Lassman C, et al. CXCL10 regulates liver innate immune response against ischemia and reperfusion injury. Hepatology. 2008;47:207–14.
61. Saraiva M, O'Garra A. The regulation of IL-10 production by immune cells. Nat Rev Immunol. 2010;10:170–81.
62. Zhang X, Majlessi L, Deriaud E, Leclerc C, Lo-Man R. Coactivation of Syk kinase and MyD88 adaptor protein pathways by bacteria promotes regulatory properties of neutrophils. Immunity. 2009;31:761–71.
63. Bamboat ZM, Ocuin LM, Balachandran VP, Obaid H, Plitas G, DeMatteo RP. Conventional DCs reduce liver ischemia/reperfusion injury in mice via IL-10 secretion. J Clin Invest. 2010;120:559–69.
64. Tsung A, Zheng N, Jeyabalan G, Izuishi K, Klune JR, Geller DA, et al. Increasing numbers of hepatic dendritic cells promote HMGB1-mediated ischemia-reperfusion injury. J Leukoc Biol. 2007;81:119–28.
65. Suzuki S, Toledo-Pereyra LH, Rodriguez FJ, Cejalvo D. Neutrophil infiltration as an important factor in liver ischemia and reperfusion injury. Modulating effects of FK506 and cyclosporine. Transplantation. 1993;55:1265–72.
66. Zwacka RM, Zhang Y, Halldorson J, Schlossberg H, Dudus L, Engelhardt JF. CD4(+) T-lymphocytes mediate ischemia/reperfusion-induced inflammatory responses in mouse liver. J Clin Invest. 1997;100:279–89.
67. Rabb H, Daniels F, O'Donnell M, Haq M, Saba SR, Keane W, et al. Pathophysiological role of T lymphocytes in renal ischemia-reperfusion injury in mice. Am J Physiol Renal Physiol. 2000;279:F525–31.
68. Shen XD, Ke B, Zhai Y, Amersi F, Gao F, Anselmo DM, et al. CD154-CD40 T-cell costimulation pathway is required in the mechanism of hepatic ischemia/reperfusion injury, and its blockade facilitates and depends on heme oxygenase-1 mediated cytoprotection. Transplantation. 2002;74:315–9.
69. Burne MJ, Daniels F, El Ghandour A, Mauiyyedi S, Colvin RB, O'Donnell MP, et al. Identification of the CD4(+) T cell as a major pathogenic factor in ischemic acute renal failure. J Clin Invest. 2001;108:1283–90.
70. Takada M, Chandraker A, Nadeau KC, Sayegh MH, Tilney NL. The role of the B7 costimulatory pathway in experimental cold ischemia/reperfusion injury. J Clin Invest. 1997;100:1199–203.

71. Shen XD, Ke B, Zhai Y, Gao F, Anselmo D, Lassman CR, et al. Stat4 and Stat6 signaling in hepatic ischemia/reperfusion injury in mice: HO-1 dependence of Stat4 disruption-mediated cytoprotection. Hepatology. 2003;37:296–303.
72. Kono H, Fujii H, Ogiku M, Hosomura N, Amemiya H, Tsuchiya M, et al. Role of IL-17A in neutrophil recruitment and hepatic injury after warm ischemia-reperfusion mice. J Immunol. 2011;187:4818–25.
73. Wolk K, Witte E, Witte K, Warszawska K, Sabat R. Biology of interleukin-22. Semin Immunopathol. 2010;32:17–31.
74. Ma HL, Liang S, Li J, Napierata L, Brown T, Benoit S, et al. IL-22 is required for Th17 cell-mediated pathology in a mouse model of psoriasis-like skin inflammation. J Clin Invest. 2008;118:597–607.
75. Ikeuchi H, Kuroiwa T, Hiramatsu N, Kaneko Y, Hiromura K, Ueki K, et al. Expression of interleukin-22 in rheumatoid arthritis: potential role as a proinflammatory cytokine. Arthritis Rheum. 2005;52:1037–46.
76. Wolk K, Witte E, Hoffmann U, Doecke WD, Endesfelder S, Asadullah K, et al. IL-22 induces lipopolysaccharide-binding protein in hepatocytes: a potential systemic role of IL-22 in Crohn's disease. J Immunol. 2007;178:5973–81.
77. Aujla SJ, Chan YR, Zheng M, Fei M, Askew DJ, Pociask DA, et al. IL-22 mediates mucosal host defense against Gram-negative bacterial pneumonia. Nat Med. 2008;14:275–81.
78. Zenewicz LA, Yancopoulos GD, Valenzuela DM, Murphy AJ, Stevens S, Flavell RA. Innate and adaptive Interleukin-22 protects mice from inflammatory bowel disease. Immunity. 2008;29:947–57.
79. Radaeva S, Sun R, Pan HN, Hong F, Gao B. Interleukin-22 (IL-22) plays a protective role in T cell-mediated murine hepatitis: IL-22 is a survival factor for hepatocytes via STAT3 activation. Hepatology. 2004;39:1332–42.
80. Zenewicz LA, Yancopoulos GD, Valenzuela DM, Murphy AJ, Karow M, Flavell RA. IL-22 but not IL-17 provides protection to hepatocytes during acute liver inflammation. Immunity. 2007;27:647–59.
81. Zhai Y, Qiao B, Gao F, Shen X, Vardanian A, Busuttil RW, et al. Type I, but not Type II, Interferon is critical in liver injury induced after ischemia and reperfusion. Hepatology. 2008;47:199–206.
82. Wolk K, Sabat R. Interleukin 22: a novel T- and NK-cell derived cytokine that regulates the biology of tissue cells. Cytokine Growth Factor Rev. 2006;17:367–80.
83. Chestovich PJ, Uchida Y, Chang W, Ajalat M, Lassman C, Sabat R, et al. Interleukin-22: implications for liver ischemia-reperfusion injury. Transplantation. 2012;93:485–92.
84. Kuchroo VK, Meyers JH, Umetsu DT, DeKruyff RH. TIM family of genes in immunity and tolerance. Adv Immunol. 2006;91:227–49.
85. Uchida Y, Ke B, Freitas MC, Ji H, Zhao D, Benjamin ER, et al. The emerging role of T cell immunoglobulin mucin-1 in the mechanism of liver ischemia and reperfusion injury in the mouse. Hepatology. 2010;51:1363–72.
86. Rong S, Park JK, Kirsch T, Yagita H, Akiba H, Boenisch O, et al. The TIM-1:TIM-4 pathway enhances renal ischemia-reperfusion injury. J Am Soc Nephro. 2011;22:484–95.
87. Uchida Y, Ke B, Freitas MC, Yagita H, Akiba H, Busuttil RW, et al. T-cell immunoglobulin mucin-3 determines severity of liver ischemia/reperfusion injury in mice in a TLR4-dependent manner. Gastroenterology. 2010;139:2195–206.
88. Ji H, Shen X, Gao F, Ke B, Freitas MC, Uchida Y, et al. Programmed death-1/B7-H1 negative costimulation protects mouse liver against ischemia and reperfusion injury. Hepatology. 2010;52:1380–9.
89. Ueki S, Castellaneta A, Yoshida O, Ozaki K, Zhang M, Kimura S, et al. Hepatic B7 Homolog 1 expression is essential for controlling cold ischemia/reperfusion injury after mouse liver transplantation. Hepatology. 2011;54:216–28.
90. Shen X, Wang Y, Gao F, Ren F, Busuttil RW, Kupiec-Weglinski JW, et al. CD4 T cells promote tissue inflammation via CD40 signaling without de novo activation in a murine model of liver ischemia/reperfusion injury. Hepatology. 2009;50:1537–46.

91. Lappas CM, Day YJ, Marshall MA, Engelhard VH, Linden J. Adenosine A2A receptor activation reduces hepatic ischemia reperfusion injury by inhibiting CD1d-dependent NKT cell activation. J Exp Med. 2006;203:2639–48.
92. Arrenberg P, Maricic I, Kumar V. Sulfatide-mediated activation of type II natural killer T cells prevents hepatic ischemic reperfusion injury in mice. Gastroenterology. 2011;140:646–55.
93. Beldi G, Banz Y, Kroemer A, Sun X, Wu Y, Graubardt N, et al. Deletion of CD39 on natural killer cells attenuates hepatic ischemia/reperfusion injury in mice. Hepatology. 2010;51:1702–11
94. Seneschal J, Clark RA, Gehad A, Baecher-Allan CM, Kupper TS. Human epidermal Langerhans cells maintain immune homeostasis in skin by activating skin resident regulatory T cells. Immunity. 2012;36:873–84.
95. Seré K, Baek JH, Ober-Blöbaum J, Müller-Newen G, Tacke F, Yokota Y, et al. Two distinct types of Langerhans cells populate the skin during steady state and inflammation. Immunity. 2012; 37:905–16.
96. Greter M, Lelios I, Pelczar P, Hoeffel G, Price J, Leboeuf M, et al. Stroma-derived interleukin-34 controls the development and maintenance of langerhans cells and the maintenance of microglia. Immunity. 2012;37:1050–60.
97. Shklovskaya E, O'Sullivan BJ, Ng LG, Roediger B, Thomas R, Weninger W, et al. Langerhans cells are precommitted to immune tolerance induction. Proc Natl Acad Sci U S A. 2011;108:18049–54.
98. Gomez de Agüero M, Vocanson M, Hacini-Rachinel F, Taillardet M, Sparwasser T, Kissenpfennig A, et al. Langerhans cells protect from allergic contact dermatitis in mice by tolerizing CD8(+) T cells and activating Foxp3(+) regulatory T cells. J Clin Invest. 2012;122:1700–11.

Index

G. Brandacher (ed.), *The Science of Reconstructive Transplantation*,
Stem Cell Biology and Regenerative Medicine, DOI 10.1007/978-1-4939-2071-6

CPSIA information can be obtained at www.ICGtesting.com
Printed in the USA
LVOW01*1727010615

440727LV00001B/8/P